T0259411

Handbook of
Visual Communications

Series in Telecommunications

Handbook of Visual Communications
Edited by
Hseuh-Ming Hang
Department of Electronics Engineering and Microelectronics
  and Information Systems Research Center
National Chiao Tung University
Hsinchu, Taiwan, ROC

John W. Woods
ECSE Department
Rensselaer Polytechnic Institute
Troy, New York

Other Books in the Series

Ali N. Akansu and Richard A. Haddad, *Multiresolution Signal Decomposition: Transforms, Subbands, Wavelets.* 1992.

Tsong-ho Wu and Noriaki Yoshikai, *ATM Transport and Network Integrity.* In preparation.

Handbook of
Visual Communications

Edited by

Hseuh-Ming Hang

*Department of Electronics Engineering and Microelectronics
and Information Systems Research Center
National Chiao Tung University
Hsinchu, Taiwan, ROC*

John W. Woods

*ECSE Department
Rensselaer Polytechnic Institute
Troy, New York*

Academic Press

San Diego   New York   Boston   London   Sydney   Tokyo   Toronto

Copyright © 1995 by ACADEMIC PRESS, INC.

Academic Press, Inc.
A Division of Harcourt Brace & Company
525 B Street, Suite 1900, San Diego, California 92101-4495

United Kingdom Edition published by
Academic Press Limited
24-28 Oval Road, London NW1 7DX

Library of Congress Cataloging-in-Publication Data

Handbook of visual communications / edited by Hseuh-Ming Hang, John W.
   Woods.
         p.   cm. -- (Series in telecommunication)
   Includes bibliographical references and index.
   ISBN 0-12-323050-0 (alk. paper)
   1. Image tramission.   2. Image processing--Digital techniques.
3. Video compression.   I. Hang, Hseuh-Ming.   II. Woods, John W.
(John William), date.   III. Series.
   TK5105.2.H36   1995
   621.36'7-dc20   95-15416
                                                      CIP

Contents

4   BILEVEL IMAGE CODING

D. L. Duttweiler

5   MOTION ESTIMATION FOR IMAGE SEQUENCE COMPRESSION

H.-M. Hang

Y.-M. Chou

6   VECTOR QUANTIZATION TECHNIQUES IN IMAGE COMPRESSION

A. Gersho

S. Gupta

S.-W. Wu

7   TRANSFORM CODING
R. L. de Queiroz
K. R. Rao

8   SUBBAND AND WAVELET FILTERS FOR HIGH-DEFINITION VIDEO COMPRESSION
T. Naveen
J. W. Woods

9   HIERARCHICAL CODING

F. Bosveld

R. L. Lagendijk

J. Biemond

10   MODEL-BASED CODING

K. Aizawa

11   IMAGE AND VIDEO CODING STANDARDS

R. Aravind

G. L. Cash

H.-M. Hang

B. G. Haskell

A. Puri

12   HYBRID HIGH-DEFINITION TELEVISION

Y. Ninomiya

13   VIDEO COMMUNICATIONS TECHNOLOGIES I: NARROWBAND TRANSMISSIONS

L. F. Chang

T. R. Hsing

14 VIDEO COMMUNICATIONS TECHNOLOGIES II: BROADBAND CABLE TELEVISION TRANSMISSIONS

W. I. Way

15 VLSI FOR VIDEO CODING

P. Pirsch

Contributors

Numbers in parentheses indicate the pages on which the authors' contributions begin.

K. Aizawa (341), Department of Electrical Engineering, University of Tokyo, Bunkyo-ku, Tokyo 113, Japan

R. Aravind (365), Visual Communications Research Department, AT&T Bell Laboratories, Holmdel, New Jersey 07733

J. Biemond (299), Delft University of Technology, Department of Electrical Engineering, Information Theory Group, 2600 AA Delft, The Netherlands

F. Bosveld[1] (299), Delft University of Technology, Department of Electrical Engineering, Information Theory Group, 2600 AA Delft, The Netherlands

G. L. Cash (365), Visual Communications Research Department, AT&T Bell Laboratories, Holmdel, New Jersey 07733

L. F. Chang (421), Bellcore, Red Bank, New Jersey 07701

Y.-M. Chou (147), Department of Electronics Engineering, National Chiao Tung University, Hsinchu 300, Taiwan, ROC

R. L. de Queiroz (223), Electrical Engineering Department, University of Texas at Arlington, Arlington, Texas 76019

D. L. Duttweiler (127), Visual Communications Research Department, AT&T Bell Laboratories, Holmdel, New Jersey 07733

A. Gersho (189), Department of Electrical and Computer Engineering, University of California, Santa Barbara, Santa Barbara, California 93106

S. Gupta[2] (189), Center for Information Processing Research, Department of Electrical Engineering, University of California, Santa Barbara, Santa Barbara, California 93106

H.-M. Hang (147, 365), Department of Electronics Engineering and Microelectronics and Information Systems Research Center, National Chiao Tung University, Hsinchu 300, Taiwan, ROC

B. G. Haskell (1, 365), Visual Communications Research Department, AT&T Bell Laboratories, Holmdel, New Jersey 07733

---

[1]*Present address:* Philips Sound and Vision, Advanced Development Center/BCL, Eindhoven, The Netherlands

[2]*Present address:* Compression Labs, Inc., San Jose, California 95134

T. R. Hsing (421), Bellcore, Morristown, New Jersey 07960

N. S. Jayant (73), Signal Processing Research Department, AT&T Bell Laboratories, Murray Hill, New Jersey 07974

J. D. Johnston (73), Signal Processing Research Department, AT&T Bell Laboratories, Murray Hill, New Jersey 07974

R. L. Lagendijk (299), Delft University of Technology, Department of Electrical Engineering, Information Theory Group, 2600 AA Delft, The Netherlands

T. Naveen (265), Video and Networking Division, Tektronix, Inc., Beaverton, Oregon 97077

Y. Ninomiya (393), Science and Technical Research Laboratories, NHK (Japan Broadcasting Corporation), Shibuyaku, Tokyo 150, Japan

W. A. Pearlman (13), Electrical, Computer and Systems Engineering Department, Rensselaer Polytechnic Institute, Troy, New York 12180

P. Pirsch (465), Institut für Theoretische Nachrichtentechnik under Informationsverarbeitung, Universität Hanover, D-30167 Hanover, Germany

A. Puri (365), Visual Communications Research Department, AT&T Bell Laboratories, Holmdel, New Jersey 07733

K. R. Rao (223), Electrical Engineering Department, University of Texas at Arlington, Arlington, Texas 76019

R. J. Safranek (73), Signal Processing Research Department, AT&T Bell Laboratories, Murray Hill, New Jersey 07974

W. I. Way (447), Department of Telecommunication Engineering, National Chiao Tung University, Hsinchu 300, Taiwan, ROC

J. W. Woods (265), Center for Image Processing Research, Rensselaer Polytechnic Institute, Troy, New York 12180

S.-W. Wu[3] (189), Center for Information Processing Research, Department of Electrical and Computer Engineering, University of California, Santa Barbara, Santa Barbara, California 93106

---

[3]*Present address:* Advanced Video Technology Department, AT&T Bell Laboratories, Murray Hill, New Jersey 07974

Preface

Research and product development in visual communications have advanced very rapidly in the past two decades. Not long ago, visual communication was still an academic research topic and activities were limited to a few research institutes. Thanks to recent progress in VLSI technology, low-cost desktop computers, and wideband network deployment, digital video has become widespread in the communications, computer, and media industries. Videophone, multimedia, digital satellite TV, HDTV (high-definition television), and interactive TV are some of the more common examples. It is the goal of this book to provide a comprehensive treatment of various topics in the field of visual communications.

Visual communications is a relatively new field; however, it is a combination of several traditional disciplines: image source coding, video processing, motion estimation, digital communications, computer vision, and computer networking. In order to provide complete and accurate coverage of the entire field, we invited scholars and experts from all over the world to contribute to this book. Although many techniques in visual communications are now in daily use, few books that describe the entire field have been published. Therefore, we believe this book to be quite unique.

There are 15 chapters in all. The first chapter, contributed by Barry Haskell, a well-known AT&T Bell Laboratories pioneer in this field, is a brief introduction to the subject of visual communications. Chapters 2 and 3 deal with the fundamental theory of image compression. The authors of these chapters are also known for important contributions on these subjects. Chapter 2 is written from a statistical signal processing point of view, whereas Chapter 3 is written from the human visual system viewpoint. Together they form the basis of most image compression techniques.

Chapters 4 to 10 describe various popular image compression schemes. Chapter 4 is on black–white or bilevel image communication (Chapter 4) and is contributed by Donald Duttweiler, a member of the standards committee drafting the latest bilevel image transmission standard. Chapter 5 covers motion estimation, an essential technique used in modern video compression. Chapters 6 to 10 then describe five classes of popular image compression methods: vector quantization, transform coding, subband coding, hierarchical coding, and model-based coding. The authors are all well recognized for their pioneering work on these topics.

Chapters 6 to 9 cover traditional waveform coding schemes, of which transform coding has been adopted for current international video transmission standards for its balanced performance in compression efficiency, input video robustness, and system complexity. The basic vector quantization structure is close to the optimal compression scheme predicted by the information theory and has the virtue of simple implementation. In reality, video signals do not completely satisfy all the idealized assumptions in information theory. The more sophisticated structures of subband coding and hierarchical coding often provide subjectively superior pictures. In addition, these techniques offer compressed data with multiple priorities and thus are suitable for multilayer transmission and database retrieval systems. Model-based coding, described in Chapter 10, is a relatively new approach. Although its concept was suggested many years ago, only recently has this idea been fully implemented. It is one of the promising techniques for the next generation of video compression standards for very-low bit rate applications.

Realizing the needs of the global communications industry, international organizations have made tremendous efforts over the past 10 years to standardize digital video communications. We invited AT&T Bell Laboratories senior researchers, who are heavily involved in the standards activities, to contribute Chapter 11 on video standards. High-definition television is a buzz word in the news media. However, the only commercial broadcast HDTV that can be received today is the MUSE system—a hybrid-type (not purely digital) TV system—in Japan. Yuichi Ninomiya, who led the team defining this system, has provided a chapter on hybrid HDTV.

A complete communication system includes both the terminal and the communications link (network). Whereas the earlier chapters emphasized the terminal side, i.e., image compression, Chapters 13 and 14, contributed by senior researchers from Bellcore, emphasize the network issues of video transmission. All the algorithm and system designs must be implemented in hardware in order for benefits to be derived from this new technology. The high-speed, high-density, low-cost VLSI technology is the key that makes the era of digital video possible. It is our great pleasure to have Peter Pirsch, who has years of experience in this area, contribute the final chapter on VLSI design.

We thank all the contributors to this book. Without them this book could never exist. And indeed it is their efforts that make this book valuable. Visual communications, as an active R&D field, is still progressing. We see new products being brought out, new systems being designed, and new standards being added weekly. Hence, if at all possible, we hope this book can be updated every few years to bring state-of-the-art knowledge to new readers. Finally, we acknowledge the patience and guidance of several Academic Press editors who helped give birth to this book.

H.-M. Hang
J. W. Woods

Chapter 1
Video Data Compression

B.G. Haskell

Visual Communications Research Department
AT&T Bell Laboratories
Holmdel, New Jersey

1.1 Introduction

A considerable effort has been underway for some time to develop inexpensive transmission techniques that take advantage of recent advances in electronic technology as well as expected future developments. Most of the attention has been focused on digital systems because, as is well known, noise does not accumulate in digital regenerators as it does in analog amplifiers and, in addition, signal processing is much easier in a digital format.

Progress is being made on two fronts. First, the present high *cost per bit* of transmitting a digital data stream has generated interest in a number of methods that are currently being evaluated for cost reduction. While these methods have general applications and are not confined to a data stream produced by a video signal source, it is important to remember that video bit rates tend to be considerably higher than those required for voice or data transmission. The most promising techniques for more economical digital transmission include optical fibers, digital satellite, broadband ISDN, and digital transmission over the air, among others.

The second front on which progress is being made involves reducing the *number of bits* that have to be transmitted in a video communication system. Bit-rate reduction is accomplished by eliminating, as much as possible, the substantial

Handbook of Visual Communications

amount of redundant information that exists in a video signal as it leaves the camera. The amount of signal processing required to reduce the redundancy determines the economic feasibility of using this method in a given system. The savings that accrue from lowering the transmission bit rate must more than offset the cost of the required signal processing if redundancy reduction is to be economical.

Present costs of digital logic and digital memory are low enough to make this type of signal processing economically very attractive for use in long distance videoconferencing links over existing facilities. Furthermore, it is expected that the cost of digital logic and memory will continue to decline. Therefore, it is conjectured by those knowledgeable in the field that signal processing for bit-rate reduction will have an important part to play in all video systems, and in many cases, it could become the overriding factor determining economic feasibility.

To transmit video information at the minimum bit rate for a given quality of reproduction, it is necessary to exploit our understanding of many branches of science. Ideally the engineer should have an appreciation of motion pictures, colorimetry, human vision, signal theory, display devices, and so on. As might be expected any individual can have only a smattering of knowledge on such a diverse range of topics, and a specialist in any one topic will readily confess to a certain amount of ignorance even in his or her chosen field. As engineers we are concerned with complex stimuli and their human perception, as well as the final utilization of the perceived information. Knowledge of these is often unavailable or sketchy, forcing us to design encoders based on a relatively primitive understanding of the problem. The limits of bit-rate compression will be approached, we believe, only as our knowledge of stimuli, perception, and utilization increases.

Thus, in opening a discussion of video bit-rate compression we are very aware of our own limitations. Our modest objective of defining the state of the art is, we are well aware, open to the criticisms of oversimplification, serious omissions, and factual disagreement. As for where the subject is heading and its inherent

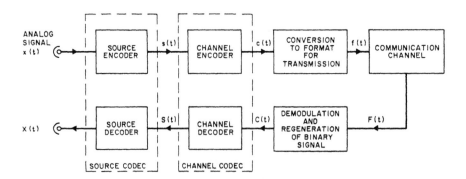

Figure 1.1   Digital communication link for the transmission of audio or pictorial information.

limitations, we confess myopia and will not be surprised by a discovery that could not have been extrapolated from existing thinking and known ignorances. But first let us set the stage for our discussion. The conventional representation of a digital communication link for the transmission of audio or pictorial information is shown in Fig. 1.1. The function of the source encoder is to operate on an analog of audio or pictures, $x(t)$, and to convert it into a stream of binary digits, $s(t)$. The source decoder at the receiver accepts a binary signal $S(t)$ and produces a continuous signal $X(t)$. It may not be necessary to ensure $X(t) = x(t)$, but what does matter is that after transduction, e.g., loudspeaker or TV tube, $X(t)$ should be perceived as $x(t)$, subject to an acceptable quality criterion. Although $x(t)$ does not always have to be identical to $X(t)$, system engineers prefer $s(t) = S(t)$; i.e., the channel appears ideal. Most practical channels contain dispersion, nonlinearities, additive noise, multipath fading, interference from other channels, and so on. These imperfections are overcome largely by preprocessing and postprocessing the binary signals $s(t)$ and $S(t)$ by the channel codec and terminal equipment. The transmitting terminal equipment operates on $c(t)$ to produce (perhaps by conversion to multilevel, modulation, filtering, etc.) a signal $f(t)$ that is suitable for combating the imperfections of the communication channel. The signal $F(t)$ that emerges from the channel may differ considerably from $f(t)$. After demodulation, a binary signal $C(t)$ is regenerated using adaptive equalization of the channel and adaptive detection strategies. The binary signal $C(t)$ is then channel and source decoded to produce $S(t)$.

The purpose of this book is to discuss mostly source encoding. However, Fig. 1.1 demonstrates that $S(t)$ is dependent on the channel terminal equipment, the channel codec, and of course, the channel. Thus, encoding picture signals is not merely a source encoding problem, but may include the complete communication system. For example, if the channel is known to result in a high bit error rate (ber), then the effect on the recovered signal $X(t)$ may be mitigated by altering the modulation and regeneration strategies, increasing the length of the check bits in the channel coding words, altering the source encoding algorithm, or combinations of all of these. The conventional arrangement of source and channel codecs may be altered, even merged. Postprocessing of $X(t)$ can also be successfully employed.

Thus, we are interested in the source codec, its algorithms, how they relate to the signals it encodes, how the bit rate can be reduced by exploiting the source signal statistics and properties of human perception, the variety of quality criteria, the codec complexity, and above all, how these phenomena are interrelated and can be traded to approach an optimum design.

We therefore present a discussion of picture sources and our scant knowledge of the salient properties of human perception. Armed with this we describe the current state of the art in waveform and parameter coding and conclude with directions for the future, guessing at where we believe some ultimate limitations may be found.

1.1.1 Picture Sources

Video processing or transmission systems typically start with a two-dimensional distribution of light intensity. Thus, three-dimensional scenes must first be projected onto a two-dimensional plane by an optical imaging system. Color pictures can usually be represented by three such light intensity distributions in three primary bands of wavelengths. If moving objects are to be accommodated, the light intensity must change with time.

The two-dimensional light intensity distribution is then usually raster scanned to produce a one-dimensional waveform. Facsimile involves single pictures, while in television the scene is repetitively raster scanned (usually with interlace to avoid flicker). Black/white pictures, e.g., printed or handwritten text, line drawings, weather maps, produce a two-level or binary waveform.

Color pictures produce three such waveforms corresponding to the three primaries. These are then usually converted by linear combination into a luminance (monochrome brightness) component and two chrominance (hue and saturation) components. Multiplexing methods for further combining these components into a single composite waveform are well known and widely used; however, the luminance component usually takes up most of the channel capacity.

1.1.2 The Eye and Seeing

The eye is the organ of sight, having at its rear an inner nervous coating known as the retina. Rays of light pass through the cornea, aqueous humor, lens, and vitreous body to form an image on the retina. The central area of the retina, known as the fovea, provides high resolution and good color vision in about 1 degree of solid angle. The images on the retinas are sent along two optic nerves, one for each eye, until they meet at the optic chiasma, where half the fibers of each nerve diverge to opposite sides of the brain. This enables observations in three dimensions.

The eye behaves as a two-dimensional low-pass filter for spatial patterns, with a high-frequency cutoff of about 60 cycles per degree of foveal vision and significant attenuation below about 0.5 cycle. Thus, high spatial frequencies in the image are not seen and need not be transmitted. The eye also acts as a temporal bandpass filter having a high-frequency cutoff between 50 and 70 Hz depending on viewing conditions. Flicker is more disturbing at high luminances and low spatial frequencies.

Noise and distortion are less visible at high-luminance levels than at middle- and low-luminance values, again depending on viewing conditions such as overall scene brightness and ambient room lighting. High- and low-frequency noise is less visible than mid-frequency noise. Distortions are also less visible near luminance transitions, such as occur at boundaries of objects in a scene. This is termed *spatial masking*, since the transitions *mask* the distortions.

Temporal masking also occurs. For example, shortly after a television scene change, the viewer is relatively insensitive to distortion and loss of resolution. This is also true of objects in a scene that are moving in an erratic and unpredictable fashion. However, if a viewer is able to track a moving object, then resolution and distortion requirements are the same as for stationary areas of a picture.

1.1.3 Subjective Assessment of Quality

As the variety of encoding algorithms increases so do the types of degradation perceived. If perception were thoroughly understood, the quality of reproduction of a particular video encoding strategy could be ascertained by objective measurements of signal parameters. The current situation is one of ad hoc objective measurements, each trying to relate subjective observations with each new encoding algorithm. Old methods of signal-to-noise ratio (SNR), spectral distance measures, pulse shapes, etc., are frequently inadequate. To postulate a new objective measure, subjective testing must be done. Here tests are made on a small sample of the population, and by statistical methods the effect on the entire population is estimated. Subjective testing is controversial. Should simple grading, bad to excellent in five steps, or multidimensional analysis be used? What form should the test take: word text, carefully assembled sentences, natural dialog, type of picture detail, amount of motion, etc.? However, what is even more in dispute is relating subjective testing results to objective measurements. Our inability to do this is a serious impediment both to communication between research scientists and to source encoding itself. Only when perception is properly understood will we have accurate objective measures. However, the day when we can, with confidence, objectively evaluate a *new* impairment without recourse to subjective testing seems very remote.

1.1.4 Statistical Redundancy and Subjective Redundancy

If an information source such as a television camera produces *statistically redundant* data—that is, information that could just as well have been derived from past data—then a saving in transmission bit rate can result if the redundant information is removed prior to transmission. In most cases, this requires, at the transmitter, a capability for storing some of the past source output so that a decision can be made as to what is and what is not redundant in the present source output. Memory of past information is also required at the receiver so that the redundance can be rederived and inserted back into the data stream in order to reproduce the original signal.

For example, in a television picture successive picture points (picture elements, or *pels* for short) along a line are very much alike, and redundancy reduction can be achieved by sending pel-to-pel differences instead of the pels themselves.

The differences are small most of the time and large only occasionally. Thus, an average bit-rate saving can be obtained by using short binary words to represent the more probable, small differences and longer binary words to represent the infrequent, large differences. In successive frames a pel also changes very little on the average.

Statistical redundancy is not the only form of redundancy in a video signal. There is also considerable *subjective redundancy;* that is, information that is produced by the source, but that is not necessary for subjectively acceptable picture quality at the receiver. For example, it is well known that viewers are less sensitive to degradations near edges; i.e., large brightness transitions, in the picture. Also, viewers require less reproduced resolution for moving objects in a picture than for stationary objects. Thus, in applications where exact reproduction of the video source output is not necessary as long as the displayed picture is subjectively pleasing, a further reduction in transmission bit rate can be achieved by removing subjective redundancy.

For example, an pel-differential PCM coder need not transmit large differences as accurately as small differences because of the viewer's insensitivity to distortion near large brightness transitions. Thus, prior to transmission, large pel differences can be quantized more coarsely than small differences, thereby reducing the number of levels that must be transmitted.

Videotelephone format pictures can be transmitted with 16-level quantization of pel differences (4 bits per pel). The bit rate can be reduced further by using multilength binary words for transmission; however, a buffer memory is then needed to transmit the resulting irregular data rate over a constant bit-rate digital channel.

In order to reduce frame-to-frame redundancy a memory or delay capable of storing an entire frame of video information is needed. At present, this requirement is the main factor determining the economic feasibility of frame-to-frame signal processing. However, it is expected that costs for digital storage will continue to decline, thereby making this type of signal processing even more attractive in the years to come.

One method of removing frame-to-frame redundancy is simply to reduce the number of frames that are transmitted per second. At the receiver, frames are repeated as in motion picture projection to avoid flicker in the display. This technique takes advantage of the fact that frame display rates must be greater than about 50 Hz to eliminate objectionable flicker, whereas something between 20 and 30 Hz is all that is required for rendition of normal motion, and less than 15 Hz for rendition of low-speed movement. A 50% reduction in bit rate can thus be obtained by transmitting only 15 frames per second and displaying each frame twice. However, jerkiness is then noticeable if the scene contains objects moving at moderate or rapid speed.

In most systems interlaced scanning already takes advantage of these phenomena to some extent. Odd numbered lines are sent during one half-frame period (field 1) and even numbered lines during the other half-frame period (field 2). For example, broadcast television systems in the United States transmit 30 frames per second using 2:1 interlace (60 fields per second).

1.2 Waveform Encoding

In waveform coding, a continuous *analog* signal $x(t)$ is encoded into a stream of bits by the source encoder, and from these bits, a decoder produces a recovered signal $X(t)$. The design objective in waveform encoding is that, for a given bit rate, $X(t)$ should be as close a replica of $x(t)$ as possible. Since many $x(t)$'s can produce the same $X(t)$ the difference, $n(t) = x(t) - X(t)$, cannot, in general, be zero all the time. This *quantization noise* is a fundamental limitation of finite bit-rate coding.

For example, in pulse code modulation (PCM), $x(t)$ is sampled at the Nyquist rate, and each pel is represented by a binary number, i.e., quantized. The decoder converts the binary numbers back to analog and low-pass filters them to obtain $X(t)$. The bit rate is the product of the sampling rate and the binary word length, the latter fixing the accuracy of conversion.

The signals from most TV cameras are already companded, i.e., a compressed nonlinear function of scene luminance. Eight-bit uniform quantization of this companded signal gives imperceptible quantization noise in most cases. Single pictures, e.g., photographs, typically require 1 bit less quantization accuracy than television, where the quantization noise is time varying and, therefore, much more visible. For black/white images only 1-bit quantization is required.

If perceptible quantization noise can be tolerated, then coarser quantization can be used and the bit rate reduced. The addition of random, or pseudo-random, noise prior to quantization, called *dithering*, changes the quantization error from being conditional on the input signal to approximately white noise and gives improved subjective results. With dithering, the saving is typically less than 3 bits per pel.

Successive pels of video are often highly correlated. In addition, periodicities exist in the waveform that lead to high correlations between pels that are separated, in some cases, by many sampling epochs. Predictive coding (also called *differential PCM* or *DPCM*) exploits these correlations by using previously transmitted quantized pels to form a prediction of the current pel to be encoded. The difference between the actual pel value and its prediction is quantized, binary encoded, and transmitted. The decoder is able to form the same prediction as the encoder (in the absence of transmission errors) because it has access to the same quantized pels. By adding the received quantized difference to the prediction the decoded quantized pel is obtained, and the signal is recovered by low-pass filtering.

The predictor may be a linear, nonlinear, or an adaptive function of previously encoded pels. Similarly the quantizer may be uniform, nonuniform, or adaptive. Using the previously quantized pel value as a prediction, companded 5-bit quantization achieves imperceptible distortions in domestic television, while 4-bit adaptive quantization achieves toll quality speech. By increasing the sampling rate, two-level or 1-bit quantization, called *delta modulation*, can be used to give a very simple implementation. However, for a given quality, bit rates are usually higher than with multilevel DPCM employing Nyquist rate sampling.

The predictor enables the variance and correlation of the difference signal to be significantly less than that of the original signal, enabling the quantization noise to be reduced for a given number of quantization levels. Further gains can be made by *entropy coding* the quantized difference signal; i.e., assigning short code words to small, but frequently occurring, values and longer code words to the seldomly occurring large values. However, with entropy coding, the bit rate depends on the input signal, and unless protective measures are taken, there is a chance of some signals producing a bit rate that exceeds the channel capacity, causing severe distortion in the recovered signal. Entropy encoding offers substantial improvements for picture signals and low bit-rate speech signals.

In some cases it pays to represent *groups* of pels with a single code word. For example, with black/white graphics and text, long strings of identical bits occur both with and without predictive encoding. Considerable savings occur from coding such strings with a single binary word, called *run-length coding*. Entropy coding yields further gains. In television, interframe codecs use previous frame prediction in nonmoving areas of the picture. These areas are efficiently encoded as groups of pels.

DPCM performance can be improved significantly by using adaptive predictors, adaptive quantizers, or both. Adaptive predictors attempt to optimize the prediction depending on the local waveform shape. In video, interframe coders typically use previous frame prediction in nonmoving areas; however, in moving areas linear combinations of pels in both the previous and present frame may be used as a prediction. By adapting the moving-area predictor to the speed and direction of motion, further improvement is achieved. ADPCM (adaptive DPCM) systems are limited by the predictor making predictions from pels corrupted by quantization noise. Therefore, the design of the predictor should take into account the characteristics of the quantizer and vice versa.

Adaptive quantization (AQ) greatly increases the effective number of quantization levels and hence the dynamic range of the signal that can be accommodated for a given bit rate and quality of encoding. In adaptive quantization, the step size is typically computed once for every block of pels.

While AQs for speech are essentially concertinalike, AQs used in picture encoding usually span the range of the input signal but adaptively discard some of their levels as a function of the video signal. Sharp transitions in the waveforms need not be represented as accurately as when variations are relatively slow. Visibility of quantization noise is markedly less at edges of objects than in flat, low-detail areas. Such subjective phenomena enable adaptive quantization to save a bit or more per pel.

Correlations and periodicities can also be exploited by transform coding. With this approach the pels to be coded are first partitioned into blocks. Pels within a block need not be contiguous in time, such as in television where groups of pels may be chosen from adjacent lines and adjacent frames in order to make up a block. Each block is then linearly transformed into another domain, e.g., frequency, having the desirable property that signal energy is concentrated in relatively few transform coefficients compared with the number of pels in the original block. Furthermore, all of the coefficients need not be quantized to the same accuracy to achieve a given quality of reproduction. By encoding only the significant coefficients with an accuracy dependent on human perception considerable bit-rate reductions are possible.

With adaptive transform coding (ATC), the coefficients selected for transmission change from block to block, as does the quantization strategy for each coefficient. For single pictures, many blocks contain little or no picture detail; i.e., they contain energy only at low frequencies and require only a few bits for encoding. Other blocks contain high-frequency components and produce more coding bits. Pictures containing an average amount of detail can be intraframe ATC encoded at around 2 bits per pel with imperceptible distortion, i.e., "excellent" quality, and 1.5 bits per pel with perceptible but not annoying distortion, i.e., "good" quality. In television three-dimensional ATC operates on blocks from several frames yielding large reductions in bit rate, particularly in nonmoving areas of the picture.

Hybrid encoding involves transform coding of blocks of pels followed by block-to-block DPCM encoding of the resulting coefficients. Used in picture encoding, its performance is similar to transform coding with larger blocks; however, implementation is simpler. Interframe adaptive hybrid coding of pictures with low movement has yielded bit rates of 1 bit per pel with "excellent" quality and 0.5 bit per pel with "good" quality.

ATC, ADPCM using entropy coding, and interframe coders all have the property that, for a fixed quality, the bit rate depends very much on the input waveform. This is undesirable for communication channels with a fixed channel capacity. Buffers can be used to accommodate the variable bit-rate generation to the constant bit-rate transmission. However, they introduce delay that may be intolerable in certain two-way communications. This may be reduced by sacrificing quality during periods of excessive bit-rate generation. Interframe coders take this approach by reducing moving-area picture quality during periods of rapid movement. However, it must be emphasized that all constant bit-rate waveform codecs produce a variable quality either on a block basis, as with ATC, or on a per pel basis, as with DPCM.

Interframe ADPCM will eventually advance to the point where object motion is tracked extremely well, including translation, rotation, and shape changes. Adaptive filters and quantizers will optimize the displayed resolution (temporal and spatial) and quantization noise to the subjective requirements of the viewer. With these techniques, camera motion (zooming and panning) will have little effect on overall bit rate.

In an interframe encoder with entropy coding, the long-term average and the short-term peak bit rates differ considerably. For purposes of digital recording this is of little consequence. However, for present day real-time communication, where data peaks cannot be *buffered out* via the use of large memories and long delays, either excess channel capacity has to be provided or picture quality has to be compromised. In the near future, there will be considerably more video traffic making it feasible for many video sources to share the same communication channel. Advantage can then be taken of the fact that simultaneous data peaks in several sources rarely occur, and the allocated *per source* channel capacity can be made much closer to the long-term average data rate without introducing long delays due to buffering.

Ultimately, such channel sharing arrangements (equivalent to packet switching with variable length packets) appear to be the only way that real-time video can take advantage of highly adaptive waveform coders that produce low bit rates for low-detail or low movement pictures, but require higher rates otherwise. While it may be true that the *average* picture has average detail and average movement, few systems will be successful unless they can accommodate the full range of pictures that the *average* viewer finds interesting.

1.3 Parameter Coding

In parameter coding of speech the signal is analyzed in terms of a model of the vocal mechanism and the parameters of the model transmitted. The receiver uses the parameters to synthesize a speech signal that is perceptually similar to the original speech. There is no equivalent of this in picture encoding because of the difference in the nature of the sources. However, image parameters, such as high-detail/low-detail indicators, positions, and orientations of edges, speed, and direction of moving objects in successive TV frames, and so on, can be used in image coding. These parameters are employed both in the prediction processes of DPCM and the quantization processes of both DPCM and ATC.

For black/white text consisting of typed characters, optical character recognition (OCR) algorithms identify each character and its position as two parameters that are then encoded and transmitted. This gives large compression gains compared to waveform encoding. Typically a standard $8\frac{1}{2} \times 11$ inch typewritten page (80×66 characters) can be transmitted with 8 bits per character; i.e., about 0.01 bit/pel. Black/white graphics can also be handled using OCR, but the character alphabet must be constructed adaptively for each document or class of documents to be coded. This usually necessitates the transmission of side information. Test documents have been encoded at about 0.025 bit/pel.

For pictures having a continuous gray scale, one parametric approach is to decompose the picture into edge information (boundaries of objects) and texture information (everything else). The texture is then assumed to be a random process for which optimum rate-distortion coding strategies apply, and the edges are coded efficiently using black/white graphics techniques.

Parameter coding has the prospect of much further reductions in bit rate compared to waveform coding. We consider parameter coding to include recognition of one or more global attributes of the image that enable more efficient coding while still achieving the quality objective.

For black/white text, the ultimate in parameter coding is character recognition. However, this approach rapidly merges with the science of computer generated graphics and the study of specialized graphics languages. Generally, the more specialized is the language, the greater the efficiency of representation. Many specialized graphics languages currently exist. For example, integrated circuit layouts never reside in computer memory in point-by-point form. Instead, they are built up from basic blocks according to instructions written in graphics language.

The graphics language approach will ultimately benefit the encoding of gray scale pictures as well. Complete specification of video scenes by a graphics language will probably not be possible, except for specialized situations like cartoons. However, important features such as boundaries and locations of objects are essentially graphical information, and it ought to be possible to represent them as such with very efficient codes. Motion of objects as well as shape changes should also be representable by fairly efficient parameter codes.

Replication of the detail, shading, etc., within boundaries of objects requires additional *fix-up* data to be transmitted. The *amount* of fix-up needed depends on the application. In some cases, e.g., surveillance, very little is necessary; however, in broadcast television full cosmetic restoration must be maintained. The *fidelity* of the fix-up depends on context. For example, the texture in a grassy field may well be replaceable with random noise having similar statistics, whereas detail in a human face may have to be replicated exactly.

Trying to estimate the ultimate bit rate achievable by parameter coding of video suffers the problem that different pictures produce different bit rates, and different applications require different fidelities of reproduction. However, the graphical parameters should require on average about 0.08 bit per pel for a single picture and perhaps a quarter of that (or less) for moving video depending on the amount of motion rendition required. The bit rate needed for fix-up is much more elusive due to the large variation in pictures and fidelity requirements. We guess it should range downward from 1 bit per pel for excellent quality, single pictures and a quarter of that (or less) for moving video, where interframe redundancy can be exploited. Only time will tell how close these estimates come to practicality.

Chapter 2

Information Theory and Image Coding

W. A. Pearlman

Electrical, Computer and Systems Engineering Department
Rensselaer Polytechnic Institute
Troy, New York

This chapter presents a tutorial exposition of principles and techniques of data compression, applied mainly but not exclusively to images. The orientation is toward methods that are founded through the tenets of information theory. Toward this end, the relevant theorems of the source coding branch of information theory, called *rate–distortion theory*, are cited and explained, omitting formal proofs. Then the various methods motivated by these theorems are presented. They include optimal coding, scalar (PCM) quantization, the role of entropy coding and entropy constraints, vector coding, transform coding, predictive (DPCM) coding, and subband coding. Within this framework are explained the operational details of a functional system. For example, general analytical formulas are derived for allocations of rate among transform and subband elements. Gain formulas, some of which are new or new generalizations, are derived to compare the performances of different systems.

2.1 Introduction

Source coding began with the initial development of information theory by Shannon in 1948 [1] and continues to this day to be influenced and stimulated by advances in this theory. Information theory sets the framework and the language, motivates the methods of coding, provides the means to analyze the methods, and establishes the ultimate bounds in performance for all methods. No study of image coding is complete without a basic knowledge and understanding of information theory.

In this chapter, we shall present several methods for coding images or other data sources, but intertwined with the motivating information theoretic principles and bounds on performance. The chapter is not meant to be a primer on information theory, so theorems and propositions will be presented without proof. The reader is referred to one of the many excellent textbooks on information theory, such as Gallager's [2], for a deeper treatment with proof. As the purpose here is to present coding methods and their performance, information theory is invoked only as needed for this purpose. Ideally, the reader will derive from this chapter both knowledge of coding methods and an appreciation and understanding of the underlying information theory.

The chapter begins with definitions of entropy and the presentation of the noiseless coding theorem and its converse, which states roughly that the minimum possible rate for perfect reconstruction of a discrete-amplitude source is its entropy. Means of achieving this minimal rate are then cited. The chapter then moves quickly to continuous-amplitude sources and rate–distortion theory, the branch of information theory that treats the coding of such sources. Optimal code structures are briefly mentioned as a reference point for the following section on scalar quantization, which describes simpler methods, such as nonuniform and uniform quantization with and without entropy constraints. Next, various means of coding sources with memory are treated: vector quantization, transform coding, predictive (DPCM) coding, and subband coding. When possible, gain formulas, some of which are new, are derived to compare performance against simpler schemes.

Although images are two-dimensional, the notation in this chapter will be one-dimensional, as a linear ordering of the two-dimensional picture element array can usually be assumed. In the case of transforms, the extension from one to two dimensions is assumed to be separable unless otherwise indicated.

2.2 Noiseless Source Coding

Consider an information source that emits a vector random variable of dimension N denoted by $\mathbf{X} = (\mathbf{X}_1, \mathbf{X}_2 \ldots \mathbf{X}_N)$ according to a probability law characterized by a probability mass function or probability density function $q_{\mathbf{X}}(\mathbf{x})$, depending on whether the random vector takes on continuous or discrete values of $\mathbf{X}$ in N-dimensional Euclidean space. Each vector element X_i, $i = 1, 2 \ldots N$, is called a *source* letter or *symbol*. Assume the source is stationary so that the probability function is the same for any N and length N vector emitted at any time. When the source is discrete in amplitude, the entropy of the vector can be defined as

$$H(\mathbf{X}) = - \sum_{\text{all } \mathbf{x}} q_{\mathbf{X}}(\mathbf{x}) \log_2 q_{\mathbf{X}}(\mathbf{x}),$$

where the sum is over all values $\mathbf{x}$ of $\mathbf{X}$. The logarithmic base of 2 provides an information measure in bits and will be understood as the base in all logarithms that follow. It is often more meaningful to refer to entropy per source letter, defined to be

$$H_N(\mathbf{X}) = \frac{1}{N} H(\mathbf{X}).$$

The source is said to be memoryless when the individual components of the vector, called the *source letters*, are statistically independent; i.e.,

$$q_{\mathbf{X}}(\mathbf{x}) = q_{X_1}(x_1) q_{X_2}(x_2) \ldots q_{X_N}(x_N)$$
$$= q_X(x_1) q_X(x_2) \ldots q_X(x_N).$$

The last equality, which removes the dependence of the probability distribution on time, follows from stationarity.

Suppose that the source X_i is memoryless and discrete. It emits at any time values (letters) from a countable set (alphabet); i.e.,

$$x_i \in \{a_1, a_2, \ldots, a_K\}, \qquad K \text{ finite or infinite,}$$
$$i = 1, 2, \ldots, N,$$

with respective probabilities $P(a_1), P(a_2) \ldots P(a_K)$. The entropy or average uncertainty of the source in bits per source letter is

$$\frac{1}{N} H(\mathbf{X}) \equiv H(X) = \sum_{k=1}^{K} P(a_k) \log \frac{1}{P(a_k)}, \tag{2.1}$$

where the base of the logarithm is 2 unless otherwise specified.

Through the nonnegativity of $-\log P(a_k)$ for all k and the inequality $\ln x \leq x - 1$, it is easy to prove that

$$H(X) \geq 0$$

and

$$H(X) \leq \log K, \tag{2.2}$$

with equality in the latter if and only if the probabilities $P(a_k)$ are equal. When the source is not memoryless, it is fairly obvious that the entropy $H(\mathbf{X})$ of the vector $\mathbf{X}$ follows (2.2) when K is interpreted as the number of values of $\mathbf{X}$ with nonzero probability. It can also be shown that

$$H_N(\mathbf{X}) \equiv \frac{1}{N} H(\mathbf{X}) \le \frac{1}{N} \sum_{i=1}^{N} H(X_i) = H(X), \qquad (2.3)$$

which means that the uncertainty per source letter is reduced when there is memory or dependence between the individual letters. Furthermore, as N tends toward infinity, $H_N(\mathbf{X})$ goes monotically down to a limit $H_\infty(\mathbf{X})$. The following source coding theorems can now be stated:

Theorem 2.1   *For any $\epsilon > 0$, $\delta > 0$, there exists N sufficiently large that a vector of N source letters can be put into one-to-one correspondence with binary sequences of length $L = N[H_\infty(\mathbf{X}) + \epsilon]$ except for a set of source sequences occurring with probability less than δ. Conversely, if $\frac{L}{N} < H_\infty(\mathbf{X})$, the set of source sequences having no binary code words, approaches 1 as N grows sufficiently large.*

Note that, when the source is memoryless, $H_\infty(\mathbf{X}) = H(X)$. The ramification of this theorem is that we can select the $K = 2^{N[H_\infty(\mathbf{X})+\epsilon]}$ vectors from the source that occur with probability greater than $1 - \delta$ and index each of them with a unique binary code word of length $L = N[H_\infty(\mathbf{X}) + \epsilon]$. If we transmit the binary index of one of these vectors to some destination where the same correspondences between the K indices and vectors are stored, then the original source vector is perfectly reconstructed. When the source emits a vector that is not among the K indexed ones, an erasure sequence is transmitted with no recovery possible at the destination. The probability of this error event is less than δ. The set of K binary sequences is called a code with rate in bits per source letter of $R = \frac{1}{N} \log_2 K = H_\infty(\mathbf{X}) + \epsilon$. The converse of the theorem means that $H_\infty(\mathbf{X})$ is the smallest possible rate in bits per source letter for a code that enables perfect reconstruction of the source vectors at the destination.

Consider now the case of a memoryless source. If one is willing to transmit binary code word sequences of variable length, one can theoretically eliminate the error event associated with fixed length code word sequences. In practice, however, one needs to utilize a fixed length buffer that may overflow or become empty with a finite probability when operating for a finite length of time. If we ignore the buffering problems by assuming an infinite buffer, the idea is to choose for $\mathbf{X} = \mathbf{x}$ a binary code word sequence of length $L(\mathbf{x})$ such that

$$\log \frac{1}{q_\mathbf{X}(\mathbf{x})} \le L(\mathbf{x}) < \log \frac{1}{q_\mathbf{X}(\mathbf{x})} + 1. \qquad (2.4)$$

Averaging this inequality, dividing by N, and using the memoryless property of the source, we obtain

$$H(X) \leq \frac{\bar{L}}{N} < H(X) + \frac{1}{N}. \tag{2.5}$$

As N approaches infinity, the average number of binary digits per source letter $\bar{L}/N$ or rate of the code approaches $H(X)$, which parallels the result for fixed length codes.

When variable length code words are strung together and transmitted, there is no assurance that they can be uniquely separated and decoded at the destination unless an extra mark digit is inserted between each code word. As insertion of an extra digit for each code word is uneconomical, one seeks codes (sets of code words) that are uniquely decodable when strung together and of minimum average length. The Huffman code [3] is the most well-known example of such codes.

For sources with memory, we can still choose a binary code word of length $L(\mathbf{x})$ satisfying

$$\log \frac{1}{q_{\mathbf{X}}(\mathbf{x})} \leq L(\mathbf{x}) < \log \frac{1}{q_{\mathbf{X}}(\mathbf{x})} + 1$$

for the vector $\mathbf{X}$. When we average over the source probability distribution and divide by N, we obtain the vector analog of (2.5) in the form

$$H_N(\mathbf{X}) \leq \frac{\bar{L}}{N} < H_N(\mathbf{X}) + \frac{1}{N}.$$

The advantage is that when the source has memory

$$H_N(\mathbf{X}) < H(X),$$

the average length of the code is smaller. As N becomes larger, the advantage in coding N-tuples becomes progressively greater. However, since every vector value must have a unique code word, the number of code words grows exponentially in the dimension N. Another possibility for exploiting memory in the source is to consider conditional probability distributions. Let $\mathbf{X}_k$ and $\mathbf{X}_l$ be two successive N-dimensional vectors (to be called N-vectors) from the source. The conditional probability distribution of $\mathbf{X}_k$ given $\mathbf{X}_l$ is

$$q_{k/l}(\mathbf{X}_k/\mathbf{X}_l) = \frac{q_{\mathbf{X}}(\mathbf{x}_k, \mathbf{x}_l)}{q_{\mathbf{X}}(\mathbf{x}_l)},$$

where $q_{\mathbf{X}}(\mathbf{x}_k, \mathbf{x}_l)$ is the joint probability distribution of the two vectors and $q_{\mathbf{X}}(\mathbf{x}_l)$ is nonzero. The conditional entropy of $\mathbf{X}_k$ given $\mathbf{X}_l$ is

$$H(\mathbf{X}_k/\mathbf{X}_l) = \sum_{\text{all } \mathbf{x}_k, \mathbf{x}_l} q_{\mathbf{X}}(\mathbf{x}_k, \mathbf{x}_l) \log \frac{1}{q_{k/l}(\mathbf{x}_k/\mathbf{x}_l)}.$$

One of the central facts of information theory is that conditioning cannot increase information required to specify a source. The precise statement in the context here is

$$H(\mathbf{X}_k/\mathbf{X}_l) \leq H(\mathbf{X}_k).$$

Therefore, if binary code words are chosen for $\mathbf{x}_k$ with lengths such that

$$\log \frac{1}{q_{k/l}(\mathbf{x}_k/\mathbf{x}_l)} \leq L(\mathbf{x}_k) < \log \frac{1}{q_{k/l}(\mathbf{x}_k/\mathbf{x}_l)} + 1.$$

the average length per source symbol over the N-vector random variables $\mathbf{X}_k$ and $\mathbf{X}_l$ are

$$H_N(\mathbf{X}_k/\mathbf{X}_l) \leq \frac{\bar{L}}{N} < H_N(\mathbf{X}_k/\mathbf{X}_l) + \frac{1}{N},$$

where the subscript N denotes usual division of $H(\ )$ by N. Since

$$H_N(\mathbf{X}_k/\mathbf{X}_l) \leq H_N(\mathbf{X}_k) = H_N(\mathbf{X}),$$

the use of conditioning has produced a code of lower average length unless $\mathbf{X}_k$ and $\mathbf{X}_l$ are independent. Now for a given $\mathbf{X}_l = \mathbf{x}_l$, a different code word must be assigned for each possible value of $\mathbf{x}_k$. The performance advantage of conditioning decreases as N grows larger, since both $H_N(\mathbf{X}_k)$ and $H_N(\mathbf{X}_k/\mathbf{X}_l)$ approach $H_\infty(\mathbf{X})$ in the limit. The objective of an efficient distortionless source coding scheme is to operate as closely as possible, within complexity constraints, to the entropy rate of the source. Thus far, considerations of complexity have prevented acceptance of conditional entropy coding schemes.

2.3 Continuous-Amplitude Sources

2.3.1 Differential Entropy

When the source letters are continuous in amplitude, the entropy rate $H_\infty(\mathbf{X})$ or $H(X)$ is infinite. It is useful, however, to define a quantity called *differential entropy* of a source, in which probability mass distributions are replaced by densities and sums by integrals. For a vector source $\mathbf{X}$ with joint probability density function $q_\mathbf{X}(\mathbf{X})$, the differential entropy is defined as

$$h(\mathbf{X}) = E\left[\log \frac{1}{q_\mathbf{X}(\mathbf{X})}\right] = \int q_\mathbf{X}(\mathbf{x}) \log \frac{1}{q_\mathbf{X}(\mathbf{x})} d\mathbf{x}.$$

The differential entropy has two failings with regard to entropy: (1) it can be negative; and (2) it changes with a linear scaling of the vector or any of its components. It does, however, reflect a dispersive property of the distribution $q_X(\mathbf{x})$ and is therefore sometimes called *dispersion*.

A memoryless source X with probability density function $q_X(x)$ has a differential entropy defined to be

$$h(X) = -\int q_X(x) \log q_X(x) dx. \tag{2.6}$$

When the source is Gaussian with zero mean and variance σ^2, it has the probability density

$$q_X(x) = \frac{1}{\sqrt{2\pi\sigma^2}} \exp\left\{-\frac{x^2}{2\sigma^2}\right\}, \qquad -\infty < x < \infty. \tag{2.7}$$

Its differential entropy is evaluated to be

$$h(X) = \frac{1}{2} \log(2\pi e \sigma^2). \tag{2.8}$$

One can also show for any source X with variance σ^2,

$$h(X) \leq \frac{1}{2} \log(2\pi e \sigma^2), \tag{2.9}$$

with equality if and only if the source is Gaussian. The Gaussian source has the largest dispersion of all sources having the same variance. A similar conclusion holds for vector sources with and without memory, but we shall defer the discussion of this subject until later when vector sources are studied in more detail.

2.3.2 Source Coding with a Fidelity Criterion

When the real-valued source letters are not discrete, but continuous, amplitude, the entropy rate is generally infinite. Perfect reconstruction is therefore impossible with a finite code rate. To reproduce the source vector of length N at a remote point, a certain amount of distortion must be accepted. First, a measure of distortion is defined between the source vector realization $\mathbf{x}$ and its corresponding reproduction $\mathbf{y}$, denoted by $d_N(\mathbf{x}, \mathbf{y})$. At the source is a list of K possible reproduction vectors $\{\mathbf{y}_1, \mathbf{y}_2, \ldots, \mathbf{y}_K\}$ of length N called the *codebook* or *dictionary* C. When $\mathbf{x}$ is emitted from the source, the codebook C is searched for the reproduction vector $\mathbf{y}_m$ that has the least distortion; i.e.,

$$d_N(\mathbf{x}, \mathbf{y}_m) \leq d_N(\mathbf{x}, \mathbf{y}_k), \qquad \text{for all } k \neq m,$$

and the binary index of m is transmitted to the destination. The rate R of the code in bits per source letter is therefore

$$R = \frac{1}{N} \log_2 K.$$

As the same codebook and indexing are stored at the destination in a table, the look-up in the table produces the same reproduction vector and same distortion found at the source through search if the channel is assumed to be distortion free. A diagram of the system model is shown in Fig. 2.1. Since the source vector is a random variable (of N dimensions), the average distortion obtained for a given codebook C can in principle be calculated with a known source probability distribution. Through the search of the codebook, the minimum distortion reproduction vector $\mathbf{y}_m$ is a function of the source vector $\mathbf{x}$ and is also a random variable.

The average distortion per source letter of the code C is therefore

$$E\left[\frac{1}{N} d_N(\mathbf{X}, \mathbf{y}_m(\mathbf{X}))/C\right]$$

where the expectation is calculated for the given source distribution $q_\mathbf{X}(\mathbf{x})$. The optimal code C is the one that produces the least average distortion for the same rate R; i.e.,

$$\min_C E\left[\frac{1}{N} d_N(\mathbf{X}, \mathbf{y}_m(\mathbf{X}))/C\right],$$

where the minimization is over all codebooks C of rate R.

To find an optimal code through an exhaustive search over an infinite number of codes of size $K = 2^{RN}$ is truly an impossible task. Furthermore, optimal

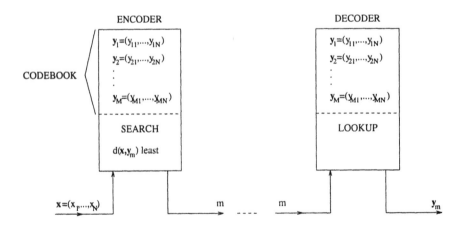

Transmission of m : ≤ log$_2$M bits/vector

or log$_2$M/N bits per sample

Figure 2.1   A block or vector coding system model.

performance is obtained in the limit of large N, as it can be proven that as N approaches infinity, $E\left[\frac{1}{N} d_N(\mathbf{X}, \mathbf{y}_m(\mathbf{X}))/C\right]$ approaches a limit in a monotone and nonincreasing fashion. We must therefore resort to another approach to find the minimum average distortion of a code of rate R. This approach is through a theory called *rate–distortion theory* and actually yields a method of code construction that is "highly probable" to achieve optimal performance. We shall now introduce this theory to make the above statements more precise.

2.3.3 Rate–Distortion Theory

Consider a continuous-amplitude vector source ensemble $\mathbf{X}$ and a reproduction vector ensemble $\mathbf{Y}$, with a stochastic relationship between them governed by the conditional probability density function $p_{\mathbf{Y}/\mathbf{X}}(\mathbf{y}/\mathbf{x})$ of a reproduction vector $\mathbf{Y} = \mathbf{y}$ given a source vector $\mathbf{X} = \mathbf{x}$. The conditional probability density function $p_{\mathbf{Y}/\mathbf{X}}(\mathbf{y}/\mathbf{x})$ is called a *test channel*, which is depicted in Fig. 2.2.

This test channel description relating output to input, being probabilistic, is not coding, in which the equivalent channel is entirely deterministic. A theorem to be stated later will establish the relationship to coding. The average distortion per source letter can be calculated for the joint density function $q_{\mathbf{X}}(\mathbf{x}) \, p_{\mathbf{Y}/\mathbf{X}}(\mathbf{y}/\mathbf{x})$ for the test channel joint input–output ensemble $\mathbf{XY}$ as

$$E\left[\frac{1}{N}d_N(\mathbf{X}, \mathbf{Y})\right] = \int \int \frac{1}{N}d_N(\mathbf{x}, \mathbf{y})q_{\mathbf{X}}(\mathbf{x})p_{\mathbf{Y}/\mathbf{X}}(\mathbf{y}/\mathbf{x})d\mathbf{x}d\mathbf{y}. \qquad (2.10)$$

The (average mutual) information per letter between the input and output ensembles is

$$\frac{1}{N}I_N(\mathbf{X}; \mathbf{Y}) = E\left[\frac{1}{N} \log \frac{p_{\mathbf{Y}/\mathbf{X}}(\mathbf{y}/\mathbf{x})}{w_{\mathbf{Y}}(\mathbf{y})}\right] \qquad (2.11)$$

$$= \frac{1}{N} \int \int q_{\mathbf{X}}(\mathbf{x})p_{\mathbf{Y}/\mathbf{X}}(\mathbf{y}/\mathbf{x}) \log \frac{p_{\mathbf{Y}/\mathbf{X}}(\mathbf{y}/\mathbf{x})}{w_{\mathbf{Y}}(\mathbf{y})}d\mathbf{x}d\mathbf{y}$$

where $w_{\mathbf{Y}}(\mathbf{y}) = \int q_{\mathbf{X}}(\mathbf{x})p_{\mathbf{Y}/\mathbf{X}}(\mathbf{y}/\mathbf{x})d\mathbf{x}$ is the probability density function of the output vector random variable $\mathbf{Y}$. The source vector probability distribution $q_{\mathbf{X}}(\mathbf{x})$

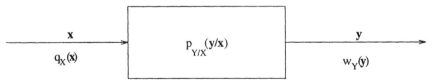

Figure 2.2   A test channel.

is considered to be known, so that the information $I_N(\mathbf{X}; \mathbf{Y})$ is a function of the test channel and is hence denoted as $I_N(p)$.

Consider now the problem of minimizing the average distortion per letter over all test channels $p_{\mathbf{Y}/\mathbf{X}}(\mathbf{y}/\mathbf{x})$, giving an average mutual information per letter no greater than some fixed rate $R \geq 0$. The mathematical statement is

$$D_N(R) = \inf_p \left\{ E\left[\frac{1}{N} d_N(\mathbf{X}, \mathbf{Y}) \right] : \frac{1}{N} I(p) \leq R \right\}. \tag{2.12a}$$

The result is an average distortion $D_N(R)$, which is a function of R and depends on N. The function $D_N(R)$ is called the *N-tuple distortion–rate function*. The distortion-rate function $D(R)$ is

$$D(R) = \inf_N D_N(R) = \lim_{N \to \infty} D_N(R). \tag{2.12b}$$

The corresponding inverse functions, $R_N(D)$ and $R(D)$, are stated mathematically as

$$R_N(D) = \inf_p \left\{ \frac{1}{N} I_N(p) : E\left[\frac{1}{N} d_N(\mathbf{X}, \mathbf{Y}) \leq D \right] \right\}, \tag{2.13a}$$

$$R(D) = \inf_N R_N(D) = \lim_{N \to \infty} R_N(D), \tag{2.13b}$$

and are called the *N-tuple rate–distortion function* and the *rate–distortion function*, respectively.

We now restrict the distortion measure to be based on a single-letter distortion $d(x, y)$ between corresponding vector components according to

$$d_N(\mathbf{x}, \mathbf{y}) = \sum_{i=1}^{N} d(x_i, y_i).$$

The following theorems connect these functions to coding.

Theorem 2.2 Distortion–Rate Theorem and Its Converse   *For stationary and ergodic sources (under certain technical conditions) and a single-letter based distortion measure, for any $\epsilon > 0$ and N sufficiently large, there exists a code C of rate R such that*

$$E\left[\frac{1}{N} d_N(\mathbf{X}, \mathbf{y}_m(\mathbf{X})) \mid C \right] < D(R) + \epsilon.$$

Furthermore, there exists no code of rate R such that

$$E\left[\frac{1}{N} d_N(\mathbf{X}, \mathbf{y}_m(\mathbf{X})) \mid C \right] < D(R).$$

The distortion–rate function $D(R)$ is an unbeatable lower bound on average distortion for a code of rate R.

The inverse statement of the above theorem is also common.

Theorem 2.3 Rate–Distortion Theorem and Its Converse   *For stationary and ergodic sources (under certain technical conditions) and a single-letter based distortion measure, given $D \geq 0$ and $\epsilon > 0$, for N sufficiently large there exists a code of rate R such that*

$$R < R(D) + \epsilon.$$

Furthermore, there exists no code of rate

$$R < R(D).$$

The rate–distortion function $R(D)$ is an unbeatable lower bound on rate for a code with average distortion D.

These two two theorems are likewise valid for discrete-alphabet sources.

2.3.4 Rate–Distortion Functions

Memoryless Sources   For memoryless sources and a single-letter based distortion measure, the rate–distortion (or distortion–rate) function can be stated more simply, since

$$E\left[\frac{1}{N} d_N(\mathbf{X}, \mathbf{Y})\right] = E[d(X, Y)],$$

where X and Y are scalar input and output random variables of a scalar test channel $p_{Y/X}(y/x)$, and average mutual information may be calculated for this test channel (instead of the vector one) as

$$I(p) = E\left[\log \frac{p_{Y/X}(y/x)}{w_Y(y)}\right], \tag{2.14}$$

where $w_Y(y) = \int q_X(x) p_{Y/X}(y/x) dx$ is the test channel output probability density. The rate–distortion function is

$$R(D) = \inf_p \{I(p) : E[d(X, Y)] \leq D\}. \tag{2.15}$$

In some cases $R(D)$ can be evaluated or bounded analytically. Otherwise, there is a computational algorithm by Blahut [4] for evaluating the points of $R(D)$. $R(D)$ is a monotone, nonincreasing, convex $\bigcup$ (concave) function of D. Some important special cases, where exact formulas or analytical bounds are known, follow.

Case 1 $R(D)$ for a Gaussian source X with zero mean and variance σ^2 and squared error distortion measure.

$$q_X(x) = \frac{1}{\sqrt{2\pi\sigma^2}} \exp\left\{-\frac{x^2}{2\sigma^2}\right\}, \qquad -\infty < x < +\infty,$$

$$d(x, y) = (x - y)^2,$$

$$R(D) = \max\left\{0, \frac{1}{2}\log\frac{\sigma^2}{D}\right\} \qquad \text{bits/source letter or}$$

$$D(R) = \sigma^2\, 2^{-2R}, \quad R \geq 0.$$

Note the two extremal points $D = 0$, $R = \infty$ and $R = 0$, $D = \sigma^2$. The first one says that infinite rate is required for zero mean squared error, as expected. The second one says that a mean squared error of σ^2 can be achieved for a zero rate. This is done by setting any input x to zero, the mean value. The reception of all zeros is certain, so no information rate is required to convey it.

Case 2   *An upper bound to R(D)*   Given a source of unknown density and variance σ^2 and squared error distortion measure, the rate–distortion function is overbounded as

$$R(D) \leq \max\left\{0, \frac{1}{2}\log\frac{\sigma^2}{D}\right\}. \qquad (2.16)$$

The upper bound is the rate–distortion function of the Gaussian source. This means that the Gaussian source of the same variance is the most difficult to encode, since it requires the largest rate to obtain the same distortion.

Case 3   A lower bound to $R(D)$.
A lower bound to $R(D)$ for squared error distortion is

$$R(D) \geq h(X) - \frac{1}{2}\log(2\pi e D), \qquad (2.17)$$

where $h(X)$ is the differential entropy of the source.

2.3.5 Optimal Code Construction

The rate–distortion or (distortion–rate) theorem guarantees the existence of optimal codes but does not tell us how to construct these codes. However, a methodology in the proof of the theorems will produce an optimal code with probability approaching 1. Consider the memoryless source case and denote by $p_o(y/x)$

the test channel that achieves the rate–distortion function in (2.13). The output probability density of this optimal test channel is

$$w_o(y) = \int q_X(x) p_o(y/x) dx. \qquad (2.18)$$

Choose variates y_{ij} independently and at random from the probability density $w_o(y)$ until K vectors of length N are obtained $(j = 1, 2, \ldots, N; i = 1, 2, \ldots, K)$. The reproduction codebook is

$$
\begin{aligned}
\mathbf{y}_1 &= (y_{11}, y_{12}, \ldots, y_{1N}) \\
\mathbf{y}_2 &= (y_{21}, y_{22}, \ldots, y_{2N}) \\
&\vdots \\
\mathbf{y}_K &= (y_{K1}, y_{K2}, \ldots, y_{KN})
\end{aligned}
$$

where the rate of the code is $R = N^{-1} \log_2 K$. For any $\epsilon > 0$ and sufficiently large N, the code will have average distortion no more than $D(R) + \epsilon$ with probability close to unity. As pointed out previously, for large N, the storage of $NK = N2^{RN}$ real variates and the $N2^{RN}$ multiplications and additions to find the minimum distortion $\mathbf{y}_m$ for a given source vector $\mathbf{x}$ is an enormous burden of storage and complexity. Fortunately, codebooks may be stored along code word-dependent structures such as trees and trellises, which relieve some of the burdens of storage and complexity, while still maintaining optimality. The rate R_T of a code residing on a tree or trellis, with branching factor (number of branches emanating from each node) α and number of reproduction letters on each branch β, is $R_T = \beta^{-1} \log \alpha$. It will not be our task here to explain the workings of such codes, only to point out their existence.

2.4 Scalar Quantization

2.4.1 Structure of Quantization

A much less complex alternative to optimal coding is an encoding procedure called *quantization*, which is illustrated in Fig. 2.3.

Consider a stationary, memoryless source X with probability density function (X of continuous–amplitude) $q(x)$.[1] To encode $N = 1$ sample at a time, assume a set of reproduction letters (codebook) $\{y_1, y_2 \ldots y_K\}$ and a distortion measure $d(x, y_k)$, $k = 1, 2 \ldots K$ for any x. As before, given x, find y_k such

---

[1] Hereafter, the subscript X in the density $q_X(x)$ will be dropped when the random variable is obvious.

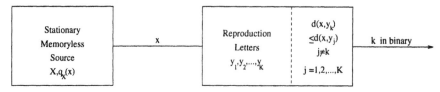

Figure 2.3   Model of scalar quantization.

that $d(x, y_k) \leq d(x, y_l)$, all $l \neq k$ and emit the binary expansion of k, which re-quires $\log_2 K$ binary digits. For a scalar source, this procedure is called *pulse code modulation*, or PCM.

Because we have defined a distortion measure and a set of reproduction symbols beforehand, one really need not compute the distortions $d(x, y_k)$ after $X = x$ is emitted from the source. One can specify beforehand intervals

$$I_k = \{x : d(x, y_k) \leq d(x, y_l), \text{ all } l \neq k\}, \qquad k = 1, 2 \ldots K, \qquad (2.19)$$

which is the set of x values closest in distortion measure to y_k for every $k = 1, 2 \ldots K$. For a distortion measure that is a monotone, non-decreasing function f of the distance between x and y_k, i.e.,

$$d(x, y_k) = f(|x - y_k|),$$

the values of I_k are nonoverlapping intervals. The endpoints of these intervals we denote as x_{k-1} and x_k:

$$I_k = [x_{k-1}, x_k) = \{x : x_{k-1} \leq x < x_k\}.$$

The definitions of these points and intervals are depicted in Fig. 2.4.

Upon emitting $X = x$, we accept as our reproduction of x the value y_k if $x \in I_k$. If x is in $I_k = [x_{k-1}, x_k)$, then x is *mapped* to y_k. Such a mapping is called *quantization*. Using functional notation,

$$y_k = Q(x) \quad \text{if} \quad x \in I_k \qquad (\text{or } x_{k-1} \leq x < x_k). \qquad (2.20)$$

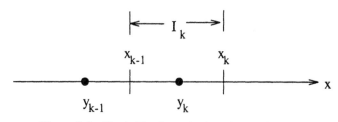

Figure 2.4   Threshold and reproduction points and intervals.

The reproduction letters $\{y_1, y_2 \ldots y_K\}$ are called *quantizer levels*, and the interval endpoints $\{x_0, x_1, x_2 \ldots x_{K-1}, x_K\}$ are called *decision thresholds*.

A graph of the function of quantization is a staircase function with the level of the kth step y_k with ends at x_{k-1} and x_k, as shown in Fig. 2.5. If there is a k such that $y_k = 0$, the quantizer is called *midtread*. If there is no k such that $y_k = 0$ and a k such that $x_k = 0$, the quantizer is called *midrise*.

The quantization error (noise) is measured by the average of $d(x, Q(x))$; i.e.,

$$D = E[d(x, Q(x))] = \int_{-\infty}^{\infty} d(x, Q(x)) q(x) dx.$$

Since $Q(x) = y_k$ if $x_{k-1} \le x < x_k$,

$$D = \sum_{k=1}^{K} \int_{x_{k-1}}^{x_k} d(x, y_k) q(x) dx. \tag{2.21}$$

When $x_0 = -\infty$ and $x_K = +\infty$, the portions of D in the two end intervals $(-\infty, x_1)$ and $[x_{K-1}, +\infty)$ are called *overload error* or *distortion*, while the contributions to D in the other regions are called *granular distortion*. Note that, since the errors in different regions add, the overload and granular errors add. The actual error $d(x, y_k)$ in the overload regions is unbounded but, when weighted by the probability density, should contribute very little to D. The actual error $d(x, y_k)$ in the interior region is bounded and usually small.

2.4.2 Optimum Nonuniform Quantization

Since the decision thresholds are determined through the distortion measure once the quantizer levels $\{y_1, y_2 \ldots y_K\}$ are specified, the problem is to find the set of levels $\{y_1, y_2 \ldots y_K\}$ such that $D = \sum_{k=1}^{K} \int_{x_{k-1}}^{x_k} d(x, y_k) q(x) dx$ is minimized.

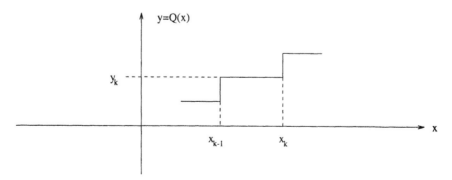

Figure 2.5   Graphical form of quantizer function.

To seek a stationary point, we can first fix the $\{y_k\}_{k=1}^{K}$ set and find the best set of decision thresholds. When $d(x, y_k) = f(|x - y_k|)$ with f a monotone and non-decreasing function of its argument, the search for $I_k = [x_{k-1}, x_k)$ such that $d(x, y_k) \leq d(x, y_l)$ for all $l \neq k$ must place x_k halfway between y_k and y_{k+1}; i.e.,

$$x_k = \frac{y_{k+1} + y_k}{2}, \qquad k = 1, 2 \ldots K - 1. \qquad (2.22)$$

Then I_k is the set of points that are closest to y_k in distance. You may check this result by $\frac{\partial D}{\partial x_k} = 0$ (holding everything else constant). To find the best $\{y_k\}$ a necessary condition is that $\frac{\partial D}{\partial y_k} = 0, \quad k = 1, 2 \ldots K.$

$$\frac{\partial D}{\partial y_k} = 0 = \int_{x_{k-1}}^{x_k} \frac{\partial}{\partial y_k}(d(x, y_k))q(x)dx. \qquad (2.23)$$

Let us now assume the distortion measure is squared error: $d(x, y_k) = (x - y_k)^2$. Substituting into (2.23), we obtain

$$\int_{x_{k-1}}^{x_k} \frac{\partial}{\partial y_k}(x - y_k)^2 q(x)dx = 0,$$

$$-2 \int_{x_{k-1}}^{x_k} (x - y_k)q(x)dx = 0,$$

$$y_k = \frac{\int_{x_{k-1}}^{x_k} xq(x)dx}{\int_{x_{k-1}}^{x_k} q(x)dx}, \qquad k = 1, 2 \ldots K. \qquad (2.24)$$

Therefore, y_k is the "centroid" of the probability density in the decision interval between x_{k-1} and x_k for every k. It may also be interpreted as a conditional mean; i.e.,

$$y_k = E[X/X \in [x_{k-1}, x_k)], \qquad k = 1, 2 \ldots K.$$

This result is not unexpected when one looks at the possible actions of the decoder. Assuming uncorrupted reception of the code word, the decoder knows only the range interval I_k, so sets the quantization point to the centroid of the region, which is the previous conditional mean, to minimize the mean squared error. Are these conditions sufficient for a minimum? Not always. They are if the Hessian matrix

$$\mathcal{H} = \left\{ \frac{\partial^2 D}{\partial y_k \partial y_l} \right\}_{k,l=1}^{K}$$

is positive definite. A simpler condition by Fleischer [5] is that if $\log q(x)$ is concave $\left(\frac{\partial^2 (\log q(x))}{\partial x^2} < 0 \right)$ and distortion is squared error, then the previous minimum is a global minimum. The distortion function for this condition was extended to a convex and increasing weighting function of absolute error by Trushkin [6].

The following algorithms compute $\{y_k\}$ and $\{x_k\}$:

I. Lloyd (Method I) Algorithm [7]

1. Choose set $\{x_1, x_2 \ldots x_{K-1}\}$ arbitrarily ($x_0 = x_{min}$, $x_K = x_{max}$).
2. Compute centroids y_k by (2.23) or (2.24).
3. Compute new set of thresholds by (2.22).
4. Set stopping criterion on decrease of D.
   If not met, go back to 2.
   If met, stop.

II. Lloyd–Max Algorithm ([7, 8])

1. Choose y_1, arbitrarily ($x_0 = x_{min}$). Set $k = 1$.
2. Calculate x_k by y_k= centroid of $q(x)$ in $[x_{k-1}, x_k)$.
3. Calculate y_{k+1} by $x_k = (y_{k+1} + y_k)/2$.
4. If $k < K$, set $k = k + 1$ and go back to 2.
   If $k = K$, compute y_c= centroid of $q(x)$ in $[x_{K-1}, x_K)$.
   If $|y_K - y_c| > \epsilon$ (error tolerance), return to 1 for a new choice of y_1, otherwise, stop.

2.4.3 Optimum Uniform Quantization

Since the number of solution parameters can be rather large for optimum nonuniform quantization, one often seeks a suboptimal solution with far fewer parameters. One such category of solutions is the uniform quantizer, where the decision thresholds, quantizer levels, or both are uniformly spaced. An additional benefit of uniform quantization is that it is implemented readily and simply in hardware. Later we shall see that uniform quantization exceeds the performance of nonuniform quantization when performance is measured by a different criterion.

Consider now the class of quantizers whose quantizer levels are restricted to a uniform spacing Δ according to

$$y_k = a + (k - 1)\Delta, \qquad k = 1, 2 \ldots K.$$

Since the thresholds must satisfy the midpoint condition (2.22),

$$x_k = a + (k - \frac{1}{2})\Delta, \qquad k = 1 \ldots K - 1.$$

The parameters a and Δ are chosen to minimize D. Such a quantizer is called an *optimum uniform quantizer* for *fixed K*.

Often a is chosen such that $y_k = 0$ for some k. Or, for even-symmetric densities, the choice is

$$a = -\frac{K-1}{2}\Delta, \qquad K \text{ odd, midtread}$$
$$K \text{ even, midrise.}$$

For *one-sided densities*, optimization of a is important only for very small K. Otherwise, take $a = \Delta/2$, so that $x_0 = 0$, in which case

$$\begin{aligned}
y_k &= (k - \tfrac{1}{2})\Delta, &\quad k &= 1, 2 \ldots K, \\
x_k &= k\Delta, &\quad k &= 0, 1 \ldots K-1, &\quad x_K = x_{max}.
\end{aligned}$$

Another, less restrictive class of uniform quantizers is called *uniform threshold quantizers*, whereby the thresholds are set to be uniformly spaced and the quantizer levels are allowed their optimum points of the centroids of the intervals. The principles that follow are true for both kinds of uniform quantizers, but will be described specifically for the more restrictive class where the thresholds and quantizer levels are uniformly spaced.

The distortion D is evaluated by substitution of the uniformly spaced values of x_k and y_k into (2.21). Here $\frac{\partial D}{\partial a} = 0$ and $\frac{\partial D}{\partial \Delta} = 0$ are necessary for a minimum. The positive definiteness of the matrix $\mathcal{H}$ is easy to check in this case to show sufficiency of the zero derivatives for a minimum. If you fix a, you need only check that $\frac{\partial^2 D}{\partial \Delta^2} > 0$. The solution for a and Δ (or Δ only) is done by computer usually through Newton–Raphson techniques.

Fig. 2.6 shows the mean quantization error D versus K, the number of output levels, for the Gaussian density in (2.7) with variance $\sigma^2 = 1$. The error criterion is squared error. The optimum non-uniform and uniform quantizers for fixed K are compared for $K = 1$ to 128.

As expected from the additional restriction of uniform spacing of the levels, the uniform quantizer always has more error for a given K than that of the nonuniform quantizer. The trend of these results is true for any probability density.

2.4.4 Entropy Coding of Quantizer Outputs

The code rate for the previous quantization scheme is $R = \log_2 K$ bits/sample. Let us now consider a block of N quantized source samples and call them $(z_1, z_2 \ldots z_N)$. Each z_n, $n = 1, 2 \ldots N$ is one of the quantizer levels $\{y_1, y_2 \ldots y_K\}$. The paradigm is shown in Fig. 2.7.

The output of the quantizer can be considered as a memoryless, discrete-alphabet source with letter probabilities:

$$P(y_1), P(y_2) \ldots P(y_K),$$

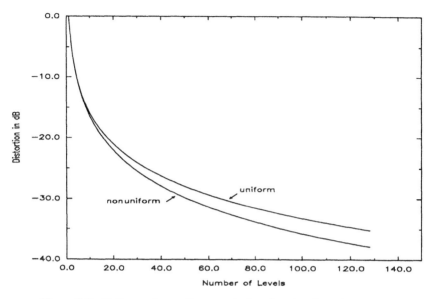

Figure 2.6 Uniform and nonuniform quantization of a unit variance Gaussian density.

$$P(y_k) = \Pr\{x_{k-1} \leq X < x_k\} = \int_{x_{k-1}}^{x_k} q(x)dx, \qquad k = 1, 2 \dots K.$$

This new source can therefore be encoded without error at a rate arbitrarily close to its entropy of

$$H = \sum_{k=1}^{K} P(y_k) \log_2 \frac{1}{P(y_k)} \quad \text{bits/sample,}$$

say, by a Huffman code on blocks of N outputs for N sufficiently large. Such further encoding of the quantized source is called *entropy coding*. As no additional error or distortion is introduced, the error is that in the quantization alone. Therefore, with entropy coding, the distortion D and the entropy H are the important quantities of

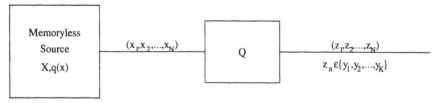

Figure 2.7 Quantization of a sequence from the source.

performance in the quantization. The number of quantizer levels K is no longer a determining factor. Note that $H \leq \log_2 K$, the previous rate without entropy coding.

If one looks at Max's tables [8] for the Gaussian density and squared error or the corresponding uncoded and coded curves in Fig. 2.8, the entropy H is always less than $\log_2 K$, since the level probabilities are not equal, and surprisingly, at a given distortion, the entropy of the uniform quantizer is less than that of the nonuniform quantizer. This latter result seems to be true for any density and squared error distortion measure. In Fig. 2.8 the distortion in dB ($10 \log_{10} D$) is plotted against the rate in bits for nonuniform and uniform quantizers, with and without entropy coding, for the Gaussian density ($\sigma = 1$) and squared error distortion. The "uncoded" curves have a rate of $\log_2 K$, whereas the "coded" curves have a rate of $H = -\sum_{k=1}^{K} P(y_k) \log_2 P(y_k)$ bits/sample. For a given rate, the coded curves give the lower distortions, whether the quantizer is uniform or non-uniform. However, as the rate grows larger, the uniform quantizer outperforms the nonuniform quantizer when the outputs are entropy coded.

So, with the view of entropy coding the quantizer outputs, we should look for a different measure of quantizer performance. Instead of constraining or fixing the number of levels, we should constrain the entropy H to be no greater than some value H_o and adjust the quantizer levels to minimize D. Equivalently, we can fix

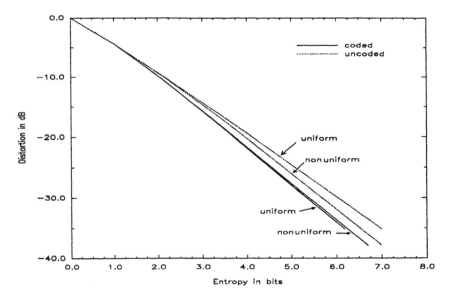

Figure 2.8 Comparison of uncoded and entropy-coded quantization for nonuniform and uniform quantizer levels. The source is the unit variance Gaussian and the distortion is squared error.

$D \le D_o$ for some distortion D_o and adjust the quantizer levels to minimize H. Since the uniform quantizer is superior in entropy coding, we need to consider only a uniform quantizer and vary the level spacing Δ (for symmetric densities).

Here is the method of Goblick and Holsinger [9] to accomplish this.

1. Choose K, the number of levels.
2. Choose Δ, the uniform quantizer spacing.
3. Calculate D and H.
4. Return to 2 (to choose a new Δ). If there are enough points to trace a curve, go to 5.
5. Return to 1 (to choose a new K).

Goblick and Holsinger plotted a figure similar to that of Fig. 2.9 for a Gaussian density with squared error distortion. For each K a curve is traced where the lower part is a straight line when H is plotted versus D on a log scale. Every curve eventually coincides with the same straight line as distortion D increases and entropy H decreases. Higher rates and lower distortion are obtained for the higher values of K. The lower envelope of these curves, which is that straight line,

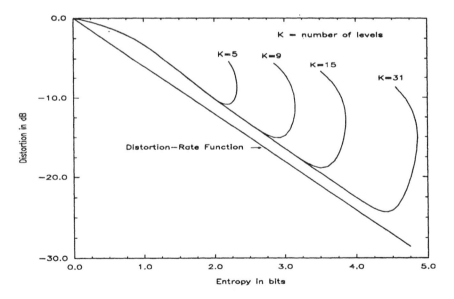

Figure 2.9 Mean squared error versus entropy for different numbers of levels in uniform quantization.

is the minimum entropy for a given distortion. This straight line envelope is well described by the following equation for rates above 1.75 bits:

$$R^*(D) = 1/4 + R(D) \quad \text{bits/sample}$$

where $R(D) = \frac{1}{2} \log_2 \frac{\sigma^2}{D}$ is the rate–distortion function of the Gaussian source with squared error for $\sigma^2 > D$. Therefore, there exists a uniform quantizer, whose rate for a given distortion is only $\frac{1}{4}$ bit larger than that of the best possible encoder (for this source and distortion measure). In another paper, Gish and Pierce [10] proved that for any source density and squared error, a uniform quantizer is asymptotically optimal (for large rates) and its entropy is only .255 bit per sample larger than the corresponding point of the rate–distortion function $R(D)$.

There are more sophisticated algorithms in the literature than the one just described for minimizing D with respect to the entropy constraint $H \leq H_o$. For those who want to delve further into the subject and see results for other memoryless sources, see [11] and [12]. We shall describe such an algorithm later in the chapter when we present vector quantization.

2.5 Vector Coding

2.5.1 Optimal Performance Bounds

Rate–Distortion theory tells us that a source, whether or not it has memory, can be more efficiently coded if we encode long blocks of data values, the larger the better. This is the interpretation of (2.13) stating that $\lim_{N \to \infty} R_N(D) = R(D)$. The function $R_N(D)$ is the N-tuple rate–distortion function and gives the least possible rate for independent encoding of blocks of N-tuples from the source with average distortion no greater than D. As N increases, $R_N(D)$ grows closer to $R(D)$ for stationary, block-ergodic sources. In fact, the following bound by Wyner and Ziv [13] for the difference between $R_N(D)$ and $R(D)$,

$$R_N(D) - R(D) \leq h_N(\mathbf{X}) - h_\infty(\mathbf{X}) \qquad (2.25)$$
$$= \lim_{n \to \infty} \left[\frac{1}{N} I(X_1, X_2 \ldots X_N; X_{N+1} \ldots X_{N+n}) \right],$$

is informative concerning the decrease of rate with N. This rate difference is bounded by the average mutual information per sample that an N-tuple conveys about the infinite remainder of the sequence. As N grows larger, this information must decrease. The bound holds with equality when the average distortion is below some critical distortion D_c, the value of which depends on the particular source and distortion measure. Optimal coding of N-tuples from a source with memory, just as with scalars, requires searches over long sequences of N-tuples to best match a long data stream and is computationally burdensome. In Section 2.5.2, after presenting the rate–distortion function of the Gaussian source with squared error distortion, we describe a method for constructing a codebook for this source whose reproduction vectors lie along the branches of a tree. A suboptimal approach, again paralleling scalar coding, is to make an instantaneous decision on the reproduction of a single N-tuple. This approach is called vector quantization and is only optimal in the limit of infinite N.

2.5.2 Vector (Block) Quantization

Consider N values emitted from a data source $\mathbf{x} = (\mathbf{x}_1, \mathbf{x}_2 \ldots \mathbf{x}_N)$, a codebook for K reproduction N-tuples $\mathbf{y}_1, \mathbf{y}_2 \ldots \mathbf{y}_K$, and a distortion measure $d(\mathbf{x}, \mathbf{y}_m)$ between the source vector $\mathbf{x}$ and any member of the codebook $\mathbf{y}_m$, $m = 1, 2 \ldots K$. The encoder searches for $\mathbf{y}_m$ such that $d(\mathbf{x}, \mathbf{y}_m) \leq d(\mathbf{x}, \mathbf{y}_l)$ for all $l \neq m$ and emits the binary index of m. The rate of the code is then $R = \frac{1}{N} \log_2 K$ bits per source sample. The objective is to choose the codebook of rate R that minimizes the average distortion $E[d(\mathbf{x}, \mathbf{y}_m)]$. This procedure is the same as that depicted in Fig. 2.1, and such an encoder is called a *vector* or *block quantizer*. This quantizer induces a rule Q that maps $\mathbf{X}$ to $\mathbf{y}_m$. In lieu of a search, given the codebook $Y = \{\mathbf{y}_1, \mathbf{y}_2 \ldots \mathbf{y}_K\}$ and the distortion, we can determine a set of decision regions $\mathcal{R}_m$ such that for $\mathbf{x}$ in $\mathcal{R}_m$, $Q(\mathbf{x}) = \mathbf{y}_m$; i.e., $\mathbf{x} \in \mathcal{R}_m$ implies $d(\mathbf{x}, \mathbf{y}_m) \leq d(\mathbf{x}, \mathbf{y}_l)$ for all $l \neq m$ or $\mathcal{R}_m = Q^{-1}(\mathbf{y}_m) = \{\mathbf{x} \in \mathcal{R}^N : Q(\mathbf{x}) = \mathbf{y}_m\}$. The regions $\mathcal{R}_m$ partition Euclidean N-space:

$$\bigcup_{m=1}^{K} \mathcal{R}_m = \mathcal{R}^N, \qquad \mathcal{R}_m \cap \mathcal{R}_l = \emptyset, \quad l \neq m,$$

for an arbitrary tie-breaking rule. The encoder identifies $\mathcal{R}_m$ for the given $\mathbf{X}$ and sends the index m in binary. At the decoder m is mapped uniquely to $\mathbf{y}_m$ for the reproduction. The following two conditions are necessary for a quantizer to be optimum. Gray, Kieffer, and Linde [14] have shown these conditions are sufficient for a local optimum.

1. Given a codebook $\mathbf{y}_1, \mathbf{y}_2 \ldots \mathbf{y}_K$ the decision regions must be such that

$$\mathcal{R}_m = \{\mathbf{x} : d(\mathbf{x}, \mathbf{y}_m) \leq d(\mathbf{x}, \mathbf{y}_l) \quad \text{all } l \neq m\}, \qquad m = 1, 2 \ldots K.$$

   These regions are called *Voronoi* regions.
2. Given any set of K decision regions $\mathcal{R}_1, \mathcal{R}_2 \ldots \mathcal{R}_K$ the reproduction points $\mathbf{y}_m$, $m = 1, 2 \ldots K$, are such that

$$E[d(\mathbf{x}, \mathbf{y}_m)/\mathbf{x} \in \mathcal{R}_m] = \min_{\mathbf{v}} \ E[d(\mathbf{x}, \mathbf{v})/\mathbf{x} \in \mathcal{R}_m].$$

Clearly the decoder knows only that X in $\mathcal{R}_m$ and must pick the point $\mathbf{v}$ in $\mathcal{R}_m$ that minimizes the expected distortion. This is easily seen since the quantizer distortion can be expressed as:

$$E[d(\mathbf{x}, Q(\mathbf{X}))/\{\mathcal{R}_m\}] =$$
$$\sum_{m=1}^{K} \underbrace{E[d(\mathbf{x}, \mathbf{v})/\mathbf{x} \in \mathcal{R}_m]}_{\text{minimize for each } m \text{ over } \mathbf{v}} \Pr\{\mathbf{x} \in \mathcal{R}_m\}.$$

For squared error $d(\mathbf{x}, \mathbf{v}) = \|\mathbf{x} - \mathbf{v}\|^2$ and $\mathbf{y}_m = E[\mathbf{x}/\mathbf{x} \in \mathcal{R}_m]$.

Proof Assume some $\mathbf{X} \neq \mathbf{y}_m$ gives lower error:

$$xxx\, E[\|\mathbf{x} - \mathbf{v}\|^2/\mathbf{x} \in \mathcal{R}_m] = E[\|(\mathbf{x} - \mathbf{y}_m) + (\mathbf{y}_m - \mathbf{v})\|^2/\mathbf{x} \in \mathcal{R}_m]$$
$$= E[\|\mathbf{x} - \mathbf{y}_m\|^2/\mathbf{x} \in \mathcal{R}_m] - 0$$
$$+ E[\|\mathbf{y}_m - \mathbf{v}\|^2/\mathbf{x} \in \mathcal{R}_m]$$
$$\geq E[\|\mathbf{x} - \mathbf{y}_m\|^2] \quad \text{with equality} \quad \mathbf{y}_m = \mathbf{v}.$$

The contradiction proves that $\mathbf{y}_m$ minimizes the mean squared error.

For a squared error, $\mathbf{y}_m$ is "centroid" of $\mathcal{R}_m$. Condition 2 is called the *centroid condition* for any distortion measure.

Also, we can show for any partition $\{\mathcal{R}_m\}$ and optimal output points $\mathbf{y}_m = E[\mathbf{x}/\mathbf{x} \in \mathcal{R}_m]$ for squared error distortion, the mean squared error:

$$e^2 = E[\|\mathbf{x} - Q(\mathbf{x})\|^2] = E[\|\mathbf{x}\|^2] - E[\|\mathbf{y}_m\|^2],$$

which means that the average squared value or power in the reproduction ensemble is lower than that in the source. This result contradicts the often used model of the quantization noise vector being independent and additive to the source vector.

Proof

$$e^2 = E[\|\mathbf{x} - Q(\mathbf{x})\|^2] = E[(\mathbf{x} - Q(\mathbf{x}))\mathbf{x}^T] - E[(\mathbf{x} - Q(\mathbf{x}))Q^T(\mathbf{x})].$$

Since $\mathbf{y}_m = Q(\mathbf{x}) = E[\mathbf{x}/\mathbf{x} \in \mathcal{R}_m]$, the second term equals 0. Expanding the first term,

$$e^2 = E[\|\mathbf{x}\|^2] - E[(Q(\mathbf{x}))\mathbf{x}^T]$$
$$= E[\|\mathbf{x}\|^2] - \sum_{m=1}^{M} E[\mathbf{x}^T\mathbf{y}_m/\mathbf{x} \in \mathcal{R}_m]\Pr\{\mathbf{x} \in \mathcal{R}_m\}$$
$$= E[\|\mathbf{x}\|^2] - \sum_{m=1}^{M} \|\mathbf{y}_m\|^2\Pr\{\mathbf{x} \in \mathcal{R}_m\}$$
$$= E[\|\mathbf{x}\|^2] - E[\|Q(\mathbf{x})\|^2].$$

There is an algorithm by Linde, Buzo, and Gray [15, 16] (LBG algorithm) based on the two necessary conditions for designing a quantizer codebook and decision regions for a known source probability distribution $q(\mathbf{x})$. It is a generalization to vectors of the Lloyd Method I [7] for scalar quantizers and is hence often called the *generalized Lloyd algorithm* (GLA).

0. Initialization: Choose some reproduction alphabet (codebook) $Y^0 = \{\mathbf{y}_1^{(0)}, \mathbf{y}_2^{(0)} \dots \mathbf{y}_K^{(0)}\}$.

1. Given Y^i find minimum distortion decision regions $\mathcal{R}_1^{(i)}, \mathcal{R}_2^{(i)} \dots \mathcal{R}_K^{(i)}$ such that

$$\mathcal{R}_m^{(i)} = \{\mathbf{x} : d(\mathbf{x}, \mathbf{y}_m^{(i)}) \leq d(\mathbf{x}, \mathbf{y}_l^{(i)}), \quad \text{all } l \neq m\}.$$

Compute resulting average distortion

$$D^{(i)} = \sum_{m=1}^{K} E[d(\mathbf{x}, \mathbf{y}_m^{(i)})/\mathbf{x} \in \mathcal{R}_m^{(i)}]\Pr\{\mathbf{x} \in \mathcal{R}_m^{(i)}\}.$$

2. If $(D^{(i-1)} - D^{(i)})/D^{(i)} < \epsilon$ for some $\epsilon > 0$, stop.

3. For decision regions $\mathcal{R}_1^{(i)}, \mathcal{R}_2^{(i)} \dots \mathcal{R}_K^{(i)}$ in (1), find centroids (optimum reproduction points) $\mathbf{y}_m^{(i+1)}$ such that

$$E[d(\mathbf{x}, \mathbf{y}_m^{(i+1)})/\mathbf{x} \in \mathcal{R}_m^{(i)}] = \min_{\mathbf{v}} E[d(\mathbf{x}, \mathbf{v})/\mathbf{x} \in \mathcal{R}_m^{(i)}],$$

(for squared error $\mathbf{y}_m^{(i+1)} = E[\mathbf{x}/\mathbf{x} \in \mathcal{R}_m^{(i)}]$)

$$Y^{i+1} = \{\mathbf{y}_1^{(i+1)}, \mathbf{y}_2^{(i+1)} \dots \mathbf{y}_K^{(i+1)}\}.$$

4. $i \to i + 1$. Go to step 1.

This algorithm is a fixed point algorithm for Y and $\{\mathcal{R}_m\}$; that is,

$$Y^{i+1} = f(Y^i) \qquad \text{and} \qquad \{\mathcal{R}_m^{(i+1)}\} = g(\{\mathcal{R}_m^i\})$$

for some functions f and g. Note that $D_{i-1} \geq D_i$, so each iteration cannot increase the error. Since D^i is nonincreasing and nonnegative, it must have some limit D_∞. Gray et al. [14] show that $\{\mathcal{R}_m^{(\infty)}\}$ and Y^∞ have the proper relationship of centroids of own partition and that there is convergence to a fixed point.

A necessary condition for a quantizer to be optimal is that it be a fixed point quantizer. It is locally optimum if no probability is on the boundary of each decision region. For a continuous probability distribution, this presents no problem. However, for a discrete distribution, there may be a problem.

For an unknown source probability distribution, use a sequence of source vectors to train the quantizer and insert them into steps 1 and 3, using numerical averages instead of expectations or ensemble averages. We obtain partitions and centroids with respect to the sequence of training vectors (called a *training sequence*). Then we use those partitions and centroids for other data judged to have similar statistics (i.e., *outside* the training sequence). We shall state the algorithm using the training sequence (TS).

0. *Initialization.*   Set $\epsilon \geq 0$, and an initial codebook $Y^0 = \{\mathbf{y}_1^{(0)}, \mathbf{y}_2^{(0)} \dots \mathbf{y}_K^{(0)}\}$. Given a TS $\tilde{\mathbf{x}}_j,$ $j = 1, 2 \dots n$ of N-tuples.

$$D_{-1} = \infty.$$
$$i = 0.$$

1. Given Y^i, find minimum distortion decision (Voronoi) regions of TS $\mathcal{R}_1^i, \mathcal{R}_2^i \dots \mathcal{R}_K^{(i)}$ such that, if $\tilde{\mathbf{x}}_j \in \mathcal{R}_m^i$ then all $d(\tilde{\mathbf{x}}_j, \mathbf{y}_m^i) \leq d(\tilde{\mathbf{x}}_j, \mathbf{y}_l^i)$ for all $l \neq m$. Calculate distortion per sample:

$$D^i = \frac{1}{Nn} \sum_{m=1}^{K} \sum_{\tilde{\mathbf{x}}_j \in \mathcal{R}_m^i} d(\tilde{\mathbf{x}}_j, \mathbf{y}_m^i).$$

2. If $(D_{i-1} - D_i)/D_i < \epsilon$, stop. If not, go to step 3.
3. Find optimal reproduction alphabet for TS partitions. For $m = 1, 2, \ldots, K$ and for all $\tilde{\mathbf{x}}_j \in \mathcal{R}_m^i$, find $\mathbf{v} \in \mathcal{R}_m$ such that

$$\frac{1}{\|\mathcal{R}_m^i\|} \sum_{\tilde{\mathbf{x}}_j \in \mathcal{R}_m^i} d(\tilde{\mathbf{x}}_j, \mathbf{v}) \leq \frac{1}{\|\mathcal{R}_m^i\|} \sum_{\tilde{\mathbf{x}}_j \in \mathcal{R}_m^i} d(\tilde{\mathbf{x}}_j, \mathbf{u})   \text{ for all } \mathbf{u} \neq \mathbf{v} \text{ in } \mathcal{R}_m^i$$

where $\|\mathcal{R}_m^i\|$ denotes the number of $\tilde{\mathbf{x}}_j$'s in $\mathcal{R}_m^i$, and set $\mathbf{v} = \mathbf{y}_m^{(i+1)}$. (For squared error, $\mathbf{y}_m^{(i+1)} = \frac{1}{\|\mathcal{R}_m^i\|} \sum_{\tilde{\mathbf{x}}_j \in \mathcal{R}_m^i} \tilde{\mathbf{x}}_j$.) Set $i = i + 1$, and go to step 1.

Problems arise in initialization. A reasonable start is to use a product quantizer, which is the scalar quantizer level for each dimension. Another method called the *splitting method* [16], starts with rate per vector $R = 0$ and increments the rate by 1 at each stage until the desired rate is reached. It is accomplished by successive splitting into two the reproduction vectors obtained at a given rate from the preceding algorithm and using them to initialize the algorithm for the next higher rate.

2.5.3 Entropy Constrained Vector Quantization

As with the Lloyd method in scalar quantization, the generalized Lloyd method for vector quantization optimizes a criterion that is inappropriate for optimal coding in the true rate versus distortion sense. When the indices of the reproduction vectors are considered as a new source, a subsequent entropy coding can achieve a rate arbitrarily close to the entropy of this index source, called the *index entropy*. It is therefore appropriate to reformulate the objective to minimize the average distortion subject to a constraint on the index entropy, which is the entropy of the reproduction vector ensemble. To accomplish this objective, we first define an objective function as the average of an impurity function $j(\mathbf{x}, \mathbf{y})$; i.e.,

$$J = E[j(\mathbf{x}, \mathbf{y})],$$
$$j(\mathbf{x}, \mathbf{y}) = d(\mathbf{x}, \mathbf{y}) - \lambda \log P(\mathbf{y}), \qquad (2.26)$$

with $\mathbf{y} = Q(\mathbf{x})$.

Note that

$$J = D + \lambda H(\mathbf{Y}),$$

where D and $H(\mathbf{Y})$ are the average distortion per vector and entropy per vector, respectively.

The Lagrange parameter $\lambda > 0$ is evaluated from the entropy constraint,

$$H(\mathbf{Y}) \le R, \qquad (2.27)$$

where the solution in the space is assumed to lie on the boundary $H(\mathbf{Y}) = R$, due to the presupposition of a convex distortion function.

For a given set of K reproduction vectors $C = \{\mathbf{y}_1, \mathbf{y}_2 \ldots \mathbf{y}_K\}$ with respective probabilities $\mathcal{P} = \{P_1, P_2 \ldots P_K\}$ and a given $\lambda > 0$, the impurity function $j(\mathbf{x}, \mathbf{y})$ partitions the input space into a set of (Voronoi) regions

$$\mathcal{R}_m = \{\mathbf{x} : j(\mathbf{x}, \mathbf{y}_m) \le j(\mathbf{x}, \mathbf{y}_l), \quad \text{all } l \ne m\}, \qquad m = 1, 2 \ldots K. \quad (2.28)$$

This partition is one of two necessary conditions to achieve a minimum of $J = E[j(\mathbf{x}, \mathbf{y})]$. The other condition is the "centroid" condition that the set of reproduction vectors $C = \{\mathbf{y}_1, \mathbf{y}_2, \ldots, \mathbf{y}_K\}$ satisfy

$$E[j(\mathbf{x}, \mathbf{y}_m)/\mathbf{x} \in \mathcal{R}_m] = \min_{\mathbf{v}} E[j(\mathbf{x}, \mathbf{v})/\mathbf{x} \in \mathcal{R}_m], \qquad (2.29)$$

paralleling the previous development with $d(\mathbf{x}, \mathbf{y})$. This second condition can be simplified, since

$$E[j(\mathbf{x}, \mathbf{v})/\mathbf{x} \in \mathcal{R}_m] = E[d(\mathbf{x}, \mathbf{v})/\mathbf{x} \in \mathcal{R}_m] - \lambda E[\log P(\mathbf{v})/\mathbf{x} \in \mathcal{R}_m].$$
$$(2.30)$$

Once $\mathbf{x}$ is known to be in a given region, the probability of the reproduction vector is the probability of that region. Therefore, in searching for the $\mathbf{v} \in \mathcal{R}_m$ that minimizes (2.30), only the first term varies while the second term stays constant. We need then to search for the centroid of the decision region $\mathcal{R}_m$ only with respect to the distortion measure $d(\mathbf{x}, \mathbf{v})$. Therefore, we need to modify the GLA only in the initialization step 0 and the Voronoi regions definition in step 1 and stopping criterion in step 2 to obtain the entropy-constrained vector quantization (ECVQ) algorithm, originated by Chou, Lookabaugh, and Gray [17]. The modified steps are:

0′. Choose codebook of reproduction vectors $Y^{(0)} = \{\mathbf{y}_1^{(0)}, \mathbf{y}_2^{(0)} \ldots \mathbf{y}_K^{(0)}\}$ with their respective probabilities $\mathcal{P}^{(0)} = \{P_1^{(0)}, P_2^{(0)} \ldots P_K^{(0)}\}$. Set some $\lambda > 0$ and $i = 0$.

1′. Given $Y^{(i)}$ and $\mathcal{P}^{(i)}$, calculate the decision regions

$$\mathcal{R}_m^{(i)} = \{\mathbf{x} : d(\mathbf{x}, \mathbf{y}_m^{(i)}) - \lambda \log P_m^{(i)} \leq d(\mathbf{x}, \mathbf{y}_l^{(i)}) - \lambda \log P_l^{(i)}, \quad \text{for all } l \neq m\},$$
$$m = 1, 2 \ldots K.$$

Calculate resulting average distortion

$$D^{(i)} = \sum_{m=1}^{K} E[d(\mathbf{x}, \mathbf{y}_m^{(i)})/\mathbf{x} \in \mathcal{R}_m^{(i)}] P_m^{(i)}$$

and entropy

$$H^{(i)} = -\sum_{m=1}^{K} P_m^{(i)} \log P_m^{(i)},$$
$$J^{(i)} = D^{(i)} + \lambda H^{(i)}.$$

2′. If $(J^{(i-1)} - J^{(i)})/J^{(i)} < \epsilon$ for some suitable $\epsilon > 0$, stop.

The entropy of the last iteration is computed in step 1′. Another value of λ is now selected, and the algorithm is rerun to get another entropy value. To obtain a code rate close to these index entropy rates, the indices of the code vectors found in the last iteration for a given λ must be entropy coded.

The initialization of the algorithm requires a set of probabilities in addition to the set of reproduction vectors of the unconstrained GLA. It also requires a guess for the value of λ that achieves the desired rate. Therefore, a good starting point is $\lambda = 0$, which reduces to the unconstrained algorithm where the entropy per vector is allowed to reach its maximum of $\log M$. A set of probabilities is not required in the initialization step for $\lambda = 0$. Once the final sets of code vectors and probabilities are found for $\lambda = 0$, they are used to initialize the algorithm for the next λ, which is a small increment from 0. This process is repeated for several increments of λ until all desired points of the distortion–rate curve are found. As λ increases, a greater penalty is exacted for entropy in the objective function, so the corresponding entropy decreases while distortion increases. The limit of $\lambda = \infty$ gives an entropy or rate of 0 and a distortion equal to the variance per vector. We note finally that this algorithm may be utilized with any vector dimension barring complexity constraints and, in particular, for one dimension. One may therefore use this algorithm to design optimal scalar entropy-constrained quantizers. There may not be much benefit in doing so, since scalar uniform quantizers are so close to optimum for nearly all rates.

2.6 Transform Coding

2.6.1 Introduction

Instead of trying to encode the outputs of a source directly, it is often advantageous to transform the source mathematically into a set of equivalent values that are more easily or efficiently coded. First we present the information theoretical framework behind such a transformation and then introduce practical realizations of this framework.

2.6.2 Optimal Coding of Stationary Gaussian Sources

Let the vector source ensemble $\mathbf{X}$ be Gaussian with zero mean and N-dimensional covariance matrix $\Phi_N = E[\mathbf{XX}^T]$. This (Toeplitz) matrix is real, symmetric, and nonnegative definite. There exists a set of N orthonormal eigenvectors $\mathbf{e}_1, \mathbf{e}_2, \ldots, \mathbf{e}_N$ and a corresponding set of N nonnegative, not necessarily distinct, eigenvalues $\lambda_1, \lambda_2, \ldots, \lambda_N$ for Φ_N. A given source sequence $\mathbf{X} = \mathbf{x}$ can be represented with respect to the eigenvector basis by

$$\mathbf{x} = P\,\tilde{\mathbf{x}}, \quad P = [\mathbf{e}_1|\mathbf{e}_2|\ldots|\mathbf{e}_N] \tag{2.31a}$$

where P is a unitary matrix ($P^{-1} = P^T$).

The vector $\tilde{\mathbf{x}}$ is called the *transform* of the source vector $\mathbf{x}$. The covariance matrix of the transform vector ensemble $\tilde{\mathbf{X}}$ is

$$\Lambda = E[\tilde{\mathbf{X}}\tilde{\mathbf{X}}^T] = E[P^{-1}\mathbf{XX}^T P]$$

$$= P^{-1}\Phi_N P = \begin{bmatrix} \lambda_1 & & & \\ & \lambda_2 & & 0 \\ & & \ddots & \\ 0 & & & \lambda_N \end{bmatrix}. \tag{2.31b}$$

The components of $\tilde{\mathbf{X}}$ are uncorrelated with variances equal to the eigenvalues. This transformation of $\mathbf{x}$ to $\tilde{\mathbf{X}}$ is called the (discrete) Karhunen–Loeve transform (KLT). Moreover, since the source vector ensemble is assumed to be Gaussian, the ensemble $\tilde{\mathbf{X}}$ is also Gaussian with independent components.

When we consider the problem of source encoding $\mathbf{X}$ with the single-letter squared error distortion measure $d(x, y) = (x - y)^2$, so that $d(\mathbf{x}, \mathbf{y}) = |\mathbf{x} - \mathbf{y}|^2$, it is equivalent to encoding $\tilde{\mathbf{X}}$ with the same distortion measure $d(\tilde{\mathbf{x}}, \tilde{\mathbf{y}}) = |\tilde{\mathbf{x}} - \tilde{\mathbf{y}}|^2$, because average mutual information and average squared error are preserved in the unitary transformation of $\mathbf{X}$ to $\tilde{\mathbf{X}}$. The N-tuple rate–distortion (or distortion–rate) function is solved for the transform ensemble $\tilde{\mathbf{X}}$ and is found to be expressed through a positive parameter θ by

$$R_N = \frac{1}{N} \sum_{n=1}^{N} \max\left\{0, \frac{1}{2}\log\frac{\lambda_n}{\theta}\right\}, \tag{2.32a}$$

$$D_N = \frac{1}{N} \sum_{n=1}^{N} \min\left\{\theta, \lambda_n\right\}. \tag{2.32b}$$

As N approaches infinity, the eigenvalues approach in a pointwise fashion a sampling of the discrete-time source power spectral density $S_X(\omega) = \sum_{n=-\infty}^{\infty} \phi(n)e^{-j\omega n}$, where $\{\phi(n)\}$ is the autocorrelation sequence of the source [18], and the formulas above tend toward the limit of the rate–distortion (or distortion–rate) function:

$$R_\theta = \frac{1}{2\pi} \int_{-\pi}^{\pi} \max\left\{0, \frac{1}{2}\log\frac{S_X(\omega)}{\theta}\right\}d\omega, \tag{2.33a}$$

$$D_\theta = \frac{1}{2\pi} \int_{-\pi}^{\pi} \min\left\{\theta, S_X(\omega)\right\}d\omega. \tag{2.33b}$$

A model of the envisioned optimal transform coding system appears in Fig. 2.10.

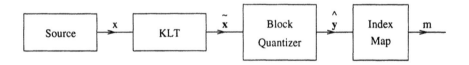

Encoder

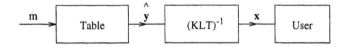

Decoder

Figure 2.10   Optimal transform coding scheme.

2.6.3 Interpretation of Formulas and Code Constructions

The interpretation of the R_N and D_N in (2.32) above is facilitated by defining the distortion sequence

$$e_n = \min \left\{ \theta, \lambda_n \right\} \tag{2.34a}$$

and the rate sequence

$$r_n = \max \left\{ 0, \frac{1}{2} \log \frac{\lambda_n}{\theta} \right\}. \tag{2.34b}$$

For those components of $\widetilde{\mathbf{X}}$, $\widetilde{X}_n$, such that $\lambda_n > \theta$, the average distortion $e_n = \theta$ and the rate is $r_n = \frac{1}{2} \log \frac{\lambda_n}{\theta}$. These components are scalar Gaussian variates of variance λ_n to be encoded with distortion θ with the minimal possible rate $\frac{1}{2} \log \frac{\lambda_n}{\theta}$. For every component whose variance does not exceed θ, the rate $r_n = 0$ and $\widetilde{X}_n$ is set to 0, its mean value, to give distortion equal to its variance λ_n. The collection of such components is hence called the *stop band*. A similar interpretation may be applied to the limiting N case where the frequency axis is continuous.

A coding theorem [2, Section 9.7] gives a code construction by which the preceding rate–distortion bounds may be achieved with high probability of success for sufficiently large N. It utilizes the random coding procedure described earlier with two modifications. First, the given source vector $\mathbf{x}$ is transformed to $\widetilde{\mathbf{x}}$ by $\mathbf{x} = P \widetilde{\mathbf{x}}$. Second, the codebook consists of transform reproduction vectors whose components are drawn independently and at random from the test channel output distributions of the components $\widetilde{Y}_n$ given by

$$w_o(\widetilde{y}_n) = \frac{1}{\sqrt{2\pi \lambda_n}} \exp \left\{ -\frac{\widetilde{y}_n^2}{2(\lambda_n - \theta)} \right\} \tag{2.35}$$

where $\lambda_n > \theta$ for any $n = 1, 2, \ldots, N$.

The nth component of the reproduction vector $\widetilde{\mathbf{y}}_m$, $m = 1, 2, \ldots, K$ is drawn from $w_o(\widetilde{y}_n)$, if $\lambda_n > \theta$. Otherwise, the component is set to 0.

This codebook construction creates an unstructured list to be searched for the $\widetilde{\mathbf{y}}_m$ that has the lowest mean squared error with respect to the given transform source vector $\widetilde{\mathbf{x}}$.

In [19] and [20], it has been shown how to construct an optimal code by depositing randomly selected reproduction words along the paths of a tree or trellis, whose branching factor (for the tree only) or number of letters per branch or both vary with level in the structure. Although optimality of the resulting code is guaranteed only for large N, the tree or trellis structure provides an opportunity for savings in storage and computation, especially when selective search procedures, which give nearly optimal performance, are invoked.

The formulas in (2.34) may be regarded as the rate r_n and mean squared error e_n to be assigned to the nth transform component in an optimal coding scheme achieving an overall rate R_N and distortion D_N in (2.32). These formulas are also guidelines for suboptimal coding schemes. For example, each component is often independently quantized with a rate determined from (2.34b). The distortion of (2.34a) cannot be achieved, however, unless $r_n = 0$ when such quantization is an optimal scheme. The rate distribution formula, however, can be modified some-what to take into account the distortion-versus-rate characteristic of the particular quantization scheme. The result is still fairly close to (2.34b) for most cases, be-cause of the exponential dependence of distortion with rate. Many schemes of quantization, however, restrict the rate r_n to be an integer number of bits. Such a confinement to integer rates imparts yet another deviation from optimality. In practice, block, tree, or trellis codes can achieve to almost any required accuracy, the rate–distribution formula in (2.34b) and distortions much closer to (2.34a) than those of quantization.

An optimal codebook for a stationary Gaussian source with squared error dis-tortion can be constructed on a regular tree with reproduction words for the original source word $\mathbf{x}$ [21]. The reproduction sequences along the tree must possess the statistics of test channel output achieving the rate–distortion function. These statistics are Gaussian with power spectral density

$$S_Y(\omega) = \max \left\{ 0, \ S_X(\omega) - \theta \right\}, \tag{2.36}$$

where θ is the parameter value solving (2.33a) for the desired rate $R = R_T$ of the tree code. The way to construct such a tree is to first populate a subsidiary tree of rate R_T with independent, Gaussian, unit variance letters. The sequence along every path through the subsidiary (or " white") tree is fed to a causal filter $H(\omega) = \sqrt{S_Y(\omega)}$ and then deposited along the corresponding path through the final code tree (the "colored tree"). The latter tree is the one searched for the best reproduction word (in the squared error sense) for the given source word $\mathbf{x}$. The corresponding path is then released to the channel and received at the destination, wherein the same white tree and filter reside. Since the path maps of the white and colored tree are in unique correspondence, perfect reception of the path map allows tracing through the white tree to find the white sequence to feed to the filter, whose output is the same reproduction word found by search at the source. The first use of this kind of system in coding of images is found in [22].

We have treated primarily optimal methods for finite N to show that the N-tuple rate–distortion bounds can be approached. These searched stochastic code methods are not generally well known and are not used in practice for images. They have been much more prevalent in the speech coding literature since their introduction by Anderson and Bodie in 1974 [23] (see also [24]). The stochastic block code procedure just described, however, is currently used in code excited linear prediction (CELP) speech coding systems [25]. The common block techniques in image coding are forms of vector quantization, which is suboptimal for finite N and optimal only in the limit of large N. Vector quantization for large N is computationally not feasible at this time.

2.6.4 Transform Quantization

Introduction A logical suboptimal coding scheme is to independently quantize the independent components of the transformed source vector instead of coding it as a block. The Gaussian assumption is then relaxed so that the transform components are modeled by other probability distributions, but they are still uncorrelated through the Karhunen–Loeve transform of (2.31). Such a scheme is called *transform quantization* in our context but is often referred to in the literature as *transform coding*. We will call it *transform quantization*.

Bit Allocation to Transform Elements The transform coefficients are assumed to have the same generic probability distribution, but with different variances given by the eigenvalues in (2.31b). Therefore, the number of bits or rate allocated to each coefficient will vary, and the Gaussian formula in (2.34b) is no longer strictly correct. We now derive the correct bit allocation formula.

Consider a single transformed source component $\tilde{x}_n$ to be quantized to $\tilde{y}_n$ as depicted in Fig. 2.11.

Every $\tilde{x}_n$, $n = 1, 2 \ldots N$, is quantized independently with rate r_n so that the total rate adds to NR; i.e.,

$$R = \frac{1}{N} \sum_{n=1}^{N} r_n \quad \text{bits/source sample.} \tag{2.37}$$

The rates add because of the independent quantization of the components. Each r_n is different because of the different variances λ_n of $\tilde{x}_n$.

Let $d_{Q_n}(r_n)$ denote the mean squared error of the nth quantizer Q_n with rate r_n. The mean squared error for independently quantizing the uncorrelated components of $\tilde{u}$ is then

$$D = \frac{1}{N} \sum_{k=1}^{N} d_{Q_n}(r_n) \quad \begin{array}{l} \text{mean squared error} \\ \text{per source sample.} \end{array} \tag{2.38}$$

The problem is now to find the rates r_n, $n = 1, 2 \ldots N$, assigned to the different quantizers. Once found, either a quantizer with $K_n = 2^{r_n}$ levels or an entropy-coded quantizer with more than K_n levels may be used to achieve rate r_n. For the former quantizer, the rate r_n must be an integer for direct binary

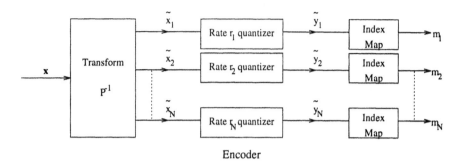

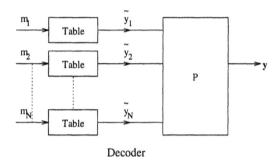

Figure 2.11   A transform quantization system.

transmission, while for the latter only the average of variable length binary code words equals r_n. We shall ignore any requirements for integer valued r_n.

Define $\rho(r)$ as the mean squared error of a unit variance Gaussian quantizer with rate $r \geq 0$ (number of levels 2^r). The distortion of the nth quantizer is then

$$d_{Q_n}(r_n) = \lambda_n \rho(r_n),$$

and the mean squared error per letter of the N quantizers is

$$D = \frac{1}{N} \sum_{n=1}^{N} \lambda_n \rho(r_n). \tag{2.39}$$

The rates r_n must satisfy

$$R = \frac{1}{N} \sum_{n=1}^{N} r_n \tag{2.40}$$

or

$$1 = \sum_{n=1}^{N} \frac{r_n}{NR} = \sum_{n=1}^{N} \alpha_n, \qquad \alpha_n = \frac{r_n}{NR}.$$

Since $0 \le \alpha_n \le 1$ and $\sum_{n=1}^{N} \alpha_n = 1$, the $\alpha = (\alpha_1, \alpha_2 \ldots \alpha_N)$ is a convex space (by definition).

We assume now that the basic quantizer characteristic $\rho(r)$ is a convex $\cup$ function of r $\left(\frac{\partial^2 \rho(r)}{\partial r^2} > 0 \right)$. Expressed in terms of α_n,

$$D = \frac{1}{N} \sum_{n=1}^{N} \lambda_n \rho(\alpha_n).$$

Because a linear combination of convex $\cup$ functions is also convex $\cup$, D is a convex $\cup$ function on a convex space. For such a function the following theorem is applicable.

Theorem Kuhn–Tucker [3, Chapter 5] *If $f(\alpha)$ is convex $\cup$ over the convex space α, then*

$$\frac{\partial f(\alpha)}{\partial \alpha_n} = S \qquad \textit{all n s.t.} \alpha_n > 0$$

$$\frac{\partial f(\alpha)}{\partial \alpha_n} \ge S \qquad \textit{all n s.t.} \alpha_n = 0$$

are necessary and sufficient conditions for a minimum point of f.

An accepted model for the mean squared error versus rate characteristic of a scalar quantizer of a unit variance random variable is the monotically nonincreasing function

$$\rho(r) = g(r)2^{-ar}, \qquad r \ge 0 \tag{2.41}$$

where $g(r)$ is an algebraic function of r, such that $g(0) = 1$ and a is a constant no greater than 2 (see [26]). For larger rates r or an entropy-coded quantizer, $a = 2$ is well justified. Monotonicity implies that $\rho(r) \le \rho(0) = 1$. We make the further approximation that, since $g(r)$ is a much more slowly varying function than the exponential 2^{-ar}, we shall regard it as a constant in the range of interest. The substitution of this quantization model into the Kuhn–Tucker theorem is shown in the steps that follow. In these steps we have set $g(r_n) = g_n$, a constant depending on n, to reflect the slow change of the function g compared to that of the exponential function in (2.41).

$$D(\alpha) = \frac{1}{N} \sum_{n=1}^{N} \lambda_n g_n 2^{-aR\alpha_n}, \qquad g_n = g(r_n),$$

$$\frac{\partial D(\alpha)}{\partial \alpha_n} = -\frac{g_n \lambda_n}{N}(\ln 2)(aRN)2^{-aNR\alpha_n} = -(a \ln 2)R\lambda_n \rho(\alpha_n),$$

$$\alpha_n > 0, \qquad -(a \ln 2)R\lambda_n \rho(\alpha_n) = S,$$

$$\lambda_n \rho(\alpha_n) = -\frac{S/R}{a \ln 2} = \theta,$$

$$\alpha_n = 0, \qquad -a(\ln 2)R\lambda_n \rho(0) \geq S,$$

$$\lambda_n \leq -\frac{S/R}{2\ln 2} = \theta \quad (\rho(0) = 1).$$

Splitting D into components where $\alpha_n > 0$ and $\alpha_n = 0$,

$$D = \frac{1}{N}\left[\sum_{n:\alpha_n > 0} \theta + \sum_{n:\alpha_n = 0} \lambda_n \right],$$

when $\alpha_n > 0$, $\rho(\alpha_n) < 1$, $\lambda_n > \theta$ and when $\alpha_n = 0$, $\lambda_n \leq \theta$. Therefore, the average distortion may be expressed as

$$D = \frac{1}{N}\left[\sum_{n:\lambda_n > \theta} \theta + \sum_{n:\lambda_n \leq \theta} \lambda_n \right] = \frac{1}{N} \sum_{n=1}^{N} \min\left\{\theta, \lambda_n\right\}. \qquad (2.42)$$

Since all the nonzero values of α_n sum to 1,

$$1 = \sum_{n:\alpha_n > 0} \alpha_n,$$

and for $\alpha_n > 0$, $\lambda_n g_n 2^{-aNR\alpha_n} = \theta$. Substituting the solution for α_n,

$$\alpha_n = \frac{1}{aNR} \log_2 \frac{\lambda_n g_n}{\theta}, \qquad \alpha_n > 0 \Rightarrow \lambda_n > \theta.$$

$$1 = \sum_{n:\alpha_n > 0} \frac{1}{aNR} \log_2 \frac{\lambda_n g_n}{\theta} = \frac{1}{NR} \sum_{n:\lambda_n > \theta} \frac{1}{a} \log \frac{\lambda_n g_n}{\theta},$$

or

$$R = \frac{1}{N} \sum_{n:\lambda_n > \theta} \frac{1}{a} \log \frac{\lambda_n g_n}{\theta} = \frac{1}{N} \sum_{n=1}^{N} \max\left\{0, \frac{1}{a} \log \frac{\lambda_n g_n}{\theta}\right\}. \qquad (2.43)$$

The rate assignment is $r_n = \frac{1}{a} \log_2 \frac{\lambda_n g_n}{\theta}$ and $r_n = 0$, $\lambda_n \leq \theta$. The multiplicative factor $g_n = g(r_n)$ is an implicit function of the rate and is often ignored or set to a constant value for all r_n. Note that when $a = 2$ and $g(r) = 1$, the quantizer characteristic is that of the memoryless Gaussian/squared error rate–distortion function. The D and R in (2.42) and (2.43) then become the rate–distortion function of the Gaussian source with memory in (2.32). This means that optimal coding of the scalar transform components with the optimal rate assignment produces an optimal code.

Transform Coding Gain If each component of the original source vector $\mathbf{X} = (x_1, x_2 \ldots x_N)$ is independently quantized with the same rate R by a scalar quantizer with distortion–rate characteristic in (2.35), the mean squared error per sample is

$$D_{\mathrm{PCM}} = \sigma_x^2 \rho(R) \qquad (\sigma_x^2 = \phi(0)).$$

But $\sigma_x^2 = \frac{1}{N} \sum\limits_{n=1}^{N} \lambda_n$, so that

$$D_{\mathrm{PCM}} = \left(\frac{1}{N} \sum_{n=1}^{N} \lambda_n \right) \rho(R).$$

Let us compare the $D = D_{\mathrm{TQ}}$ of transform quantization to that of D_{PCM} for the special case when the rate R is such that

$$\lambda_n > \theta \qquad \text{for all} \quad n = 1, 2 \ldots N,$$

with $\rho(r) = g(r)2^{-ar} \quad (a \le 2)$.

$$R = \frac{1}{N} \sum_{n=1}^{N} \frac{1}{a} \log \frac{\lambda_n g_n}{\theta} = \frac{1}{aN} \log \frac{\prod\limits_{n=1}^{N} \lambda_n g_n}{\theta^N},$$

and

$$D_{\mathrm{TQ}} = \theta = \left(\prod_{n=1}^{N} g_n \lambda_n \right)^{1/N} 2^{-aR}.$$

The transform quantization (TQ) gain over PCM is defined to be

$$G_{\mathrm{TQ/PCM}} = \frac{D_{\mathrm{PCM}}}{D_{\mathrm{TQ}}}$$

and equals

$$G_{\mathrm{TQ/PCM}} = \frac{g(R)}{\left[\prod\limits_{n=1}^{N} g(r_n) \right]^{1/N}} \cdot \frac{\frac{1}{N} \sum\limits_{n=1}^{N} \lambda_n}{\left[\prod\limits_{n=1}^{N} \lambda_n \right]^{1/N}}. \tag{2.44}$$

Although not strictly correct, except for optimal scalar coding, the function $g(\ )$ is often set to a constant to obtain the ideal coding gain:

$$G^o_{TQ/PCM} = \frac{\frac{1}{N}\sum\limits_{n=1}^{N}\lambda_n}{\left[\prod\limits_{n=1}^{N}\lambda_n\right]^{1/N}}, \qquad (2.45)$$

which by the arithmetic–geometric mean inequality [27] that follows,

$$\left(\prod_{n=1}^{N}\lambda_n\right)^{1/N} \le \frac{1}{N}\sum_{n=1}^{N}\lambda_n,$$

is no less than 1 and equals 1 if and only if the values of λ_n are equal; i.e., the process is white. If one does not ignore the variation of $g(\ )$ with rate and assumes $g(\ )$ is convex downward, then $g(R) \ge \left[\prod_{n=1}^{N} g(r_n)\right]^{1/N}$ and $G_{TQ/PCM} \ge 1$ also in this case. Otherwise, the ideal coding gain $G^o_{TQ/PCM}$ of (2.39) must be tempered by the additional factor in (2.44). The formulas are applicable when the rate R is high enough that $\lambda_n > \theta$ for all n. This means that all the components of the transform receive nonzero rate; i.e., they are being quantized and not set directly to 0. The smallest rate for that $\lambda_n > \theta$ for all n is called the critical rate R_c. The θ which produces that rate, $\theta_c = \min\limits_{n} \lambda_n$, is equal to the distortion and is called the *critical distortion* $\theta_c = D_c$.

Suboptimal Transforms   The discrete KLT requires knowledge of the covariance function of the source and a solution for the eigenvectors of the $N \times N$ covariance matrix. In general, especially for large N, the solution and the transform are computationally burdensome procedures with no fast algorithms for their execution. Therefore, one uses instead a source-independent transform with a fast execution algorithm. The hope is that the transform, although suboptimal, will give nearly as good a performance as the optimal KLT.

One such transform that comes to mind is the unitary discrete Fourier transform,

$$\tilde{x}(n) = \frac{1}{\sqrt{N}}\sum_{k=0}^{N-1} x(k)e^{-j2\pi kn/N}, \qquad (2.46a)$$

$$x(k) = \frac{1}{\sqrt{N}}\sum_{n=0}^{N-1} \tilde{x}(n)e^{+j2\pi kn/N}, \quad k = 0, 1 \ldots N - 1,$$

or in vector-matrix form,

$$\tilde{\mathbf{X}} = F_N\mathbf{X}, \qquad [F_N]_{n,k} = \frac{1}{\sqrt{N}}\exp\{-j2\pi nk/N\}. \qquad (2.46b)$$

The matrix F_N is unitary, that is, $(F_N^*)^T = F_N^{-1}$, and is *circulant*, because its rows and columns are circular shifts of each other. The covariance matrix $\Lambda_{\widetilde{x}}$ of the transform vector $\widetilde{\mathbf{X}}$, related to the covariance matrix Λ_x of the source vector $\mathbf{X}$ by $\Lambda_{\widetilde{x}} = F_N \Lambda_x F_N^{-1}$, is diagonal if and only if $\phi_x(|k - l|)$ is periodic with period N.

The discrete Fourier transform (DFT) has a fast computational algorithm (FFT) and is asymptotically optimal in the strong sense that the variance of each $\tilde{x}(n)$ converges pointwise to a value of the spectral density that in turn equals one of the eigenvalues of the covariance matrix [18].

Although for finite N, the elements $\tilde{x}(n)$ are not uncorrelated, we encode (quantize) them independently as if they were independent. Because the correlations between the components are ignored, we obtain inferior performance to that of the KLT. As N grows larger, the performance of DFT quantization approaches that of KLT quantization. Bounds on the difference in performance between optimal DFT coding and optimal KLT coding have been derived by Pearl [28] and Pearlman [29]. One of the seeming disadvantages of the DFT is that it produces complex coefficients, but because of the induced Hermitian symmetry for a real input sequence ($\tilde{x}^*(n) = \tilde{x}(N - n)$), there are only N real non-redundant quantities to be encoded.

Another important transform is the (unitary) discrete cosine transform (DCT) given by

$$\tilde{x}(n) = \begin{cases} \dfrac{1}{\sqrt{N}} \displaystyle\sum_{k=0}^{N-1} x(k), & n = 0 \\[2mm] \dfrac{2}{\sqrt{N}} \sum_{k=0}^{N-1} x(k) \cos \dfrac{n\pi}{2N}(2k + 1), & n = 1, 2 \ldots N - 1. \end{cases} \tag{2.47a}$$

The inverse is the same form with k and n interchanged. The matrix transformation is

$$C = \{C_{nk}\}_{n,k=0}^{N-1}, \qquad C_{nk} = \begin{cases} \dfrac{1}{\sqrt{N}} \quad n = 0 \\ \qquad k = 0, 1 \ldots N - 1 \\[2mm] \sqrt{\dfrac{2}{N}} \cos \dfrac{n\pi}{2N}(2k + 1) \\ \qquad n = 1, 2 \ldots N - 1 \\ \qquad k = 0, 1 \ldots N - 1 \end{cases} \tag{2.47b}$$

$$C^{-1} = C^T = C.$$

The transform elements are real.

This transform is asymptotically optimal in a distributional sense; i.e., for a function f, $f(E[|\tilde{x}(n)|^2])$, $n = 0, 1 \ldots N - 1$, converges to $f(\lambda_0, \lambda_1 \ldots \lambda_{N-1})$ as $N \to \infty$, if the eigenvalues are bounded.

Yemini and Pearl [30] proved the asymptotic equivalence of DCT and KLT for all finite-order Markov processes. Ahmed, Natarajan, and Rao [31] gave empirical evidence that DCT performance is close to KLT even for small values of N in the case of Markov-1 signals.

You can evaluate the DCT through the DFT, which has a fast algorithm. First express the DCT as

$$\tilde{x}(n) = \sqrt{\frac{2}{N}} \, \text{Re} \left\{ e^{-j\frac{n\pi}{2N}} \alpha(n) \sum_{k=0}^{N-1} x(k) e^{-j\frac{2\pi kn}{2N}} \right\},$$

$$\alpha(n) = \begin{cases} \frac{1}{\sqrt{2}}, & n = 0 \\ 1, & n \neq 0. \end{cases}$$

Note that the summation is a $2N$-point DFT with $x(k) = 0$, $N \leq k \leq 2N - 1$. The $2N$-point DFT is at a slight disadvantage along with the exponential (or sine–cosine) multipliers. If the sequence is reordered as

$$\left. \begin{aligned} w(k) &= x(2k) \\ w(N - 1 - k) &= x(2k + 1) \end{aligned} \right\} \quad k = 0, 1 \dots \frac{N}{2} - 1,$$

$$\tilde{x}(n) = \sqrt{\frac{2}{N}} \, \alpha(n) \text{Re} \left\{ e^{j\frac{\pi n}{2N}} \sum_{k=0}^{N-1} w(k) e^{j2\pi nk/N} \right\},$$

the summation is an N-point DFT of the reordered sequence [32]. Similar procedures apply to the inverses.

Another transform often used is the Haadamard–Walsh transform. The transform matrix consists just of 1's and -1's. Therefore, the transform is formed by a series of additions and subtractions, making it very fast to compute. It has worse performance than the DFT and DCT and is not asymptotically optimal.

Implementation Issues   To implement a transform coding scheme for images, several issues must be addressed. The first concerns the statistical model of the image process and the possibility of making the model flexible or adaptive; the second concerns the transmission of overhead information required for the receiver to reconstruct the image from the message bit stream; and the third concerns the realization of the bit allocation described in the preceding formulas and the operation of the decoder.

Modelling   In real applications of transform quantization the source sequence is not Gaussian. Therefore, the transform sequence is also not Gaussian. However, if N is large, the probability distribution of the transform sequence is approximately jointly Gaussian for almost any transform.

The justification for this statement comes from the multidimensional central limit theorem. The scalar central limit theorem says roughly that weighted sums of N random variables (not necessarily independent), appropriately scaled, approach Gaussian in the limit as $N \to \infty$. Transforms are weighted sums of random variables. So Gaussian quantizers of transform coefficients are often justified. For small N and certain data processes, such as speech and images, empirical evidence has shown that DCT coefficients are better modeled as Laplacian or Gamma (except for the 0th component).

The assumption of a fixed model for an image process is not realistic. The image is subdivided into an array of smaller subblocks (usually 8×8 or 16×16) for two reasons: one, to save computation time versus a full image transform; and two, so that the smaller blocks can be individually treated according to their local statistical or variational properties. An example of the latter is to set the same average distortion target for each block and allow the rate to vary among the blocks to meet this target. One can still take a statistical approach of fitting a parametric autocorrelation model to each block, use the model for allocating the bit among the transform coefficients, and transmit the parameters to inform the receiver of the model. The number of such parameters must be small enough to keep the overhead information rate relatively small and large enough to produce an effective model. Such a scheme is described in [33].

When the Gaussian assumption is relaxed, the KLT still produces uncorrelated coefficients, but uncorrelated coefficients are not necessarily independent. In most realizations, the coefficients are still independently scalar quantized, although some attempts have been made to vector quantize individual segments of transform coefficients. Except for the lowest order dc coefficient, the coefficients are usually modeled as Laplacian. The autocorrelation or spectral model provides the variance of these coefficients.

Bit Allocation and Entropy Coding   The bit allocation is dependent on the rate versus distortion characteristic in the quantization of each coefficient. The most efficient quantization is an optimal entropy-constrained quantizer. To realize the promised rate, the quantizer outputs must be entropy coded. Even if a fixed level quantizer is used, a lower rate can be achieved when the quantizer outputs are entropy-coded. When entropy coding is used, the binary code word's length is variable, depending on the quantizer level's probability, giving a noninteger average number of bits per sample. Therefore, there does not need to be a constraint in the bit allocation of an integer number of bits per sample. The encoding to an integer number of bits can be actualized through a quantizer constrained in number of levels or in entropy. As uniform quantizers are nearly optimal, one can simply specify a different step size to set the desired entropy of rate for each coefficient. A particularly convenient kind of entropy coding for transform coefficients is the arithmetic code [34]. Each coefficient that receives nonzero rate has a different probability table for its quantization levels, and the string of such coefficients is encoded with one long binary code word. A Huffman code can also be used, but not as conveniently when each coefficient has a different variance and quantizer step size and with a sacrifice in performance.

Bit Map Transmission and Threshold Coding   In the scheme just described, the locations of which coefficients are encoded and which are set to 0, the probabilities of the quantizer outputs, and their step sizes are all conveyed by the transmission of model parameters and the overall rate or average distortion target. The receiver then executes the same bit allocation procedure as that in the encoder. There are adaptive methods that convey only a bit map, which is an array of integers equal to the number of bits allocated to each coefficient. To transmit this map efficiently, the two-dimensional transform block is scanned in a zigzag or cross-diagonal manner, which is likely to produce a monotonically nonincreasing set of bit map integers. Runs of zeros, which become more probable as the overall rate is lowered, are efficiently coded by a run length code. Some schemes set the bit map adaptively by encoding only those coefficients that exceed an experimentally determined threshold, allocating to them a given number of bits according to their magnitude. These so-called threshold codes are highly adaptive and obviate the statistical model but typically use more overhead information than model-based schemes.

2.7 Predictive Coding

2.7.1 Introduction

As with transform coding, predictive coding is another technique that exploits the memory of a source. The basic idea is to encode the difference or residual between the source value and a prediction of its value from previous source values. These values of the source or its past can be scalar or vector values. If the source has memory, the prediction residual will have lower variance than the source itself. For the same number of bits, the mean squared error when encoding the residual will be lower than that obtained when encoding the source directly. Certain principles must be followed in order to make the idea of predictive coding a workable one. Here we present these principles for the case of scalar predictive quantization, which is called *differential pulse code modulation*, or DPCM.

2.7.2 Linear Prediction

Let $\ldots, x_{-1}, x_0, x_1 \ldots x_n$ be a stationary, zero mean source sequence. The linear prediction of x_n from the past p values is

$$\widehat{x}_n = -\sum_{k=1}^{p} a_k x_{n-k},$$

where $a_1, a_2 \ldots a_p$ are the prediction coefficients. The well-known solution for the vector of prediction coefficients $\mathbf{a}^o = (a_1^o, a_2^o \ldots a_p^o)^T$ that minimizes the mean squared error $E[|\widehat{x}_n - x_n|^2]$ follows the projection principle that the error $x_n - \widehat{x}_n$ be orthogonal to the data or past values; i.e.,

$$E[(x_n - \widehat{x}_n)x_{n-k}] = 0, \quad k = 1, 2 \ldots p,$$

which results in the matrix (Yule–Walker) equation

$$\Phi_p \mathbf{a}^o = \phi$$

where $\phi = (\phi(1), \phi(2) \ldots \phi(p))^T, \phi(l) = E[x_n x_{n-l}], \quad l = 1, 2 \ldots p,$ and $\Phi_p = \{\phi(|k - l|)\}_{k,l=0}^{p-1}$ is the pth order autocorrelation matrix of the source. The minimum mean squared error can be expressed as

$$e_{\min}^2 = \min \; E[|x_n - \widehat{x}_n|^2] = \phi(0) \; - \; \phi^T \mathbf{a}^o$$
$$= \phi(0) \; - \; \phi^T \Phi_p^{-1} \phi. \tag{2.48}$$

Assuming that Φ_p is strictly positive definite assures that Φ_p^{-1} is also positive definite, so that $\phi^t \Phi_p^{-1} \phi > 0$ for any vector ϕ (by definition). Therefore, the minimum error variance $e_{\min}^2$ is always less than $\phi(0)$, the variance of the source.

2.7.3 DPCM System Description

The fact that the variance of the prediction residual is smaller than that of the source motivates its use in a workable coding system. The obstacle is that a prediction is made from past values of the source and that these values are not available at the destination, since any reconstruction of past values must contain some error if the source has continuous amplitude. If one does proceed with prediction from true values at the source by always holding the past p values in memory and attempts reconstruction at the destination, the reconstruction error will accumulate as decoding progresses. One must therefore predict using reconstructed values at the source that are the same as those at the destination. The DPCM system illustrated in Fig. 2.12 accomplishes this objective.

In this system the difference e_n between the current source value x_n and a prediction $\widehat{x}_n$ is quantized to $\widehat{e}_n$, whose binary code is transmitted to the destination. The prediction $\widehat{x}_n$ has been formed from a linear combination of the past p reconstructed values of the source $\widetilde{x}_{n-1}, \widetilde{x}_{n-2} \ldots \widetilde{x}_{n-p}$, which in turn can be derived only from past values of $\widehat{e}_n$. Therefore, the current reconstruction $\widetilde{x}_n$ is the sum of its prediction $\widehat{x}_n$ and the quantized prediction error $\widehat{e}_n$. With this arrangement of feeding the prediction value $\widehat{x}_n$ back to add to the quantized prediction residual (feedback around the quantizer), the error in the reconstruction $\widetilde{x}_n$ is equal to the error in the quantization of e_n. That is,

$$x_n - \widetilde{x}_n = (x_n - \widehat{x}_n) - (\widetilde{x}_n - \widehat{x}_n)$$
$$= e_n - \widehat{e}_n.$$

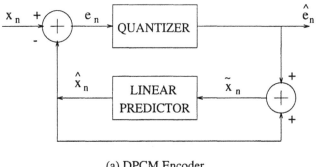

(a) DPCM Encoder

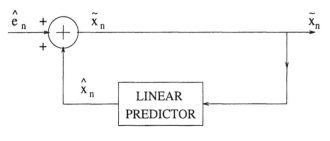

(a) DPCM Decoder

Figure 2.12 DPCM encoder and decoder.

The decoder can form the reconstructions $\tilde{x}_n$ from $\hat{e}_n$ as does the encoder. The operation is shown also in Fig. 2.12.

2.7.4 DPCM Coding Error and Gain

The sequence of prediction residuals is a stationary scalar source whose values are independently quantized for minimum mean squared error. The distortion versus rate function is proportional to the source variance, which is the mean squared value of the prediction residual. It is not $e_{\min}^2$ in (2.48), because in DPCM the prediction takes place on reconstructed past values, not actual past values. If we make the approximation that the predictor coefficients are designed for actual past values using the source covariances $E[x_n x_{n-l}]$, $l = 0, 1 \ldots p$, and that the prediction $\hat{x}_n$ from the quantized reconstructions is approximately the same as that from actual past values, then $e_{\min}^2$ in (2.48) is the variance of the prediction residual. This approximation becomes more realistic at high rates when the quantization error is small and the reconstructions become close to their actual values. Therefore, the approximation is

$$E[\hat{e}_n^2] \approx \phi(0) - \phi^{\mathrm{T}}\Phi_p^{-1}\phi.$$

The gain in DPCM coding over PCM coding may be defined as

$$G_{\mathrm{DPCM/PCM}} = \frac{D_{\mathrm{PCM}}}{D_{\mathrm{DPCM}}}.$$

Since the distortions (mean squared errors) of both PCM and DPCM have the same unit variance distortion versus rate function, such as that in (2.41), the gain is just the ratio of their respective source variances, which is

$$G_{\mathrm{DPCM/PCM}} = \frac{\phi(0)}{\phi(0) - \phi^{\mathrm{T}}\Phi_p^{-1}\phi}.$$

This gain exceeds one for a source with memory and is always optimistic due to the approximation of assuming prediction from nonquantized values.

2.8 Subband Coding

2.8.1 Introduction

We now treat another kind of transformation upon the source and explore the encoding of its output. This transformation is a decomposition of the source into contiguous spatial frequency subbands. The encoding takes place on the narrow-band spatial processes that are the outputs of the decomposition. The particular properties of these processes lead to some interesting consequences in coding. As before, we shall explore the information theory of optimal coding before treating the theory of practical coding. We begin with a mathematical description of the subband decomposition.

2.8.2 Subband Decompositions of a Source

Consider now a subband decomposition, wherein the source sequence is fed to a bank of bandpass filters that are contiguous in frequency so that the set of output signals can be recombined without degradation to produce the original signal. We shall assume that the contiguous filters are ideal with zero attenuation in the passband and infinite attenuation in the stop-band and that they cover the full frequency range of the input (Fig. 2.13). The assumption of ideal filters is not a consequential limitation, as filters called *quadrature mirror filters* (QMFs) exist that when used in a contiguous filter bank approximate well the characteristics of ideal ones.

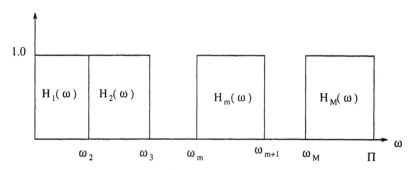

Figure 2.13   Subband filter transfer functions.

When the input waveform to the bank of M ideal filters is a one-dimensional sequence, the output of any one filter whose lower cutoff frequency is an integer multiple of its bandwidth W_m, for any $m = 1, 2, \ldots, M$, is subsampled by a factor equal to $V_m = \pi / W_m$ and is now a full-band sequence in the frequency range from $-\pi$ to π referenced to the new, lower sampling frequency. This combination of filtering and subsampling, called *decimation*, is depicted in Fig. 2.14. We shall assume that the integer bandwidth to lower frequency relationship holds for all filters in the bank, so that all outputs are decimated by the appropriate factor. These outputs are called the *subband signals* or *waveforms*, and their aggregate number of samples equals that in the original input waveform. The original input can be reconstructed exactly from the subband signals. The sampling rate of each subband signal is increased to that of the original input by filling in the appropriate number of zero samples, and the zero-filled waveform is fed into an ideal filter with gain equal to the subsampling factor covering the original passband (called *interpolation*) (see Fig. 2.15). The sum of these interpolated subband signals equals the original input signal.

When the input is a sampled image, subband images are produced by an analogous two-dimensional filter bank of contiguous, nonoverlapping ideal filters covering the two-dimensional frequency range. In two dimensions, there is much more flexibility in the sampling pattern (e.g., rectangular or hexagonal) and the passband shapes (e.g., rectangular, hexagonal, or hexagonal sections). We assume that each filter domain is such that a subsampling pattern exists that replicates the filter characteristic so that it appears at the central low-frequency band and becomes full band when referenced to its subsampling pattern. The most common example is a sampling of the original image on a rectangular grid and locating every filter domain so that the coordinates of its lower frequency in each dimension are integer multiples of the bandwidths belonging to their respective dimensions.

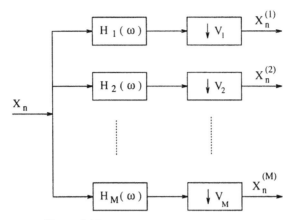

Figure 2.14   Analysis of source into subbands.

Although we shall use the paradigm of rectangular sampling grids for our subsequent analyses, the methods will extend to other kinds of sampling grids as well, such as hexagonal.

A particular type of subband decomposition that has attracted a great deal of attention lately is the pyramid or multiresolution decomposition. We say a pyramid is obtained when the low-frequency band image is recursively split into four equal area subbands, which are then subsampled by a factor of 2 in each dimension. So

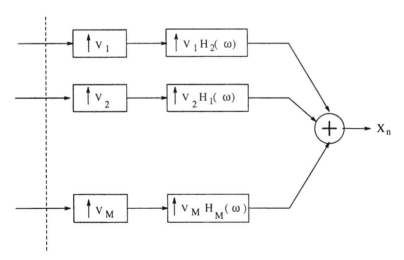

Figure 2.15   Synthesis of source from subbands.

we begin with the original image, filter into four subbands of equal size, subsample by a factor of 2 in each dimension, and then keep repeating the procedure on the low-pass subband until a 2×2 subband image terminates the process. When subband images belonging to different splitting levels are stacked one on top of another by decreasing size with the largest at the bottom, the figure resembles a pyramid. Actually, we have four pyramids, one to each distinct subband. As we travel from the bottom to the top of any one pyramid, we regress by successive factors of 2 in resolution for each dimension.

2.8.3 Rate–Distortion Theory for Subband Coding

We now apply the rate–distortion theory of Section 2.3.3 to subband coding, as was originally done in [35]. The one-dimensional case will be treated in detail first and the generalization to two-dimensional sampling grids will be indicated. Assume a one-dimensional stationary Gaussian sequence with power spectral density $S_X(\omega)$ has been decomposed with ideal filters into subbands with positive lower edge frequencies $\omega_1, \omega_2, \ldots, \omega_M$ with $\omega_1 = 0$ as shown in Fig. 2.13. The frequency range of the mth subband is denoted by $I_m = [\omega_m, \omega_{m+1}) \cup (-\omega_{m+1}, -\omega_m]$, $\omega_{M+1} = \pi$, $m = 1, 2, \ldots, M$, and its bandwith by $W_m = \omega_{m+1} - \omega_m = \pi/V_m$. We require that

$$\bigcup_{m=1}^{M} I_m = [-\pi, \pi] \quad \text{and} \quad I_m I_n = \phi \quad \text{for } m \neq n \qquad (2.49)$$

and that V_m be a rational number. The ideal filtering of the mth subband filter produces a Gaussian random sequence with power spectrum

$$S_m(\omega) = \begin{cases} S_X(\omega), & \omega \in I_m \\ 0, & \omega \notin I_m. \end{cases}$$

Resampling by a factor of V_m results in a random subsequence $\mathbf{x}^{(m)}$ with power spectrum

$$S^{(m)}(\omega) = \frac{1}{V_m} S_X\left(\frac{\omega}{V_m} + \omega_m \operatorname{sgn}(\omega)\right) \qquad (2.50)$$

in the frequency interval $-\pi < \omega \leq \pi$ referenced to the new sampling frequency. According to (2.33), the rate–distortion function of each Gaussian subband for the squared error distortion measure is given by

$$R_{\theta_m}^{(m)} = \frac{1}{2\pi} \int_{-\pi}^{\pi} \max\left\{0, \frac{1}{2} \log \frac{S^{(m)}(\omega)}{\theta_m}\right\} d\omega, \qquad (2.51a)$$

$$D_{\theta_m}^{(m)} = \frac{1}{2\pi} \int_{-\pi}^{\pi} \min\left\{\theta_m, S^{(m)}(\omega)\right\} d\omega, \quad m = 1, 2, \ldots, M. \quad (2.51b)$$

The $1/V_m$ factor in $S^{(m)}(\omega)$ equalizes the subband variances before and after resampling so that the mean squared error per sample of the reconstructed process after recombining the subbands is

$$D = \sum_{m=1}^{M} D_{\theta_m}^{(m)}. \quad (2.52)$$

(The per-sample scaling of the subband distortions in zero filling is cancelled by the gain factor V_m in the interpolation filters.) The contribution of the rate $R_\theta^{(m)}$ to the overall rate per sample R of the reconstructed process must be reduced by a factor of V_m to reflect the fraction of the samples in the subband. Therefore,

$$R = \sum_{m=1}^{M} \frac{1}{V_m} R_{\theta_m}^{(m)}. \quad (2.53)$$

If we substitute $\theta_m = \theta/V_m$, where θ is the parameter solving the rate–distortion function for the original process, we obtain by straightforward manipulations

$$D = D_\theta \quad \text{and}$$
$$R = R_\theta,$$

where D_θ and R_θ are the distortion and rate of the rate–distortion function of the source process $\mathbf{X}$. The meaning of this result is that if ideally filtered subbands of the process $\mathbf{X}$ are optimally encoded and the rate and distortion are distributed among the subbands according to (2.51) with $\theta_m = \theta/V_m$, the encoding method is optimal.

The result extends readily to two dimensions for sampling on a rectangular grid. The analogous rate and distortion allocation to the mth subband is

$$R_{\theta_m}^{(m)} = \left(\frac{1}{2\pi}\right)^2 \int_{-\pi}^{\pi} \int_{-\pi}^{\pi} \max\left\{0, \frac{1}{2}\log\frac{S^{(m)}(\omega_1, \omega_2)}{\theta_m}\right\} d\omega_1 d\omega_2, \quad (2.54a)$$

$$D_{\theta_m}^{(m)} = \left(\frac{1}{2\pi}\right)^2 \int_{-\pi}^{\pi} \int_{-\pi}^{\pi} \min\left\{\theta_m, S^{(m)}(\omega_1, \omega_2)\right\} d\omega_1 d\omega_2, \quad (2.54b)$$

for $m = 1, 2, \ldots, M^2$, with $\theta_m = \theta/V_m$ and V_m being the resampling factor of the subband. Two common implementations are equal-size subbands, $V_m = M^2$, $m = 1, 2, \ldots, M^2$ and recursive splitting and 2×2 subsampling of the low-frequency subband to form a pyramid, $V_m \in \{2^{2r} : r = 1, 2, \ldots, M_R\}$, where M_R is the highest level of the splitting corresponding to the lowest resolution. Incidentally for hexagonal grids, the formulas are of the same form except that the integration is over the smallest hexagon that circumscribes a circle of radius π. Since this hexagon has area $2\sqrt{3}\pi^2$, the reciprocal factor preceding the integral changes from $(2\pi)^2$, the rectangular frequency area, to this hexagonal area.

The optimal performance specified by the preceding rate–distortion formulas can be achieved only in the limit as the size of the source process and hence the subband processes grow to infinity. Later in this chapter, we shall consider the implications of coding the finite size source processes encountered in practice.

2.8.4 Practical Subband Coding

Introduction   Constraints of complexity and limitations of modeling lead to avoidance of complicated coding schemes that attempt to reach close to the rate–distortion theoretic limit. Once a finite and realizable complexity is accepted, a subband coding scheme becomes attractive because it achieves measurable and theoretical performance advantages over the same scheme applied directly to the full band source. We shall examine PCM and DPCM in this light.

Rate Allocation to Subbands   The first step in coding is allocating a given number of bits among the subbands to achieve the minimum distortion. Distortion is again defined to be mean squared error. Let there be N samples from the image source and a given code rate of R bits/sample. If b_m bits are given to the mth subband, then

$$NR = \sum_{m=1}^{M} b_m \tag{2.55}$$

because the subbands are statistically independent for a Gaussian source.

Since the fullband source is stationary, any given subband has samples with the same variance σ_m^2, $m = 1, 2, \ldots, M$. Define $d_m(b)$ to be the mean squared error obtained for coding a subband of unit variance per sample with b bits. Then the total mean squared error after combining the subbands is

$$ND = \sum_{m=1}^{M} \sigma_m^2 d_m(b_m) \tag{2.56}$$

where D is the average distortion per sample.

Note now the analogy to the rate and distortion equations (2.38) and (2.39), which were obtained for quantization of transform coefficients. With the replacement of r_k, rate per sample, by b_m, rate per subband, and λ_k, the eigenvalue, by σ_m^2, the subband variance, the solution by the Kuhn–Tucker theorem becomes

$$\sigma_m^2 d_m'(0) \le \frac{S}{R}, \qquad b_m = 0,$$

$$\sigma_m^2 d_m'(b_m) = \frac{S}{R}, \qquad b_m > 0, \tag{2.57}$$

where the prime indicates derivative with respect to the argument of $d_m(\ )$. Note that this derivative is negative.

There are also algorithmic methods for allocating bits among the subbands, which are useful when analytic expressions for $d_m(b)$ are not available. Many of them, however, constrain the number of bits per sample to be an integer. Here we do not impose this restriction, because we may want to consider general subband coders such as entropy-coded quantization, vector quantization, tree coding, or trellis coding, all of which do not require integer rates. The algorithm in [36] does not impose any restrictions on the rate or type of subband coders. It iteratively seeks a point on the lower convex hull in the distortion–rate plane given rate and distortion points of coders in different subbands. An elegant way to implement the same solution, as pointed out in [37], is to use the Breiman–Friedman–Olshen–Stone (BFOS) algorithm [39] to prune an unary branching tree with each path attached to the root corresponding to a subband. The rate and distortion of a quantizer are associated with each node such that the lower the rate, the nearer to the root node. All interior nodes of a path are examined to find the one with the greatest marginal return from the leaf node and the intervening branches are then pruned. All paths are examined in turn and the process is repeated until the desired rate is reached. For further explanation, see [38] and [40]. The rate–distortion point is guaranteed to lie on the convex hull. The analytical solution given here also achieves a point on the convex hull of all rate–distortion points (R,D) satisfying (2.55) and (2.56).

Subband Predictive Coding Gain Thus far, no advantageous properties of subband coding have been revealed, only that optimality can be preserved by coding subbands. Assume now that every sample of every subband is to be quantized independently with the same quantizer and the length of its representative binary code word either the base 2 logarithm of the number of levels or the entropy. The distortion (mean squared error) versus rate characteristic of such a scalar quantizer of a unit variance sample can again be modeled as in (2.41).

Since the quantization of the coefficients in any subband is independent and the subband processes are stationary, the quantization noise at the input of the upsampling and interpolation filters is stationary and white. The variance per sample of a subband quantization noise process is therefore $\sigma_m^2 \rho(b_m/n_m)$ and the total quantization mean squared error (mse) is $n_m \sigma_m^2 \rho(b_m/n_m)$. The upsampling leaves this mse unchanged, but the factor of V_m reduction in bandwidth and power gain V_m^2 of the (ideal) interpolation filter contribute a net gain of V_m to the mse. Therefore, prior to the final re-combining of the subbands, the reconstruction mse of the subband becomes $V_m n_m \sigma_m^2 \rho(b_m/n_m) = N\sigma_m^2 \rho(b_m/n_m)$. Thus, we take $d_m(b_m) = N\rho(b_m/n_m)$ in (2.56) to obtain

$$g_m(\sigma_m^2/n_m)2^{-ab_m/n_m} = \theta, \qquad \text{for all} \quad b_m > 0,$$
$$\sigma_m^2/n_m \le \theta, \qquad \text{for all} \quad b_m = 0, \qquad (2.58)$$

with $\theta \equiv -S/a(\ln 2)NR$. Again, we have assumed $g_m = g(b_m/n_m)$ is constant in the differentiation and $g(b_m/n_m) \le g(0) = 1$. Solving for b_m in terms of θ, we obtain for $r_m = b_m/n_m$, the rate per sample,

$$r_m = \begin{cases} 1/a \log[(\sigma_m^2 g_m/n_m)/\theta], & \sigma_m^2/n_m > \theta \\ 0, & \sigma_m^2/n_m \le \theta \end{cases} \qquad (2.59)$$

where a is usually taken to be 2. Solving for θ using the rate constraint of (2.55) and letting $J_c = \{m : \sigma_m^2/n_m > \theta\}$ be the index set of nonzero rate subbands, we obtain the subband rate allocations

$$r_m = \begin{cases} NR/N_c + 1/a \log[(\sigma_m^2 g_m/n_m)/(\pi_{k \in J_c}(\sigma_k^2 g_k/n_k)^{n_k})^{1/N_c}], & m \in J_c \\ 0, & m \notin J_c \end{cases} \qquad (2.60)$$

where $N_c = \sum_{m \in J_c} n_m$ is the total number of samples in the nonzero rate subbands and the number of samples in the mth subband is $n_m = NW_m/\pi$. For this rate allocation the mse after reconstruction is

$$D_{\text{SB/PCM}} = \sum_{m=1}^{M} \sigma_m^2 g_m 2^{-ar_m} = \sum_{m \in J_c} n_m \theta + \sum_{m \notin J_c} \sigma_m^2 \qquad (2.61)$$

where θ is given through the rate constraint.

Consider now the special case of rate R sufficiently large that $\sigma_m^2/n_m > \theta$ for all m. The rate per sample for the mth subband r_m, after solving for θ,

$$\theta = \sigma_{\text{WGM}}^2 2^{-aR}, \qquad \sigma_{\text{WGM}}^2 = \prod_{m=1}^{M} (\sigma_m^2 g_m/n_m)^{n_m/N},$$

becomes

$$r_m = R + \frac{1}{a} \log \frac{\sigma_m^2 g_m/n_m}{\sigma_{\text{WGM}}^2}, \qquad (2.62a)$$

and the mse after reconstruction

$$D_{\text{SB/PCM}} = N\sigma_{\text{WGM}}^2 2^{-aR}. \qquad (2.62b)$$

When the samples of the full band source are directly and independently coded using the same technique, the average distortion is

$$D_{\text{FB/PCM}} = g(R)\sigma_X^2 2^{-aR} = g(R)\left(\sum_{m=1}^{M} \sigma_m^2\right) 2^{-aR}. \tag{2.63}$$

The subband coding gain for PCM is defined to be

$$G_{\text{SB/PCM}} = \frac{D_{\text{FB/PCM}}}{D_{\text{SB/PCM}}}, \tag{2.64}$$

the ratio of the full band to subband mse's. When the rate R is sufficiently large that $r_m > 0$ for all $m = 1, 2, \ldots, M$, this gain, by substitution of (2.63) and (2.62b) is expressed as

$$G_{\text{SB/PCM}} = \frac{g(R)\sum_{m=1}^{M} \sigma_m^2}{N\sigma_{\text{WGM}}^2} = \frac{\sigma_x^2}{\prod_{m=1}^{M} (\sigma_m^2 V_m)^{1/V_m}} \times \frac{g(R)}{\prod_{m=1}^{M} (g_m)^{1/V_m}}. \tag{2.65}$$

If the function $g(\ )$ is constant, which is the case for ideal scalar coding, then the ideal gain is

$$G_{\text{SB/PCM}}^o = \frac{\sigma_x^2}{\prod_{m=1}^{M} (\sigma_m^2 V_m)^{1/V_m}}. \tag{2.66}$$

Noting that $\sum_{m=1}^{M} (1/V_m) = 1$, the convexity of the logarithm yields

$$\sum_{m=1}^{M} \frac{1}{V_m} \log(\sigma_m^2 V_m) \leq \log\left(\sum_{m=1}^{M} \frac{1}{V_m} \cdot \sigma_m^2 V_m\right) = \log \sigma_X^2, \tag{2.67}$$

with equality if and only if $\sigma_m^2 V_m$ is the same for all subbands. Therefore,

$$G_{\text{SB/PCM}}^o \geq 1, \tag{2.68}$$

with equality if and only if $\sigma_m^2 V_m$ is a constant independent of m. Since V_m is inversely proportional to bandwidth W_m, the equality condition is satisfied if the input process is white, even for subbands of different bandwidths.

If $g(\ )$ is a convex downward function of rate, then $g(R) \geq \prod_{m=1}^{M} (g_m)^{1/V_M}$, and the previous ideal gain is lower bound. Otherwise, the ideal gain must be tempered by the ratio in (2.65), which is close to 1 in most cases. Therefore, PCM coding of subbands is advantageous over direct PCM coding of the full band signal even in the case of unequal size subbands. This result is a generalization of previous formulations [26], which treated only the case of equal size subbands and ignored the effect of the variation of the value of $g(\ )$.

Differential Predictive Coding   In general, the subbands do not have flat-topped power spectra and therefore are not memoryless. One way to take advantage of this memory to reduce the encoding rate is to encode each subband by DPCM. We assume again that the prediction coefficients, which are optimized for past values of the samples, are suboptimal and approach optimality only in the limit of large encoding rate. This large rate approximation for predictor design is much more tenable for subband coding than for full band coding, because the subbands are coded at rates that are higher than their contributions to the overall rate by a factor equal to the resampling factor V_m. Moreover, the flatter spectra of the subbands enable more accurate predictors of lower order than that for the full band image.

In a DPCM coder, as shown in Fig. 2.16, the mean squared reconstruction error per sample is equal to the mean squared quantization error per sample. In the mth subband, the difference between the input sample $x_n^{(m)}$ and its prediction $\hat{x}_n^{(m)}$ is scalar quantized to a value $\hat{e}_n^{(m)}$ with rate $r_m > 0$ for all m to yield a mean squared error by (2.41) of

$$E[(e_n^{(m)} - \hat{e}_n^{(m)})^2] = \sigma_{e,m}^2 g(r_m) 2^{-a r_m}, \tag{2.69}$$

and this error is equal to the reconstruction error; i.e.,

$$E[(e_n^{(m)} - \hat{e}_n^{(m)})^2] = E[(x_n^{(m)} - \tilde{x}_n^{(m)})^2]. \tag{2.70}$$

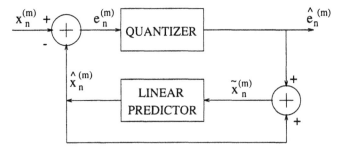

Figure 2.16   DPCM encoder for the mth subband.

The variance of the quantizer input, $\sigma_{e,m}^2$, is the variance of the prediction error and $g(r_m)$ is again considered to be approximately a constant g_m in each subband. The total reconstruction mse per sample $D_{\mathrm{SB/DPCM}}$ is therefore

$$D_{\mathrm{SB/DPCM}} = \sum_{m=1}^{M} E[(x_n^{(m)} - \tilde{x}_n^{(m)})^2] = \sum_{m=1}^{M} g_m \sigma_{e,m}^2 2^{-ar_m}. \tag{2.71}$$

The optimal rate allocation $\{r_m\}$ for a given rate R depends on the variances of the elements being quantized, and (2.62b) is therefore applicable with $\sigma_{e,m}^2$ replacing σ_m^2 and $\sigma_{e,\mathrm{WGM}}^2 = \prod_{m=1}^{M}(\sigma_{e,m}^2 g_m/n_m)^{n_m/N}$ replacing σ_{WGM}^2. Substituting these correspondences into expression (2.71) for $D_{\mathrm{SB/DPCM}}$ results in

$$D_{\mathrm{SB/DPCM}} = N\sigma_{e,\mathrm{WGM}}^2 \, 2^{-aR}, \tag{2.72}$$

where the mse is equally distributed across the M subbands.

When the rate R is sufficiently low, some subbands may be assigned zero rate, making the preceding formula no longer applicable. In this case, we use the rate allocation equation (2.60) and distortion equation (2.61) with $\sigma_m^2 = \sigma_{e,m}^2$ to obtain

$$D_{\mathrm{SB/DPCM}} = \left[\theta \left(\sum_{m \in J_c} n_m \right) + \sum_{m \notin J_c} \sigma_{e,m}^2 \right] \tag{2.73}$$

with θ determined by the overall rate constraint,

$$R = \frac{1}{N} \sum_{m \in J_c} n_m r_m, \tag{2.74a}$$

and

$$J_c = \{m : \sigma_{e,m}^2/n_m > \theta\}. \tag{2.74b}$$

The corresponding full band DPCM coding for a given rate R results in the reconstruction mse of

$$D_{\mathrm{FB/DPCM}} = g(R)\sigma_e^2 2^{-aR} \tag{2.75}$$

where σ_e^2 is the variance of e, the quantizer input, which is the prediction error. To assess whether there is any benefit of DPCM coding of subbands over DPCM coding of the full band signal, consider the ratio of their distortions as the gain

$$G_{\mathrm{SB/DPCM}} = \frac{D_{\mathrm{FB/DPCM}}}{D_{\mathrm{SB/DPCM}}}. \tag{2.76}$$

In the high-rate case this gain becomes

$$G_{SB/DPCM} = \frac{\sigma_e^2 g(R)}{N\sigma_{e,WGM}^2} \qquad (2.77)$$

$$= \frac{\sigma_e^2}{\displaystyle\prod_{m=1}^{M} \left(V_m \sigma_{e,m}^2\right)^{1/V_m}} \times \frac{g(R)}{\displaystyle\prod_{m=1}^{M} \left(g_m\right)^{1/V_m}}.$$

Again, for ideal coding of prediction errors or $g(\ )$ constant, the ideal gain is

$$G_{SB/DPCM}^o = \frac{\sigma_e^2}{\displaystyle\prod_{m=1}^{M} \left(V_m \sigma_{e,m}^2\right)^{1/V_m}}. \qquad (2.78)$$

Pearlman [41] has shown that with optimal prediction and N approaching infinity the previous ideal gain approaches 1. Woods and O'Neil [42] had previously reached the same conclusion for subbands of equal bandwidths. This limiting case is unrealistic, because it presupposes infinite order prediction in all subbands and in the full band source. A realistic assumption is that the prediction orders are finite. S. Rao and Pearlman [43] have proven that this gain is greater than 1 in the cases when the subband prediction orders are the same as the full band prediction order and when the sum of the full band prediction orders equals the full band prediction order. The gain is 1 if the source process is white. The reason for a subband coding gain is that every subband process has less memory than the full band process and can be predicted more accurately with a lower order predictor. This makes the error process more nearly white so that it is encoded more efficiently with independent scalar quantization.

2.9 Conclusions

This chapter has presented several coding methods used for image compression and has tried to cite or use information theory as a motivation or tool of analysis. Adherence to the theory will serve the student or user well in the field of image coding. The theory, however, has its limitations, because of simplistic statistical models and distortion criteria. The stationary Gaussian source and squared error distortion criterion are prominent in the theory, because they allow the derivation of closed-form formulas and provide the basis for understanding the origin of a technique. Techniques with more realistic statistical models and distortion measures are usually adaptations of the fundamental technique found by Gaussian/squared error analysis. Images are non-Gaussian, nonstationary, and are not viewed by a human observer with squared error as a distortion measure. Therefore, adaptation methods are very important for achieving the required results. This chapter was written to lay the basic foundation for later chapters, where many of these adaptive coding techniques will be treated in great detail.

Acknowledgments

The first draft of this work was written when the author was on sabbatical as Lady Davis Visiting Professor in the Department of Electrical Engineering at the Technion–Israel Institute of Technology. The author acknowledges and appreciates the partial support of the Lady Davis Foundation and the National Science Foundation, Grant Number NCR-9004758. The author wishes to thank Anet Berg of the Technion and Charmaine Darmetko for typing the manuscript, and Elizabeth Haight and Sudhakar Rao for their help in generating the figures. The author is indebted to to his former and current graduate students, especially Sanjiv Nanda, Mahesh Balakrishnan, R. Padmanabha Rao, Sudhakar Rao, and Diego de Garrido, whose perspectives and encouragement helped shape the work.

References

[1] C. E. Shannon, "A mathematical theory of communication." *Bell Syst. Tech. J.*, vol. 27, pp. 379–423 and 623–656, July and Oct. 1948. Also C. E. Shannon and W. Weaver, *The Mathematical Theory of Communication*. Urbana: University of Illinois Press, 1949.

[2] R. G. Gallager, *Information Theory and Reliable Communication*. New York: J. Wiley and Sons, Inc., 1968.

[3] D. A. Huffman, "A method for construction of minimum redundancy codes." *Proc. IRE*, vol. 40, pp. 1098–1101, Sept. 1952.

[4] R. E. Blahut, "Computation of channel capacity and rate–distortion function." *IEEE Trans. Inform. Theory*, vol. IT-18, pp. 460–473, July 1972.

[5] P. E. Fleischer, "Sufficient conditions for achieving minimum distortion in a quantizer." *IEEE Int. Conv. Rec., Part I*, pp. 104–111, 1964.

[6] A. V. Trushkin, "Sufficient conditions for uniqueness of a locally optimal quantizer for a class of convex error weighting functions." *IEEE Trans. Inform. Theory*, vol. IT-28, pp. 187–198, March 1982.

[7] S. P. Lloyd, "Least squares quantization in PCM." *IEEE Trans. Inform. Theory*, vol. IT-28, pp. 129–137, March 1982.

[8] J. Max, "Quantizing for minimum distortion." *IRE Trans. Inform. Theory*, vol. 6, pp. 7–12, March 1960.

[9] T. J. Goblick, Jr., and J. L. Holsinger, "Analog source digitization: A comparison of theory and practice." *IEEE Trans. Inform. Theory*, vol. IT-13, pp. 323–326, April 1967.

[10] H. Gish and J. N. Pierce, "Asymptotically efficient quantizing." *IEEE Trans. Inform. Theory*, vol. IT-14, pp. 676–683, Sept. 1968.

[11] A. N. Netravali and R. Saigal, "Optimum quantizer design using a fixed-point algorithm." *Bell Syst. Tech. J.*, vol. 55, pp. 1423–1435, Nov. 1976.

[12] N. Farvardin and J. W. Modestino, "Optimum quantizer performance for a class of non-Gaussian memoryless sources." *IEEE Trans. Inform. Theory*, vol. IT-30, pp. 485–497, May 1984.

[13] A. D. Wyner and J. Ziv, "Bounds on the rate–distortion function for stationary sources with memory." *IEEE Trans. Inform. Theory*, vol IT-17, pp. 508–513, Sept. 1971.

[14] R. M. Gray, J. C. Kieffer, and Y. Linde, "Locally optimal block quantizer design." *Inform. and Control*, vol. 45, pp. 178–198, May 1980.

[15] Y. Linde, A. Buzo, and R. M. Gray, "An algorithm for vector quantizer design." *IEEE Trans. Commun.*, vol. COM-28, pp. 84–95, Jan. 1980.

[16] R. M. Gray, "Vector quantization." *IEEE ASSP Magazine*, vol IT-1, pp. 4–29, April 1984.

[17] P. A. Chou, T. Lookabaugh, and R. M. Gray, "Entropy-constrained vector quantization." *IEEE Trans. Acoust., Speech, Signal Processing*, vol IT-ASSP-37, pp. 31–42, Jan. 1989.

[18] W. A. Fuller, *Introduction to Statistical Time Series*. New York: J. Wiley and Sons, 1976.

[19] W. A. Pearlman and P. Jakatdar, "A transform tree code for stationary Gaussian sources." *IEEE Trans. Inform. Theory*, vol IT-31, pp. 761–768, 1985.

[20] B. Mazor and W. A. Pearlman, "A trellis code construction and coding theorem for stationary Gaussian sources." *IEEE Trans. Inform. Theory*, vol IT-29, pp. 924–930, Nov. 1983.

[21] B. Mazor and W. A. Pearlman, "A tree coding theorem for stationary Gaussian sources and the squared-error distortion measure." *IEEE Trans. Inform. Theory*, vol IT-32, pp. 156–165, March 1986.

[22] J. W. Modestino, V. Bhaskaran, and J. B. Anderson, "Tree encoding of images in the presence of channel errors." *IEEE Trans. Inform. Theory*, vol IT-27, pp. 677–697, Nov. 1981.

[23] J. B. Anderson and J. B. Brodie, "Tree encoding of speech." *IEEE Trans. Inform. Theory*, vol IT-21, pp. 379–387, July 1975.

[24] H. G. Fehn and P. Noll, "Multipath search coding of stationary signals with applications to speech." *IEEE Trans. Commun.*, vol IT-COM-30, pp. 687–701, April 1982.

[25] P. Kroon and E. F. Deprettere, "A class of analysis-by-synthesis predictive coders for high quality speech coding at rates between 4.8 and 16 kbits/s." *IEEE J. Selected Area Commun.*, vol IT-6, pp. 353–363, Feb. 1988.

[26] N. S. Jayant and P. Noll, *Digital Coding of Waveforms*. Englewood Cliffs, NJ: Prentice-Hall, 1984.

[27] R. Bellman, *Introduction to Matrix Analysis*, 2d ed. New York: McGraw-Hill, 1970.

[28] J. Pearl, "On coding and filtering stationary signals by discrete Fourier transform." *IEEE Trans. Inform. Theory*, vol IT-19, pp. 229–232, May 1973.

[29] W. A. Pearlman, "A limit on optimal performance degradation in fixed-rate coding of the discrete Fourier transform." *IEEE Trans. Inform. Theory*, vol IT-22, pp. 485–488, July 1976.

[30] Y. Yemini and J. Pearl, "Asymptotic properties of discrete unitary transforms." *IEEE Trans. Pattern Anal. Machine Intell.*, vol IT-PAMI-1, pp. 366–371, Oct. 1979.

[31] N. Ahmed, T. R. Natarajan, and K. R. Rao, "On image processing and a discrete cosine transform." *IEEE Trans. Comput.*, vol IT-23, pp. 90–93, Jan. 1974.

[32] M. J. Narasimha and A. M. Peterson, "On the computation of the discrete cosine transform." *IEEE Trans. Commun.*, vol IT-COM-26, pp. 934–936, June 1978.

[33] W. A. Pearlman, "Adaptive cosine transform image coding with constant block distortion." *IEEE Trans. Commun.*, vol IT-COM-38, pp. 698–703, May 1990.

[34] I. J. Witten, R. M. Neal, and J. G. Cleary, "Arithmetic coding for data compression." *Commun. ACM*, vol IT-30, pp. 520–540, June 1987.

[35] S. Nanda and W. A. Pearlman, "Tree coding of image subbands." *IEEE Trans. Image Processing*, vol IT-1, pp. 133–147, April 1992.

[36] P. H. Westerink, "An optimal bit allocation for subband coding." in *Proc. IEEE Int. Conf. Acoust., Speech, Signal Processing*, New York, April 1988, pp. 757–760.

[37] P. Chou, T. Lookabaugh, and R. M. Gray, "Optimal pruning with applications to tree-structured source coding and modeling." *IEEE Trans. Inform. Theory*, vol IT-35, pp. 299–315, March 1989.

[38] E. A. Riskin, "Optimal bit allocation via the generalized BFOS algorithm." *IEEE Trans. Inform. Theory*, vol IT-37, pp. 400–402, March 1991.

[39] L. Breiman, J. H. Friedman, R. A. Olshen, and C. J. Stone, *Classification and Regression Trees.* Monterey, CA: Wadsworth, 1984.

[40] B. Mahesh and W. A. Pearlman, "Multiple-rate structured vector quantization of image pyramids." *J. Visual Commun. Image Representation*, vol IT-2, pp. 103–113, Jan. 1991.

[41] W. A. Pearlman, "Performance bounds for subband coding." In *Subband Image Coding*, Ed. J. W. Woods. Norwell, MA: Kluwer Academic Publishers, 1991, Chapter 1.

[42] J. W. Woods and S. D. O'Neil, "Subband coding of images." *IEEE Trans. Acoust., Speech, Signal Processing*, vol IT-ASSP-34, pp. 1278–1286, Oct. 1986.

[43] S. Rao and W. A. Pearlman, "Analysis of linear prediction, coding, and spectral estimation from subbands." Submitted to *IEEE Trans. Inform. Theory*.

Chapter 3

Image Compression Based on Models of Human Vision*

Nikil Jayant, James Johnston, and Robert Safranek
Signal Processing Research Department
AT&T Bell Laboratories
Murray Hill, New Jersey

The problem of *image compression* is to achieve a low bit rate in the digital representation of an input image or video signal with minimum perceived loss of picture quality. Since the ultimate criterion of quality is that judged or measured by the human receiver, it is important that the compression (or *coding*) algorithm minimize perceptually meaningful measures of signal distortion, rather than more traditional and tractable criteria such as the mean squared difference between the waveforms at the input and output of the coding system.

This chapter develops the notion of *perceptual coding* based on the concept of *distortion masking* by the signal being compressed and describes how the field has progressed as a result of advances in classical coding theory, modeling of human vision, and digital signal processing. We propose that fundamental limits in the science can be expressed by the semi-quantitative concepts of *perceptual entropy* and the *perceptual distortion–rate function*, and we examine current compression technology with respect to that framework. We conclude with a summary of future challenges and research directions.

---

*An altered version of this material, entitled "Signal Compression Based on Models of Human Perception," was published in *Proceedings of the IEEE*, October 1994.

3.1 Introduction

The problem of image compression is to achieve a low bit rate in the digital representation of an input image or video signal with minimum perceived loss of picture quality. The function of *compression* is often referred to as *low bit rate coding*, or *coding*, for short. In coding signals such as image and video, the ultimate criterion of signal quality is that judged or measured by the human receiver. As we seek lower bit rates in the digital representations of these signals, it is imperative that we design the coding (or compression) algorithm to minimize perceptually meaningful measures of image distortion, rather than more traditional and more tractable criteria such as the mean squared difference between the intensity waveforms at the input and output of the coding system.

Central to the preceding idea is the notion of distortion masking or noise masking, whereby the distortion (or noise) that is inevitably introduced in the coding process, *if properly distributed or shaped*, is masked by the input signal itself. The masking can be partial or total, leading either to increased quality compared to a system without noise shaping or perfect image quality that is equivalent to that of the uncoded signal. In either case, the masking occurs because of the inability of the human visual mechanism to distinguish two signal components (one belonging to the image, one belonging to the noise) in the same spectral, temporal, or spatial locality. An important effect of this perceptual limitation is that the perceptibility of noise can be zero even if the objectively measured local signal-to-noise ratio is modest or low. If the noise distribution is such that the masking effect is successfully invoked at all points in frequency, time, and space, the result will be a global coding operation of high, or perfect signal quality.

Ideally, the noise level at all points in the signal space is exactly at the level of *just noticeable distortion* (JND). This corresponds to perfect signal quality at the lowest possible bit rate. The signal is now neither overcoded nor undercoded. This bit rate is a fundamental limit to which we can compress the signal with zero (perceived) distortion. We call this limit the *perceptual entropy*. A signal coding algorithm that is based on the criterion of minimizing perceptual error is called a *perceptual coding algorithm*.

The basic notion of maximizing perceived image quality rather than the more tractable mean squared error is not novel. However, significant progress is now being made in the sophistication, degree, and dynamics of perceptual coding. Models of human vision, while still imperfect, are leading to more accurate models for JND and noise masking. In parallel, the arithmetic capabilities of digital signal processors have increased to the point where the computational complexity of perceptual coding can be supported in practical hardware.

While perceptual coding is important for various signals such as speech, audio, and image, it is most significant for signal classes that lack a good source model. Speech signals do have a universal and reliable production model, based on our knowledge of the human vocal apparatus. The source model leads directly to efficient models for removing signal redundancy. These models are utilized in techniques such as linear predictive coding to reduce the bit rate while maintaining a specified level of signal quality. In general, audio, image, and video signals do not have a similar single or universal *source* model. As a result, more of the burden for bit rate reduction falls on the *sink* model—the notion of shaping the noise into components of distortion that are perceptually unnoticed, the paradigm of perceptual coding.

The promises of perceptual coding are significant. But several issues need to be addressed to achieve its maximum potential. Traditional perceptual models for image coding have been generally based on global knowledge. As a result, they are not locally optimal, but on the other hand fairly robust across variations such as those in input signal or viewing distance. The new generation of signal-dependent or dynamic models based on local properties are more powerful in the context of input nonstationarity, but they can also be more sensitive to the accuracy of analysis and the correctness of the mapping from the signal analysis to the JND model for quantization. Designing dynamic perceptual systems that combine peak performance with robustness is a formidable research problem.

The following sections of the chapter are organized as follows. Section 3.2 is an essay on signal compression. This includes a description of four dimensions of performance: signal quality, bit rate, algorithm complexity, and communication delay. It is followed by a summary of applications, standards, and technology goals. Section 3.2 also discusses the tools of the trade: reduction of redundancy in the input signal, removal of irrelevant information in the operation of quantization, and signal enhancement by postfiltering. Focus in this chapter will be on the second operation, the utilization of *perceptual irrelevancy* for realizing high quality at low bit rates.

Section 3.3 provides a very brief and pragmatic discussion of the properties of visual signals and the physics of the human visual mechanism, as utilized in our examples of perceptual coding.

Section 3.4 describes how certain basic building blocks in signal compression algorithms are qualitatively matched to the properties and methods of human perception. These building blocks, filter banks, for example, provide an advantageous framework for utilizing signal redundancy as well as for the subsequent utilization of the irrelevancy principle by the quantizing system.

Section 3.5 discusses amplitude quantization. This is the part of the compression algorithm where the flexibilities afforded by the perceptual mechanism are utilized in conserving the bit rate in the digital representation of the signal being coded. This section of the chapter describes the evolution of perceptual coding, beginning with primitive examples of non-minimum mean squared error quantizers and concluding with systems driven by the perceptual notions of just noticeable distortion and minimally noticeable distortion.

Sections 3.6 and 3.7 discuss perceptual coding algorithms that have proven useful in the compression of image and video signals and point out the impact of these algorithms on current technology and international standards for image compression.

Section 3.8 is a summary of research directions. It describes the need to incorporate notions of perceptual coding in existing standards as well as in evolving technology. It also describes research advances that will be needed to make perceptual coding less of an art and establish it as a robust and well-understood scientific discipline.

3.2 Overview of Signal Compression

The material in this section is borrowed from a recent article by one of the authors [1]. The first two sections of that article are reproduced here as background for the subsequent discussion of perceptual coding in the remainder of this chapter. Readers well-versed in the theory of signal compression and its application to image signals may choose to skip this section and proceed directly to Section 3.3.

3.2.1 Signal Compression

Signal coding is the process of representing an information signal in a way that realizes a desired communications objective such as analog-to-digital conversion, low bit rate transmission, or message encryption. In the literature, the terms *source coding, digital coding, data compression, bandwidth compression,* and *signal compression* are all used to connote the function of achieving a compact digital representation of a signal, including the important subclass of analog signals such as speech, audio, and image. The subject of this chapter is the art and science of image and video compression. When the terms *coding, encoding,* and *decoding* are used in this chapter, they will all refer to the specific common objective of compression. An important theme of our discussion is the human receiver at the end of the communication process (Fig. 3.1).

It is useful to discuss the role of signal compression (source coding) in digital communication. While the source coder attempts to minimize the necessary bit rate for faithfully representing the input signal, the *mo*dulator-*dem*odulator (modem) seeks to maximize the bit rate that can be supported in a given channel or

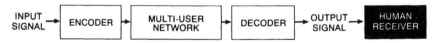

Figure 3.1   Digital coding for signal compression.

storage medium without causing an unacceptable level p_e of bit error probability. The bit rate in source coding is measured in bits per sample or bits per second (bps). In modulation, the rate is measured in bits per second per Hertz (bps/Hz). Channel coding adds redundancy to the encoder bit stream for the purpose of error protection. In so-called coded modulation systems, the operations of channel coding and modulation are combined for greater overall efficiency. The processes of source and channel coding, as well as the process of multiuser networking, can sometimes be integrated to increase the efficiency of digital communication.

The capability of signal compression has been central to the technologies of robust long-distance communication, high-quality signal storage, and message encryption. Compression continues to be a key technology in communications in spite of the promise of optical transmission media of relatively unlimited bandwidth. This is because of our continued and, in fact, increasing need to use bandlimited media such as radio and satellite links, packet networks, and bit-rate-limited storage media such as CD-ROMs and miniaturized memory modules.

3.2.2 Background

The information-theoretical foundations of signal compression date back to the seminal work of Shannon [2, 3]. His mathematical exposition defined the information content or *entropy* of a source and showed that the source could be coded with zero error if the encoder used a transmission rate equal to or greater than the entropy, and if the encoder used a long processing delay, tending in general to infinity. In the special case of the infinite-alphabet or analog source, the encoding error tends to approach zero only at an infinite bit rate. However, in practice, the error is close enough to zero at finite rates. In the case of a finite-alphabet or discrete-amplitude source, the entropy is finite, and the bit rate needed for zero encoding error is finite as well. An important example of a finite-entropy source is an analog signal stored in a computer as a sequence of discrete amplitudes. The raw (uncompressed) bit rate of such a signal is typically 8 or 24 bits per sample, respectively, for a gray-level image or a color image with three 8-bit components. The entropy, or the minimum bit rate for zero encoding error, will be typically smaller because of the statistical redundancy in the input sequence. For a gray-level image, this *statistical entropy* may be on the order of 5 bits per sample.

The inadequacies of the classical source coding theory are twofold. First, the theory is nonconstructive, offering bounds on distortion–rate performance rather than techniques for achieving these targets. However, the classical theory teaches us important qualitative recipes such as delayed encoding, as in vector quantization with a large vector dimension or block length. Second, the source model used in the classical theory does not capture what are now recognized as fundamental nuances in audio and visual signal processing. These include the fact that the input signal is non-Gaussian, nonstationary and in general has a complex and intractable power spectrum; and further, the observation that the human receiver does not employ a mean squared error criterion in judging the similarity of a coded image to the uncoded original. As a result of these complications, some of the observations of classical source coding do not carry over in an obvious way to image compression as discussed in this chapter. One such result is that the source entropy measured with a perceptual distortion criterion is different from, and generally much lower than, the classical statistical entropy measured with a mean squared error criterion for coding distortion (the 5 bits per sample entropy in the previous paragraph). Another classical result that needs to be reexamined is the thesis that, in principle, the processes of source and channel coding can be separated without loss of optimality. This thesis would hold very nicely for the digital communication of data sequences, but does not necessarily suggest an optimal solution to the complex problem of communicating visual signals over a noisy channel with high perceptual fidelity, robustness, and bit rate efficiency.

The technology and literature of image and video compression have therefore evolved independently, with basic and valuable inspiration from Shannon's theory [2, 3] and the rate–distortion theory that followed it [4, 5, 6], but also with a great deal of innovative engineering on the part of scientists closely familiar with the signals in question [7–20].

3.2.3 The Dimensions of Performance in Signal Compression

The generic problem in signal compression is to minimize the bit rate in the digital representation of the signal while maintaining required levels of signal quality, complexity of implementation, and communication delay. We will now provide brief descriptions of these parameters of performance.

Signal Quality   Perceived signal quality is sometimes measured on a 5-point scale that is well-known as the *mean opinion score* or *mos* scale. The five points of quality are associated with a set of standardized adjectival descriptions: *bad*, *poor*, *fair*, *good*, and *excellent*, and every example of an input being evaluated is assigned one of these levels in the course of a subjective test. An alternative methodology for subjective testing is based on an inverted scale that categorizes levels of *impairment* [21, 22] (*very annoying*, *annoying*, *slightly annoying*, *perceptible but not annoying*, and *imperceptible*). Our quantitative discussion of image and video quality will only be impressionistic, however, given the multidimensionality of the problem (dependence of subjective quality on input scene, picture resolution, image size, and viewing distance) and given the general lack of formal quality assessments in recent image coding literature.

Bit Rate   We measure the bit rate of the digital representation in *bits per sample*, *bits per pixel* (bpp), or *bits per second* (bps), depending on context, where *pixel* (sometimes shortened to *pel*) refers to a *picture el*ement, or an image sample. The rate in bits per second is merely the product of the sampling rate and the number of bits per sample. The sampling rate is typically slightly higher than about twice the respective signal bandwidth, as required by the Nyquist sampling theorem [11].

Table 3.1 defines commonly used grades of video in terms of sampling rate in pixels per second (pps) or Hertz (Hz). The sampling rates for the CIF, CCIR, and HDTV formats defined in the table are 3, 12, and 60 MHz. Respective Nyquist bandwidths are approximately 1.5, 6, and 30 MHz, although bandwidth limiting of image and video signals is in general less formal than the bandlimiting operations used for speech and audio signals. The HDTV format in the table is merely a specific example, one of several alternative formats currently in consideration.

The sampling rates in Table 3.1 refer to luminance information. Overheads for including color information are system dependent. In the CIF format, the color overhead is 50% in sampling rate, corresponding to a 50% subsampling relative to luminance in each of the horizontal and vertical directions, and a total of two chrominance components. Higher degrees of subsampling are sometimes used, leading to overall color overheads lower than 50%. In the line-interlaced CCIR format, the subsampling of color is performed only in the horizontal direction, and the final overhead in sampling rate due to color is 100%.

Complexity   The complexity of a coding algorithm is the computational effort required to implement the encoding and decoding processes in signal processing hardware, and it is typically measured in terms of arithmetic capability and memory requirement. Coding algorithms of significant complexity are currently being implemented in real-time, some of them on single-chip processors. Other, related measures of coding complexity are the physical size of the encoder, decoder, or codec (en*cod*er plus *dec*oder), their cost (in dollars), and the power consumption (in watts or milliwatts, mW), a particularly important criterion for portable systems.

Table 3.1   Digital television formats.[a]

Format	Spatio–temporal resolution	Sampling rate
CIF	$360 \times 288 \times 30 =$	3 Mpps
CCIR	$720 \times 576 \times 30 =$	12 Mpps
HDTV	$1280 \times 720 \times 60 =$	60 Mpps

[a]CIF, common intermediate format; CCIR, International Consultative Committee for Radio; HDTV, one example of a high-definition television format.

The technology of digital signal processing (DSP) has been evolving quite rapidly. One aspect of this progress can be measured in terms of the millions of instructions per second (mips) that can be accommodated on a single general-purpose processor, as a function of time. The evolution is exponential, with no evidence of saturation in the near-term [23]. In the five-year period from 1990 to 1995, the typical per-chip capability is expected to increase 10-fold, from about 25 mips to about 250 mips per chip. Supporting this evolution in arithmetic capability is a parallel advance in memory capability. The significance of these advances is that sophisticated compression algorithms that demand increasing levels of complexity will be supported only by DSP technology, in the form of single-chip, multi-channel implementations of the relatively less complex algorithms, and realistic parallel-processing machines for the more complex techniques. Power dissipation and cost are also expected to decrease steadily, making DSP technology increasingly useful for personal portable devices and for other high-volume consumer applications.

Communication Delay   Increasing complexity in a coding algorithm is usually associated with increased processing delays in the encoder and decoder. Although improved DSP capability can be used as an argument in favor of more sophisticated algorithms, the need to constrain communication delay should not be underemphasized. This need places important practical restrictions on the permissible sophistication of a signal compression algorithm. Depending on the communication environment, the permissible total delay for one-way communication (coding plus decoding delay) can be as low as about 1 msec (as in network telephony under conditions of no echo control) and as high as about 500 msec (as in very low bit rate videotelephony (or *videophony*) where the delay performance is severely compromised in the interest of obtaining a received picture good enough for communication). Communication delay is largely irrelevant for applications involving one-way communication (as in television broadcasting) or storage and message-forwarding. However, even in these applications, processing delay needs to be controlled to the extent it affects the latency of the coder in the context of input scene changes and station changes in a digital receiver.

3.2.4 Coding and Digital Communication

Figure 3.2 describes performance criteria in digital communication, by recapitulating the four dimensions of coder performance, explained specifically for source coding in Section 3.2.3. These dimensions of performance apply to channel coding and modulation as well, although the units of quality and bit rate are different. Respective units, defined either in Section 3.2.3 or in the second paragraph of Section 3.2.1, appear along the *quality* and *efficiency* axes in Fig. 3.2. Along each axis, the left and right entries refer respectively to source and channel coding. The units of *complexity* and *delay* are identical for source and channel coding, although those parameters are used for different reasons in the two cases. Processing delay is used in source coding to remove signal redundancy. In channel coding, delay may be used for adding error protection bits and for processes such as interleaving for the randomization of burst errors.

The axes in Fig. 3.2 define a four-dimensional space in which some regions are theoretically allowable and some regions are desirable for specific communication applications. Researchers in source and channel coding attempt to describe the allowable regions and the tradeoffs as quantitatively as possible. The focus of this chapter is on the domain of source coding. In particular, we shall comment extensively on the quantitative relationship between compressed signal quality and bit rate.

3.2.5 Applications of Image Compression

Figure 3.3 depicts various applications of image and video compression. The vertical axis in Fig. 3.3 has no special meaning. The numbers on the horizontal axis are bit rates *after* compression. The labels in Fig. 3.3 represent, in an approximate fashion, current capabilities. The bit rates spanned by these labels, and in some cases, the bit rates on which the labels are centered, represent the rates at which compressed signals render the corresponding application practical. As our capabilities in compression improve, the labels in Fig. 3.3 tend to shift to the left. In the following paragraphs, we provide very brief summaries of current capabilities in the compression of image signals.

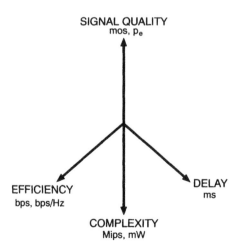

Figure 3.2   The dimensions of coder performance.

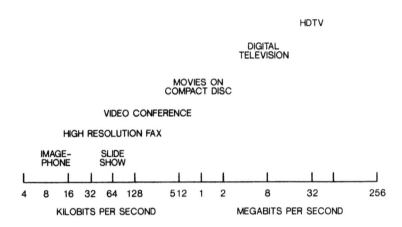

Figure 3.3   Applications of image compression.

Still Images   A 500 × 500 pixel color image, with the uncompressed format of 24 bits per pixel (bpp), will require about 100 sec of transmission time over a 64 kbps link. With 0.25 bpp coding, the transmission time is about 1 sec, a number that would be deemed excellent for an interactive "slide show" [24]. Current technology for coding a 500 × 500 image is capable of providing good picture quality at 0.25 bpp for a wide class of color images, assuming a viewing distance of about six times the picture height [25, 26]. For most images, increasing the bit rate to 1 bpp provides excellent and, in some cases, perceptually perfect image quality. The corresponding transmission time over a 64 kbps link is 4 s. Likewise, high-resolution facsimile typically takes several seconds of transmission time over a 64 kbps channel even after the use of powerful techniques for fax compression [27, 28].

Techniques for *progressive transmission* [28, 26, 29] involve a first stage of coding characterized by a low bit rate and rapid picture access, followed if needed, by additional stages of transmission that upgrade the picture quality. Progressive transmission is ideal for applications such as telebrowsing. It is also appropriate for applications where one expects display modalities (terminals and printers) of varying resolutions. The price paid for this flexibility is that progressive systems provide a slightly lower signal quality at a given bit rate compared to a single bit rate algorithm tuned to that specific rate.

The image-phone application in Fig. 3.3, which assumes the use of a telephone line and a 9.6 kbps modem, involves pictures of very low spatial and temporal resolution. A typical resolution would be 100 × 100 pixels per frame, and about 1 to 5 frames per second. With an even lower temporal resolution such as less than 1 frame per second, the system degenerates to a sequence-of-stills service, sometimes referred to as *freeze-frame video*.

Digital Video   Comfortable videoconferencing requires CIF resolution (360 ×
288 pixels per frame, Table 3.1) or at least quarter-CIF resolution (180 × 144 pixels
per frame). Input temporal resolutions are usually submultiples of 30 frames per
second, say 15 or even 10, for bit rates lower than about 1.5 Mbps. With CIF
resolution and a bit rate of 1.5 Mbps, the communications quality of the service
is generally agreed to be high. With quarter-CIF or somewhat lower resolution,
and with correspondingly lower values of temporal resolution, it is possible to
achieve lower bit rates such as 48 or 112 kbps. But the video quality is useful
only if one accepts low levels of sharpness in the output picture and very low
levels of motion activity in the input scene, as in the head-and-shoulders view
of a single person—an environment sometimes referred to as video*telephony*,
rather than video*conferencing*. The bit rates of 48 and 112 kbps are appropriate
for ISDN systems with total bit rates of 64 and 128 kbps, respectively, and a
bit rate of 16 kbps for voice transmission. The bit rate of 384 kbps is a very
interesting number in the current state of the technology. At this bit rate, it is
possible to provide a fair, if not high, level of picture quality in the coding of a
videoconference scene.

   CD-ROM media have a net throughput rate of about 1.5 Mbps for source data
and a total bit capacity on the order of a gigabit. If video can be compressed to
about 1 Mbps, a CD-ROM device could store and play out about an hour or more of
the video signal, together with compressed stereo sound. This capability is central
to various emerging applications of CD-ROM multimedia, including the specific
example of a movie on an audio compact disk [30]. The additional capability of
selecting a still-image snapshot of a desired part of the image sequence leads to the
concept of *addressable video*. This is an important feature in emerging systems
for video storage.

   Uncompressed high-definition television has a bit rate of over a gigabit per
second (the product of a sampling rate on the order of 60 MHz, as in Table 3.1,
and the representation of three color components with a total of 24 bits per sample).
Compression of the HDTV signal to a bit rate on the order of a few tens of Mbps
will create several important opportunities for HDTV broadcasting. In particular,
a bit rate in the range of 20 Mbps will bring the service into the realm of a 6 MHz
transmission channel [31], implying the capability of simulcasting the HDTV
version of a program in vacant slots of an NTSC channel set. Transmission rates
higher than 20 Mbps are appropriate for higher quality transmissions over satellite
and broadband ISDN channels, cable transmission, and applications of HDTV for
movie production.

3.2.6 Compression Standards

The need for interoperating different realizations of signal encoding devices (transmitters) and signal decoding devices (receivers) has led to the formulation of

several international and national standards for compression algorithms. A recent article [1] provides a brief summary of compression standards for image [27, 28, 26] and video [30, 32, 33]. Additional information provided in that article includes typical applications, typical levels of signal quality, and the approximate date of formulation of the standard.

The recent explosion in standards activity has had an important impact on research and development in the field. Standards have led to increased focus in applied research. They have sometimes stimulated highly productive new research as well. But they have also elevated the threshold of performance that a novel research algorithm needs to exceed before it is widely accepted, given that the supplanting of a recently endorsed standard is generally difficult and expensive.

Several applications of signal compression are decoder, intensive in the sense that users need access only to a decoder, the encoding being a one-time operation by the provider of the service. Examples are multimedia and HDTV decoders. In recognition of this, corresponding standards have specified the decoder algorithms and bit stream syntax rather than the encoder. In these cases, compatible enhancements to the standard are possible in the encoding module, as well as in optional modules of pre- and postprocessing: prefiltering at the encoder and postfiltering at the decoder.

3.2.7 Quality of the Compressed Signal

We noted earlier that several dimensions define the performance of a coding system. If we ignore the dimensions of algorithmic complexity and communication delay for the moment, coder improvements can be demonstrated in two ways: by measuring signal quality improvement at a specified bit rate, or by realizing a specified level of signal quality at a lower bit rate. Depending on the application, one of these approaches would be more relevant than the other. For example, in the problem of coding HDTV at 15 to 20 Mbps, the bit rates are defined by important generic applications, and the goal of coding research is to enhance picture quality at those rates.

So far, our discussions of bit rate have been in terms of kilobits per second and megabits per second. All of these numbers can be converted to equivalent numbers in bits per sample based on sampling rates such as the illustrative numbers in Table 3.1. For example, the 15 Mbps rate for 60 MHz-sampled HDTV corresponds to 0.25 bit per sample. In our ensuing definition of technology targets in signal compression, we shall use the normalized unit of bits per sample in the interest of a unified perspective for audio and visual signals.

3.2.8 Technology Targets

Video signals are easier to compress than still images on a bit-per-sample basis. This is attributable to the well-known redundancy in video information, particularly in the temporal domain of the signal.

In the category of still images, facsimile documents constitute a special subclass, if we agree to regard text and line graphics, rather than gray-level photographs, as typical fax documents. A halftoned (black/white) document is generally highly compressible. The bit rate for the lossless coding of a fax document can be as low as 0.1 bpp or lower.

As we seek to advance the state of the art as depicted earlier in Fig. 3.3, it is useful to talk about bit rate targets at which one expects image signals to be digitized with a quality rating such as 4.0 or higher on a 5-point scale. Without loss of generality or realism, our targets can be described by the goal of 4.5 quality at 0.25 bpp. We are closer to this goal in typical video examples than in typical still-image examples. It is also possible that the 4.5-quality goal at rates down to 0.25 bpp is impossible to achieve in some cases, regardless of coder complexity or processing delay, because of fundamental limits imposed by information theory and the acuity of the human perceptual system. But it is fair to ask the question: as we seek to approach these (sometimes unattainable) levels of high quality at low bit rates, what techniques are most likely to succeed?

3.2.9 Tools of the Trade: Perceptual Coding

Three fundamental operations are common to low bit rate signal coding: reduction of signal redundancy in the input signal, removal of irrelevant information in the operation of quantization, and signal enhancement by postfiltering. Of these, postprocessing is generally considered to be a process outside of the coding operations em per se, although the benefits of performing the process can be very significant. Prefiltering can likewise increase the performance of a compression algorithm. This is accomplished in video coding, for example, by the reduction of camera noise in the input image or by the insertion of an explicit bandlimiting filter. The remainder of this chapter will focus on the two operations that are intrinsic to signal coding: *removal of redundancy* and *reduction of irrelevancy*.

Almost all sampled signals in coding are redundant because Nyquist sampling typically tends to preserve some degree of intersample correlation. This is reflected in the form of a nonflat power spectrum. Greater degrees of nonflatness, as resulting from a low-pass function for signal energy versus frequency or from periodicities, lead to greater gains from redundancy removal. These gains are also referred to as *prediction gains* or *transform coding gains,* depending on whether the redundancy is processed in the time domain or in the frequency domain (or transform domain).

In a signal compression algorithm, the inputs to the quantizing system are typically sequences of prediction error or transform coefficients. The idea is to quantize the time components of the prediction error, or the transform coefficients, just finely enough to render the resulting distortion imperceptible, although not mathematically zero. If the available bit rate is not sufficient to realize this kind of perceptual transparency, the intent is to minimize the perceptibility of the distortion by shaping it advantageously in time or frequency, so that as many of its components as possible are masked by the input signal itself. We use the term *perceptual coding* to signify the matching of the quantizer to the human auditory or visual system, with the goal of either minimizing perceived distortion or driving it to zero where possible. These goals do *not* correspond to the maximization of signal-to-noise ratio (the minimization of mean squared error).

The parts of a coder that process redundancy and irrelevancy are sometimes separate, as in the preceding explanation. On the other hand, there are examples where the two functions cannot be easily separated. One example is a vector quantizer that combines intersample processing and quantization in a single stage of processing.

Almost all coding systems depend on the complementary interworking of the two basic operations just defined. A notable exception is a pulse code modulation (PCM) system, based on memoryless coding and quantizing algorithms, where no attempt is made to remove signal redundancy. This simple procedure is adequate for high-quality coding at bit rates in the range of 8 to 16 bits per sample, depending on the input signal. On the other hand, low bit rate coders depend heavily on more sophisticated signal analysis and redundancy removal prior to perceptually tuned quantization.

The conceptually orthogonal principles of redundancy reduction and irrelevancy removal were realized fairly early in the field of signal coding (Fig. 3.4). By the same token, the concept of perceptual coding is not novel either. What is significant is that both dimensions in Fig. 3.4 have advanced considerably over the years, and the interactions between the two dimensions are also getting better understood in coder design.

The next section describes the properties of visual signals and the psychophysics of perception, and this motivates the generic designs of algorithms that perform the dual functions of Fig. 3.4.

3.3 Visual Signal, Human Perception, and Time–Frequency Analysis

The input to the digital coder of Fig. 3.1 is the *source*, the visual signal. After compression (coding followed by decoding), the distorted low bit rate version of

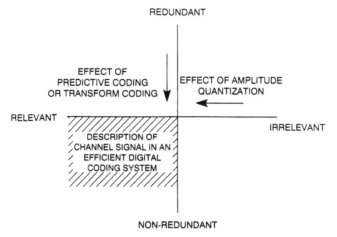

Figure 3.4   The Schouten diagram: redundancy and irrelevancy.

the signal is evaluated by the *sink* in Fig. 3.1, the human perceptual mechanism. Common to this mechanism and the coding algorithm are various signal processing and computing functions, prominent among which is time–frequency analysis. The notion of coding algorithms simulating the human visual system in their internal optimization loops is a useful one; and in fact, one could measure the sophistication of an algorithm in terms of how well it has evolved in response to the human visual model, to the extent human vision is understood.

3.3.1 Properties of Visual Signals

Visual signals are generally recognized as intensity versus space displays of image information, and intensity versus space and time displays of video scenes. These *waveforms* reveal significant information about the properties of the waveform as well as coder functions that will be needed to utilize these properties.

Nonstationarity   An important characteristic of image waveforms is the property of *nonstationarity*. Image and video signals contain a wealth of segments of flat or slowly changing intensity, as well as edges and textured regions. Coding algorithms need to render edge information faithfully, although these subsignals may represent minority subclasses from the point of view of energy or the fraction of samples or pixels involved. Adaptive algorithms for quantization, prediction, and bit allocation have given a good deal of the requisite capabilities to coding algorithms. Perceptual coding provides additional capabilities in terms of handling nonstationary inputs.

Redundancy   There are several sources of redundancy in visual signals. Examples are pixel to pixel, line to line, and frame to frame correlations. One example of a redundancy-removing algorithm is the function of motion-compensation in video (Fig. 3.5). In this procedure, interframe prediction for any given block in an image is obtained from a previous frame block providing the greatest prediction gain, and the two blocks are usually spatially displaced because of local or global motion.

Power Spectral Density   In a global or long-time-average sense, visual signals tend to have *low-pass* frequency spectra. Short-term frequency analysis, however, reveal parts of visual signals that are either *all-pass* or *high-pass*; examples are image signals with a predominance of high frequencies, a prototypical example being a checkerboard image pattern. Incidentally, a *white* or flat power spectrum corresponds to an unpredictable time signal, one for which the functional diagram of Fig. 3.4 collapses to a single dimension, the horizontal axis of quantization. In general, the nonflat spectrum of an image signal is a source of intraframe redundancy, and a generic algorithm for exploiting this redundancy is the variable bit allocation that follows linear transform analysis.

Color Signals   The use of so-called YUV and YIQ spaces (rather than the RGB space) in color image processing provides statistical as well as perceptual efficiency. In typical inputs, the UV or IQ (chroma) components have lower energies and lower bandwidths compared to the luminance component. Further, the human eye also has fewer receptors for the chroma signals. This permits the use of subsampling for the chrominance signals as well as the use of coarser quantization compared to luminance. Significant efficiencies in overall bit rate are attained as a result. In a typical color system for low bit rate video, the overall bit rate overhead for chroma can be as low as 10–25% of the total bit rate. Perceptual criteria relate to the lower sensitivities to coarse quantization, and the related observation that

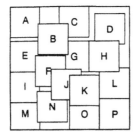

Figure 3.5   Motion compensation on a block-by-block basis.

additional liberties in chroma quantization are possible in the context of a strong change in the luminance component.

3.3.2 Models of the Human Visual System

A common model of vision incorporates a low-pass filter, a logarithmic nonlinearity, and a multichannel signal-sharpening high-pass filter. A biologically correct and complete model of the human perceptual system would incorporate descriptions of several physical phenomena including peripheral as well as higher level effects, feedback from higher to lower levels in perception, interactions between audio and visual channels, as well as elaborate descriptions of time-frequency processing and nonlinear behavior. Some of these effects are reflected in existing coder algorithms, either by design or by accident. For example, certain forms of adaptive quantization and prediction provide efficient performance in spite of inadequate response time because of temporal *noise masking*, a term that we will define presently. However, even where the matching to the human perceptual mechanism is by design, the matching is intuitive and approximate rather than formal or exact. In other words, at the time of this writing, an *optimal* perceptual coder does not exist. As a matter of fact, a biologically complete model of human perception also does not presently exist. In addition, significant intersubject variations exist in most examples of human perception.

An example of a perceptual phenomenon that is not directly reflected in compression technology is the possible feedback from higher to lower levels in perception. Examples of effects that are informally reflected in coder design are nonlinearities such as Weber's law (used as a justification for nonuniform quantization and homomorphic coders, to be elaborated later) and results about the importance of phase (used as a justification for phase-preserving filter banks, especially in spatial image processing and for avoiding spatial shifts in transitional regions of an image).

Focus in this section is on the time–frequency analyses in human vision and the desire to incorporate corresponding capabilities and flexibilities in coder modules that try to utilize the phenomenon of distortion- or noise-masking. *Masking* is a complex result of the transducing and neural components of perception. It is highly adaptive and refers to the perceptibility of one signal in the presence of another in its time or frequency vicinity. In coding, one of the signals is the input, and the second is the distortion in the low bit rate coding of it or in the communication of it over a noisy channel or a lossy network.

The basic timefrequency analyzers in the human perceptual chain are described by *bandpass filters* in vision (Fig. 3.6). The visual response in Fig. 3.6 is a function of spatial frequency (in cycles per degree, cpd) as well as temporal frequency (Hz). Bandpass filters in audiovisual perception are typically reflected in coder design and telecommunication practice in the form of "rules of thumb." For example, if the viewing distance in high-definition television is 8 feet and the image resolution is 1200×800 pixels, it is necessary that the physical image (screen) size is 5 feet in diagonal measurement, so that important high frequencies in the image correspond to values lower than say 25 cpd in the spatial response curve of Fig. 3.6. In this chapter, we shall attempt to incorporate the bandpass filter information explicitly in the calculation of the distortion-masking models in image coding.

A particularly interesting aspect of the signal processing model of the human system is nonuniform frequency processing. To incorporate this in coder design, it is necessary to use masking models with a nonuniform frequency support (Section 3.4). It is also necessary to recognize that high-frequency signals in visual information tend to have a short time or space support, while low-frequency signals tend to last longer. An efficient perceptual coder therefore needs not only to exploit properties of distortion-masking in time and frequency, but it also needs to have a time–frequency analysis front end that is sufficiently flexible to incorporate the complex phenomena of distortion masking by nonstationary input signals. All of this is in contrast to the classical redundancy-removing coder driven purely by considerations of *minimum mean squared error* (mmse), mmse bit allocation, or mmse noise shaping matched to the input spectrum (rather than to the composite of input spectrum and perceptual spectrum).

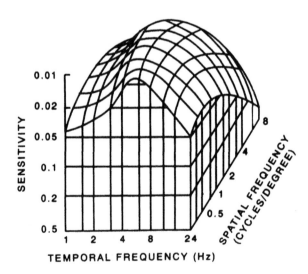

Figure 3.6   Spatio–temporal sensitivity of the human visual system (after [15]).

3.4 Filterbanks and Transforms in Image Processing

Predictive methods as well as subband and transform techniques have been used as bases for frequency-dependent signal coding. A recently described predictive coding system uses the spatio–temporal model of Fig. 3.6 to optimize the compression algorithm, based on both spatial and temporal criteria [34]. Subband filter banks and transforms provide direct control of frequency or frequency-related components in the input signal.

Fig. 3.7 depicts subband decompositions in one, two, and three dimensions. It is assumed that the filter banks used for these decompositions satisfy requirements on reconstruction error, processing delay, and interband interference. In typical examples of 2-D and 3-D filter banks, the structures are realized by separable, identical operations in the multiple dimensions. Quadrature mirror filter banks (QMFs) and related structures are very efficient designs in that the filters in the bank are of reasonably low order to permit practical implementation (typically, finite impulse response (FIR) structures with 16, 32, or 48 taps) and containment of artifacts in spatial and temporal dimensions. The interband aliasing due to over-lapping filter responses is controlled in a way that permits subsequent cancellation of it in the synthesis filter. In the absence of quantization, the analysis and synthesis filter banks constitute a unity or near-unity operation [35]. Efficient signal compression results when subband signals are quantized with subband-specific bit allocation and quantization, based on input power spectrum and the model of perception.

Figure 3.8 exemplifies transform coding [11]. The image blocks in Fig. 3.8 are the 64 basis images of an 8 × 8 2-D transform. The 2-D discrete cosine transform (DCT) decomposes the image into components that are analogous to the vertical and horizontal frequency components in the 2-D filter-bank of Fig. 3.7b. In the absence of quantization, the 2-D DCT and the identical operation of the inverse DCT at the receiver (2-D IDCT) constitute a unity operation. Efficient signal compression results when the transform coefficients are quantized with coefficient-dependent bit allocation and quantization.

In subband and transform coders, the number of frequency components (sub-bands or coefficients) is a compromise between prediction gain (due to variable bit rate coding of the nonflat spectrum of the redundant signal) and practical considerations such as complexity and processing delay.

The coefficients describing the QMF analysis filter are reused with a trivial modification in the QMF synthesis filter. The DCT and IDCT equations are like-wise very similar. The operations describing the QMF filter, the DCT, and the *modified* DCT (MDCT) are described by the following equations.

(a)

(b)

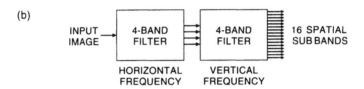

(c)

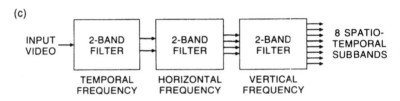

Figure 3.7   Subband decompositions in (a) one, (b) two, and (c) three dimensions of frequency.

QMF:

$$h_\ell(n) = h_u(n) = 0, \qquad \text{for } 0 > n \geq N.$$

$$h_\ell(n) = h_\ell(N - 1 - n), \qquad n = 0, 1 \ldots N/2 - 1.$$
$$h_u(n) = -h_u(N - 1 - n), \qquad n = 0, 1 \ldots N/2 - 1.$$

$$h_u(n) = (-1)^n h_\ell(n).$$
$$\left|H_\ell(e^{j\omega})\right|^2 + \left|H_u(e^{j\omega})\right|^2 = 1.$$

DCT:

$$F(u) = \sqrt{\frac{2}{N}}\alpha(u) \sum_{n=0}^{N-1} x(n) \cos \frac{(2n + 1)u\pi}{2N}, \qquad u = 0, 1 \ldots N - 1.$$

$$\alpha(0) = 1/\sqrt{2}, \qquad \alpha(u) = 1, \qquad u \neq 0.$$

IDCT:

$$x(n) = \sqrt{\frac{2}{N}} \sum_{k=0}^{N-1} \alpha(u)F(u) \cos \frac{(2n + 1)u\pi}{2N}, \qquad n = 0, 1 \ldots N - 1.$$

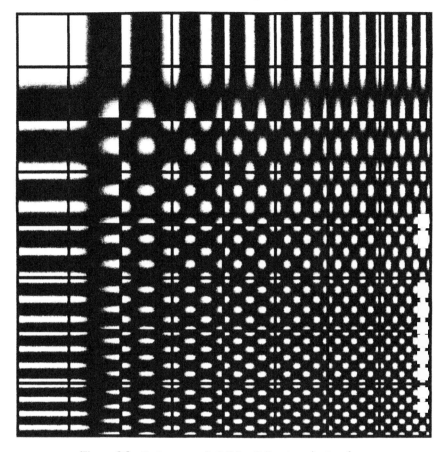

Figure 3.8 Basis vectors of a 2-D 8 × 8 discrete cosine transform.

2-D DCT:

$$F(u, v) = \frac{2}{N}\alpha(u)\alpha(v)\left[\sum_{m=0}^{N-1}\sum_{n=0}^{N-1} x(m, n)\cos\frac{(2m + 1)u\pi}{2N}\cos\frac{(2n + 1)v\pi}{2N}\right],$$

$$\alpha(u) = \alpha(v) = 1/\sqrt{2}, \qquad \alpha(u) = \alpha(v) = 1, \qquad \text{if } u \neq 0, \qquad v \neq 0.$$

2-D IDCT:

$$X(m, n) = \frac{2}{N}\left[\sum_{u=0}^{N-1}\sum_{v=0}^{N-1}\alpha(u)\alpha(v)F(u, v)\cos\frac{(2m + 1)u\pi}{2N}\cos\frac{(2n + 1)u\pi}{2N}\right].$$

MDCT:

$$F(u) = \sum_{n=0}^{2N-1} h(n)x(n) \cos \left[\frac{\pi}{2N}(2u+1)(2n+1+N) \right],$$
$$u = 0, 1, \ldots, 2N - 1.$$

$$h^2(N - 1 - i) + h^2(i) = 2,$$
$$h^2(N + 1) + h^2(2N - 1 - i)^2 = 2, \quad 0 \leq 1 < N/2.$$

The QMF system (Fig. 3.9a) is based on the division of a frequency band into contiguous but overlapping subbands. The broken line shows the characteristic of a high-pass filter that is the mirror image of the solid line low-pass filter characteristic. The extent of overlap is a decreasing function of the number of filter taps. But the allowing of a nonzero overlap simplifies filter design. Frequency aliasing is caused in QMF analysis by sampling each of the two bands (lower and upper bands) in the QMF split at twice the *nominal* bandwidth (rather than twice the actual, say −90 dB, bandwidth). However, with a special design of QMF filters, the process of QMF synthesis provides cancellation of this aliasing if the quantization noise inserted in the system is zero [36, 11, 37, 35].

The MDCT system [38] is a dual of the QMF approach (with $N = 2$ in the QMF equations) in that it permits an overlap between successive transform blocks in the time domain (as determined by the window function $h(n)$ of Fig. 3.9b), but decimates the resulting sequence to maintain the original sampling rate. In place of the frequency aliasing in the QMF system, the MDCT exhibits time-domain

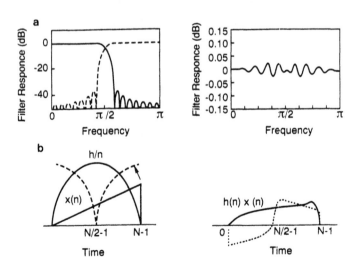

Figure 3.9   (a) Quadrature mirror filter and (b) overlapped transform as in MDCT.

aliasing. But this is cancelled by the inverse MDCT process in the receiver due to the design of the DCT basis vectors and the analysis window. The overlap in the MDCT is 50% and the decimation rate is 2:1 [39]. The lapped orthogonal transform is an MDCT system. While a 50% frequency domain overlap is possible in QMF design in principle, it is untypical and undesirable.

Although QMF and MDCT systems are essentially dual, they exhibit different properties in the context of a specific overall coder. For example, operations such as signal anticipation and temporal bit allocation can be used to some extent to control quantizing distortion and the consequent phenomenon of uncancelled time-domain aliasing in the MDCT system. This in turn permits the use of the 50% time overlap, which provides a smooth handling of the time (or space) process as well as a simple decimator design. In QMF design on the other hand, one still prefers, in the current state of the art, to employ a small spectral overlap, within the constraints on filter complexity. This implies a relatively discontinuous handling of the frequency process. Finally, in both MDCT and QMF, the available time–frequency behavior is a compromise between a match to sustained, stationary inputs and the ability to handle input discontinuities in time or frequency.

Effect of Quantization Error   The assumption of zero aliasing is a conspicuous simplification in almost all of subband coding literature. Techniques such as quadrature mirror filtering and the modified discrete cosine transform can provide perfect cancellation of aliasing in the absence of quantization errors. This ideal solution never occurs in a practical coding situation, and the assumption of the ideal case becomes increasingly inappropriate at lower bit rates because of a corresponding increase of quantization error. Recent work has given us a fairly good understanding of filter banks that provide zero or near-zero reconstruction error in the absence of quantization. However, the design of a filter bank that minimizes the combined effect of quantizing and aliasing errors is an unsolved problem. Here again, a rigorous optimization is extremely intractable in our current state of knowledge, but we sorely need at least partial solutions.

A somewhat orthogonal, though not unrelated, research area is that of efficient time–frequency analysis. Considerations of signal nonstationarity and perceptual distortion criteria have resulted in increasingly sophisticated demands on signal analysis. In particular, techniques that provide flexible combinations of time support and bandwidth represent a powerful generic tool for efficient coding. The discrete Fourier transform and a uniform-bandwidth quadrature mirror filter bank are well-understood and widely used analysis tools. But in their simplest forms, they lack the flexibility for time–frequency analysis mentioned previously. QMF trees with unequal bandwidth branches, as well as subband-DFT hybrids, are relatively newer structures with more flexible features. So are wavelet filters [40, 41].

Wavelets   Unlike the basis vectors of a DFT (sinusoids and cosinusoids of various frequencies and constant time support), the wavelet filter structure is characterized by a shorter time support at higher frequencies and a longer time support at lower frequencies, a direct result of a dilating operation that is a basic component of wavelet design [41]. The time–frequency characteristic of a wavelet filterbank is a natural match to some of the properties of audiovisual information: high-frequency events often occur for a short time and stand to benefit from a finer resolution in time analysis, while low-frequency events are often sustained in time and require less frequent sampling in time. The wavelet approach, especially if used in a time-varying framework, may therefore offer powerful forms of adaptive analysis in a more basic sense than a nonuniform frequency-band QMF system or a variable window-length MDCT system. Wavelet transforms provide the additional feature of perfect reconstruction, a property not generally offered in conventional methods of analysis. Conventional methods, however, do offer the property of *almost-perfect* reconstruction, which is adequate for many low bit rate applications.

Wavelet filtering is a promising analytical tool and applications of it are also beginning to emerge. What is still lacking, however, is a thorough understanding of what wavelets can do for coding that the more sophisticated examples of conventional analyses cannot; and as we seek to apply wavelet (or nonwavelet) tools to low bit rate coding, attention must necessarily shift to the yet-untouched problems of uncancelled aliasing and the computationally intensive but extremely important notion of a signal-adaptive filter bank.

Subband and transform coders provide a natural framework for variable rate coding, embedded coding, unequal bit allocation, and unequal channel error protection. In an embedded coder, the coded signal is partitioned into *essential* information and *enhancement* information. When an embedded coding system is used in an ATM packet network, the enhancement layers are much more likely to be discarded in the event of network congestion.

The left top corner in Fig. 3.10a represents one example of a partition for *essential* 2-D DCT coefficients. The shaded area in Fig. 3.10b is an example of *essential* spatio–temporal subbands.

Frequency-domain coders (subband, transform, and related multiresolution systems) also offer an excellent framework for *progressive transmission*. In an application like telebrowsing, it may be useful to obtain a rough version of an image first, using minimal communication resources; and to request a finer version only if an image needs to be scrutinized further. In this application, progressive transmission of frequency content is subjectively more useful than progressive top-to-bottom transmission of pixel rows (Fig. 3.11).

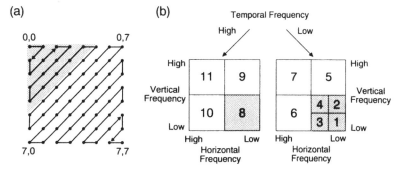

Figure 3.10   Embedded coding in (a) transform and (b) subband frameworks.

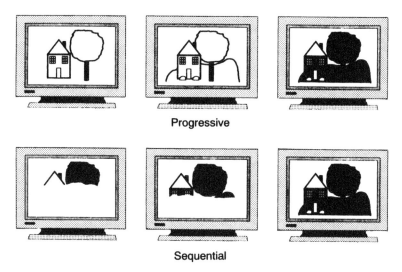

Figure 3.11   Progressive versus sequential transmission.

3.5 Quantization

The bit rate in the digital representation of an analog signal is determined as the bit rate of the (primary) quantizing system in a coder, together with any extra information that the coder may need to signify control parameters such as motion vectors in video coding. Such side information is also digitized by means of a (secondary) quantizing system.

The (primary and secondary) quantizing systems consist of either a scalar or a vector quantizer, followed (especially in the case of scalar quantization) by an entropy coder such as a Huffman or Ziv–Lempel coder that exploits residual redundancies *after* quantization. Since the entropy coder is a *noiseless* or *information-lossless* operation in a mathematical sense, the quantizer is the part of the overall coder where perceptual coding is explicitly realized.

In this section, we trace the history of quantization in a nonexhaustive manner and point out the increasing incorporation of perceptual cues as coding systems have become increasingly sophisticated and efficient.

A scalar quantizer is a memoryless device by definition. However, the statistical and perceptual effects of quantization are not adequately described by instantaneous properties. For example, the power spectral density of quantization noise, or equivalently, its autocorrelation function, is a function of the quantizer as well as the input signal. The perceptual impact of quantization noise depends on its spectral distribution and as we see later in discussions of masking, it depends on its relationship to the spectral distribution of the input signal as well. The ratio of the areas under the two spectral densities is the overall *signal to noise ratio* (SNR), and this is a partially useful descriptor of performance for the case of uncorrelated noise. But with low bit rate signals using noise-shaping, the SNR is not only inadequate but can also be misleading in terms of perceptual significance.

The PCM Quantizer   Analog signals are stored on the computer as R-bit PCM codes. For example, R may be 8 bits per sample for the luminance and chrominance components of color images. If these PCM codes are considered to be high-quality representations, it means that the additional resolution provided by an $(R + \Delta R)$-bit PCM system ($\Delta R > 0$) is perceptually irrelevant (while if $\Delta R < 0$, the signal representation may be considered inadequate). Thus, if the coder is constrained to be no more complex than a memoryless quantizer, the R-bit quantizer provides perceptually critical coding of the input.

Nonuniform and Midtread Quantizers   Although the theory of mmse quantization is extensive, quantizer designs are often guided by non-mmse criteria that make more intuitive sense from a perceptual viewpoint. One example is a nonuniform quantization algorithm for a differential PCM image coding system where the quantizer characteristics are designed to maximize the masking of distortion by luminance changes (the input to the differential coder). A better known and more widely practiced example is that of a midtread quantizer with an odd number $(N - 1)$ of output levels, including 0, sometimes with an extrawide dead zone near 0 [30, 26]. These features, while providing a nonminimum mean squared error, avoid the perceptually unpleasant oscillations between the smallest positive and the smallest negative output levels. Examples of such quantizers are the midtread quantizers with a dead zone for quantizing DCT coefficients in interframe video coding.

The Preference of Overload to Granularity in Quantization Another example of a non-mmse algorithm is one where quantizer design (quantizer step size, characteristic, or adaptation algorithm) favors a greater predominance of overload distortion compared to granular noise. This is well-understood in delta modulator designs which favor significant amounts of slope-overload distortion [11]. Such designs are suboptimal from an mse viewpoint but provide the best balance of distortion types from a perceptual viewpoint. To use the formal language of later parts of this section, the reason why the human perceptual system is relatively sensitive to granular noise is that the near-zero-slope input is a poor masker of the high-frequency noise caused by the succession of positive and negative steps in the delta modulator output.

The Use of Dither in Low Bit Rate Quantization In low bit rate PCM coders, the quantizing distortion tends to be highly structured, rather than random. This is reflected by significant input-distortion correlations and the resulting phenomenon of *contouring* distortion (as in the image coding example of Fig. 3.12). Dithering is a technique where random noise (typically with a small amplitude support) is added to the input *prior to quantization.* The random noise can be subsequently removed from the quantizer output in an ideally synchronized encoder–decoder system, or as is usual in image coding, there may be no such correction at all. In the latter case, the SNR of the overall system is unchanged. In the former, the SNR actually decreases because of dithering. But in both cases, dithering helps to break up signal-dependent patterns in the distortion and makes the coarse quantizer more palatable.

Figure 3.12 The destructuring of distortion by the use of dithering in the quantization of an image signal (after [17]).

In the image example of Fig. 3.12, the bit rate is 3 bits per sample. The left picture has no dithering. In the right image, the inclusion of dithering increases the mean square value of the distortion, but the break-up of the contours in the image provides a better chance for the input image to mask the distortion or at least decrease its visibility.

Dithering is an integral part of analog-to-digital converters and image halftoning. It is a precursor of more advanced systems for noise shaping.

Weighted Distortion   The SNR criterion of quantizer performance can be made perceptually more meaningful by weighting it in some ways (Fig. 3.13). For example, distortion components can be weighted by local input amplitude to reflect sensitivities to errors in low-amplitude input segments. Such weighting can be used to show the desirability of midtread quantizers and local average error metrics [8].

Still another example of error weighting places heavier emphasis on interblock differences and a relatively lower emphasis on intra-block distortions in a vector quantizer with memory. Such error weighting leads to code-vector selections that compromise the intrablock SNR, but create a perceptually more pleasing low bit rate image characterized by smooth interblock transitions. In the example of Fig. 3.14, the tendency is to select codevectors that maximize smooth vertical transitions from a vertical neighbor and smooth horizontal transitions from a horizontal neighbor, with both neighbors being previously quantized with a vector quantizer with the same codebook that is available to the current input.

Perhaps the most important weighting of distortion in the generic system of Fig. 3.13 is a suitable form of time–frequency weighting. This is not surprising,

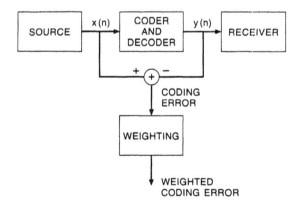

Figure 3.13   Weighted distortion in signal coding.

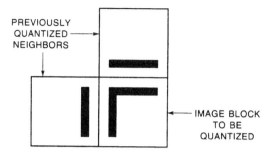

PREVIOUSLY
QUANTIZED
NEIGHBORS

IMAGE BLOCK
TO BE
QUANTIZED

Figure 3.14   Adaptive vector quantization with an interblock distortion metric.

given the discussion in Sections 3.3 and 3.4. Frequency-weighted distortion measures have been particularly valuable, for transparent as well as nontransparent coding.

The Homomorphic Model for Perceptual Coding   In a sense, this paradigm is an extension of the distortion-weighting principle. However, rather than weighting the distortion, the system weights the input and transforms it into a (*perceptually flat*) domain where unweighted error is useful. As a result, the quantizer itself can be very simple in structure and in terms of cues for optimization. A simple example of input weighting is the logarithmic nonlinearity following Weber's law. A generic homomorphic, psychovisual coder [42, 19] is shown in Fig. 3.15, and it includes a gamma correction factor to allow for camera nonlinearity. What is significant in the system is that the input and distortion (introduced in the quantizer Q) go through different weightings before reaching the eye; and emphasis is on optimizing the HVS model rather than the quantizer Q. Perhaps equally significant is the fact that the paradigm of Fig. 3.15 is a one-signal model, rather than a two-signal masking model; as such, it does not provide a specific mechanism for driving local distortion components to zero. Rather, it tends to minimize noise visibility by the use of a global perceptual model. This global model is also a static one, although dynamic extensions have been proposed recently. Finally, the homomorphic model is not necessarily tuned to the characteristics of a specific compression algorithm.

We next propose a model that is dynamic, local, coder tuned, and naturally suited to the explicit use of what we know about distortion masking in audition and vision.

A Paradigm for Perceptually Lossless Low Bit Rate Coding   Figure 3.16 defines a perceptual coding methodology that has recently produced very promising results in image coding. The methodology provides a framework for perceptually

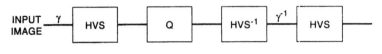

Q: QUANTIZER   HVS: HUMAN VISUAL SYSTEM

Figure 3.15   The homomorphic model for incorporating the human visual system (HVS) in coding (after [42, 19]).

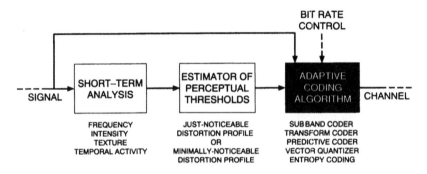

Figure 3.16   A paradigm for adaptive perceptual coding.

lossless coding at the lowest possible bit rate within the constraints of a well-known, mature coding algorithm. It also provides a framework for perceptually optimum performance when the available bit rate is too low to provide transparent compression.

The first box in Fig. 3.16 performs a short-time or spatio–temporally local analysis of the input visual signal and extracts properties such as frequency, intensity, texture, and temporal activity. These local properties are then used in the second stage of the process to derive a perceptual distortion threshold. This threshold can be a function of time, space, or frequency. It expresses a critical distortion profile in the sense that, if the distortion caused by the compression algorithm is at or below the threshold at all points in time, space, or frequency, the degradation in signal quality is imperceptible. The critical distortion profile will be called the *just noticeable distortion* (JND). A suprathreshold generalization of it, suitable for nontransparent (but still perceptually optimum) coding will be called the *minimally noticeable distortion* (MND).

The mapping from local properties to the JND profile is performed in real time in general and, by necessity, in a compression algorithm for two-way communication. The knowledge needed for the mapping is derived, on the other hand, from off-line experimentation with a large number of subjects performing a coder-specific perceptual task (as in the perceptual image coding algorithm of Section 3.6). Alternatively, the knowledge for the dynamic JND derivation may come from adapting results from psychophysical literature to the particular coding algorithm in empirical procedures. In either method, intersubject variations are addressed by defining a sensitivity to represent $P\%$ of the test population. Values of P may be 50, 75, or 95, for example, depending on how conservative the algorithm is desired to be. In the perceptual experiments described in this chapter, we use the 95% criterion. Other sources of variation come into play in describing perceptual effects, such as the effect of viewing distance in image coding. The method of Fig. 3.16 is therefore a generic methodology rather than a specific design.

MND profiles are also derived from off-line information. An alternative, implicit way of deriving them, especially at bit rates that are high enough to permit near-transparency, is to upshift the JND profile until the desired bit rate is achieved, at the cost of undercoding some parts of the input signal.

Given the JND or MND information, the rest of the coding algorithm is in principle straightforward, and it is the function of the third box of Fig. 3.16. A typical operation at this point is *adaptive bit allocation* that is steered by the JND or MND cues. The short-term analysis implied in these functions precedes the quantization function in the third box of Fig. 3.16, and this analysis can be either the same as that in the first box of Fig. 3.16 or different.

The overall result of the paradigm in Fig. 3.16 will be a *variable bit rate, constant-quality* algorithm. If needed, feedback from a bit buffer can be used to perturb the algorithm into a *constant bit rate, variable-quality* method in which, ideally, most of the signal will be slightly overcoded and some of the signal will be slightly undercoded. The desired constant bit rate is realized typically by an iterative process that can contribute significantly to the complexity of the perceptual coder.

In the critically coded, variable bit rate mode that is the only known *optimal* mode of the system of Fig. 3.16, the total number of bits needed to encode the signal is really a fundamental limit: the lowest rate at which transparent coding is possible. We call this the *perceptual entropy* in deference to the information-theoretic terminology of entropy. Two qualifications are needed in this context. First, these entropy estimates are only as good as the subjective data used to estimate JND profiles; we hope that, as our knowledge about noise masking improves, lower entropies can result. Second, the perceptual entropy is input specific, and it varies from segment to segment in a nonstationary signal. It is useful in these cases to talk about the long time-averaged perceptual entropy, as well as histograms of segmental perceptual entropy. Figure 3.17 is a sample of such a histogram. The bit rate information in Fig. 3.17 is based on real quantizers with real Huffman coders following them. In a constant bit rate coding system designed for transparent reproduction of the signal, the encoder tends to operate at the peak of the histogram rather than at the average of perceptual entropy. The results of Fig. 3.17 are based on a database of over 100 color images. For this database, the

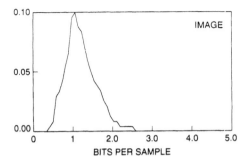

Figure 3.17   Histogram of perceptual entropy

average value of perceptual entropy is about 1 bpp. The corresponding number for video is expected to be significantly lower, as discussed earlier.

Several, perhaps most, coders used for compressing visual information can be regarded as special cases of the perceptual coder of Fig. 3.16. For example, elimination of the first two boxes leads to well-known classic algorithms. The most degenerate case of Fig. 3.16 is a memoryless PCM coder.

3.6 Perceptual Image Coding

Most of the previous work in applying human perception to image compression has utilized the human visual system's *frequency sensitivity* as described by the *modulation transfer function* (MTF). This function describes the human eye's sensitivity to sine wave gratings at various frequencies. The psychovisual experiments reported, for example by [43, 44] have proposed a commonly used model for the MTF. Since the MTF is defined for sine wave inputs, it is not directly usable for DCT-based coders. The work in [45] develops a transformation that accounts for the difference between the Fourier and DCT bases. Similar procedures could be used to transform the model to other frameworks for frequency analysis. From these models, given that the minimum viewing distance is fixed, it is possible to determine a static JND threshold for each frequency band. These thresholds can be used for quantization and bit allocation [46, 26]. The models can be extended to include *contrast sensitivity* as well. The homomorphic models mentioned earlier also attempt to utilize these properties. Contrast sensitivity can be included either implicitly, by developing functions of masking threshold vs local brightness, or explicitly, by using homomorphic systems or so-called *equi-luminant color spaces* [47].

The main problem with the aforementioned systems is that they do not go far enough in utilizing the masking properties of the human visual system. Frequency sensitivity is a global property dependent on only the image size and viewing conditions. Contrast sensitivity has been exploited mainly via pre- and postprocessing as in the homomorphic models. What is needed is a more dynamic model that allows for finer control of the quantization process. Frequency-dependent masking functions in the visual system [48] depend on the local frequency content or texture of the image and therefore have the desired property of *local control.*

The starting point for the locally adaptive perceptual coder is the two-dimensional subband decomposition of Fig. 13.7b. Image blocks in each of 16 subbands are further analyzed to determine local values of brightness and texture (as in the first block of Fig. 3.16).

Distortion sensitivity profiles are then derived as functions of frequency (subband number), brightness, and texture [25]. This is done by using the basic experimental paradigm of Fig. 3.18, in which a subject views the variable-power noise square at a specified distance and determines the noise power for which the noise square is *just noticeable* against a specified background of flat texture. (To be precise, the random noise is added to a frequency band in question, and the corrupted images are passed through the synthesis filter bank.) In the cited experiment, the noise had approximately the same mean value as the background and its density function was uniform. The experiment is repeated at various background brightness levels to determine the *brightness sensitivity* (Fig. 3.19). When the mean value of the noise square is the same as that of the background, the noise square tends to be most visible against a mid-gray background. The *frequency sensitivity* is inferred from the eye-sensitivity diagram, the MTF (Fig. 3.20, a result of the more general Fig. 13.6); alternatively, the frequency sensitivity is experimentally measured by repeating the basic subjective experiment for each of the 16 subbands. Table 3.2 shows the experimentally measured frequency sensitivity for a viewing

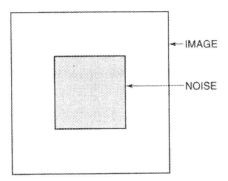

Figure 3.18 An experiment to evaluate the visibility of image distortion.

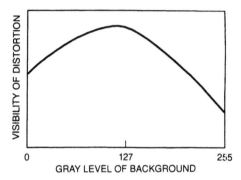

Figure 3.19   Distortion visibility as a function of background brightness.

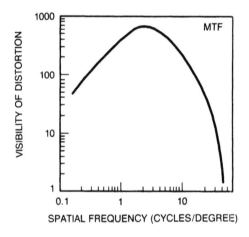

Figure 3.20   Distortion visibility as a function of spatial frequency.

distance of six times the picture height and for a mid-gray background of flat texture. Finally, the *texture sensitivity* is determined by an empirical calculation in which texture is defined as a local intrablock variance with a local dominant frequency and the MTF is (again) used to establish a relation between visibility of distortion and texture.

Fig. 3.21a depicts a constant level of noise (rms value of 8 for a peak input of 255) added to the center third of the lowest subband of a ramp input whose brightness changes continuously from white (at the left) to black (at the right). Fig. 3.21a demonstrates that distortion visibility is greatest against a mid-gray background. Fig. 3.21b depicts a constant texture background with white noise of rms value 8 added to the center third of the right half of the picture and absent in

Figure 3.21   Illustration of distortion masking by (a) background brightness and (b) background texture.

Table 3.2   Just noticeable noise levels in
2-D subband coding with 16 subbands.[a]

Subbands	Horizontal frequency			
	0.25	0.40	2.0	6.0
Vertical	0.50	1.0	4.0	8.0
frequency	2.0	3.0	4.0	6.0
	3.0	6.0	10.0	11.0

[a] Entries are rms values, for a constant mid-
grey background level of 127; rows and
columns represent subbands of horizontal and
vertical frequency.

the other half. Fig. 3.21b demonstrates the expected phenomenon that distortion visibility is low when the background has a strong texture.

Data from these generic experiments are used to derive JND estimates as spatial profiles for each of the 16 subbands of the image to be coded. These JND profiles lead in turn to spatially adaptive bit allocation in each subband. With a typical image, several spatial blocks are allocated zero bits, and in fact a large number of subbands see zero bit allocation to *all* of their spatial blocks.

These effects are illustrated in Fig. 3.22. Part (a) is a 512 × 512 pixel head and shoulders image. Part (b) is the image of the first subband (lowest horizontal and vertical frequencies). Part (c) is the JND image of (b): dark areas in (c) represent parts of (a) where the JND is high, implying the possibility of coarse quantization; white areas in (c) represent parts of (a) where the JND is low; and intermediate intensities in (c) signify intermediate values of JND. Corresponding results in terms of variable bit allocation (including zero bit allocation) are shown in part (d): white blocks in (d) signify that for corresponding regions in the full-band input (a), 4 out of 16 subbands are retained; light-gray blocks in (d) signify the retention of 3 subbands; dark-gray blocks in (d) signify the retention of 2 subbands; and black blocks in (d) signify that only 1 out of 16 subbands (almost always the lowest frequency subband) is retained for corresponding input blocks. As a result of the JND-paradigm, the input (a) can be compressed in a perceptually lossless fashion at a bit rate of about 1.1 bits per pixel, assuming a viewing distance of six times the picture height.

Figure 3.23A shows the result of perceptual subband coding of the 8-bit image (top left) at rates of 0.5 bpp (top right), 0.33 bpp (bottom left), and 0.25 bpp (bottom right).

Figure 3.23B shows the result of perceptual subband coding of a 24-bit color input (704 × 480 pixels) at rates of 1.0, 0.67, and 0.34 bit per pixel. Distortion is noticeable only at the very lowest bit rate. On a 64 kbps transmission line, the

Figure 3.22   Perceptual subband coding: (a) 512 × 512 input, (b) lowest frequency subband, (c) spatial JND profile for (b), and (d) spatial profile of the number of subbands retained (out of 16).

time needed to transmit the 24-bit original is about 130 seconds. With 0.67 bit per pixel coding, the transmission time is 3.6 seconds.

The ISO-JPEG Image Coding Standard   This system has recently been adopted as an international standard for the storage and transmission of still images [49, 26]. To achieve interoperability, the standard specifies the information contained in the compressed bit stream and a decoder architecture that can reconstruct an image from the data in the bit stream. However, the exact implementation of the encoder

Figure 3.23A   Perceptual subband coding of a gray-level image: 8-bit original (a) and compressed image at 0.5 bpp (b), 0.33 bpp (c), and 0.25 bpp (d).

is not standardized. The only requirement on the encoder is that it should generate a compliant bit stream. This provides an opportunity to incorporate the ideas presented in the previous section into a JPEG encoder.

The challenge in improving these standards-based codecs is to generate a compliant bit stream that produces an image that is perceptually equivalent to the

Figure 3.23B   Perceptual subband coding of a color image: (a) 24-bit original and compressed image at (b) 1.00, (c) 0.67, and (d) 0.34 bpp.

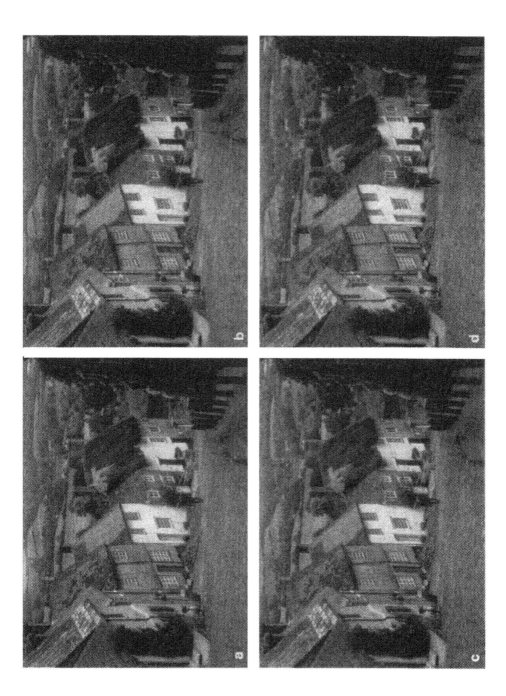

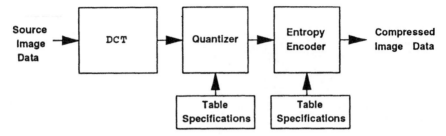

Figure 3.24   The baseline sequential JPEG encoder consists of three major components.

one generated by the baseline system, but with a higher compression ratio. This results in a lower encoded bit rate without loss in perceptual quality.

Figure 3.24 illustrates the basic operations of the baseline sequential JPEG encoder. They consist of a frequency analysis block, which utilizes an 8×8 DCT, a quantizer, which consists of 64 uniform quantizers, one for each DCT coefficient, and an entropy coder. The only user-settable parameters of this system are the step sizes of the uniform quantizers and the type of entropy coder.

Perceptual criteria can be incorporated into this system in two ways. In the first, a quantization matrix that has been designed using a model of the human visual system can be specified. Such an approach is presented in [50]. They determined the visibility of DCT basis functions for a variety of viewing distances and lighting conditions through subjective experimentation. This visibility threshold information was used to generate a model from which custom quantization tables can be computed.

Since this method produces a single quantization matrix that depends only on the global properties of the viewing environment and not on the content of the image, it will be suboptimal for those inputs that have high local masking thresholds.

Perceptual JPEG   To take advantage of these local elevations in masking threshold, a second technique is useful. To adapt the encoder to these local variations in masking threshold, the texture model described in the previous section is applied to each block of the input. This results in a masking threshold elevation due to the data in that block. From this information and the global quantization matrix, a custom set of masking thresholds can be determined for each block in the input image. If the magnitude of a DCT coefficient is less than its corresponding threshold value, it can be set to zero with no subjective loss in quality. This operation also produces a bit stream that is compliant with the the JPEG specification. The net result is that by setting to zero or "prequantizing" coefficients that are lower than the local perceptual masking thresholds, it is possible to generate a smaller compressed image with no loss in subjective quality (Fig. 3.25).

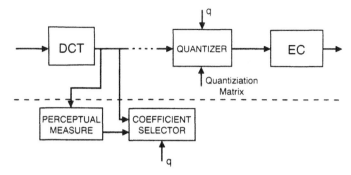

Figure 3.25   The perceptual JPEG encoder adds components to compute a perceptual model and perform prequantization to those in the baseline encoder.

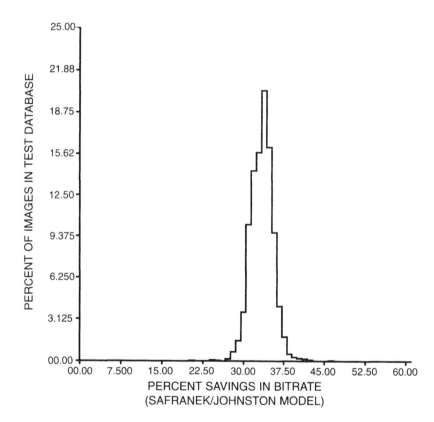

Figure 3.26   Shown here is a histogram of the bitrate savings obtained by P×PJEG compared to baseline JPEG.

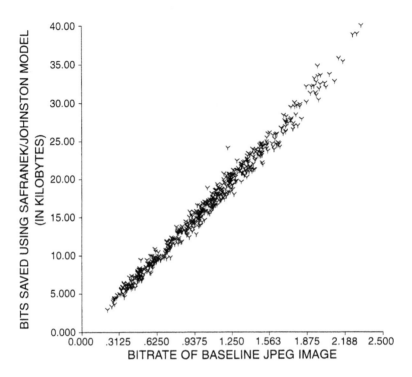

Figure 3.27   This scatter plot shows that P×PJEG achieves its greatest gains for images that do not compress well using baseline JPEG.

Figures 3.26 and 3.27 illustrate the gains that can be achieved using this procedure on database of about 3000 images. The reference system was base-line sequential JPEG using the default quantization matrices. The test system (P×PJEG—Perceptual JPEG), used the perceptually optimal quantization matri-ces, and the perceptual pre-quantization technique. The median gain in bit rate using perceptual pre-quantization is around 33%. It should be noted that larger bit rate savings are obtained for images that are less compressible with baseline JPEG. This occurs because these images tend to have many textured areas where the masking thresholds are elevated. In contrast, images that compress well with baseline JPEG tend to have less texture and therefore less opportunity to apply prequantization. This effect is illustrated in Fig. 3.28.

Perceptual Coding of Facsimile   The low bit rates provided by perceptual im-age coding suggest a possible new approach to the transmission of gray-level

Figure 3.28 Comparison of JPEG image coding algorithm (a) at 2.1 bpp, without preprocessor and (b) at 1.6 bpp, with perceptual preprocessor.

facsimile. Rather than transmitting a gray-level photograph by first halftoning it and transmitting it as black/white document of high spatial resolution, one could transmit the picture *as a gray-level image* (using sub-band coding) and do the halftoning *at the receiver*. In particular, the following two alternatives have been specifically compared: conventional fax transmission of a 1536 × 1536 halftoned image, using standard black/white coding techniques; and the new method of transmitting a 512 × 512 gray-level image and using halftoning at the receiver. The compression algorithms were arranged to be lossless in both cases, with a perceptual losslessness criterion in the subband method. The gray-level transmission method resulted in a 50% decrease of bit rate. More significantly, the availability of gray-level information (in the new technique) permits the use of new model-based techniques for efficient halftoning at the receiver. The models used include the HVS model and the model of the printer itself. Figure 3.29 illustrates the effect. With the same (300 dot per inch) printer, the left image is the result of conventional halftoning. The (sharper) right image is the result of model-based halftoning at the receiver, a precondition for which is the transmission of the picture as a gray-level image [51, 52].

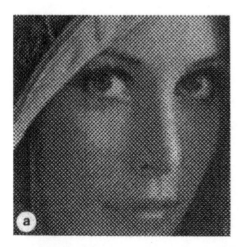 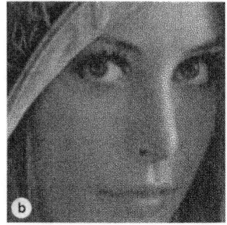

Figure 3.29   (a) Gray-level transmission of fax document using perceptual coding at 0.5 bpp and (b) model-based halftoning at the receiver (after [51] and [52]).

3.7 Perceptual Coding of Video

This is an application area where the use of perceptual cues are least understood. We have only begun to use the JND methodology of Fig. 3.16 for quantizer design. On a different level, even prior to quantization, we are still using very informal criteria for fundamental designs such as choice of temporal (and spatial) resolutions and the degree of chrominance subsampling.

In the following, we shall comment on two fundamentally different approaches to video coding and examine the inroads made by perceptual designs in each case.

Motion Compensation Followed by 2-D DCT   This is a hybrid approach to 3-D coding where a differential interframe process is followed by frequency-domain coding of the uncompensated interframe error (Fig. 3.30). This method is the basis for several international (CCITT and ISO-MPEG) standards for videoconferencing and addressable video [30, 32].

A low bit rate HDTV coder based on perceptual coding of the interframe error has been recently described. The methodology used here was similar to that of Fig. 3.16, but the extension was approximate in that the structure of the interframe error was very different from the uniform-noise model used in the JND experiment of Fig. 3.18, and the HDTV viewing distance was smaller than the six times picture height model of that experiment. These effects were offset by the fact that the JND could now be a function of the time dimension as well, and the

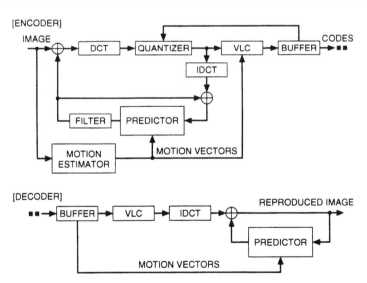

Figure 3.30   Block diagram of video coder using motion compensation followed by 2-D DCT and entropy coding.

benefits of temporal masking were now available. Figure 3.31 explains an example of temporal masking. In the context of newly uncovered background (due to the passing truck), there is a small latency period (say the frame called $t = 1$) during which the newly exposed areas do not have to be perfectly reconstructed.

HDTV images (1280×720 pixels, 60 frames per sec) were compressed to 17 Mbps using this procedure, to facilitate simulcasting over 6 MHz NTSC channels [31]. The rate of 17 Mbps is not adequate to provide perceptually transparent coding of all HDTV inputs. As a result, the system oscillates between the JND and MND modes of Fig. 3.16, with a constant bit rate constraint in both cases, and no claim is made of overall optimality.

Three-Dimensional Subband Coding   Figure 3.32 shows the result of decomposing a *beach and flowers* scene using the 3-D subband analysis described in Fig. 3.10b. Most of the picture energy is in subband 1, which contains the low-frequency information in the horizontal, vertical, and temporal frequency dimensions. Other subbands have much lower energy. Subband 8 contains high temporal and low spatial frequencies and therefore acts as a motion detector.

Energy-based adaptive bit allocation provides a classical approach to algorithm optimization. Although straightforward in principle, such optimization is very difficult because of the dynamics of the problem: all energies vary in three dimensions—horizontal, vertical, and temporal, and even if an optimal bit allocation algorithm is defined, there is the additional problem of transmitting the bit allocation information to the decoder.

Perceptual optimization of the coder is even more difficult. In recent experimentation with the system for videoconferencing at medium bit rates (such as 384 kbps), empirical perceptual designs have been proposed for the following functions: bit allocation based partly on energy (favoring subband 1) and partly

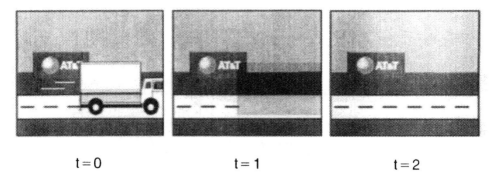

$t = 0$ $t = 1$ $t = 2$

Figure 3.31   Explanation of temporal noise masking.

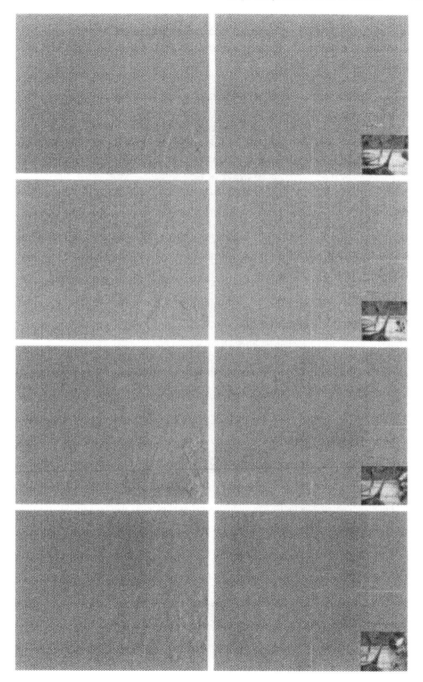

Figure 3.32 3-D subband decomposition of *Beach and Flowers* (after [53, 54]).

on motion fidelity (favoring subband 8); adaptive exchange of spatial and temporal resolution in subband 1; and perceptually efficient quantization of the higher frequency subbands [53, 54].

The higher frequency subbands in 3-D coding are dominated by edgelike information of low energy and high perceptual value for edge intelligibility and motion rendition. An efficient method for reproducing the edgelike information at very low average bit rates is the technique of *geometric vector quantization.* In the simplified example of Fig. 3.33, the codebook consists of codevectors that can reproduce horizontal, vertical, and diagonal edges, as well as a null vector for the representation of areas of constant or near-constant gray level. Very low bit rates result because of the high probability and spatial clustering of the null vector output.

At the bit rate of 384 kbps, and with inputs in the CIF format (360×240 pixels, 15 frames per sec), the 3-D coder operates well below the transparency level; the JND principle of Fig. 3.16 does not apply; and the MND cues used in the system are at best informal and empirical.

3.8 Research Directions

The technology of signal compression has seen significant advances in recent years, but we still see a gap between current capabilities and technology targets that we feel are attainable (for example, high-quality coding of unrestricted images

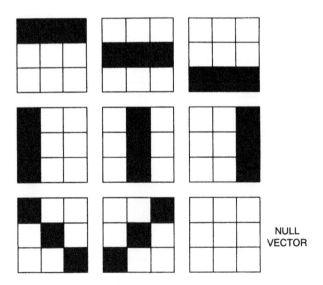

Figure 3.33   Geometric vector quantization.

at 0.25 bpp). Several research directions are needed to attain these goals. Perceptual coding is one of them. This concept has made a steady and ever-increasing impact on the status of signal compression. In the future, to enhance the role of perceptual coding and thus that of image compression in general, we need to address several research challenges, some of which are described in the following.

Algorithms for perceptual coding need to become more robust, scalable, and portable. One should be able to address different signal types (luminance and chrominance), different coder types (motion compensation and 3-D coding, sub-band and transform coding), and different signal environments (interlaced versus progressive scanning, very-clean versus camera-noise-limited inputs, near- versus far-distance viewing), without having to reinvent an empirical perceptual model for every situation. If the perceptual algorithms are made more robust in these ways, there is also a much better chance of incorporating them in rapidly evolving technology for new services and coding standards.

The role of perception in various modules of coding needs to be better understood. In other words, perceptual metrics need to be integrated better into algorithms for motion compensation, spatio–temporal 3-D coding, vector quantization, pre- and postprocessing, methods of time–frequency analysis including multiresolution, pyramid, and wavelet techniques, and finally, emerging models for signal production (such as the wire-frame model for images of the human face [1]).

In vision, the theories of noise masking need to be better unified, and in particular, temporal masking needs to be better understood and applied.

The continuum of distortion in the MND model of Fig. 3.16 needs significant additional work. At this time, it is much easier to establish the JND point (the rate $R_p(0)$ for zero perceptual distortion) than to define the way perceived distortion $D_p(R)$ varies as a function of R in the so-called supra-threshold region of nonzero D. This is shown in Fig. 3.34. The exact form of $D_p(R)$ is really unknown although we know that it is displaced well to the left of the mse-based distortion-rate curve; this curve has a zero distortion rate $R_{rmse}(0)$ that is significantly higher than the JND point $R_p(0)$. The homomorphic model may prove to be useful in methodologies for determining the shape of the perceptual distortion–rate curve. Even if the $D_p(R)$ function is available, the incorporation of this knowledge in an actual encoder with a nonstationary input is a formidable problem. Part of solving this problem is the design of an efficient buffer control mechanism that provides the bit rate control in Fig. 3.16.

With increased understanding of perceptual quality metrics (in the internal workings of a coding algorithm), one could also expect better methodologies for evaluating the subjective quality of the final signal (at the output of the decoder). In particular, we can expect more progress in our search for a subjectively meaningful objective measure of overall quality, and we can perhaps minimize the need for time-consuming and intricate subjective tests.

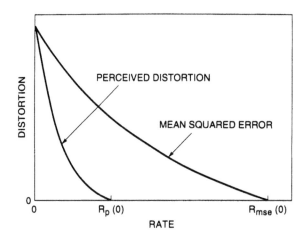

Figure 3.34   Distortion–rate function with mean squared and perceptual error metrics.

Finally, as perceptual coding evolves into more of a science (rather than the art it currently is), the lessons and the tools from this field will ideally extend to tasks well beyond signal compression itself. One such problem is that of measuring and maximizing the quality of service in a multiuser network in the presence of various classes of signal degradation: coding distortion, bit error effects, packet losses, and delay. Another challenge is that of measuring and maximizing the quality of composite signals in a multimedia system: the perceived overall quality of an audiovisual signal, as opposed to the individual quality levels of its audio and visual components; and the related bigger problem of evaluating and improving the perceived quality of telepresence in the sophisticated communication systems of the next generation.

Acknowledgments

The authors thank their colleague Thrasyvolous Pappas for his kind help in generating the pictures in Figs. 3.22, 3.23, and 3.29 of this chapter.

References

[1] N. S. Jayant, "Signal compression: Technology targets and research directions." *IEEE J. Selected Areas Commun.*, Special Issue on Speech and Image Coding, June 1992.

[2] C. E. Shannon, "A mathematical theory of communication." *Bell Syst. Tech. J.*, vol. 27, pp. 623–656, 1948.

[3] C. E. Shannon, "Coding theorems for a discrete source with a fidelity criterion." *IRE Nat. Conv. Rec., Part 4*, pp. 142–163, 1959.

[4] T. Berger, *Rate Distortion Theory*. Englewood Cliffs, NJ: Prentice–Hall, 1971.

[5] R. G. Gallager, *Information Theory and Reliable Communication*. New York: McGraw–Hill, 1965.

[6] D. Huffman, "A method for the construction of minimum redundancy codes." *Proc. IRE*, pp. 1098–1101, Sept. 1952.

[7] C. C. Cutler, "Differential quantization for communication signals." U.S. Patent No. 2,605,361, July 1952.

[8] C. C. Cutler, "Delayed encoding: stabilizer for adaptive coders." *IEEE Trans. Commun.*, pp. 898–904, Dec. 1971.

[9] R. C. Gonzalez and P. Wintz, *Digital Image Processing*. Reading, MA: Addison-Wesley, 1977.

[10] R. M. Gray, "Vector quantization." *IEEE ASSP Magazine*, pp. 4–29, 1984.

[11] N. S. Jayant and P. Noll, *Digital Coding of Waveforms: Principles and Applications to Speech and Video*. Englewood, NJ: Prentice Hall, 1984.

[12] M. Kunt, A. Ikonmopoulos, and M. Kocher, "Second-generation image coding technique." *Proc. IEEE*, pp. 549–574, April 1985.

[13] J. Max, "Quantizing for minimum distortion." *IRE Trans. Inform. Theory*, pp. 7–12, March 1960.

[14] H. G. Musmann, P. Pirsch, and H.-J. Grallert, "Advances in picture coding." *Proc. IEEE*, pp. 523–548, April 1985.

[15] A. N. Netravali and B. G. Haskell, *Digital Pictures: Representation and Compression*. New York: Plenum Press, 1988.

[16] W. Pratt, *Digital Image Processing*. New York: J. Wiley and Sons, 1978.

[17] L. G. Roberts, "Picture coding using pseudo-random noise." *IRE Trans. Inform. Theory*, pp. 145–154, Feb. 1962.

[18] W. F. Schreiber, "Psychophysics and the improvement of television picture quality." *SMPTE J.*, pp. 717–725, Aug. 1984.

[19] T. G. Stockham, "Image processing in the context of a visual model." *Proc. IEEE*, pp. 828–842, July 1972.

[20] P. A. Wintz, "Transform picture coding." *Proc. IEEE*, pp. 809–820, July 1972.

[21] CCIR, *Method for the Subjective Assessment of the Quality of Television Pictures*, Rec. 500-1, 1978.

[22] J. S. Goodman and D. E. Pearson, "Multidimensional scaling of multiply-impaired television pictures." *IEEE Trans. Syst., Man, Cybern.*, pp. 353–356, June 1979.

[23] S. S. Magan, "Trends in DSP system design." In *Short course on Digital Signal Processing*, IEEE Int. Electron. Device Meeting, Dec. 1989.

[24] S. R. Quackenbush, "Hardware implementation of a color image decoder for remote database access." In *Proc. ICASSP*, 1990.

[25] R. J. Safranek and J. D. Johnston, "A perceptually tuned sub-band image coder with image-dependent quantization and post-quantization data compression." In *Proc. ICASSP*, 1989.

[26] G. K. Wallace, "The JPEG still picture compression standard." *Commun. ACM*, pp. 31–43, April 1991.

[27] International Telephone and Telegraph Consultative Committee, *Facsimile Coding Schemes and Coding Control Functions for Group 4 Facsimile Apparatus*, Red Book, Fascicle VII.3 Rec. T.6, 1980.

[28] ISO, *Coded Representation of Picture and Audio Information—Pro-gressive Bi-Level Image Compression Standard*, ISO/IEC Draft, Dec. 1990.

[29] L. Wang and M. Goldberg, "Progressive image transmission using vector quantization on images in pyramid form." *IEEE Trans. Commun.*, pp. 1339–1349, 1989.

[30] D. LeGall, "MPEG: A video compression standard for multimedia applications." *Commun. ACM*, pp. 47–58, April 1991.

[31] A. N. Netravali, E. Petajan, S. Knauer, K. Mathews, R. J. Safranek, and P. Westerink, "A high quality digital HDTV codec." *IEEE Trans. Consumer Electron.*, pp. 320–330, Aug. 1991.

[32] M. Liou, "Overview of the px64 kbits/s video coding standard." *Commun. ACM*, pp. 60–63, April 1991.

[33] Y. Ninomiya, "HDTV broadcasting systems." *IEEE Commun. Mag.*, pp. 15–23, Aug. 1991.

[34] B. Girod, "Psychovisual aspects of image communication." *Signal Processing*, pp. 239–251, 1992.

[35] P. P. Vaidyanathan, "Quadrature mirror filter banks, m-band extensions and perfect-reconstruction techniques." *IEEE ASSP Mag.*, pp. 4–20, 1987.

[36] R. V. Cox, "The design of uniformly and nonuniformly spaced pseudoquadrature mirror filters." *IEEE Trans. Acoust., Speech, Signal Processing*, pp. 1090–1096, 1986.

[37] J. D. Johnston, "A filter family designed for use in quadrature mirror filter banks." In *Proc. ICASSP*, 1980.

[38] J. Princen, A. Johnson, and A. Bradley, "Sub-band transform coding using filterbank designs based on time-domain aliasing cancellation." In *Proc. ICASSP*, 1987.

[39] H. S. Malvar, *Signal Processing with Lapped Transforms*. Norwood, MA: Artech House, 1992.

[40] I. Daubechies, "Orthonormal bases on compactly supported wavelets." *Commun. Pure Appl. Math.*, pp. 909–996, 1988.

[41] P. Rioul and M. Vetterli, "Wavelets and signal processing." *IEEE Signal Processing Mag.*, pp. 14–38, Oct. 1991.

[42] D. J. Granrath, "The role of human visual models in image processing." *Proc. IEEE*, pp. 552–561, May 1981.

[43] F. W. Campbell and J. G. Robson, "Application of Fourier analysis to the visibility of gratings." *J. Physiology*, pp. 551–566, 1968.

[44] J. L. Mannos and D. J. Sakrison, "The effects of a visual fidelity criterion on the encoding of images." *IEEE Trans. Inform. Theory*, July 1974.

[45] N. B. Nill, "A visual model weighted cosine transform for image compression and quality assessment." *IEEE Trans. Commun.*, pp. 551, June 1985.

[46] M. G. Perkins and T. Lookabaugh, "A psychophysically justified bit allocation algorithm for subband image coding systems." In *Proc. ICASSP*, 1989.

[47] R. E. Van Dyck and S. A. Rajala, "Subband/VQ coding in perceptually uniform color spaces." In *Proc. ICASSP*, 1992.

[48] C. F. Stromeyer and B. Julesz, "Spatial frequency masking in vision: Critical bands and spread of masking." *J. Opt. Soc. Amer.*, pp. 1221–1232, Oct. 1972.

[49] W. B. Pennebaker and J. L. Mitchell, *JPEG Still Image Data Compression Standard*. New York: Van Norstrand Rienhold, 1993.

[50] H. A. Peterson, A. J. Ahumada, Jr., and A. B. Watson, "Improved detection model for DCT coefficient quantization." In *SPIE Conf. Human Vision, Visual Processing, and Digital Display IV*, Feb. 1993, vol. 1913, pp. 191–201.

[51] D. L. Neuhoff and T. N. Pappas, "Perceptual coding of images for halftone display." *IEEE Trans. Image Processing*, July 1994.

[52] T. N. Pappas and D. L. Neuhoff, "Printer models and error diffusion." *IEEE Trans. Image Processing*, Jan. 1995.

[53] C. I. Podilchuk and N. Farvardin, "Perceptually based low bit rate video coding." in *Proc. ICASSP*, 1991.

[54] C. I. Podilchuk, N. S. Jayant, and P. Noll, "Sparse codebooks for the quantization of non-dominant sub-bands in image coding." in *Proc. ICASSP*, 1990.

Chapter 4

Bilevel Image Coding

D. L. Duttweiler
Visual Communications Research Department
AT&T Bell Laboratories
Holmdel, New Jersey

4.1 Introduction

The terms *bilevel*, *two tone*, and *black and white* are used more or less interchangeably to denote images in which pixels take on only one of two colors. Generally the two colors are black and white, but from an image compression standpoint just what the two colors are is of no consequence. The adjective *black and white* is more common, but the broader adjectives *two tone* and *bilevel* are arguably more technically precise.

Algorithms for bilevel image coding are generally *bit preserving* or *lossless*. These synonyms denote that the decoded image is digitally identical to the encoded image. This losslessness of most bilevel image coding algorithms is in marked contrast to the general *lossy* nature of algorithms for coding gray-scale images, color images, and video. For these other types of image coding, the goal is usually only to preserve as much as possible the subjective quality of the image. Presumably this dichotomy has evolved because it is rather easy to get significant (say a factor of 20 or greater) compressions of bilevel images under a lossless constraint, whereas gray-scale, color, and video can be significantly compressed only under the more relaxed constraint of only preserving subjective quality.

Although bilevel coding algorithms are generally lossless, the actual final compression achieved in any particular application can generally be improved by "cleaning up" source images before coding them. Such preprocessing might reasonably remove isolated pixels of one color in a field of the other, or it might straighten ragged lines. Despite the benefit to compression, however, the bilevel image coding world generally considers preprocessing like this to be image enhancement rather than a part of coding.

Eliminating subjective quality as an issue in comparing coding algorithms is a great simplification, but certainly does not end all discussion of algorithm worth. The compressions achieved by different algorithms are generally sensitive to just what type of bilevel images are being coded. Applications with differing image types will see differing relative compression performance. Some algorithms are less complex and more readily implemented in hardware or can execute faster in software. Some algorithms recover gracefully from a small number of errors in the compressed data stream, while others fall apart after just one.

Bilevel images can be usefully classified into a few broad categories. In many applications, all the images to be coded are *text* images. A typewritten page is the archetypical text image, but the terminology is broader and includes images with glyphs from multiple fonts and of multiple point sizes.

Line-art images are typified by block diagrams. In additions to glyphs of various sizes and fonts, line-art images contain lines, rectangles, and other filled or unfilled geometric shapes.

Bilevel images by definition contain pixels of only two colors, but this does not mean that they cannot render gray-scale. Gray scale can be very satisfactorily rendered by rapidly varying pixel colors within small regions. If the display resolution in dots per inch is sufficiently high and the image is viewed at sufficient distance, the eye will integrate intensity over regions many pixels in size and perceive a shade of grey (assuming for simplicity that the two colors available are black and white). Such a process is used routinely in printing.

There are many techniques that can be used to map gray-scale intensities to rapidly varying patterns of pixels in a bilevel image. Ulichney's book [1] provides an excellent discussion. For purposes of image coding, all the various techniques for rendering gray-scale can be usefully divided into two classes based on whether or not the bilevel images generated contain strong periodicities. The *periodic* gray-scale-rendering techniques include screening, ordered dither, and in general any techniques that deterministically map gray-scale *superpixels* to a cluster of bilevel pixels. Aperiodic gray-scale-rendering techniques include error diffusion, the blue-noise technique described in Ulichney's book, and the model-based technique of Pappas and Neuhoff [2].

The easiest images to compress are text and line-art images. Even the dot over an *i* is many pixels in diameter when scanned at normal scanning densities. Hence there is available for exploitation a strong correlation between the colors of neighboring pixels. Reasonable, but generally not as good, compression is achieved by some algorithms in coding gray-scale renderings containing periodicities. Not all coding techniques can achieve this, however, and in fact the dominant techniques being used today (G3 and G4 facsimile) do not. Adaptive algorithms, adjusting at a minimum to the fact that the image is a gray-scale rendering and, better yet, also to its periodicity, are needed. The hardest images to compress are aperiodic gray-scale renderings. The best algorithms might provide a factor of 2 compression. Many algorithms will actually create "compressed" files larger than the source file.

Many, perhaps too many, bilevel image coding algorithms have been proposed and analyzed in the technical literature. It would be impossible without expanding this chapter to a book in itself to begin to cover them all. To maintain manageability, coverage will be restricted to the standardized bilevel image coding techniques, that is, the CCITT G3 and G4 facsimile coding algorithms and the newer CCITT/ISO JBIG algorithm. The G3 and G4 algorithms have found broad acceptance not only for document transmission as in facsimile, but also in document storage and retrieval systems. The JBIG standard is much newer and not yet widely used. Its advantages over the G3/G4 algorithms are better compression, especially on gray-scale renderings, and possible parameterization for progressive (multiresolution) coding. The JBIG downside is increased complexity.

Many current applications for bilevel image coding often have as an alternative to raster image compression, the possibility of capturing the document in an electronic form such as Postscript [3] or the input to some word processing program. Once the high-level description has been translated to a raster, it will never be possible to get back to a file size as small as that of the original abstract characterization. The next section steps back from the mainline objective of describing the G3/G4 algorithms (Section 4.3) and the JBIG algorithm (Section 4.4) to develop rough estimates of the relative efficiencies of these two approaches.

4.2 Compressed Rasters versus Page Description Language

The most widely used test image for bilevel image coding is CCITT test image number 1, the "Slerexe" letter explaining facsimile transmission. Various scannings of the original paper image are unfortunately in existence. The one we have is a 200 dot per inch scanning, 1728 pixels wide, and 2376 pixels high. An image generated from this particular scanning appears as Fig. 4.1.

For the purposes of this section, a high-level capture of this letter is also needed. Fig. 4.2 shows the results of one attempt at creating one. This particular image started as a *troff* source file. *Troff* is an older word processing language that

THE SLEREXE COMPANY LIMITED

SAPORS LANE · BOOLE · DORSET · BH 25 8 ER

TELEPHONE BOOLE (945 13) 51617 · TELEX 123456

Our Ref. 350/PJC/EAC   18th January, 1972.

Dr. P.N. Cundall,
Mining Surveys Ltd.,
Holroyd Road,
Reading,
Berks.

Dear Pete,

   Permit me to introduce you to the facility of facsimile
transmission.

   In facsimile a photocell is caused to perform a raster scan over
the subject copy. The variations of print density on the document
cause the photocell to generate an analogous electrical video signal.
This signal is used to modulate a carrier, which is transmitted to a
remote destination over a radio or cable communications link.

   At the remote terminal, demodulation reconstructs the video
signal, which is used to modulate the density of print produced by a
printing device. This device is scanning in a raster scan synchronised
with that at the transmitting terminal. As a result, a facsimile
copy of the subject document is produced.

   Probably you have uses for this facility in your organisation.

                                Yours sincerely,

                                Phil.

                                P.J. CROSS
                                Group Leader - Facsimile Research

Registered in England: No. 2088
Registered Office: 60 Vicara Lane, Ilford. Essex.

Figure 4.1   The original CCITT test image number 1.

THE SLEREXE COMPANY LIMITED

SAPORS LANE - BOOLE - DORSET - BH 25 8 ER

TELEPHONE BOOLE (94513) 51617 - TELEX 123456

Our Ref. 350/PJC/EAC   18th January, 1972

Dr. P.N. Cundall,
Mining Surveys Ltd.,
Holroyd Road,
Reading,
Berks.

Dear Pete,

Permit me to introduce you to the facility of facsimile
transmission.

In facsimile a photocell is caused to perform a raster scan over
the subject copy. The variations of print density on the document
cause the photocell to generate an analogous electrical video signal.
This signal is used to modulate a carrier, which is transmitted to a
remote destination over a radio or cable communications link.

At the remote terminal, demodulation reconstructs the video
signal, which is used to modulate the density of print produced by a
printing device. This device is scanning in a raster scan synchronised
with that at the transmitting terminal. As a result, a facsimile
copy of the subject document is produced.

Probably you have uses for this facility in your organisation.

Yours sincerely,

P.J. CROSS
Group Leader - Facsimile Research

Registered in England: No. 2038
Registered Office: 60 Vicars Lane, Illford. Essex.

Figure 4.2   A word processor imitation of CCITT test image number 1.

originated on Unix^{TM} systems around 1975. Since most of the *troff* source is text as opposed to formating directives, none of the results to be given should be particularly sensitive to the particular choice of word processing language.

Two noteworthy differences between Figs. 4.1 and 4.2 are that in Fig. 4.2 the company logo is missing and the signature is missing. *Troff* like most other word processing languages has hooks for capturing bit maps in an electronic document, but using them begins to undo the comparison we are trying to achieve. The presence of this problem is in itself a comment on one of the main problems in trying to capture documents electronically.

Table 4.1 gives byte counts for the various ways in which Figs. 4.1 and 4.2 can be captured.

The *troff* source for Fig. 4.2 is a 1368 byte ASCII file. Because the relative frequencies of bytes within this file differ (bytes with values greater than 128 never occur at all) and because there is correlation between adjacent bytes, this electronic source is readily compressed by standard file compression algorithms. Ziv–Lempel compression [4] is one of the more common and effective such algorithms. It compresses the original source to 964 bytes.

Postscript has become the de facto page description language. Most word processors either automatically create Postscript files as an intermediary enroute to a raster image or at least can be requested to do so. Because of this, an interesting alternative to capturing word processor input is capturing Postscript. Doing so eliminates the need to choose some particular word processing software.

The Postscript intermediate created by our word processing software includes a standard header file of commonly used Postscript subroutines, some, but not all, of which are called from the main body of the Postscript description. Depending on the application, byte counts with or without this header file are more appropriate

Table 4.1   File sizes for different capture formats.

Capture	Original CCITT #1	Electronic Rendition
troff		1368
troff compressed by Ziv–Lempel		964
Postscript (with header)		11215
Postscript (with header) compressed by Ziv–Lempel		6469
Postscript (no header)		6756
Postscript (no header) compressed by Ziv–Lempel		3659
raster	513216	513216
Raster compressed by Ziv–Lempel	32364	27808
Raster compressed by G4	18103	15121
Raster compressed by JBIG	14715	10597

for comparison. Both are given in Table 4.1 along with the byte counts of their Ziv–Lempel compressions.

The raster of the original Slerexe letter and its electronic imitation both contain $1728 \times 2376/8 = 513,216$ bytes. Ziv–Lempel compression is general purpose and not tuned for bilevel image compression, but it is simple to apply and achieves remarkably good compression despite lacking forehand knowledge of the image dimensions or even that the byte sequence represents an image. Presumably Ziv–Lempel compression does this well in this instance because there are numerous zero bytes occurring whenever 8 pixels in a row are all white. G4 compression is the most efficient of the G3/G4 techniques and does about a factor of 2 better than Ziv–Lempel. JBIG gains another 20 %.

All the compressed rasters are smaller when the raster is the imitation rather than true Slerexe letter. Presumably this is largely attributable to the absence of the logo and signature in the imitation. The compressions of the imitation may also be beneficial because the input image is digitally generated and hence has no scanning noise or ragged edges on character glyphs.

The data of Table 4.1 are for a particular image and two (*troff* and Postscript) particular page description languages. Furthermore, even with the choice of a page description language fixed, there is no unique defining file. Many are possible and they will generally have differing byte counts. Because of problems like this, all that can be expected from comparisons like that in Table 4.1 is rough rule-of-thumb guidelines. For this example, the high-level capture with a word processor language offers about an order of magnitude compression advantage over a compressed-raster capture. A Postscript capture has about a factor of 2 advantage over the compressed-raster capture.

Byte count is far from the only issue in choosing between high-level capture and compressed-raster capture. High-level capture allows keyword searching, which may be essential in some applications. Compressed rasters have the advantage of unrestrained applicability alluded to earlier while noting the absence of the logo and signature in Fig. 4.2. The fact that they require less processing to be converted to a viewable image can also be a benefit. Both approaches will coexist for some time.

4.3 Group 3 and Group 4 Coding

By far the most widely used algorithms for bilevel image coding are the CCITT Group 3 (G3) and Group 4 facsimile algorithms [5]. Group 3 facsimile actually defines two coding algorithms. One goes by the alternative names of *modified Huffman* (MH) and *G3 one-dimensional* (G3-1D). The other is called either *modified READ* (MR) or *G3 two-dimensional* (G3-2D). The term *READ* is itself an acronym for row element address designate. The Group 4 (G4) algorithm is almost identical to the G3-2D algorithm and is sometimes referred to as *modified-modified READ* (MMR).

Purists object to using "G3" and "G4" to refer to coding algorithms. CCITT Group 3 facsimile apparatus is defined by much more than just an image coding algorithm. Image resolution, transmission bit rates, modem signaling constellations, and handshaking protocols are also defined. The actual CCITT Recommendation in which G3 coding (both one- and two- dimensional) is defined is T.4 [6]. G4 coding is defined in Recommendation T.6 [7]. Hence the names *T.4* and *T.6* are also in common use.

4.3.1 One-Dimensional Group 3 Coding

The easiest code to describe or implement is the G3-1D code. Each line of the raster is coded independently of the others. Suppose a line to be coded has 16 pixels of color

$$W W W B B W W W W B W W W W W W$$

where W denotes a white pixel and B denotes a black pixel (more generally, W denotes a pixel of the background color and B denotes a pixel of the foreground color). In G3-1D coding, one first maps this line to the sequence

$$WR(3), BR(2), WR(4), BR(1), WR(6)$$

of run lengths, where $WR(n)$ denotes a white run of length n pixels and $BR(n)$ similarly denotes black runs. Each of these run lengths is then mapped by look-up in a fixed table to a particular Huffman code [8, 9]. The needed translations here are:

$$WR(3) = 1000,$$
$$BR(2) = 11,$$
$$MWR(4) = 1011,$$
$$BR(1) = 010,$$
$$WR(6) = 1110.$$

These Huffman codes are concatenated and an

$$EOL = 000000000001$$

appended, so that the final coding for the line becomes

$$1000111011010111 0000000000001.$$

For this example, G3-1D coding "compresses" the original 16 pixel (bit) line to 29 bits. What has gone wrong? One thing is simply that the line is atypically short and the appended 12-bit EOL marker is significant rather than inconsequential as it usually is. More subtly and significantly, another thing that has gone wrong, is that the simple short choppy line of this example is not well matched to the G3-1D Huffman tables. These tables are optimized for coding typewritten lines scanned at 200 dots per inch (dpi). The relative probabilities of occurrence of the various runs dictate the length of the codes used to denote them. Typical lines when scanned at 200 dpi often generate much longer runs than those in this example, and because of their relatively common occurrence, such run lengths have to be given reasonably short codes. Doing so uses up code space and forces the Huffman codes for extremely short runs such as those of this example to be longer than they would be in a code table tuned for short choppy lines.

G3-1D coding is actually slightly more complicated than the simple picture described so far. Most lines will begin with a run of white, the presumed foreground color. But this need not be so for all lines of all images. The code tables for coding white runs and black runs do not share the same code space, and therefore decoders need somehow to be kept in synchrony. The trick is to have a Huffman code for a white run of zero length available and code this first whenever a line begins with black.

Another slight complication is introduced by *makeup* codes. The run length Huffman codes discussed so far are more fully referred to as *terminating* Huffman codes. It is not practical, nor desirable from an implementation standpoint, to define codes for all possible run lengths (0 to 1728 at standard 200 dpi scanning resolution). Instead what is defined are terminating codes for white and black runs of lengths between 0 and 63 pixels, plus makeup codes for runs of $n \times 64$ pixels for multipliers n between 1 and 27. If an actual run is greater than or equal to 64 pixels in length, what is transmitted is the concatenation of a makeup code and a terminating code. The makeup codes and terminating codes share the same code space, so decoders can maintain synchrony. When a makeup code is received, a terminating code for the same color is known to be the following Huffman code rather than a code describing the length of the following run.

From an information theoretic point of view, the EOL markers incorporated into the line coding are redundant. Their presence is justified by the fact that they make it possible to gracefully recover from infrequent errors. Decoders always resynchronize on EOL markers. Because of this, an isolated error does not cause image errors except to the right of where it occurs on the particular line. Because G3 facsimile equipment is used on analog lines at high bit rates, errors are an unfortunate reality on at least a small percentage of calls, and error recover like this is important.

4.3.2 Two-Dimensional Group 3 Coding

Two-dimensional coding generally offers at least a factor of 2 advantages over one-dimensional coding. Because two-dimensional coding is not as sensitive to an assumed scanning density as is one-dimensional coding, the two-dimensional advantage can be even larger at unusually large or small scanning densities.

A free parameter in G3-2D coding is the K factor. Common choices for K are 4 and infinity. In G3-2D coding, the top line and every K^{th} line thereafter are coded one-dimensionally, while the intervening lines are coded two-dimensionally. Since one-dimensional coding makes no reference to prior lines, choosing K to be finite allows error recovery after at most K lines. The most efficient coding is achieved with K equal to infinity. Such a choice is entirely appropriate if errors are not likely, as is the case when lower level protocols correct errors either by retransmission or error correcting codes or when the application is a document retrieval application and there is no transmission with its attendant possibility of error.

One-dimensional line coding in G3-1D and G3-2D is nearly identical. The only difference is that the termination code

$$EOL0 = 0000000000010$$

is used to terminate a one-dimensionally coded line during G3-2D coding. The appended 0 informs the decoder that the next line to be coded will be two-dimensionally coded. Two-dimensionally coded lines are terminated in

$$EOL1 = 0000000000011$$

or $EOL0$ depending on whether the next line is to be one-dimensionally or two-dimensionally coded, respectively.

In two-dimensionally coding a line, the basic idea is to code differences between the line being coded and the line above it. As in the one-dimensional coding of a line, Huffman codes specify lengths of runs to be laid down from left to right. However, now the Huffman codes indicate run length not simply by specifying a length but rather by locating run terminations in the current line relative to run terminations in the line above it.

Figure 4.3 defines some key pixels along two image lines, the bottom line being the *coding* line and the top being the *reference* line.[1] The a_0 pixel is the pixel on the current line immediately following the last pixel to be laid down by a Huffman code. The a_1 pixel is the first pixel following the a_0 pixel and of opposite color. The a_2 pixel begins the run following the run of the a_1 pixel. The b_1 pixel is the first pixel on the reference line that is (strictly) to the right of a_0, is of color opposite to a_0, and is changing in the sense that it differs in color from its left neighbor. The b_2 pixel begins the run following the run of the b_1 pixel.

---

[1] This figure and the next two to follow duplicate figures in Recommendation T.4 itself so that passing between this narrative and that in T.4 is as easy as possible. Figure 4.6 might be expected to continue this commonality, but actually has been drawn different from its T.4 counterpart, which is unfortunately rather confusing.

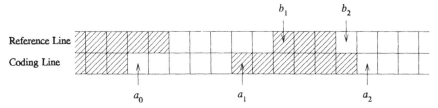

Figure 4.3   Definitions of key pixels in two-dimensional coding of a line (redrawn from [6]).

If as is shown in Fig. 4.4, b_2 lies (strictly) to the left of a_1, the next Huffman code to be transmitted is a

$$PASS = 0001$$

Huffman code, which instructs a decoder to lay down pixels of the color of a_0 up to but not including the pixel a_0' shown in Fig. 4.4. Following the processing of this code, a_0 is repositioned to a_0' and a_1, a_2, b_1, and b_2 are reidentified.

If b_2 is not to the left of a_1 (the trigger for entering pass mode), exactly what is transmitted depends on how far apart a_1 and b_1 are. If as shown in Fig. 4.5, they are 3 or fewer pixels apart, one of seven (±3, ±2, ±1, 0) vertical mode Huffman codes exactly specifies this relationship. Pixels up to but not including a_1 are laid down and a_0 is repositioned to a_1. When crossing vertical lines in an image, a_1

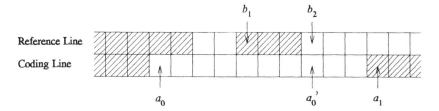

Figure 4.4   Sample configuration to trigger pass mode (redrawn from [6]).

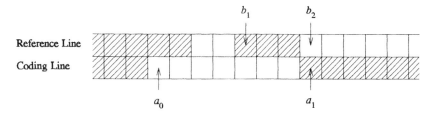

Figure 4.5   Sample configuration to trigger vertical mode (redrawn from [6]).

lies under b_1. This is a very common situation and the vertical mode code for a 0 offset is allotted half the coding space (it is a single 1 bit) so that only 1 bit need be transmitted.

The remaining possibility is that b_2 is not to the left of a_1 and that a_1 and b_1 are more than 3 pixels part. An example of this situation is shown in Fig. 4.6. When it arises, the first code transmitted is a

$$\text{HORIZONTAL} = 001$$

Huffman code, which in itself lays down no pixels, but serves only to indicate entry into horizontal mode. This horizontal marker is followed by specifications of the run lengths $a_1 - a_0$ and $a_2 - a_1$. Each of these specifications is most commonly just one of the terminating one-dimensional Huffman codes, but will be a makeup-code/terminating-code pair for runs of 64 or more pixels.

Beginning-of-line, end-of-line, and end-of-page considerations add complication, but this detail is not particularly interesting and is well explained in Recommendation T.4 itself.

4.3.3 Group 4 Coding

G4 coding is almost identical to G3-2D coding with K equal to infinity. The only difference is that the EOL0 and EOL1 line terminating codes are not appended to every line coding. As noted before, these codes coupled with a finite value for K make a graceful recovery from errors possible. Since G4 assumes a digital transmission medium and hence errors are not expected, setting K to infinity and deleting line terminating codes is entirely appropriate.

4.4 Joint Bilevel Imaging Group Coding

The bilevel coding algorithm most recently standardized is the Joint Bilevel Imaging Group (JBIG) algorithm. JBIG was chartered in 1988 to establish a standard

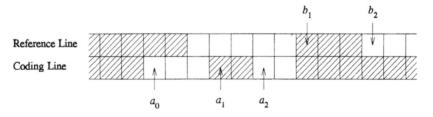

Figure 4.6   Sample configuration to trigger horizontal mode.

for the progressive coding of bilevel images. The *joint* in its name reflects the fact that it reports to both ISO (specifically, ISO-IEC/JTC1/SC29/WG1) and ITU-T (specifically, ITU-T/SGVIII/Q16). ITU-T is the successor organization to CCITT.

JBIG coding is more complex than G3/G4 coding, but offers two compensating advantages. One is superior compression. On text or line-art images, JBIG achieves about 20% greater compression than G4, the most efficient of the G3/G4 algorithms. The JBIG compression advantage is much greater, generally ranging between a factor of 2 and a factor of 10, on bilevel images rendering gray scale. The JBIG algorithm is adaptive and indirectly recognizes and adjusts to input images rendering gray scale. In addition, special features are built in to recognize and exploit any periodicities present in gray-scale renderings.

The second advantage of JBIG coding is that, if desired, it can be parameterized for progressive coding.

4.4.1 What Is Progressive Coding?

Progressive, or synonymously, *multi resolution*, coding captures images as a compression of a low-resolution rendition plus a sequence of *delta* files that each provide another level of resolution enhancement. In the case of JBIG, each delta file contains the information needed to double both the vertical and horizontal resolution.

Progressive coding is attractive in database applications where output devices of varying resolution capability (say PC screens, workstation screens, and laser printers) are all to be served. Only the bottom-layer coding and as many delta files as are of use are transmitted and processsed. In contrast, if images are stored nonprogressively, it is necessary to either store compressions at all resolutions or to store compressions at only the highest resolution but require output devices to first decode to high resolution and then map down to display resolution. The first alternative wastes storage capacity. The second wastes transmission capacity and processing power.

Another application for progressive coding is in image browsing over medium rate communication links. A low-resolution rendition of an image can be made available quickly and then followed by as much resolution enhancement as is desired. A user on seeing at low resolution that the image being developed is not that desired, can quickly interrupt its transmission and move on to the next. This advantage for progressive coding accrues only on medium rate links, roughly those with speeds between 9.6 kbit/sec and 64 kbit/sec when bilevel images are being retrieved. On slower links, no user would have the patience required for image browsing no matter what the form of presentation. On higher speed links, the image comes so fast relative to human reaction times that how it develops is immaterial.

A third potential application for progressive coding is on packet networks on which packets can or must be priority classified as droppable or nondroppable. The packets carrying the information for the final resolution doubling would be sent at low priority and if the network were to drop them, the only penalty would be a slightly less sharp image. No entire regions would be lost or destroyed.

The JBIG specification uses the symbol D to denote the number of resolution doublings (delta files) that are to be available. The parameter D is unrestricted and may be chosen as 0 when progression is of no benefit. Making this choice generally improves compression by about 5%, but surprisingly, can degrade compression by as much as 5% on gray-scale renderings created by periodicity generating algorithms.

Common choices for D when progression is wanted are 4, 5, and 6. If the original is a 400 dpi scanning of an 8.5 × 11 inch page and D is chosen as 5, the bottom-layer image will be 107 × 138 pixels in size. Normally such a small image would be blown up by pixel replication (or perhaps something more sophisticated) before display. If instead, however, it is displayed dot for dot on something like a 1000 × 1000 pixel display, it can serve nicely as an icon.

4.4.2 Functional Clocks

The JBIG algorithm is complex and definitively describing it here is not appropriate. Such detail is best obtained from the specification itself [10]. As in the discussion of the G3/G4 algorithms, the goal here is to only convey the essence of the algorithm.

Conceptually a JBIG encoder can be decomposed (see Fig. 4.7) into a chain of D differential-layer encoders followed by a bottom-layer encoder. In Fig. 4.7, I_d denotes the image at layer d and C_d denotes its coding. A hardware implementation would in all likelihood time share one physical differential-layer encoder, but for heuristic purposes the decomposition of Fig. 4.7 is helpful. Each differential-layer encoder can be decomposed into the functional blocks shown in Fig. 4.8. The bottom-layer encoder has the somewhat simpler decomposition shown in Fig. 4.9.

Resolution reduction   The resolution reduction block in a differential-layer encoder accepts the high-resolution image I_d and creates the low-resolution image I_{d-1} with, as nearly as possible,[2] half as many rows and half as many columns. A simple way of performing this resolution reduction would be by subsampling; that is, discarding every other row and every other column. Subsampling is simple, but the low-resolution images it creates are poorer in subjective quality than is necessitated by their diminished resolution alone. On images with line art, whole lines can be lost if they happen to lie on rows or columns being discarded. Gray-scale

---

[2]There need not in general be even numbers of rows and columns in I_d.

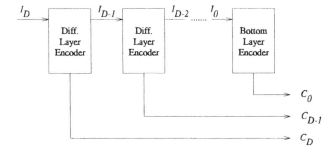

Figure 4.7   Decomposition of a JBIG encoder.

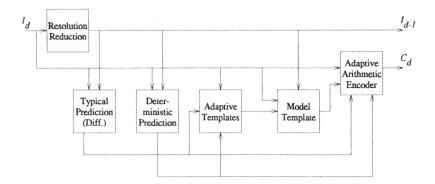

Figure 4.8   Functional blocks in a JBIG differential-layer encoder.

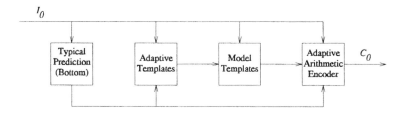

Figure 4.9   Functional blocks in a JBIG bottom-layer encoder.

renderings with periodicities frequently display aliasing artifacts caused by beats between the superpixel frequency and the subsampling by 2.

JBIG's resolution reduction algorithm works remarkably well for all image types—text, line art, and gray-scale renderings. It is a table-based algorithm. The low-resolution image is created pixel by pixel in the usual raster scan order, that is, top to bottom and left to right. The color of any given low-resolution pixel is dictated by the colors of nine particular high-resolution neighbors and three causally positioned low-resolution neighbors.

Arithmetic coding   The remaining functional blocks in Figs. 4.8 and 4.9 implement the compression algorithm. The heart of this functionality in both the bottom-layer coder and the differential-layer coder is an adaptive arithmetic coder [11]. Arithmetic coders are distinguished from other entropy coders such as Huffman coders and Ziv–Lempel coders in that, conceptually at least, they map a string of symbols to be coded into a real number x on the unit interval [0.0, 1.0]. In the JBIG application, the string of symbols to be coded is the pixels of the image I_d presented in raster scan order. What is transmitted or stored instead of the image I_d is a binary representation of x.

The particular number x to which the input sequence maps is determined by the recursive probability interval subdivision of the Elias coder [12]. Figure 4.10 shows an example of such interval division through an initial sequence 0,1,0,0 to be coded.

The portion of the unit interval on which x is known to lie after coding an initial sequence of symbols is known as the *current coding interval*. For each binary input the current coding interval is divided into two subintervals with sizes proportional to estimates of the relative probabilities of symbol-value occurrences. The new current coding interval becomes the portion of the old coding interval associated with the symbol value actually occurring. The fact that most frequently the symbol to be coded will be the more probable and that by design this symbol codes into the larger of the two subintervals means that most of the time the coding interval is not reduced in size by as much as one half and an input symbol is coded without generating as much as 1 bit in the binary expansion of x. It is true that whenever

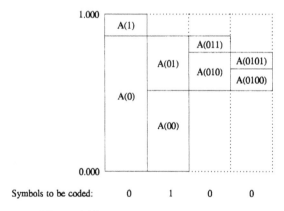

Figure 4.10   Example of interval subdivision.

the less probable symbol does happen to occur more than 1 output bit is generated, but the central result of information theory is that on the average there is still less than 1 bit generated for each input symbol.

The Elias coder was conceived soon after Shannon's seminal paper gave birth to information theory in 1948 [13]. It was of little practical use, however, until ways were found in the mid-1970s to perform all the necessary arithmetic with finite precision arithmetic and in a pipelined fashion so that a decoder could start its processing without having to wait for an encoder to finish its processing [14–17].

Model templates   For each high-resolution pixel to be coded, the model-templates block provides the arithmetic coder with an integer called the *context*. For differential-layer coding this integer is determined by the colors of 6 particular pixels in the causal high-resolution image, by the colors of 4 particular pixels in the already available low-resolution image, and by the spatial phase of the pixel being coded. The term *spatial phase* denotes which of the four possible orientations the high-resolution pixel has with respect to its corresponding low-resolution pixel. The 6 particular high-resolution pixels and 4 particular low-resolution pixels whose colors (along with spatial phase) define the context are known as the *coding template* or *model template*.

The arithmetic coder maintains for each context an estimate of the conditional probability of the symbol given that context. Doing this well requires carefully balancing speed of adaptation against quality of adaptation [18]. The greatest coding gain is achieved when this probability estimate is both accurate and close to 0 or 1. Thus good templates have good predictive value so that when the values of the pixels in the template are known, the value of the pixel to be coded is highly predictable.

For bottom-layer coding, the coding template includes only high-resolution pixels. There are no low-resolution pixels to incorporate nor is there an analog of the spatial-phase concept.

Adaptive Templates   The comparatively good compression obtained by JBIG on gray-scale renderings is partially attributable to the adaptation within the arithmetic coder. The adaptive-template (AT) block adds a final factor of 2 improvement on gray-scale renderings with periodicities. AT looks for periodicities in the image and on finding a strong indication for one changes the template so that as nearly as possible a pixel offset by this periodicity is incorporated into it. Such a pixel of course has excellent predictive value.

Any changes of this sort to the template must be made only infrequently since a period of readaptation in the probability estimator ensues and compression degrades while it occurs. The JBIG algorithm for determining the if, when, and how of any template rearrangement has substantial hysteresis so that template changes occur only when there is an extremely strong indication that a new template offers substantial improvement over the current.

Differential-Layer Typical Prediction   The differential-layer typical prediction (TP) block provides some coding gain, but its primary purpose is to speed implementations. Differential-layer TP looks for regions of solid color, and when it finds that a given current high-resolution pixel for coding is in such a region, none of the processing normally done in the deterministic prediction, adaptive templates, model templates, or arithmetic coding blocks is needed. On text and line-art images differential-layer TP usually makes it possible to avoid coding about 95% of the pixels. Greyscale renderings do not generally have the large regions of continuous color and do not allow processing savings like this.

The key idea behind differential-layer TP is that if all the pixels in an 8-pixel neighborhood of a low-resolution pixel are the same color as it, then it is extremely likely that all 4 high-resolution pixels to be associated with it are that same color also. Unfortunately, it is not certain that this is so, but exceptions occur infrequently enough that it is efficient and reasonable to flag them. In particular, an encoder notes at the beginning of each high-resolution line pair whether or not a decoder would ever go wrong on that line pair if it always were to "typically expand" any low-resolution pixels it found to be within a common-color 8-pixel neighborhood into 4 high-resolution pixels of that color. The failure or success of this strategy over the line pair is coded and sent to the decoder. Note that failure here is not that some low-resolution pixel in the line pair is not within a common-color 8-pixel neighborhood, but rather that some low-resolution pixel to be associated with the line pair is within a common-color 8-pixel neighborhood but will not have associated high-resolution pixels of that color. Failure in this sense is extremely rare and on many images never occurs. When success over the line pair is coded, both the encoder and decoder skip over high-resolution pixels associated with any low-resolution pixels found to be within a common-color 8-pixel neighborhood. If failure must be coded, the only penalty is that no skipping can be done for the line pair and everything must be coded.

Bottom-Layer Typical Prediction   Bottom-layer typical prediction, like differential-layer typical prediction, tries to exploit solid color regions of the image to save processing effort. However, the algorithms are quite different. The bottom-layer algorithm is a line-skipping algorithm. A given line is said to be "typical" and all its pixels are declared *typical*, if it is identical to the line above it. Which lines are typical is again transmitted to the decoder. Both the encoder and decoder skip the coding of all pixels in typical lines and generate them instead by line duplication.

It is not possible in bottom-layer coding to skip as large a percentage of pixels as it is in differential-layer coding. On text and line-art images, savings are generally about 40%.

Deterministic Prediction   When images are reduced in resolution by the JBIG resolution reduction algorithm, it sometimes happens that the value of the particular high-resolution pixel being coded can be inferred from the pixels already known to both the encoder and decoder; that is, all the pixels in the low-resolution image and those in the high-resolution image that are causally related in a raster sense to the current pixel. When this occurs, the current pixel is said to be deterministically predictable. The deterministic prediction (DP) block flags any such pixels and inhibits their coding by the arithmetic coder. DP is a table-driven algorithm. The values of particular surrounding pixels in the low-resolution image and causal high-resolution image are used to index into a table to check for determinicity and, when it is present, obtain the deterministic prediction. DP provides about a 7% coding gain.

4.5 Conclusions

It is important to distinguish page-description-language (PDL) specifications of a bilevel image from compressed-raster specifications. PDL captures are in general more compact and because of their higher level of abstraction are amenable to keyword searching and future editing. Compressed-raster specifications have the advantage of universal applicability.

By far the most widely used algorithms for coding bilevel image rasters are the MH, MR, and MMR algorithms defined in the G3 and G4 facsimile standards. These algorithms are simple to implement and provide compression that is often judged satisfactory. The more recently standardized JBIG algorithms offer improved compression and the possibility of multiresolution capture, but are somewhat more complex and not yet widely implemented.

To date, bilevel compression algorithms have generally been lossless so that there is no coding error and decoded images are identical to originals. In general greater and greater compression can be attained via ever-increasing complexity of algorithms, but it seems likely that JBIG is at a turning point and any further significant increase in lossless compression efficiency will come at the expense of a tremendous increase in complexity. For this reason it seems likely that much future bilevel image compression activity will focus on lossy algorithms, where substantial additional compression is still attainable. A difficulty in any work of this sort, however, is that subjective quality becomes an issue and evaluating algorithms becomes difficult.

References

[1] R. Ulichney, *Digital Halftoning*. The MIT Press, 1987.

[2] T. N. Pappas and D. L. Neuhoff, "Model-based halftoning," *Proc. SPIE*, vol. 1453, Feb. 1991.

[3] Adobe Systems Inc., *Postscript Language Tutorial and Cookbook*. Addison-Wesley, 1985.

[4] J. Ziv and A. Lempel, "A universal algorithm for sequential data compression," *IEEE Trans. Inform. Theory*, vol. 30, pp. 520–540, June 1987.

[5] R. Hunter and A. H. Robinson, "International digital facsimile coding standards," *Proc. IEEE*, vol. 68, pp. 854–867, July 1980.

[6] CCITT Recommendation T.4, *Standardization of Group 3 Facsimile Apparatus for Document Transmission*, 1980 with amendments in 1984.

[7] CCITT Recommendation T.6, *Facsimile Coding Schemes and Coding Control Functions for Group 4 Facsimile Apparatus*, 1984.

[8] D. A. Huffman, "A method for the construction of minimum redundancy codes," *Proc. IRE*, vol. 40, pp. 1098–1101, Sept. 1952.

[9] R. G. Gallager, *Information Theory and Reliable Communication*. Wiley, 1968.

[10] ISO Draft International Standard 11544, *Coded Representation of Picture and Audio Information—Progressive Bi-Level Image Compression*, 1992.

[11] T. C. Bell, J. G. Cleary, and I. H. Witten, *Text Compression*. Prentice-Hall, 1990.

[12] N. Abramson, *Information Theory and Coding*. McGraw-Hill, 1963, pp. 61–62.

[13] C. E. Shannon, "A mathematical theory of communication," *Bell Syst. Tech. J.*, vol. 27, pp. 379–423 and 623–656, 1948.

[14] I. H. Witten, R. M. Neal, and J. G. Cleary, "Arithmetic coding for data compression," *Commun. ACM*, vol. 30, pp. 520–540, June 1987.

[15] F. Rubin, "Arithmetic stream coding using fixed precision registers," *IEEE Trans. Inform. Theory*, vol. 25, pp. 672–675, Nov. 1979.

[16] R. Pasco, "Source coding algorithms for fast data compression," Ph.D. thesis, Dept. of Electrical Eng., Stanford University, 1976.

[17] J. J. Rissanen, "Generalized Kraft inequality and arithmetic coding," *IBM J. Res. Develop.*, vol. 20, 1976.

[18] C. Chamzas and D. Duttweiler, "Probability estimation in arithmetic and adaptive-Huffman entropy coders," *IEEE Trans. Image Process*, vol.4, pp. 237–246, Mar. 1995.

Chapter 5

Motion Estimation for Image Sequence Compression*

H.-M. Hang
Department of Electronics Engineering and
Microelectronics and Information Systems Research Center
National Chiao Tung University
Hsinchu, Taiwan, Republic of China

and

Y.-M. Chou
Department of Electronics Engineering
National Chiao Tung University
Hsinchu, Taiwan, Republic of China

This chapter provides a comprehensive review of the motion estimation techniques that are pertinent to image sequence compression.

We begin with a general description and analysis of the motion estimation–compensation problem. Then, three popular groups of motion estimation methods are presented: (1) block matching methods, (2) differential (gradient) methods, and (3) Fourier methods. In addition to their basic operation, issues discussed are their extensions, their performance limit, their relationships with each other, and other advantages or disadvantages of these methods. A brief summary on some related topics is included at the end.

---

*This work was supported in part by the NSC grant 83-0408-E009012.

5.1 Introduction

Motion information extraction is fundamental and essential for many image processing applications such as computer vision, target tracking, industrial monitoring, and image sequence compression. Our focus in this chapter is on the motion estimation techniques that are suitable for video compression purposes. Compared with other motion estimation applications such as moving target recognition, motion estimation for image coding has some specific restrictions and requirements. For instance, because our goal is to reduce the total transmission bit rate for reconstructing images at the receiver, the motion information should occupy only a small amount of the transmission bandwidth additional to the picture contents information. Also, we may not care about the *true* motion parameters as long as the motion parameters we obtain can effectively reduce the total bit rate. Besides, the reconstructed images at the receiving end are often distorted. Therefore, if the reconstructed images are used for estimating motion information, a rather strong noise component cannot be neglected. Hence, motion estimation should be done at the transmitter and thus become a component in the total bit rate. For practical purposes, hardware complexity (now often measured in terms of VLSI design and manufacturing cost) is also an important factor in choosing a motion estimation scheme for real-time image coding systems.

Motion estimation techniques for image coding have been explored by many researchers over the past 25 years [1, 2]. These techniques can be classified, roughly, into three groups: block matching methods, differential (gradient) methods, and Fourier methods [3]. This chapter is organized as follows. In Section 5.2, we first give a general introduction to the motion estimation problem for image sequence coding. Also included in this section is a brief review on the performance limit of motion compensation in bit rate reduction. Block-based motion compensation is very popular and is generally considered robust and effective for image coding. Section 5.3 describes block matching schemes that often work in conjunction with block-based motion compensation. Fast search algorithms that reduce computational complexity of the block matching method are also summarized in Section 5.3. Section 5.4 describes another popular approach to estimating motion vectors; it is often called the *differential* method or spatio–temporal *gradient* method. The optical flow method [4] in computer vision is very similar to this differential method although these two schemes were originally derived from different viewpoints. The third group of motion estimation methods, the Fourier method, is described in Section 5.5. Although the initial concept of Fourier methods was suggested 20 years ago, it did not receive much attention for the past two decades. But this approach still provides us insights in understanding the underlying principle of motion estimation algorithms and can be used as a tool to analyze the limits of motion estimation. There are many other related topics, but due to space limitations we can only mention a few of them in Section 5.6, such as motion-compensated interpolation, 3-D motion parameter estimation, and hardware implementation.

5.2 Motion Estimation and Compensation

In the rest of this chapter we assume that the image sequences under consideration contain moving objects whose shape does not change much, at least over a short period of time. Motion estimation is thus the operation that estimates the motion parameters of moving objects in the image sequence. Frequently, the motion parameters we look for are the displacement vectors of pels between two picture frames. On the other hand, motion compensation is the operation of predicting a picture, or portion thereof, based on displaced pels of a previously transmitted frame in an image sequence.

5.2.1 Motion Estimation

In the physical world, objects are moving in the fourth-dimensional (4-D) spatio–temporal domain—three spatial coordinates and one temporal coordinate. However, images taken by a single camera are the projections of these 3-D spatial objects onto the 2-D image plane as illustrated in Fig. 5.1. A 3-D point $P(x, y, z)$ is thus represented by a 2-D point $P'(X, Y)$ on the image plane. Hence, if pels on the image plane (the outputs from an ordinary camera) are the only data we can collect, the information we have is a 3-D cube—two spatial coordinates and one temporal coordinate—as shown in Fig. 5.2. The goal of motion estimation is to extract the motion information of objects (pels) from this 3-D spatio–temporal cube. In reality, video taken by an ordinary camera is first sampled along the temporal axis. Typically, the temporal sampling rate (frame rate) is 24, 25, or 29.97

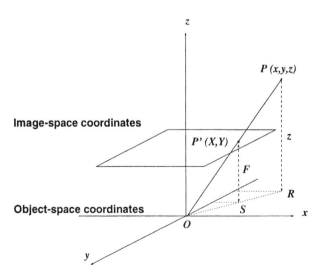

Figure 5.1   Perspective projection geometry of image formulation.

Figure 5.2 A 3-D spatio–temporal cube.

frames (pictures) per second. Spatially, every frame is sampled vertically into a number of horizontal lines. For example, the *active* portion of an NTSC TV frame (i.e., the portion containing the image) has about 483 lines. In the case of the digitized images that are almost exclusively used in this chapter, a line is sampled into a number of pels. According to the CCIR 601 standard, the active portion of an NTSC TV line consists of 720 pels.

Practically, to reduce computation and storage complexity, motion parameters of objects in a picture are estimated based on two or three nearby frames. Most of the motion estimation algorithms assume the following conditions: (1) objects are rigid bodies, hence, object deformation can be neglected for at least a few nearby frames; (2) objects move only in translational movement for, at least, a few frames; (3) illumination is spatially and temporally uniform, hence, the observed object intensities are unchanged under movement; (4) occlusion of one object by another and uncovered background are neglected [2]. Often, practical motion estimation algorithms can tolerate a small amount of inexact matches to these assumptions. Specifically, several algorithms have been invented that somewhat relieve some of the conditions.

The motion estimation problem, in fact, consists of two related sub-problems: (1) identify the *moving object* boundaries, motion segmentation, and (2) estimate the motion parameters of each moving object, motion estimation in strict sense. In our use, a *moving object* is a group of contiguous pels that share the same set of motion parameters. Hence, it does not necessarily match the ordinary meaning of object. For example, in a videophone scene, the still background may include wall, bookshelf, decorations, etc. As long as these items are stationary (sharing the same zero motion vector), they can be considered as a single object in the context of motion estimation and compensation.

The smallest object may contain only a single pel. In the case of pel-recursive estimation (described in Section 5.4), it appears that we can calculate the motion vector for every pel; however, either a minimum set of 2 pels of data or additional constraint(s) have to be used in the process of estimating the 2-D motion vector. One difficulty we encounter in using small objects (or evaluation windows) is the ambiguity problem—similar objects (image patterns) may appear at multiple locations inside a picture and may lead to incorrect displacement vectors. Also, statistically, estimates based on a small set of data are more vulnerable to random noise than those based on a larger set of data. On the other hand, if a large number of pels are treated as a single unit for estimating their motion parameters, we must first know precisely the moving object boundaries. Otherwise, we encounter the accuracy problem—pels inside an object or evaluation window do not share the same motion parameters and, therefore, the estimated motion parameters are not accurate for some or all pels in it. Alternatively, the criterion of grouping pels into moving objects, no matter which scheme is in use, must be consistent with the motion information of every pel. Hence, we are running into the chicken and egg dilemma: accurate motion information for every pel is necessary for precise moving object segmentation, and precise object boundaries are necessary for computing accurate motion parameters.

There exist practical solutions to circumvent the aforementioned motion segmentation and motion estimation dilemma. One solution is partitioning images into regular, nonoverlapped blocks, assuming that moving objects can be approximated reasonably well by regular shaped blocks. Then, a single displacement vector is estimated for the entire image block under the assumption that all the pels in the block share the same displacement vector. This assumption may not always be true because an image block may contain more than one moving object. In image sequence coding, however, prediction errors due to imperfect motion compensation are coded and transmitted. Hence, because of its simplicity and small overhead this block-based motion estimation–compensation method is widely adopted in real video coding systems. The other extreme is the pel-based method that treats every single pel as the basic motion estimation unit. Its motion vector is computed over a small neighborhood surrounding the evaluated pel. No explicit object boundaries are assumed in this approach; however, it suffers from the aforementioned ambiguity problem and, in addition, because of the small data size, it suffers from the noise problem. Detailed analysis of the above problems in various estimation schemes will be presented in the following sections.

5.2.2 Motion-Compensated Coding

The explicit use of motion compensation to improve video compression efficiency can be dated back to the late 1960s, [5] (patent filed in 1969) and [6] (Picture Coding Symposium held in 1969). A number of coding structures using motion-compensated prediction have been proposed since then [1, 2, 7–10]. Recently, this technique was adopted by the international video transmission standards committee (see Chapter 11). The effectiveness of motion compensation in coding is now being well-recognized.

Although detailed implementation of motion-compensated systems varied significantly over the past 25 years, the general concept remains unchanged. Haskell and Limb stated in [5]:

> In an encoding system for use with video signals, the velocity of a subject between two frames is estimated and used to predict the location of the subject in a succeeding frame. Differential encoding between this prediction and the actual succeeding frame are used to update the prediction at the receiver. As only the velocity and the updating difference information need be transmitted, this provides a reduction in the communication channel capacity required for video transmission.

The general structure of a motion-compensated codec is depicted in Fig. 5.3.

For the moving areas of an image, there are roughly three processing stages at the encoder: motion parameter estimation, moving pels prediction (motion compensation), and prediction-errors/original-pels compression. Only two stages are

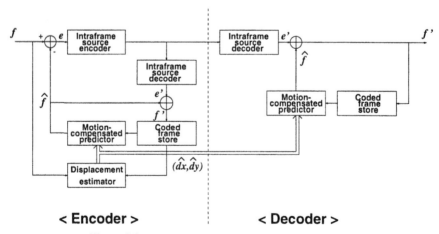

Figure 5.3   A general motion-compensated coding structure.

needed at the decoder: prediction-errors/original-pels decompression, and motion compensation. In these processes, the motion compensators at the encoder and decoder are similar. They often consist of two operations: access the *previous-frame* image data according to the estimated displacement vectors, and construct the predicted pels by passing the previous-frame image through the *prediction filter*. The data compression and decompression processes are inverse operations of each other. The most popular image data compression procedure uses discrete transforms, coefficient selection–quantization, and variable-word-length coding (see Chapters 7 and 11).

How much does motion compensation offer in bit-rate reduction? The answer is clearly data (image) dependent and codec (compression algorithm) dependent. Under certain general assumptions on the statistical characteristics of images and codecs, we could possibly derive an average performance of the ideal motion compensator for image sequence coding [11, 12]. The theoretical analysis presented in [11, 12] is briefly reviewed here. However, we may want to keep in mind that the following derivations are based on somewhat simplified and idealized assumptions of data and algorithms. The real image data and processing algorithms may differ from the ones given here.

Let $f(x, y, t)$ represent the spatio–temporal 3-D video signal (intensity, brightness) at spatial location (x, y) and time t. (Color video has three separate components that can often be treated independently.) The reconstructed video signal is denoted by $f'(x, y, t)$. The difference between the original video and the coded video is modeled as white Gaussian noise, $v(x, y, t) = f(x, y, t) - f'(x, y, t)$, independent of $f(\cdot)$. We further assume that $f(x, y, t)$ is a displaced copy of the previous-frame pel at $(x - d_x, y - d_y, t - \Delta t)$; i.e.,

$$f(x, y, t) = f(x - d_x, y - d_y, t - \Delta t).$$

In other words, the pels under consideration satisfy the aforementioned conditions—rigid body, translational movement, uniform lighting, etc.

Suppose the estimated displacement is $(\hat{d}_x, \hat{d}_y)$, and it differs from the true displacement by $(\Delta d_x, \Delta d_y) \equiv (d_x, d_y) - (\hat{d}_x, \hat{d}_y)$. A time and space invariant filter $h(x, y)$ is used to generate the motion-compensated prediction:

$$\hat{f}(x, y, t) = h(x, y) \otimes f'(x - \hat{d}_x, y - \hat{d}_y, t - \Delta t),$$

where $\otimes$ denotes spatial convolution and $h(x, y)$ is the spatial impulse response of the prediction filter. The prediction error is defined as $e(x, y, t) = f(x, y, t) - \hat{f}(x, y, t)$. Although our ultimate goal is to reduce the coded bits, according to rate–distortion theory, for a Gaussian independent source this goal leads to minimizing the variance of $e(\cdot)$, σ_e^2. Using the coding structure in Fig. 5.3, we can also derive the optimum prediction filter that achieves the minimum error variance.

A closed-form solution to the previous optimization problem can be obtained in the frequency domain. Let $\Phi_e(\omega_x, \omega_y)$ be the power spectral density of $e(x, y, t)$ at time t. Similarly, $\Phi_f(\omega_x, \omega_y)$ and $\Phi_v(\omega_x, \omega_y)$ are the power spectral densities of $f(x, y, t)$ and $v(x, y, t)$, respectively. Under the assumption that the coding error $v(\cdot)$, the image signal $f(\cdot)$, and the displacement error $(\Delta d_x, \Delta d_y)$ are jointly statistically independent, we can obtain [11, 12]

$$\Phi_e(\omega_x, \omega_y) = \Phi_f(\omega_x, \omega_y) \left(1 + |H(\omega_x, \omega_y)|^2 - 2\mathrm{Re}\{H(\omega_x, \omega_y)P(\omega_x, \omega_y)\}\right)$$
$$+ \Phi_v(\omega_x, \omega_y)|H(\omega_x, \omega_y)|^2, \tag{5.1}$$

where Re{.} denotes the real part of a complex number, $H(\omega_x, \omega_y)$ is the 2-D Fourier transform of $h(\cdot)$, and $P(\omega_x, \omega_y)$ is the 2-D Fourier transform of the continuous probability density function (pdf) of the displacement error $(\Delta d_x, \Delta d_y)$. By differentiating (5.1) with respect to $H(\omega_x, \omega_y)$, it can be shown that the prediction error variance is minimized at each frequency if $H(\omega_x, \omega_y)$ is the Wiener filter with frequency response

$$H(\omega_x, \omega_y) = P^*(\omega_x, \omega_y) \frac{\Phi_f(\omega_x, \omega_y)}{\Phi_f(\omega_x, \omega_y) + \Phi_v(\omega_x, \omega_y)}, \tag{5.2}$$

where the superscript $*$ is used to denote complex conjugate. The Wiener filter in (5.2) can be interpreted as having two components: one for reducing the coding noise $(v(\cdot))$ effect, and the other for reducing errors in the displacement estimate. To reduce coding errors, $H(\cdot)$ picks up the signal components at the frequencies where the signal power is stronger than the coding error power. Furthermore, the shape of $H(\cdot)$ also depends on the pdf of the displacement errors. A high uncertainty of displacement errors leads to a narrowband Wiener filter, which blurs the estimates.

The preceding analysis can also be employed to evaluate the performance of fractional-pel motion compensation. The exact shape of $H(\cdot)$ and the minimum prediction error variance σ_e^2 depend on both the signal characteristics and the motion estimation algorithm. Explicit results can be obtained either by assuming the statistical distributions of $f(\cdot)$, $v(\cdot)$, and $p(\cdot)$ or by conducting experiments on test images using a prechosen coding scheme. Girod [11] reported that, assuming some idealized models for $f(\cdot)$, $v(\cdot)$, and $p(\cdot)$, a motion-compensated interframe codec with the optimal Weiner filter saves about 0.8 bit/sample when compared with the 2-D intraframe codec at bit rates above 1.5 bits/sample. In terms of SNR, this is translated to a factor of about 5 dB. Recent reports on MPEG video coding (for example, [13, 14]) indicate that at 3–9 Mbits/sec, the bit rate for predictive-coded frames is less than one third the rate needed for intracoded frames. Girod [12] also reported that, based on experiments performed on several test image sequences, the use of a Wiener filter with displacement vectors of higher accuracy offers a reduction of σ_e^2 from 0.7 dB to 5.2 dB. Although it is picture dependent, $\frac{1}{2}$-pel to $\frac{1}{4}$-pel accuracy seems to be beneficial. It was also concluded that the use of a Wiener filter becomes more critical in the cases that have low coding noise but less accurate motion compensation or high coding noise but accurate motion compensation.

5.3 Block Matching Method

Jain and Jain [15] suggested an interframe coding structure using the block matching motion estimation method and proposed a fast search algorithm to reduce computation. A number of papers have been published since then improving or extending their method. There have gone roughly two directions. One direction was to reduce the computational load in calculating the motion vectors; the other was to increase the motion vector accuracy.

5.3.1 Basic Concept

Block matching is a correlation technique that searches for the best match between the current image block and candidates in a confined area of the previous frame. Figure 5.4 illustrates the basic operation of this method. In a typical use of this method, images are partitioned into nonoverlapped rectangular blocks. Each block is viewed as an independent object and it is assumed that the motion of pels within the same block is uniform. The block matching method is essentially an object recognition approach. The displacement (motion) vector is the by-product when the new location of the object (block) is identified.

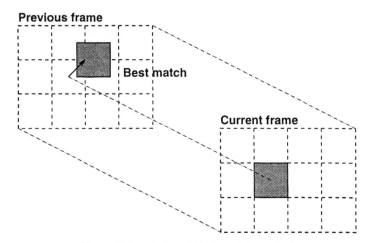

Figure 5.4 Principle of block matching method.

The size of the block affects the performance of motion estimation. Small block sizes afford good approximation to the natural object boundaries; they also provide good approximation to real motion, which is now approximated by piecewise translational movement. However, small block sizes produce a large amount of raw motion information, which increases the number of transmission bits or the required data compression complexity to condense this motion information. From a performance point of view, small blocks also suffer from the object (block) ambiguity problem and the random noise problem. Large blocks, on the other hand, may produce less accurate motion vectors since a large block may likely contain pels moving at different speeds and directions. For typical images used for entertainment and teleconferencing, overall size ranges from 240 lines by 352 pels to 1080 lines by 1920 pels (HDTV). Block sizes of 8×8 or 16×16 are generally considered adequate for these applications. The international video transmission standards, H.261, MPEG1, and MPEG2, all adopt the block size of 16×16.

The basic operation of block matching is picking up a candidate block and calculating the matching function (usually a nonnegative function of the intensity differences) between the candidate and the current block. This operation is repeated until all the candidates have gone through and then the best matched candidate is identified. The location of the best matched candidate becomes the estimated displacement vector. Several parameters are involved in the above searching process: (1) the number of candidate blocks, *search points*; (2) the matching function; and (3) the search order of candidates. All of them could have an impact on the final result.

We often first decide on the maximum range of motion vectors. This motion vector range, the *search range*, is chosen either by experiment or due to hardware constraints. Assume that the size of the image block is $N_1 \times N_2$ and that the maximum horizontal and vertical displacements are less than d_{max_x} and d_{max_y}, respectively. For the moment, we consider only integer-valued motion vectors. Except for the blocks on the picture boundaries, the size of the search region SR is $(N_1 + 2d_{max\_x})(N_2 + 2d_{max\_y})$, and therefore, the number of possible search points equals to $(2d_{max\_x} + 1)(2d_{max\_y} + 1)$ as shown by Fig. 5.5. The *exhaustive search* calculates the matching function of every candidate in the search region. Its computational load is thus proportional to the product of d_{max_x} and d_{max_y}. When the search range becomes fairly large, as in the cases of large picture sizes or interpolative coding (e.g., MPEG), both the computational complexity and the data input–output bandwidth grow very rapidly.

5.3.2 Matching Function

It is necessary to choose a proper matching function in the process of searching for the optimal point. The selection of the matching function has a direct impact on the computational complexity and the displacement vector accuracy. Let (d_1, d_2)

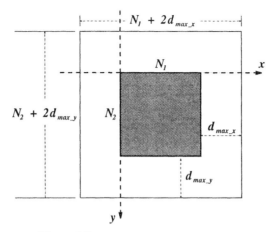

Figure 5.5   Search region in block matching.

represent a motion vector candidate inside the search region and $f(n_1, n_2, t)$ be the digitized image intensity at the integer-valued 2-D image coordinate (n_1, n_2) of the tth frame. Several popular matching functions that appear frequently in the literature follow.

1. Normalized cross-correlation function (NCF):

$$\text{NCF}(d_1, d_2) = \frac{\sum \sum f(n_1, n_2, t) f(n_1 - d_1, n_2 - d_2, t - 1)}{[\sum \sum f^2(n_1, n_2, t)]^{\frac{1}{2}} [\sum \sum f^2(n_1 - d_1, n_2 - d_2, t - 1)]^{\frac{1}{2}}}$$

2. Mean squared error (MSE):

$$\text{mse}(d_1, d_2) = \frac{1}{N_1 N_2} \sum_{n_1=0}^{N_1-1} \sum_{n_2=0}^{N_2-1} [f(n_1, n_2, t) - f(n_1 - d_1, n_2 - d_2, t - 1)]^2$$

3. Mean absolute difference (MAD):

$$\text{mad}(d_1, d_2) = \frac{1}{N_1 N_2} \sum_{n_1=0}^{N_1-1} \sum_{n_2=0}^{N_2-1} |f(n_1, n_2, t) - f(n_1 - d_1, n_2 - d_2, t - 1)|$$

4. Number of thresholded differences (NTD) [16]:

$$\text{NTD}(d_1, d_2) = \sum_{n_1=0}^{N_1-1} \sum_{n_2=0}^{N_2-1} N[f(n_1, n_2, t), f(n_1 - d_1, n_2 - d_2, t - 1)]$$

where

$$N[\alpha, \beta] = \begin{cases} 1 \text{ if } |\alpha - \beta| > T_0 \\ 0 \text{ if } |\alpha - \beta| \leq T_0 \end{cases}$$

is the counting function with threshold T_0. The absolute difference operator inside the counting function can be replaced by the squared difference operator or any other adequate threshold detector.

To estimate the motion vector, we normally maximize the value of NCF or minimize the values of the other three functions. In detection theory, if the total noise, a combination of coding error and the other factors violating our motion assumptions, can be modeled as white Gaussian, then, NCF is the optimal matching criterion. However, the white Gaussian noise assumption is not completely valid for real images. In addition, the computation requirement of NCF is enormous. Hence, the other matching functions are regarded as more practical, and they perform almost equally well for real images. Notably NTD can be adjusted to match the subjective thresholding characteristics of the human visual system. Nevertheless, mad seems to be the most popular choice in designing practical image coding systems because of its good performance and relatively simple hardware structure.

5.3.3 Fast Search Algorithms

An exhaustive search examines every search point inside the search region and thus gives the best possible match; however, a large amount of computation is required. Several fast algorithms have thus been invented to save computation at the price of slightly impaired performance. The basic principle behind these fast algorithms is breaking up the search process into a few sequential steps and choosing the next-step search direction based on the current-step result. At each step, only a small number of search points are calculated. Therefore, the total number of search points is significantly reduced. However, because the steps are performed in sequential order, an incorrect initial search direction may lead to a less favorable result. Also, the sequential search order poses a constraint on the available parallel processing structures.

Normally, a fast search algorithm starts with a rough search, computing a set of scattered search points. The distance between two nearby search points is called (search) *step size*. After the current step is completed, it then moves to the most promising search point and does another search with probably a smaller step size. This procedure is repeated until it cannot move further and the (local) optimum is reached. If the matching function is monotonic along any direction away from the optimal point, a well-designed fast algorithm can then be guaranteed to converge to the global optimal point [15, 17]. But in reality the image signal is not a simple Markov process, and it contains coding and measurement noises; therefore, the monotonic matching function assumption is often not valid and consequently fast search algorithms are often suboptimal.

The first fast search algorithm that follows is the 2D-log search scheme proposed by Jain and Jain [15]. It is an extension of the 1-D binary logarithm search. The searching procedure is illustrated by the example in Fig. 5.6. It starts from the center of the search region—the zero displacement. Each step consists of calculating five search points, as shown in Fig. 5.6. One of them is the center of a diamond-shaped search area and the other four points are on the boundaries of the search area, n pels (*step size*) away from the center. This step size is reduced to half of its current value if the best match is located at the center or located on the border of the maximum search region. Otherwise, the search step size remains the same. The search area in the next step is centered at the best matching point as a result of the current step. When the step size is reduced to 1, we reach the final step. Nine search points in the 3×3 area surrounding the last best matching point are compared. The location of the best match determines the motion vector. Assuming the maximum search range is $d_{max} \geq 2$ in both horizontal and vertical directions, the initial step size n would be equal to $\max(2, 2^{m-1})$, where $m = \lfloor \log_2 d_{max} \rfloor$ and $\lfloor z \rfloor$ denotes the largest integer less than or equal to z. For the example shown in Fig. 5.6 ($d_{max} = 7$), 6 steps and 22 calculated search points are required to reach the final destination at $(3, 7)$. This total computational cost is much smaller than the exhaustive search that requires 225 search points in this case.

Another popular fast search algorithm is the so-called three step search proposed by Koga *et al.* [18]. In their example and the example given in Fig. 5.7, the search

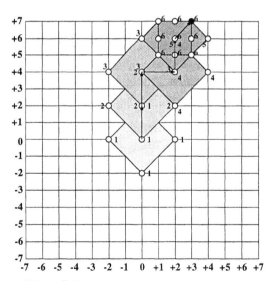

Figure 5.6   Illustration of 2D-log search procedure.

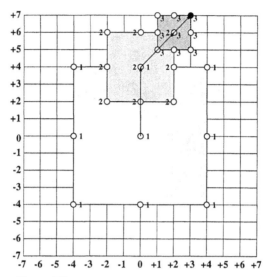

Figure 5.7   Illustration of three step search procedure.

starts with a step size equal to or slightly larger than half of the maximum search range. In each step, nine search points are compared. They consists of the central point of the square search area and eight search points located on the search area boundaries as shown in Fig. 5.7. The step size is reduced by half after each step, and the search ends with step size of 1 pel. Similar to that of the 2D-log search, this search proceeds by moving the search area center to the best matching point in the previous step. Koga *et al.* [18] did not give the general form of this search algorithm other than the $d_{max} = 6$ example in their paper, but following the described principle, we need three search steps for a maximum search range between 4 to 7 pels and four steps for a maximum range between 8 to 15 pels. For searches with 3 steps, 25 search points have to be calculated.

There are several other fast search algorithms. Limited by space, we briefly introduce only some of them. Kappagantula and Rao [19] developed a modified search algorithm that combines the preceding two schemes. In addition, a threshold function is used to terminate the searching process without reaching the final step. This is based on the observation that, as long as the matching error is less than a small threshold, the resultant motion vector would be acceptable. The one-at-a-time search suggested by Srinivasan and Rao [20] is a special case of the conjugate direction search. The basic concept is to separate a 2-D search problem into two 1-D problems. Their algorithm looks for the best matching point in one direction (axis) first, and then looks in the other direction. In each step, only two 1-D neighboring points are checked and compared with the central one. Therefore, an incorrect estimate at the beginning or middle of this process may lead to a totally different final destination. Although it often has the fewest calculating points, this one-at-a-time search is generally considered to have the least favorable performance among all the fast algorithms described in this section. Puri, Hang, and Schilling [17] described an orthogonal search algorithm of which the primary goal was to minimize the total number of search points in the worst case. The search procedure consists of horizontal search steps and vertical search steps executed alternately. It begins with three aligned but scattered search points with roughly a step size of half of the maximum search range. At the next step, centered at the previously chosen matching point, the search point pattern is altered to the perpendicular direction. The step size is reduced by half after each pair of horizontal and vertical steps. This scheme seems to provide reasonable performance with the least search points in general. In addition, a few other fast search algorithms were developed recently (for example, [21–23]).

In comparing the aforementioned fast search schemes, the attributes we look for are their abilities in entropy and prediction error reduction, number of search steps, number of search points, and noise immunity. Table 5.1 summarizes the numbers of search points and search steps of these algorithms in the best and the worst cases for $d_{max} = 7$. Fewer total search points in general would mean less computation. Fewer search steps imply a faster convergence to the final result, which also has an influence on VLSI implementation. Figure 5.8 shows the peak signal-to-noise ratio (PSNR) of *Flower Garden*, a panning image sequence (camera does translational motion), using some of the preceding search schemes. Here the block size is 16×16 and the maximum search range d_{max} is 7. The noise in this plot is the motion-compensated prediction error, assuming that the previous frame is perfectly reconstructed. The simple frame difference without motion compensation clearly has the highest mean squared error or the smallest

Table 5.1   Comparison of fast search algorithms ($d_{max} = 7$).

Search algorithm	Number of search points		Number of search steps	
	minimum	maximum	minimum	maximum
Exhaustive search	225	225	1	1
2D-log search	13	26	2	8
Three step search	25	25	3	3
Modified-log search	13	19	3	6
One-at-a-time search	5	17	2	14
Orthogonal search	13	13	6	6

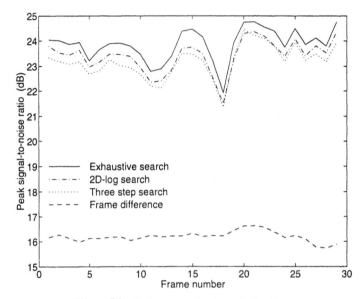

Figure 5.8   Performance of fast search algorithms.

PSNR. For typical motion pictures, block matching algorithms can significantly reduce prediction errors regardless of which search scheme is in use. So far, we have not taken into account the regularity and parallelism that play an important role in VLSI design. In hardware systems, the exhaustive search and the three step search are often favored for their good PSNR performance, their fixed and fewer number of search steps, and their identical operation in every step [24–26].

5.3.4 Variants of Block Matching Algorithms

The fast search algorithms described in the previous section are designed to reduce computation in the process of finding the best match of a block. We could also reduce computation by calculating fewer blocks of an image. In the meanwhile, another problem we would like to solve is the conflict between decreasing object ambiguity and increasing motion vector accuracy. These considerations lead to modifications and extensions of the basic block matching algorithms.

To increase search efficiency, we could place the initial search point at a location predicted from the motion vectors of the spatially or the temporally adjacent blocks [16]. A best match can often be obtained by searching a smaller region surrounding this initial point. However, an incorrect initial search point may lead to an undesirable result. To increase the robustness of this "dependent" search, the zero (stationary) motion vector is always examined. Some computational savings were reported [16].

There are other approaches to reduce computation. For example, we could first separate the moving image blocks from the stationary ones and then conduct block matching only on the moving blocks. This is because a moving or change detector can be implemented with much fewer calculations than a motion estimator. In a typical scene, about half of the picture blocks are not changing between two successive frames. Hence, a change detector that removes the stationary blocks would reduce the average computational complexity [27]. To further reduce computation, we could use only a portion of the pels inside an image block in calculating the matching function. However, the use of a simple subsampling pattern can seriously decrease the accuracy of motion vectors. Specific subsampling patterns were proposed by Liu and Zaccarin [28] to maintain roughly the same level of performance. Another technique proposed in [28] is to perform estimation only on the alternate blocks in an image; the motion vectors of the missing blocks are "interpolated" from the calculated motion vectors. Significant savings in computation are reported at the price of minor performance degradation.

We said at the beginning of this section that block size is an important factor in determining motion estimation performance. To reduce the ambiguity and noise problems caused by small-size blocks and the inaccuracy problem due to large-size blocks, the hierarchical block matching algorithm [29, 30] and the variable-block-size motion estimation algorithm [27] were invented. Their basic principles are similar and can be summarized as follows. A large block size is chosen at the beginning to obtain a rough estimate of the motion vector. Because a large-size image pattern is used in matching, the ambiguity problem—blocks of similar content—can often be eliminated. However, motion vectors estimated from large blocks are not accurate. We then refine the estimated motion vectors by decreasing the block size and the search region. A new search with a smaller block size starts from an initial motion vector that is the best matched motion vector in the previous stage. Because pels in a small block are more likely to share the same motion vector, the reduction of block size typically increases the motion vector accuracy.

In hierarchical block matching [29, 30], the current image frame is partitioned into nonoverlapped small blocks of size, say 12×12. For each partitioned block, large blocks of sizes 28×28 and 64×64 are constructed by taking windows centering at the evaluated 12×12 block. Hence, the large-size blocks overlap with the ones derived from the neighboring 12×12 blocks. The motion estimation process starts with the largest 64×64 blocks and the motion vectors are refined by using the subsequent smaller blocks. To reduce computational load, large image blocks are filtered and subsampled before the block matching process is engaged. A different block formulation is adopted by the variable-block-size motion estimation algorithm [27] because its goal is to produce a more efficient overall coding structure. Image frames are partitioned into nonoverlapped large image blocks. If the motion-compensated estimation error is higher than a threshold, this large block is not well-compensated; therefore, it is further partitioned into, say, four smaller blocks. In searching for the motion vectors of the four small blocks, the large block motion vector is used as the initial search location. This idea has also been included as a part of the Zenith and AT&T HDTV proposal [31]. These schemes and other variants of block matching algorithms improve the estimation performance but also complicate the system configurations and may thus increase implementation cost.

5.4 Differential Method

This approach assumes an analytic relationship between the spatial and temporal changes of image intensities. Early work was done by Limb and Murphy [32, 33]. Figure 5.9 illustrates the basic principle behind their motion estimation scheme. We examine only the horizontal movement for the moment. The shaded area in Fig. 5.9 represents the sum of frame differences (FD(·)) between the previous frame and the current frame. This quantity increases linearly with object speed at low speeds. Therefore, it provides a measure of the displacement vector. The displacement value can thus be approximated by dividing the shaded area by the height of the parallelogram, which is the intensity difference between the left and the right of the moving area (denoted by *MA*). This intensity difference can also be calculated by taking the sum of neighboring pel differences (HD(·) horizontally

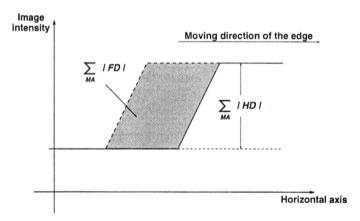

Figure 5.9   Illustration of displacement estimation scheme proposed by Limb and Murphy [33].

and VD($\cdot$) vertically) in the moving area. Extending their idea to the 2-D image plane, we obtain the estimated displacement at time t as ([33] note 2)

$$\hat{d}_1 = -\frac{\sum_{(n_1,n_2) \in MA} \sum \text{FD}(n_1, n_2) \text{ sign}(\text{HD}(n_1, n_2))}{\sum_{(n_1,n_2) \in MA} \sum |\text{HD}(n_1, n_2)|} \qquad (5.3)$$

and

$$\hat{d}_2 = -\frac{\sum_{(n_1,n_2) \in MA} \sum \text{FD}(n_1, n_2) \text{ sign}(\text{VD}(n_1, n_2))}{\sum_{(n_1,n_2) \in MA} \sum |\text{VD}(n_1, n_2)|}, \qquad (5.4)$$

where

$$\text{sign}(z) = \begin{cases} \dfrac{z}{|z|} & \text{if } z \neq 0 \\[2mm] 0 & \text{if } z = 0, \end{cases}$$

$$\text{FD}(n_1, n_2) \equiv f(n_1, n_2, t) - f(n_1, n_2, t - \Delta t) \qquad (5.5)$$

is the frame difference, and

$$\text{HD}(n_1, n_2) \equiv f(n_1, n_2, t) - f(n_1 - 1, n_2, t) \qquad (5.6)$$
$$\text{VD}(n_1, n_2) \equiv f(n_1, n_2, t) - f(n_1, n_2 - 1, t) \qquad (5.7)$$

are the horizontal and vertical differences, respectively. In these equations, $f(\cdot)$ is assumed to have values only on the discrete sampling grid (n_1, n_2, t).

5.4.1 Basic Equations

Unlike the previous heuristic formulation, an analytic, mathematical formula is derived by Cafforio and Rocca [34] using differential operators. Suppose that the continuous spatio–temporal 3-D image intensity function has the property

$$f(x, y, t) = f(x - d_1, y - d_2, t - \Delta t) \qquad (5.8)$$

over the entire moving area. We can rewrite the frame difference signal as

$$\text{FD}(x, y) = f(x, y, t) - f(x + d_1, y + d_2, t). \qquad (5.9)$$

Assuming that the image intensity function $f(\cdot)$ is differentiable at any point (x, y), for small motion vectors $\mathbf{D} = [d_1 \ d_2]^T$, the second term on the right-hand side of (5.9) has a Taylor series expansion. We thus obtain the *basic equation* for displacement estimation:

$$FD(x, y) = -\nabla_x f(x, y, t) d_1 - \nabla_y f(x, y, t) d_2 + o(x, y), \quad (5.10)$$

where $\nabla_x = \frac{\partial}{\partial x}$, $\nabla_y = \frac{\partial}{\partial y}$, and $o(\cdot)$ represents the higher order terms. We further assume that $f(\cdot)$ and $o(\cdot)$ are well-behaved random signals. The random process $f(x, y, t)$ has stochastic differentials with respect to x and y, and $o(\cdot)$ has an even probability density function and is uncorrelated with the differentials of $f(\cdot)$. Then, the optimal linear estimates of d_1 and d_2 are

$$\hat{d}_1 = -\frac{E\{(\nabla_y f)^2\}E\{FD \cdot \nabla_x f\} - E\{\nabla_x f \cdot \nabla_y f\}E\{FD \cdot \nabla_y f\}}{E\{(\nabla_x f)^2\}E\{(\nabla_y f)^2\} - E^2\{\nabla_x f \cdot \nabla_y f\}} \quad (5.11)$$

and

$$\hat{d}_2 = -\frac{E\{(\nabla_x f)^2\}E\{FD \cdot \nabla_y f\} - E\{\nabla_x f \cdot \nabla_y f\}E\{FD \cdot \nabla_x f\}}{E\{(\nabla_x f)^2\}E\{(\nabla_y f)^2\} - E^2\{\nabla_x f \cdot \nabla_y f\}}, \quad (5.12)$$

where $\nabla_x f \equiv \nabla_x f(x, y, t)$ and $\nabla_y f \equiv \nabla_y f(x, y, t)$. The preceding result can be written in a compact form:

$$\hat{D} = -\begin{bmatrix} E\{(\nabla_x f)^2\} & E\{\nabla_x f \cdot \nabla_y f\} \\ E\{\nabla_x f \cdot \nabla_y f\} & E\{(\nabla_y f)^2\} \end{bmatrix}^{-1} \begin{bmatrix} E\{FD \cdot \nabla_x f\} \\ E\{FD \cdot \nabla_y f\} \end{bmatrix}. \quad (5.13)$$

In reality, images are digitized. $FD(x, y)$ is evaluated only on discrete grid, (n_1, n_2). Hence, the two gradients in the above equation are approximated by $HD(\cdot)$ and $VD(\cdot)$ in (5.6) and (5.7), respectively. These two gradients can also be approximated by unbiased but noncausal formulas,

$$HD(n_1, n_2) = \frac{1}{2}[f(n_1 + 1, n_2, t) - f(n_1 - 1, n_2, t)]$$

and

$$VD(n_1, n_2) = \frac{1}{2}[f(n_1, n_2 + 1, t) - f(n_1, n_2 - 1, t)].$$

In addition, the ensemble averages of FD, $\nabla_x f$, and $\nabla_y f$ in (5.13) have to be replaced by their sample averages in the moving area. Suppose that there are M pels in the moving area. Then, (5.13) now becomes

$$\hat{D} = -\begin{bmatrix} \sum HD^2 & \sum(HD \cdot VD) \\ \sum(HD \cdot VD) & \sum VD^2 \end{bmatrix}^{-1} \begin{bmatrix} \sum(FD \cdot HD) \\ \sum(FD \cdot VD) \end{bmatrix}, \quad (5.14)$$

where the summation is taken over the M pels in the moving area with the same displacement.

Compared to the variances of HD(·) and VD(·), the cross term $E\{HD \cdot VD\}$ is usually small enough to ignore. Aiming at real-time implementation, if $\left(\sum(FD \cdot HD)\right)/\left(\sum HD^2\right)$ and similar terms can be approximated by

$$\frac{\sum FD(\cdot)\,\text{sign}(HD(\cdot))}{\sum |HD(\cdot)|}, \quad etc.,$$

then (5.14) can be simplified to

$$\hat{\mathbf{D}} = \begin{bmatrix} \hat{d}_1 \\ \hat{d}_2 \end{bmatrix} = - \begin{bmatrix} \dfrac{\sum\sum FD(n_1,n_2)\,\text{sign}(HD(n_1,n_2))}{\sum\sum |HD(n_1,n_2)|} \\ \dfrac{\sum\sum FD(n_1,n_2)\,\text{sign}(VD(n_1,n_2))}{\sum\sum |VD(n_1,n_2)|} \end{bmatrix}, \qquad (5.15)$$

which is identical to (5.3) and (5.4), and the summation is taken over a uniform moving area.

Equation (5.15) is much easier to compute than (5.14). So far we consider translational motion of only a rigid body. Brofferio and Rocca [35] went further by assuming a correlational model for images and, therefore, obtained explicit expressions for the displacement estimator in terms of correlation parameters. In addition, Cafforio and Rocca [36] explored the noisy cases where the preceding assumptions do not exactly hold.

5.4.2 Recursive Algorithms

Although (5.14) seems to be a good solution for estimating motion vectors, the direct use of it may engender difficulties. In deriving (5.10) using Taylor series expansion, the fundamental assumption is that the motion vector $\mathbf{D}$ is small. As $\mathbf{D}$ increases, the quality of the approximation becomes poor. To overcome this problem, the $\mathbf{D}$ value should be kept small in every use of (5.14). Also, (5.14) should be evaluated over a uniform moving area; however, we do not know the object boundaries (motion segmentation problem) before we calculate the motion vector of every pel. Some kind of bootstrapping procedure is needed. One approach is based on the observation that the spatially or the temporally nearby pels often have similar motion vectors. Hence, we start from a single pel, compute its motion vector, and then use this motion vector as the initial value for computing the motion vector of its neighboring pel. A recursive procedure is thus developed. It is claimed that this recursive procedure saves computation, too [37,38].

The first two well-known recursive algorithms were proposed by Netravali and Robbins [37]. Their first algorithm still needs multiple pels in each recursion. Let $\hat{\mathbf{D}}^{(k)} = [\hat{d}_1^{(k)}\ \hat{d}_2^{(k)}]^T$ be the estimate of the kth recursion. Our goal is to construct an algorithm that calculates an improved estimate $\hat{\mathbf{D}}^{(k+1)}$ based on the *previous* estimate $\hat{\mathbf{D}}^{(k)}$ according to

$$\hat{\mathbf{D}}^{(k+1)} = \hat{\mathbf{D}}^{(k)} + \mathbf{U}^{(k)}, \qquad (5.16)$$

where $\mathbf{U}^{(k)}$ is the update term, which becomes our focus in the following derivation. The physical recursion direction in this equation can be either along the temporal axis (from the previous frame to the current frame) or along the spatial axis (from the left pel to the right pel). Now we define the *displaced frame difference* (DFD) at time t as

$$\mathrm{DFD}(n_1, n_2, \hat{\mathbf{D}}^{(k)}) = f(n_1, n_2, t) - f(n_1 - \hat{d}_1^{(k)}, n_2 - \hat{d}_2^{(k)}, t - \Delta t). \quad (5.17)$$

Again using the translational movement assumption of (5.8) we have

$$
\begin{aligned}
\mathrm{DFD}&(n_1, n_2, \hat{\mathbf{D}}^{(k)}) \\
&= f(n_1 - d_1, n_2 - d_2, t - \Delta t) - f(n_1 - \hat{d}_1^{(k)}, n_2 - \hat{d}_2^{(k)}, t - \Delta t) \\
&= -\nabla_{n_1} f(n_1 - \hat{d}_1^{(k)}, n_2 - \hat{d}_2^{(k)}, t - \Delta t) \cdot (d_1 - \hat{d}_1^{k}) \\
&\quad -\nabla_{n_2} f(n_1 - \hat{d}_1^{(k)}, n_2 - \hat{d}_2^{(k)}, t - \Delta t) \cdot (d_2 - \hat{d}_2^{(k)}) + o(n_1, n_2).
\end{aligned}
\qquad (5.18)
$$

This equation has the same form as (5.10) except that $\hat{\mathbf{D}}$ is replaced by $\hat{\mathbf{D}} - \hat{\mathbf{D}}^{(k)}$. Following the simplification in deriving (5.15), we obtain the following recursive relationship:

$$
\hat{\mathbf{D}}^{(k+1)} = \hat{\mathbf{D}}^{(k)} - \begin{bmatrix} \dfrac{\sum\sum \mathrm{DFD}(n_1, n_2, \hat{\mathbf{D}}^{(k)}) \, \mathrm{sign}(\mathrm{HD}(n_1, n_2))}{\sum\sum \mathrm{HD}(n_1, n_2)} \\[4mm] \dfrac{\sum\sum \mathrm{DFD}(n_1, n_2, \hat{\mathbf{D}}^{(k)}) \, \mathrm{sign}(\mathrm{VD}(n_1, n_2))}{\sum\sum \mathrm{VD}(n_1, n_2)} \end{bmatrix}. \quad (5.19)
$$

Note that the summations are still carried over a uniform moving area. However, since the update term computes only the often small difference between the motion vectors at the kth and the $(k + 1)$th iterations, the result should be more accurate. Tested on two simple moving sequences with only one large moving object, it has been demonstrated by [37] that (5.19) is far more accurate than (5.15), particularly when the displacement is larger than 3 pels between two successive frames. In the preceding case, the moving area can first be identified by a simple motion detector because there is only a single moving object.

Netravali and Robbins [37] also proposed a pel-recursive scheme suggested by adaptive signal processing theory. As the name implies, each estimation iteration can be performed on only a single pel. Suppose that the squared value of motion-compensated frame difference, $\mathrm{DFD}^2(\cdot)$, is convex with respect to $\hat{\mathbf{D}}^{(k)}$ at any pel (n_1, n_2). Several well-known recursive algorithms [38] search for the best $\hat{\mathbf{D}}^{(k)}$ that minimizes an error criterion, which is now $\mathrm{DFD}^2(\cdot)$. The *steepest descent* algorithm is a popular one. As illustrated by Fig. 5.10, starting from an initial

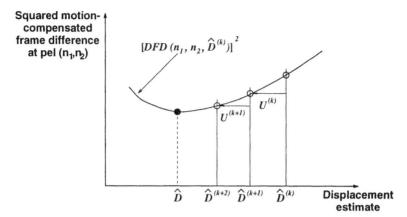

Figure 5.10 Illustration of the steepest descent algorithm.

guess, the estimator computes the best direction that reduces the error criterion fastest and then moves toward the minimum point along that direction. This can be expressed by the mathematical equations

$$\hat{\mathbf{D}}^{(k+1)} = \hat{\mathbf{D}}^{(k)} - \frac{\mu}{2}\nabla[\text{DFD}(n_1, n_2, \hat{\mathbf{D}}^{(k)})]^2$$
$$= \hat{\mathbf{D}}^{(k)} - \mu\,\text{DFD}(n_1, n_2, \hat{\mathbf{D}}^{(k)})\nabla\text{DFD}(n_1, n_2, \hat{\mathbf{D}}^{(k)}), \tag{5.20}$$

where ∇ is the 2-D gradient operator with respect to $\hat{\mathbf{D}}^{(k)}$ and μ is an update constant controlling the speed of convergence. The term $\nabla\text{DFD}(n_1, n_2, \hat{\mathbf{D}}^{(k)})$ can be evaluated by using (5.17) and substituting into (5.20) to obtain

$$\hat{\mathbf{D}}^{(k+1)} = \hat{\mathbf{D}}^{(k)} - \mu\,\text{DFD}(n_1, n_2, \hat{\mathbf{D}}^{(k)}) \cdot \mathbf{G},$$

where

$$\mathbf{G} = \begin{bmatrix} \nabla_{n_1} f(n_1 - \hat{d}_1^{(k)}, n_2 - \hat{d}_2^{(k)}, t - \Delta t) \\ \nabla_{n_2} f(n_1 - \hat{d}_1^{(k)}, n_2 - \hat{d}_2^{(k)}, t - \Delta t) \end{bmatrix}$$

is the gradient vector that can be approximated by $\text{HD}(\cdot)$ and $\text{VD}(\cdot)$. This equation can be computed, in theory, using a single pel, but to reduce noise sensitivity it would perform better when the update term is evaluated over a group of nearby pels [2, 39, 40]; that is,

$$\hat{\mathbf{D}}^{(k+1)} = \hat{\mathbf{D}}^{(k)} - \mu\left[\sum_{(n_1, n_2)\,\in\,MA}\sum W(n_1, n_2)\,\text{DFD}(n_1, n_2, \hat{\mathbf{D}}^{(k)}) \cdot \mathbf{G}\right], \tag{5.21}$$

where $W(\cdot)$ is a set of weighting factors: $W(\cdot) \geq 0$ and $\sum W(\cdot) = 1$.

Assuming the linear model of (5.18) is valid, a more sophisticated multi-pel recursive formula can be derived using least-squares estimation theory:

$$\hat{\mathbf{D}}^{(k+1)} = \hat{\mathbf{D}}^{(k)} - \mu \left[\sum_{(n_1, n_2) \in MA} \sum \mathbf{G} \cdot \mathbf{G}^T \right]^{-1}$$
$$\left[\sum_{(n_1, n_2) \in MA} \sum DFD(n_1, n_2, \hat{\mathbf{D}}^{(k)}) \cdot \mathbf{G} \right]. \tag{5.22}$$

It is reported that the performance is significantly improved when the evaluated area MA includes several pels (compared to the single pel configuration), and the multi-pel least-squares algorithm (5.22) and the weighted gradient algorithm (5.21) perform almost equally well [39, 40]. The effectiveness of differential methods in image sequence coding has been reported by many researchers such as [34–37, 39–41].

One major difference between (5.21) and (5.22) is that the update constant μ in the former is replaced by $\mu \left[\sum \mathbf{G} \cdot \mathbf{G}^T \right]^{-1}$ in the latter to achieve better adaptation to the local image statistics. In fact, there are several other forms of this update constant μ designed to improve convergence speed and estimation accuracy [1].

5.4.3 Further Development and Performance Analysis

The research group at Bell Laboratories made several extensions of the previous recursive algorithms [40, 42–44]. For example, they proposed a motion estimation algorithm executed in the transform domain [42] and algorithms with gain compensation to overcome the illumination variation problem [40, 44]. Prabhu and Netravali [45] also found that the motion vectors estimated based on the luminance component can often be used for the two chrominance components with only negligible increase in the compression bit rate.

There are many variations to the basic differential method. We can sample only a few of them here. One major research direction is to improve the stability and convergence speed of the differential (gradient) method. For example, starting from maximizing the correlation function of the displaced frame difference, Bergmann [46] arrived at an iterative formula similar to the pel-recursive formula with extra second-order terms. This rather complicated formula is reported to be more accurate than Netravali's algorithm for large displacement [46, 47]. Paquin and Dubois [48] proposed an adaptive pel-recursive algorithm that selects the initial displacement value between the temporal and the spatial estimates. Walker and Rao [49] suggested several heuristic modifications to enhance the robustness of the pel-recursive algorithm. The key is to control the *update term* in the recursive formula in accordance with the magnitude of DFD($\cdot$). It is reported that their algorithm converges in two to three iterations using spatial recursion. Aiming at practical implementation, Yamaguchi [50] computed the motion vector inside a

9×9 block using the gradient formula iteratively, and he could assure its stability. In fact, given the linear message model of (5.18) and assuming an independent random high-order term (noise), estimating the displacement vector becomes a standard linear estimation problem. The optimal solution is a Wiener filter type algorithm [51, 52].

We would like to make a few remarks about the limitations of the differential method. First of all, it is derived based on the Taylor series expansion, which is valid only when the displacement is small. This assumption can be invalid in practice when objects move more than several pels between two consecutive image frames. This situation occurs often for the typical TV and moving pictures that are sampled temporally at a rate less than 30 frames per second. The recursive schemes have been invented to relieve this restriction to a certain extent. Second, the gradient operator is sensitive to data noise—a small perturbation of data may lead to a sizable change of the estimated results. Actually, this is not due to the particular formulation of the differential method; it is inherent in the motion estimation problem and in many other early vision problems as well [53]. If we focus only on the differential method, the noise sensitive problem can be reduced by using a larger set of data or smoothness constraints, as will be described in the next section. However, the price paid for these amendments is in motion vector accuracy. These problems are also reflected in the mathematical formulas of the differential method as shown later.

Looking into (5.13), or (5.14), we may discover two cases where the differential method may fail. One case happens if the pels are located in a (spatially) smooth area so that

$$\nabla_x f(\cdot) \approx 0 \quad \text{or} \quad \nabla_y f(\cdot) \approx 0.$$

Then, the inverse matrix in (5.13) becomes singular. The other case occurs when motion is parallel to the edges of image patterns; that is,

$$\mathbf{D}^T \cdot \begin{bmatrix} \nabla_x f(\cdot) \\ \nabla_y f(\cdot) \end{bmatrix} \approx \mathbf{0}.$$

Although both the gradients and the displacement vector may be nonzero, the frame difference in (5.10) is nearly zero. Thus, the estimated displacement vector ($\approx \mathbf{0}$) is different from the true value (nonzero). It can be seen that these two cases are a consequence of the locality nature of the differential method. Hildreth called them *aperture* problems and gave some interesting examples in [54]. These problems may be solved partially by increasing the evaluated area of data, but then again, we face the dilemma of ambiguity versus accuracy. However, for image compression purposes, the goal is to reduce the transmitted bits rather than to find the *true* motion vectors. Therefore, if these situations are detected in video coding, we could simply set the motion vectors to zero without much affecting the coding efficiency.

5.4.4 Optical Flow

Although coming from a different viewpoint, the so-called optical flow (or image flow) approach in computer vision is similar to the differential approach in image compression. It is assumed that the image intensity (brightness) of a moving object remains constant so that [4]

$$f(x, y, t) = f(x + \Delta x, y + \Delta y, t + \Delta t),$$

where Δx, Δy, and Δt are small. Approximating the right-hand side by a Taylor series expansion and eliminating the common $f(\cdot)$, we obtain

$$\nabla_x f(x, y, t)\Delta x + \nabla_y f(x, y, t)\Delta y + \nabla_t f(x, y, t)\Delta t + o(x, y, t) = 0, \quad (5.23)$$

which is called the *optical flow constraint*. Dividing this equation by Δt and letting $[u \; v]^T = \left[\dfrac{\Delta x}{\Delta t} \; \dfrac{\Delta y}{\Delta t} \right]^T$ be the instantaneous velocity in the limit as $\Delta t \to 0$, we have

$$\nabla_x f(x, y, t) u + \nabla_y f(x, y, t) v + \nabla_t f(x, y, t) = 0. \quad (5.24)$$

This is equivalent to the assumption that the total derivative of $f(\cdot)$ over time is zero,

$$\frac{d}{dt} f(x, y, t) = 0,$$

and then it can be expanded in terms of the partial derivatives,

$$\nabla_x f(x, y, t)\frac{dx}{dt} + \nabla_y f(x, y, t)\frac{dy}{dt} + \nabla_t f(x, y, t) = 0. \quad (5.25)$$

Equation (5.24) poses only one constraint on the two unknown parameters u and v for each (x, y) location. One way to obtain a unique solution is to impose an additional smoothness constraint under the assumption that the motion field does not change drastically in a small neighborhood. It turns out that this *smoothness constraint* also reduces the noise effect and the singular matrix inversion problem. The specific smoothness constraint suggested by Horn and Schunck [4] is to minimize the magnitude square of the gradient of optical flow velocity:

$$\mathcal{E}_c^2 = \left(\frac{\partial u}{\partial x} \right)^2 + \left(\frac{\partial u}{\partial y} \right)^2 + \left(\frac{\partial v}{\partial x} \right)^2 + \left(\frac{\partial v}{\partial y} \right)^2.$$

It is also clear that we would like to minimize the error in computing (5.24) using the collected data:

$$\mathcal{E}_b = \nabla_x f(x, y, t) u + \nabla_y f(x, y, t) v + \nabla_t f(x, y, t).$$

Therefore, the total optimization criterion becomes

$$\mathcal{E}^2 = \int \int (\alpha^2 \mathcal{E}_c^2 + \mathcal{E}_b^2) dx dy, \tag{5.26}$$

where α^2 controls the relative cost of deviation from the motion field smoothness constraint and deviation from the optical flow constraint. Using the calculus of variations the authors obtained [4]

$$(\nabla_x f)^2 \cdot u + \nabla_x f \cdot \nabla_y f \cdot v = \alpha^2 \left[\left(\frac{\partial u}{\partial x} \right)^2 + \left(\frac{\partial u}{\partial y} \right)^2 \right] - \nabla_x f \cdot \nabla_t f,$$

$$\nabla_x f \cdot \nabla_y f \cdot u + (\nabla_y f)^2 \cdot v = \alpha^2 \left[\left(\frac{\partial v}{\partial x} \right)^2 + \left(\frac{\partial v}{\partial y} \right)^2 \right] - \nabla_y f \cdot \nabla_t f.$$

$$\tag{5.27}$$

Comparing this result with the gradient formula (5.13), the latter can be viewed as a special case of the former without the smoothness constraint ($\alpha^2 = 0$).

An estimate of $[u \ v]^T$ can be obtained by solving the pair of partial differential equations in (5.27). Using the following approximations to the Laplacians,

$$\left(\frac{\partial u}{\partial x} \right)^2 + \left(\frac{\partial u}{\partial y} \right)^2 = u - u_{ave} \quad \text{and} \quad \left(\frac{\partial v}{\partial x} \right)^2 + \left(\frac{\partial v}{\partial y} \right)^2 = v - v_{ave}.$$

Horn and Schunck [4] derived a closed-form solution:

$$u - u_{ave} = - \frac{\nabla_x f [\nabla_x f \cdot u_{ave} + \nabla_y f \cdot v_{ave} + \nabla_t f]}{\alpha^2 + (\nabla_x f)^2 + (\nabla_y f)^2},$$

$$v - v_{ave} = - \frac{\nabla_y f [\nabla_x f \cdot u_{ave} + \nabla_y f \cdot v_{ave} + \nabla_t f]}{\alpha^2 + (\nabla_x f)^2 + (\nabla_y f)^2}. \tag{5.28}$$

Replacing (u, v) and (u_{ave}, v_{ave}) in the above equation by $(u^{(k+1)}, v^{(k+1)})$ and $(u_{ave}^{(k)}, v_{ave}^{(k)})$, respectively, Horn and Schunck [4] finally obtained an iterative algorithm.

Although the optical flow constraint equation, (5.24), is motivated by an analytic assumption on the image intensity, it is shown that this constraint is also valid at both image intensity discontinuities and motion field discontinuities [55, 56]. However, the smoothness constraint has to be removed or modified to have accurate estimates of the motion field near its discontinuities. As discussed in the previous section, it is difficult to obtain accurate displacement information in nearly flat image areas. It has long been observed that motion information can best be estimated at the "gray value corners"— "cornerlike" image patterns [57]. Several approaches have been proposed to improve motion field estimation around discontinuities. For instance, the constant weighting factor (α^2) in front of the smoothness constraint is replaced by a weighting matrix that depends on the gray value changes [58–60]. Hildreth [54] suggested computing motion vectors only along the 1-D image contours whose components are tangent and perpendicular to the contours. Since motion field discontinuities often appear on object boundaries, an algorithm that consists of two stages: (1) motion boundary detection (segmentation) and (2) motion field estimation, and that iterates between them may be beneficial [61].

Many other refined differential-based motion estimation algorithms aim at performance improvement or simple implementation. Interested readers can find additional references in [62–65]. It may be worth mentioning that, similar to the hierarchical schemes in block matching motion estimation, there are several *multigrid* or *multiscale* optical flow computing schemes [66–68]. The typical approach is creating a hierarchy of images of various resolutions using, say, the Gaussian pyramid. Then, a coarse-to-fine sequential motion estimation process is undertaken. This class of algorithms has the advantages of larger range of motion vectors and less computation, which is similar to the advantages associated with hierarchical block matching.

5.5 Fourier Method

This type of motion estimation method is not as popular as the previous two approaches. This is partially because conventional Fourier-domain schemes often cannot provide very accurate motion information for multiple moving objects of small to medium sizes. However, the Fourier method still has its own distinct advantages and can provide insight on the theoretical limits of motion estimation as shall be described next.

5.5.1 Displacement in the Fourier Domain

It has long been observed that a pure translational motion corresponds to a phase shift in the frequency domain [3, 69]. Again, we assume the image intensity functions of two successive frames, in a uniform moving area, differ only due to a positional shift (d_1, d_2):

$$f(n_1, n_2, t) = f(n_1 - d_1, n_2 - d_2, t - 1). \tag{5.29}$$

Taking the 2-D discrete-time Fourier transform (DTFT) on the spatial variables (n_1, n_2), we obtain

$$F_t(w_1, w_2) = F_{t-1}(w_1, w_2) \exp(-jw_1 d_1 - jw_2 d_2), \tag{5.30}$$

where $F_t(\cdot)$ and $F_{t-1}(\cdot)$ are the 2-D Fourier transformations of the current frame and the previous frame, respectively. In other words, the displacement information is contained in the phase difference. Haskell [69] noticed this relationship but did not propose a step-by-step algorithm to estimate the displacement information from this phase shift. Let

$$F_t(w_1, w_2) = |F_t(w_1, w_2)| \exp(j \cdot \arg[F_t(w_1, w_2)]), \tag{5.31}$$

$$F_{t-1}(w_1, w_2) = |F_{t-1}(w_1, w_2)| \exp(j \cdot \arg[F_{t-1}(w_1, w_2)]). \tag{5.32}$$

Huang and Tsai [3] suggested taking the phase difference between the previous frame's transform and the current frame's transform as a consequence of (5.30),

$$\Delta\phi(\omega_1, \omega_2) = \arg[F_t(\omega_1, \omega_2)] - \arg[F_{t-1}(\omega_1, \omega_2)] = -\omega_1 d_1 - \omega_2 d_2. \tag{5.33}$$

In theory, we need to evaluate this equation at only two independent frequency points, we can then obtain the displacement (d_1, d_2). Practically, the displacement estimation should be calculated using more than a few data points to reduce "noise." Huang and Tsai [3] also pointed out that, in calculating the phase difference, there is an ambiguity of integer multiples of 2π. This ambiguity is one of the problems to be tackled in using (5.33). The preceding *frequency component* formulation has not attracted much attention over the past 15 years. The most well-known Fourier method is the so-called phase correlation.

5.5.2 Phase Correlation Algorithm

The phase correlation algorithm was first proposed to solve the image registration problem [70]. Its aim is to extract displacement information from phase components. Let $C_r(n_1, n_2)$ be the inverse discrete-time Fourier transform (IDTFT) of $\exp(j\Delta\phi(\omega_1, \omega_2))$; thus,

$$\begin{aligned} C_r(n_1, n_2) &= \text{IDTFT}[\exp(j\Delta\phi(\omega_1, \omega_2))] \\ &= \text{IDTFT}[\exp(-jw_1 d_1 - jw_2 d_2)] \\ &= \delta(n_1 - d_1, n_2 - d_2). \end{aligned} \tag{5.34}$$

That is, if noise were not present in the displacement equation, (5.29), the correlation surface $C_r(\cdot)$ would have a distinctive peak at (d_1, d_2) corresponding to the displacement value. On the other hand, $\exp(j\Delta\phi(\omega_1, \omega_2))$ can be calculated using (5.31) and (5.32),

$$\begin{aligned} \exp(j\Delta\phi(\omega_1, \omega_2)) &= \exp(j \cdot \arg[F_t(w_1, w_2)]) \cdot \exp(-j \cdot \arg[F_{t-1}(w_1, w_2)]) \\ &= \frac{F_t(w_1, w_2)}{|F_t(w_1, w_2)|} \cdot \frac{F_{t-1}^*(w_1, w_2)}{|F_{t-1}(w_1, w_2)|}. \end{aligned} \tag{5.35}$$

In other words, $C_r(\cdot)$ represents the correlation of the phase components of the transforms corresponding to the previous and the current frames. Figure 5.11 shows

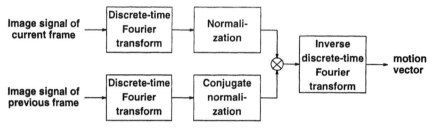

Figure 5.11   Schematic diagram of a phase correlator.

the block diagram of a basic phase correlator. An example of correlation surface of a 32 × 32 window of the *Flower Garden* sequence is shown in Fig. 5.12. Since particularly images are of finite sizes, the Fourier transforms in all the preceding equations, (5.30) to (5.35), can be replaced by the discrete Fourier transforms computed at only the discrete frequencies. Consequently, the correlation operation that takes a large amount of computation in the spatial domain now requires much less computation in the frequency domain (multiplication instead of convolution).

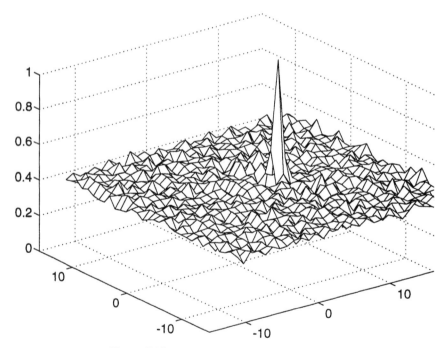

Figure 5.12   Correlation surface of *Flower Garden*.

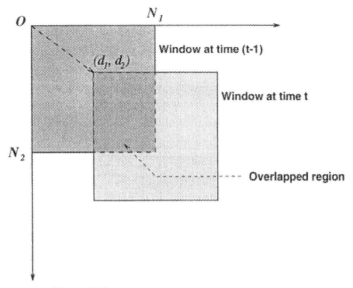

Figure 5.13   Effective window areas in phase correlation.

In addition, there exist fast Fourier transform routines that can compute the forward and the inverse discrete Fourier transforms efficiently.

The phase correlation $C_r(n_1, n_2)$ defined by (5.35) assumes that a segment in frame $(t - 1)$ is a cyclically shifted copy of one in frame t. This is approximately true when the moving object is inside the evaluation window and the background is uniform, which is rarely the case in reality. As depicted in Fig. 5.13 the pels in the nonoverlapped region, which arise due to the motion, can be treated as "noise" in the previous formulation. In the case of large displacement, the nonoverlapped region could be rather large in size, and therefore, the estimate would become inaccurate. Also, if the measurement window contains several objects with different moving vectors, there may appear several blurred peaks. In fact, we may interpret the phase correlation algorithm as a special type of block matching (cross-correlation) method, in which images are "normalized" so that their frequency components have unit amplitude but their original phase values are retained. This normalization procedure acts roughly like a high-pass filter. It amplifies the high frequency components that often contain a significant percentage of measurement noise.

5.5.3 Performance and Improvement

Figure 5.14 is a plot of the PSNR of the *Flower Garden* sequence using the phase correlation method with different window sizes. It is clear that the window size

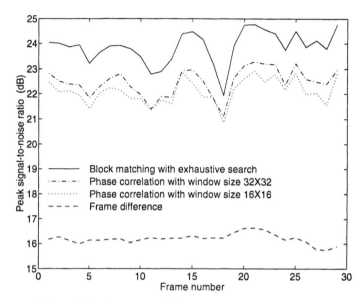

Figure 5.14   Comparison of phase correlation and block matching.

plays an important role. Compared with the exhaustive search block matching scheme (block size of 16 × 16), the phase correlation method does not perform as well. As discussed in the previous section, various "noise" components would degrade the accuracy of a simple phase correlation method.

A few improvements to phase correlation have been invented. Thomas [71] did a rather extensive study on the phase correlation method. He suggested a two-stage process. In the first stage (phase correlation), the input picture is divided into fairly large windows, say, 64 pels square. The phase correlation is performed between the corresponding blocks in two consecutive frames. Then, he searches for *several* dominant peaks on the correlation surface. Their corresponding motion vectors are the candidates for the second stage. In the second stage (vector assignment), the input picture is divided into *smaller* blocks, and for each small block every candidate motion vector is tested by matching the current frame's block to the shifted candidate block in the previous frame [71, 72]. For motion-compensated interpolation applications, the image block in the second stage is chosen to be a single pel in [71, 72]. In addition, the pels near the window borders may have the same motion vector values as the nearby windows. Hence, the correlation peaks of the nearby windows are also included as the candidate motion vectors of the current block. To reduce the ambiguity problem, similar image patterns, Thomas [71] recommended image data be (low-pass) filtered before computing the matching function and large motion vector candidates be penalized in selecting motion vectors. Ziegler [73] proposed a similar algorithm but with overlapped 64×64 windows in the first stage and four 16×16 blocks at the center of the overlapped window in the second stage. Because the vector assignment blocks are located at the center of the correlation window, the window border effect is reduced.

Thomas [71] also suggested multiplying a *windowing (weighting)* function to the input picture data to reduce noise. The effect of this windowing function is similar to the use of windowing in power spectral estimation: it reduces noise but blurs the sharpness of peaks. We could also use a weighting function in the frequency domain before taking the inverse Fourier transform to reduce noise [70]. This is approximately equivalent to performing a smoothing filter on the input pictures [71]. This filter could be signal dependent, and as pointed out in [71], it may be profitable to reduce the weights of the low-frequency components in a phase correlation calculation. Otherwise, noise may dominate those frequency components. If each frequency component is given a weight proportional to its amplitude in the input pictures, the phase correlation process becomes the cross-correlation (block matching) of the original pictures and the sharpness of peaks reduces significantly.

5.5.4 Frequency Components

So far, the phase correlation algorithms we have discussed take all the frequency components into consideration at the same time. To understand the intrinsic properties of motion estimation better we now go back to (5.33) and examine the contribution of individual components in the motion estimation process [74]. It is clear from (5.33) that

$$(\omega_1\ \omega_2) \begin{pmatrix} d_1 \\ d_2 \end{pmatrix} = -\Delta\phi(\omega_1, \omega_2). \qquad (5.36)$$

Since the input pictures are of finite sizes, it is necessary to evaluate only the frequency components at the sampling grid, multiples of $(2\pi/M_1, 2\pi/M_2)$, where M_1 and M_2 are the horizontal and vertical window sizes, respectively; that is,

$$(\omega_1, \omega_2) = (\frac{2\pi}{M_1}k_1, \frac{2\pi}{M_2}k_2), \qquad 0 \le k_1 < M_1, \qquad 0 \le k_2 < M_2.$$

Now, we want to analyze the noise effect. Assuming that the totality of noises from various sources can be modeled as a noise $v(\cdot)$ added to (5.29),

$$f(n_1, n_2, t) = f(n_1 - d_1, n_2 - d_2, t - 1) + v(n_1, n_2), \qquad (5.37)$$

and $v(\cdot)$ can be represented by a Fourier series,

$$v(n_1, n_2) = \sum\sum |V(k_1, k_2)| e^{j\left(\frac{2\pi}{M_1}k_1 n_1 + \frac{2\pi}{M_2}k_2 n_2 + \phi_v(k_1, k_2)\right)}, \tag{5.38}$$

where $\phi_v(k_1, k_2)$ is the phase of $V(k_1, k_2)$. If $f(\cdot, \cdot, t)$ and $f(\cdot, \cdot, t-1)$ are both represented by their Fourier series, (5.37) becomes

$$|F_t(\cdot)| e^{j\left(\frac{2\pi}{M_1}k_1 n_1 + \frac{2\pi}{M_2}k_2 n_2 + \phi_t(\cdot)\right)}$$

$$= |F_{t-1}(\cdot)| e^{j\left(\frac{2\pi}{M_1}k_1 n_1 + \frac{2\pi}{M_2}k_2 n_2 + \phi_{t-1}(\cdot)\right)} + |V(\cdot)| e^{j\left(\frac{2\pi}{M_1}k_1 n_1 + \frac{2\pi}{M_2}k_2 n_2 + \phi_v(\cdot)\right)}.$$

The index (k_1, k_2) is omitted for simplicity in the preceding and following equations. Separating the amplitude and the phase components, we obtain

$$|F_t| = \left[(|F_{t-1}| + |V| \cos(\phi_v - \phi_{t-1}))^2 + (|V| \sin(\phi_v - \phi_{t-1}))^2\right]^{\frac{1}{2}}, \tag{5.39}$$

$$\phi_t = \phi_{t-1} - \left(\frac{2\pi}{M_1}k_1 d_1 + \frac{2\pi}{M_2}k_2 d_2\right)$$

$$+ \arctan\left(\frac{|V| \sin(\phi_v - \phi_{t-1})}{|F_{t-1}| + |V| \cos(\phi_v - \phi_{t-1})}\right). \tag{5.40}$$

The preceding equations seem rather complicated. However, they are identical to the noise analysis in the phase modulation or the frequency modulation systems in communication [75]. For $|F_{t-1}| \gg |V|$, the noise disturbance to the phase information is less than its effect on the original signal. This is the well-known noise-reduction property of continuous phase modulation. Therefore, the displacement estimate based on phase information is less sensitive to noise than that based on the original signal provided that the signal magnitude is much higher than the noise magnitude. However, this noise-reduction situation is reversed when the noise magnitude is close to or higher than the signal magnitude. In this case, the phase information suffers more distortion than the original signal. This is the well-known "threshold effect" in continuous phase modulation [75]. The preceding analysis, therefore, tells us that we should avoid using the phase information at those frequencies for which the signal power is not much higher than the noise power.

On the other hand, our desired information, (d_1, d_2), is scaled by (k_1, k_2) in the phase component. For example, if $k_1 = 0$ and $|F_{t-1}| \gg |V|$ in (5.40), then

$$d_2 = -\frac{\frac{M_2}{2\pi}\Delta\phi}{k_2} + \frac{\frac{M_2}{2\pi}\arctan\left(\frac{|V| \sin(\phi_v)}{|F_{t-1}|}\right)}{k_2}.$$

Since the second (noise) term is divided by k_2, given the same amount of noise, the higher frequency components are more accurate in estimating motion vectors. However, the high-frequency component has short (spatial) cycles and hence it repeats itself after a short shift. This corresponds to the ambiguity problem— the true displacement may be equal to the estimated phase value plus an integer multiple of cycles. Thus far, we have arrived at three conflicting requirements: (1) noise, low-frequency components are favored for their strong power; (2) accuracy, high-frequency components are more accurate because of the division factor; and (3) ambiguity, low-frequency components are less ambiguous due to their long cycles. Consequently, the middle-frequency components seem to provide the most reliable information for motion estimation purposes. Experiments also indicate that bandpass-filtered images could produce more reliable motion estimates. Another approach is the multiresolution hierarchical estimation process. Low-resolution images are used in the initial stages to reduce noise and ambiguity problems, and high-resolution images are used in the later stages to improve accuracy.

5.6 Concluding Remarks

We have briefly described three classes of popular motion estimation algorithms used in video coding. In what follows, we will outline a few other topics that are related to video coding and motion estimation.

One important application of motion estimation is *motion-compensated interpolation*, which can also be used to reduce transmission bit rate. The basic idea is illustrated by Fig. 5.15. The pel values in the current image frame are estimated based on both the previous and future frames. This temporal interpolation process can be performed entirely at the receiver to increase the frame rate [76]. Alternately it can be performed at the encoder, and the interpolation errors would be coded and transmitted to the receiver (MPEG standards, for example). Because for most image frames the interpolation errors are rather small, it has been shown that this motion-compensated interpolative coding can save a considerable number of bits compared to the motion-compensated prediction only coding systems [77].

Some early work on motion-compensated interpolation is summarized by [1]. It was pointed out by Bergmann [1] that there are three major problems in motion-compensated interpolation: accurate motion estimation, object segmentation, and appropriate interpolation filtering. In the case of a receiver performing the temporal interpolation, the missing pictures are constructed relying on only the received frames; therefore, accurate motion estimation is critical in producing good quality pictures. Toward this end, areas covered and not covered by the moving objects in the current frame need to be segmented and processed separately. Without segmentation, a simple averaging of the previous frame pels and the future frame pels

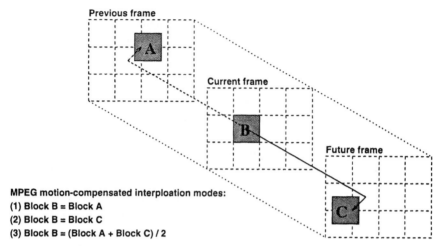

MPEG motion-compensated interploation modes:
(1) Block B = Block A
(2) Block B = Block C
(3) Block B = (Block A + Block C) / 2

Figure 5.15   Illustration of motion-compensated interpolation.

will blur object boundaries. The selection of interpolation filter is influenced by the segmentation scheme. Once we have a rather precise segmentation algorithm, an interpolation filter could adaptively select its inputs among the future frame, the previous frame, and the average of the two (MPEG standards). Better picture quality can thus be achieved.

Many motion-compensated interpolation algorithms have been proposed [29, 71, 72, 78–85]. Typically the newer schemes are improved versions of the earlier proposals with the addition of more sophisticated structures. Their motion estimation portion can be designed based on either block matching method [29, 78], differential method [79, 80], or a phase correlation approach [71, 72, 81]. Some schemes do not perform explicit object segmentation. The interpolated pels are assigned according to their associated motion vectors and their neighboring motion vectors [29, 79, 83]. Several schemes have an explicit object segmentation stage to separate the foreground (moving objects) from the background [78, 82, 84, 85]. The latest development includes global motion parameter compensation such as for camera zoom, pan, or rotation [85, 86].

So far, we have only considered motion estimation methods used by *waveform* coding techniques; i.e., those that compress and reconstruct signal waveforms without explicitly analyzing picture content. Waveform coding schemes are rather general and robust. Hence, they have been adopted as international video standards. On the other hand, there exists a separate class of video coding schemes in which *objects*, rather than waveforms, are the targets to be compressed. These schemes transmit a set of condensed parameters for each object. Although complicated, this class of coding algorithms can potentially offer a very high compression ratio.

Because of advances in high-speed computing, 3-D modeling, and 3-D motion estimation, research results on this subject have started to emerge. The recent studies split roughly into two directions. One is *model-based* coding (Chapter 10) and the other, *object-oriented* coding.

One essential element in an object-oriented coder is to estimate the 3-D motion of the moving objects [87–89]. Two types of 3-D motion models have been used. In the first case, eight mapping parameters are used to describe the 3-D motion of each 2-D rigid object [87, 90, 91]. In the second case, three translation motion parameters and three rotary motion parameters are used to describe the 3-D motion of a 3-D object [92]. A general 3-D motion estimation procedure is composed of two stages: (1) find the (feature) corresponding points between two pictures (sampled at different times or viewed from two cameras), and (2) calculate the motion parameters from the locations of the corresponding features. Depending on the type (2-D or 3-D) of features and object models, different sets of motion parameters are derived. Detailed descriptions of these approaches can be found in [93, 94]. This *feature-based* class of motion estimation algorithms is popular in the computer vision and robot vision disciplines. Since the motion vectors are estimated only at sparsely located image *features*, this approach cannot be used directly together with the conventional waveform coding, where a dense motion field is necessary to generate motion-compensated prediction errors for every pel.

A number of video coding schemes with explicit motion estimation have been proposed in the past 20 years. A partial list of these studies can be found in the survey papers [1, 95], in the monographs [2, 96], and in the paper collections [8–10]. In particular, several systems had been implemented in hardware and were available in the market. Examples of some earlier motion-compensated hardware codecs are [17, 76, 97–100]. They are all block-based motion-compensated coders. Applications of compressed digital video are currently rapidly evolving in the telecommunications, computer, and consumer electronics industries. Because of the advances in VLSI technology, it is predicted that the single video encoder chip containing motion estimation circuits will soon be realized and become very popular (Chapter 15).

References

[1] H. G. Musmann, P. Pirsch, and H.-J. Grallert, "Advances in picture coding." *Proc. IEEE*, vol. 73, pp. 523–548, April 1985.

[2] A. N. Netravali and B. G. Haskell. *Digital Pictures: Representation and Compression.* New York: Plenum Press, 1988, pp. 334–340.

[3] T. S. Huang and R. Y. Tsai, "Image sequence analysis: Motion estimation." In *Image Sequence Analysis*, ed. T. S. Huang. New York: Springer-Verlag, 1981.

[4] B. K. P. Horn and B. G. Schunck, "Determining optical flow." *Artificial Intell.*, vol. 17, pp. 185–204, 1981.

[5] B. G. Haskell and J. O. Limb, "Predictive video encoding using measured subjective velocity." U.S. Patent No. 3,632,865, Jan. 1972.

[6] F. Rocca, "Television bandwidth compression utilizing frame-to-frame correlation and movement compensation." In *Proc. 1969 Symp. Picture Bandwidth Compression*, ed. T. S. Huang and O. J. Tretiak. New York: Gordon and Breach, 1972.

[7] A. N. Netravali and J. O. Limb, "Picture coding: A review." *Proc. IEEE*, vol. 68, pp. 366–406, March 1980.

[8] K. R. Rao and R. Srinivasan, Eds., *Teleconferencing*. New York: Van Nostrand Reinhold, 1985.

[9] A. N. Netravali and B. Prasada, Eds., *Visual Communications Systems*. New York: IEEE Press, 1989.

[10] T. R. Hsing and A. G. Tescher, Eds., *Selected Papers on Visual Communication: Technology and Applications*. Bellingham, Washington: SPIE Optical Eng. Press, 1990.

[11] B. Girod, "The efficiency of motion-compensating prediction for hybrid coding of video sequences." *IEEE J. Selected Areas Commun.*, vol. 5, pp. 1140–1154, Aug. 1987.

[12] B. Girod, "Motion-compensating prediction with fractional-pel accuracy." *IEEE Trans. Commun.*, vol. COM-41, pp. 604–612, April 1993.

[13] ISO/MPEG, *Coding of Moving Pictures and Associated Audio: Part 2 Video, Annex D*, ISO-11172-2, 1992.

[14] A. H. Wong and C.-T. Chen, "A comparison of ISO MPEG1 and MPEG2 video coding standards." In *Proc. SPIE Visual Commun. Image Processing*, Boston, Nov. 1993, vol. 2094, pp. 1436–1448.

[15] J. R. Jain and A. K. Jain, "Displacement measurement and its application in interframe image coding." *IEEE Trans. Commun.*, vol. COM-29, pp. 1799–1808, Dec. 1981.

[16] A. Puri, H.-M. Hang, and D. L. Schilling, "Motion-compensated transform coding based on block motion-tracking algorithm." In *Proc. IEEE Int. Conf. Commun.*, Seattle, June 1987, pp. 5.3.1–5.3.5.

[17] A. Puri, H.-M. Hang, and D. L. Schilling, "An efficient block-matching algorithm for motion-compensated coding." In *Proc. IEEE Int. Conf. Acoust., Speech, Signal Processing*, Dallas, April 1987, pp. 25.4.1–25.4.4.

[18] T. Koga, K. Iinuma, A. Hirano, Y. Iijima, and T. Ishiguro, "Motion-compensated interframe coding for video conferencing." In *Proc. Nat. Telecommun. Conf.*, New Orleans, Nov. 1981, pp. G5.3.1–G5.3.5.

[19] S. Kappagantula and K. R. Rao, "Motion compensated predictive interframe coding." *IEEE Trans. Commun.*, vol. COM-33, pp. 1011–1015, Sept. 1985.

[20] R. Srinivasan and K. R. Rao, "Predictive coding based on efficient motion estimation." *IEEE Trans. Commun.*, vol. COM-33, pp. 888–896, Sept. 1985.

[21] H. Gharavi, "The cross-search algorithm for motion estimation." *IEEE Trans. Commun.*, vol. COM-38, pp. 950–953, July 1990.

[22] L.-G. Chen *et al.*, "An efficient parallel motion estimation algorithm for digital image processing." *IEEE Trans. Circuits Syst. Video Technol.*, vol. 3, pp. 148–157, April 1993.

[23] K. H.-K. Chow and M. L. Liou, "Genetic motion search algorithm for video compression." *IEEE Trans. Circuits Syst. Video Technol.*, vol. 3, pp. 440–445, Dec. 1993.

[24] P. Pirsch and T. Komarek, "VLSI architecture for block matching algorithm." In *Proc. SPIE Visual Commun. Image Processing*, Cambridge, MA, Nov. 1988, vol. 1001, pp. 882–891.

[25] K.-M. Yang, M-T. Sun, and L. Wu, "A family of VLSI designs for the motion compensation block-matching algorithm." *IEEE Trans. Circuits Syst.*, vol. CAS-36, pp. 1317–1325, Oct. 1989.

[26] L. DeVos, "VLSI-architectures for the hierarchical block-matching algorithm for HDTV applications." in *Proc. SPIE Visual Commun. Image Processing*, Boston, Nov. 1990, vol. 1360, pp. 398–409.

[27] A. Puri, H.-M. Hang, and D. L. Schilling, "Interframe coding with variable block-size motion compensation." In *Proc. IEEE Global Commun. Conf.*, Tokyo, Nov. 1987, pp. 2.7.1–2.7.5.

[28] B. Liu and A. Zaccarin, "New fast algorithms for the estimation of block motion vectors." *IEEE Trans. Circuits Syst. Video Technol.*, vol. 3, pp. 148–157, April 1993.

[29] M. Bierling and R. Thoma, "Motion compensating field interpolation using a hierarchically structured displacement estimator." *Signal Processing*, vol. 11, pp. 387–404, Dec. 1986.

[30] M. Bierling, "Displacement estimation by hierarchical blockmatching." In *Proc. SPIE Visual Commun. Image Processing*, Cambridge, MA, Nov. 1988, vol. 1001, pp. 942–951.

[31] Zenith and AT&T, *Digital Spectrum Compatible HDTV System Description*, Sept. 1991.

[32] J. O. Limb and J. A. Murphy, "Measuring the speed of moving objects from television signals." *IEEE Trans. Commun.*, vol. COM-23, pp. 474–478, April 1975.

[33] J. O. Limb and J. A. Murphy, "Estimating the velocity of moving images in television signals." *Comput. Graph. Image Processing*, vol. 4, pp. 311–327, 1975.

[34] C. Cafforio and F. Rocca, "Methods for measuring small displacements of television images." *IEEE Trans. Inform. Theory*, vol. IT-22, pp. 573–579, Sept. 1976.

[35] S. Brofferio and F. Rocca, "Interframe redundancy reduction of video signals generated by translating objects." *IEEE Trans. Commun.*, vol. COM-35, pp. 448–455, 1977.

[36] C. Cafforio and F. Rocca, "Tracking moving objects in television images." *Signal Processing*, vol. 1, pp. 133–140, 1979.

[37] A. N. Netravali and J. D. Robbins, "Motion-compensated television coding: Part I." *Bell Syst. Tech. J.*, vol. 58, pp. 631–670, March 1979.

[38] B. Widrow and S. D. Stearns, *Adaptive Signal Processing*. Englewood Cliffs, NJ: Prentice-Hall, 1985.

[39] A. N. Netravali and J. D. Robbins, "Motion-compensated coding: Some new results." *Bell Syst. Tech. J.*, vol. 59, pp. 1735–1745, Nov. 1980.

[40] J. D. Robbins and A. N. Netravali, "Recursive motion compensation: A review." In *Image Sequence Processing and Dynamic Scene Analysis*, ed. T. S. Huang. New York: Springer-Verlag, 1983.

[41] S. Sabri, "Movement compensated interframe prediction for NTSC color TV signals." *IEEE Trans. Commun.*, vol. COM-32, pp. 954–968, Aug. 1984.

[42] J. A. Stuller and A. N. Netravali, "Transform domain motion estimation." *Bell Syst. Tech. J.*, vol. 58, pp. 1673–1702, Sept. 1979.

[43] A. N. Netravali and J. A. Stuller, "Motion-compensated transform coding." *Bell Syst. Tech. J.*, vol. 58, pp. 1703–1718, Sept. 1979.

[44] J. A. Stuller, A. N. Netravali, and J. D. Robbins, "Interframe television coding using gain and displacement compensation." *Bell Syst. Tech. J.*, vol. 59, pp. 1227–1240, Sept. 1980.

[45] K. A. Prabhu and A. N. Netravali, "Motion-compensated component color coding." *IEEE Trans. Commun.*, vol. COM-30, pp. 2519–2527, Dec. 1982.

[46] H. C. Bergmann, "Displacement estimation based on the correlation of image segments." In *Proc. IEEE Int. Conf. Electron. Image Processing*, York, England, July 1982, pp. 215–219.

[47] H. C. Bergmann, "Analysis of different displacement estimation algorithms for digital television signals." In *Image Sequence Processing and Dynamic Scene Analysis*, ed. T. S. Huang. New York: Springer-Verlag, 1983.

[48] R. Paquin and E. Dubois, "A spatio-temporal gradient method for estimating the displacement field in time-varying imagery." *Comput. Vision, Graph., Image Processing*, vol. 21, pp. 205–221, 1983.

[49] D. R. Walker and K. R. Rao, "Improved pel-recursive motion estimation." *IEEE Trans. Commun.*, vol. COM-32, pp. 1128–1134, Oct. 1984.

[50] H. Yamaguchi, "Interactive method of movement estimation for television signals." *IEEE Trans. Commun.*, vol. COM-37, pp. 1350–1358, Dec. 1989.

[51] J. Biemond, L. Looijenga, D. E. Boekee, and R. H. J. M. Plompen, "A pel-recursive Wiener-based displacement estimation algorithm." *Signal Processing*, vol. 13, pp. 399–412, Dec. 1987.

[52] S. N. Efstratiadis and A. K. Katsaggelos, "Multi-frame pel-recursive Wiener-based displacement estimation algorithm." In *Proc. SPIE Visual Commun. Image Processing*, Philadelphia, PA, Nov. 1989, vol. 1199, pp. 51–60.

[53] M. Bertero, T. A. Poggio, and V. Torre, "Ill-posed problems in early vision," *Proc. IEEE*, vol. 76, pp. 869–889, Aug. 1988.

[54] E. C. Hildreth, "Computations underlying the measurement of visual motion." *Artificial Intell.*, vol. 23, pp. 309–354, 1984.

[55] B. G. Schunck and B. K. P. Horn, "Constraints on optical flow." In *Proc. IEEE Conf. Pattern Recogn. Image Processing*, Aug. 1981, pp. 205–210.

[56] B. G. Schunck, "The image flow constraint equation." *Comput. Vision, Graph., Image Processing*, vol. 35, pp. 20–46, 1986.

[57] H.-H. Nagel, "Displacement vectors derived from second-order intensity variations in image sequences." *Comput. Vision, Graph., Image Processing*, vol. 21, pp. 85–117, 1983.

[58] H.-H. Nagel and W. Enkelmann, "An investigation of smoothness constraints for the estimation of displacement vector fields from image sequences." *IEEE Trans. Pattern Anal. Machine Intell.*, vol. PAMI-8, pp. 565–593, Sept. 1986.

[59] H.-H. Nagel, "On the estimation of optical flow: Relations between different approaches and some new results." *Artificial Intell.*, vol. 33, pp. 299–324, 1987.

[60] J. Aisbett, "Optical flow with an intensity-weighted smoothing." *IEEE Trans. Pattern Anal. Machine Intell.*, vol. PAMI-11, pp. 512–522, May 1989.

[61] B. G. Schunck, "Image flow: Fundamentals and algorithms." In *Motion Understanding: Robot and Human Vision*, ed. W. N. Martin and J. K. Aggarwal. Norwell, MA: Kluwer, 1988.

[62] R. Chellappa and A. A. Sawchuk, Eds., *Digital Image Processing and Analysis: Digital Image Analysis*. New York: IEEE Press, 1985.

[63] J. K. Aggarwal and N. Nandhakumar, "On the computation of motion from sequences of images—A review." *Proc. IEEE*, vol. 76, pp. 917–935, Aug. 1988.

[64] W. N. Martin and J. K. Aggarwal, Eds., *Motion Understanding: Robot and Human Vision*. Norwell, MA: Kluwer, 1988.

[65] M. I. Sezan and R. L. Lagendijk, Eds., *Motion Analysis and Image Sequence Processing*. Norwell, MA: Kluwer, 1993.

[66] W. Enkelmann, "Investigations of multigrid algorithms for the estimation of optical flow fields in image sequences." *Comput. Vision, Graph., Image Processing*, vol. 43, pp. 150–177, 1988.

[67] J. Konrad and E. Dubois, "Multigrid Bayesian estimation of image motion fields using stochastic relaxation." In *Proc. IEEE Int. Conf. Comput. Vision*, Dec. 1988, pp. 354–362.

[68] M. R. Luettgen, W. C. Karl, and A. S. Willsky, "Efficient multiscale regularization with applications to the computation of optical flow." *IEEE Trans. Image Processing*, vol. 3, pp. 41–64, Jan. 1994.

[69] B. G. Haskell, "Frame-to-frame coding of television pictures using two-dimensional Fourier transforms." *IEEE Trans. Inform. Theory*, vol. IT-20, pp. 119–120, Jan. 1974.

[70] C. D. Kuglin and D. C. Hines, "The phase correlation image alignment method." In *Proc. IEEE Int. Conf. Cybern. Soc.*, San Francisco, Sept. 1975, pp. 163–165.

[71] G. A. Thomas, "Television Motion Measurement for DATV and Other Applications." British Broadcasting Corp. Res. Dept. Report No. 1987/11, Sept. 1987.

[72] G. A. Thomas, " HDTV bandwidth reduction by adaptive subsampling and motion-compensated DATV techniques." *SMPTE J.*, pp. 460–465, May 1987.

[73] M. Ziegler, "Hierarchical motion estimation using the phase correlation method in 140 Mbit/s HDTV coding." In *Signal Processing of HDTV*, ed. L. Chiariglione. Amsterdam, The Netherlands: Elseviers, 1990, pp. 131–137.

[74] H.-M. Hang, Y.-M. Chou, and T.-H. S. Chao, "Motion estimation using frequency components." In *Proc. SPIE Visual Commun. Image Processing*, Boston, Nov. 1992, vol. 1818, pp. 74–85.

[75] A. B. Carlson, *Communication Systems*, 3d. New York: McGraw-Hill, 1986.

[76] A. Furukawa, T. Koga, and K. Linuma, "Motion-adaptive interpolation for videoconference pictures." In *IEEE Int. Conf. Commun.*, Amsterdam, The Netherlands, 1984, pp. 707–710.

[77] A. Puri, R. Aravind, B. G. Haskell, and R. Leonardi, "Video coding with motion-compensated interpolation for CD-ROM applications." *Signal Processing: Image Commun.*, vol. 2, pp. 127–144, Aug. 1990.

[78] R. Thoma and M. Bierling, "Motion compensating interpolation considering covered and uncovered background." *Signal Processing: Image Commun.*, vol. 1, pp. 191–212, Oct. 1989.

[79] C. Cafforio, F. Rocca, and S. Tubaro, "Motion compensated image interpolation." *IEEE Trans. Commun.*, vol. COM-38, pp. 215–222, Feb. 1990.

[80] C. Bergeron and E. Dubois, "Gradient-based algorithms for block-oriented MAP estimation of motion and application to motion-compensated temporal interpolation." *IEEE Trans. Circuits Syst. Video Technol.*, vol. 1, pp. 72–84, March 1991.

[81] M. Götze, "Generation of motion vector fields for motion compensated interpolation of HDTV signals." in *Signal Processing of HDTV*, ed. L. Chiariglione. Amsterdam, The Netherlands: Elseviers, 1988, pp. 383–391.

[82] R. Lenz and A. Gerhard, "Image sequence coding using scene analysis and spatio-temporal interpolation." In *Image Sequence Processing and Dynamic Scene Analysis*, ed. T. S. Huang. New York: Springer-Verlag, 1983.

[83] B. Girod and R. Thoma, "Motion-compensating field interpolation from interlaced and non-interlaced grids." *Second Int. Tech. Symp. Optical and Electro-Optical Applied Science and Eng.: Image Processing Symp.*, Cannes, Dec. 1985.

[84] S. C. Brofferio, "An object-background image model for predictive video coding." *IEEE Trans. Commun.*, vol. COM-37, pp. 1391–1394, Dec. 1989.

[85] S. Tubaro and F. Rocca, "Motion field estimators and their application to image interpolation." In *Motion Analysis and Image Sequence Processing*, ed. M. I. Sezan and R. L. Lagendijk. Norwell, MA: Kluwer, 1993, pp. 153–187.

[86] M. Höetter, "Differential estimation of the global motion parameters zoom and pan." *Signal Processing*, vol. 16, pp. 249–265, March 1989.

[87] H. G. Musmann, M. Höetter, and J. Ostermann, "Object-oriented analysis-synthesis coding of moving images." *Signal Processing: Image Commun.*, vol. 1, pp. 117–138, Oct. 1989.

[88] M. Höetter, "Object-oriented analysis-synthesis coding based on moving two-dimensional objects." *Signal Processing: Image Commun.*, vol. 2, pp. 409–428, Dec. 1990.

[89] M. Höetter, "Optimization and efficiency of an object-oriented analysis-synthesis coder." *IEEE Trans. Circuits Syst. Video Technol.*, vol. 4, pp. 181–194, April 1994.

[90] M. Höetter and R. Thoma, "Image segmentation based on object oriented mapping parameter estimation." *Signal Processing*, vol. 15, pp. 315–334, Oct. 1988.

[91] N. Diehl, "Object-oriented motion estimation and segmentation in image sequences." *Signal Processing: Image Commun.*, vol. 3, pp. 23–56, 1991.

[92] J. Ostermann, "Modelling of 3D moving objects for an analysis-synthesis coder." In *Proc. SPIE/SPSE Symp. Sensing and Reconstruction of 3D Objects and Scenes*, ed. B. Girod. Santa Clara, CA, Feb. 1990. vol. 1260, pp. 240–250,

[93] T. S. Huang, "Determining three-dimensional motion and structure from two perspective views." In *Handbook of Pattern Recognition and Image Processing*, ed. T. Y. Young and K. S. Fu. New York: Academic Press, 1986.

[94] T. S. Huang and A. N. Netravali, "Motion and structure from feature correspondences: A review." *Proc. IEEE*, vol. 82, pp. 251–268, Feb. 1994.

[95] C.-T. Chen and T. R. Hsing, "Digital coding techniques for visual communications." *J. Visual Commun. Image Representation*, vol. 2, pp. 1–16, March 1991.

[96] K. R. Rao and P. Yip, *Discrete Cosine Transform*. San Diego, CA: Academic Press, 1990.

[97] T. Ishiguro and K. Linuma, "Television bandwidth compression transmission by motion-compensated interframe coding." *IEEE Commun. Mag.*, vol. 10, pp. 24–30, Nov. 1982.

[98] Y. Ninomiya and Y. Ohtsuka, "A motion compensated interframe coding scheme for television pictures." *IEEE Trans. Commun.*, vol. COM-30, pp. 201–211, Jan. 1982.

[99] T. Koga, A. Hirano, K. Iinuma, Y. Iijima, and T. Ishiguro, "A 1.5Mb/s interframe codec with motion-compensation." In *IEEE Int. Conf. Commun.*, 1983, pp. D8:7.1–D8:7.5.

[100] Y. Ninomiya and Y. Ohtsuka, "A motion compensated interframe coding scheme for NTSC color television signals." *IEEE Trans. Commun.*, vol. COM-32, pp. 328–334, March 1984.

Chapter 6

Vector Quantization Techniques in Image Compression

A. Gersho
Department of Electrical and Computer Engineering
University of California, Santa Barbara
Santa Barbara, California

S. Gupta
Compression Labs, Inc.
San Jose, California

and

S.-W. Wu
AT&T Bell Laboratories
Murray Hill, New Jersey

Vector quantization (VQ) is a powerful approach for signal compression, and a wealth of diverse techniques have been developed for exploiting the basic concept of VQ and controlling or circumventing its encoding complexity. We review and explain the basic concept of VQ, the use of structured VQ, VQ with memory, adaptive VQ, and other techniques that have useful applications for image and video compression.

Handbook of Visual Communications
189

6.1 Introduction

Algorithms for digital compression of images generally consist of two distinct components—signal processing and quantization. Signal processing is needed for the analysis or decomposition of the signal into sets of parameters or features and for the synthesis of signals from a set of quantized parameters or features. Quantization is the direct mapping of parameters or samples into a compact digital representation for transmission or storage. Several other chapters of this book on coding techniques extensively discuss signal processing methods while giving relatively brief mention of quantization. Here we focus on the use of vector quantization for simultaneously digitizing vectors of signal samples or parameter values.

We shall see that ultimately there is no simple partitioning of the task of image compression into processing and quantization. While separate treatment of each is often convenient, with increasing emphasis on high coding efficiency and optimality, the two parts tend to merge into a single challenging joint optimization problem. We shall briefly review the basic concepts of quantization and then focus on the capabilities and limitations of VQ for image compression. Most of our attention is on structurally constrained techniques, where signal processing is an intrinsic part of the task of quantization. A more extensive and comprehensive presentation of the theory and techniques of VQ may be found in the book by Gersho and Gray [1].

6.1.1 Definitions

An ordered set of k samples (or features, parameters, etc.) may be viewed as a k-dimensional vector $\mathbf{x}$; i.e., a point in k-dimensional Euclidean space $\mathcal{R}^k$. An *encoder* (specifically, a vector encoder) denoted by the operator $\mathcal{E}$ maps this vector into an index I, an integer in the finite set $\mathcal{I} \equiv \{1, 2, \cdots, N\}$. Correspondingly, a *decoder*, denoted by $\mathcal{D}$, maps an index I in the set $\mathcal{I}$ to a reproduction vector $\hat{\mathbf{x}}$ in a finite set $\mathcal{C}$ containing N distinct output or reproduction vectors, called *code vectors*, which are elements of $\mathcal{R}^k$. The ordered set $\mathcal{C} = (\mathbf{y}_1, \mathbf{y}_2, \cdots, \mathbf{y}_N)$ is called the *codebook*.

A *vector quantizer* Q of dimension k and size N is a mapping from a vector in $\mathcal{R}^k$ to a code vector in $\mathcal{C}$. Every N-point vector quantizer has an associated *partition* of $\mathcal{R}^k$ into N regions or *cells*, R_i for $i \in \mathcal{I}$. The ith cell is defined by

$$R_i = \{\mathbf{x} \in \mathcal{R}^k : Q(\mathbf{x}) = \mathbf{y}_i\} . \tag{6.1}$$

The *code rate*, or, simply, *rate* of a vector quantizer is $r = (\log_2 N)/k$, which measures the number of bits per vector component used to represent the input vector and gives an indication of the accuracy or precision that is achievable with a vector quantizer if the codebook is well-designed. This quantity is sometimes called the *resolution* of the quantizer, referring to amplitude resolution, and should not be confused with spatial resolution in image processing.

The most basic approach to image coding with VQ is to partition an image into square blocks of $p \times p$ pixels and treat each block as a vector of dimension $k = p^2$. In principle (but certainly not in practice), an entire image can be regarded as a single vector and coded with VQ. In practice, when VQ techniques are applied to image coding, the vector dimension is kept fairly small for complexity reasons, as discussed later.

6.1.2 Distortion Measures

To assess performance in any lossy compression system, a measure of distortion $d(\mathbf{x}, \mathbf{y})$ that measures the dissimilarity between two vectors $\mathbf{x}$ and $\mathbf{y}$ in $\mathcal{R}^k$ is needed. Most common is the squared error distortion measure, defined as

$$d(\mathbf{x}, \hat{\mathbf{x}}) = ||\mathbf{x} - \hat{\mathbf{x}}||^2 = \sum_{i=1}^{k} (x_i - \hat{x}_i)^2 . \tag{6.2}$$

When a random input vector $\mathbf{X}$ is quantized with a resulting reproduction $\hat{\mathbf{X}}$, the *mean squared error distortion*, or *average distortion*, is defined as

$$D = Ed(\mathbf{X}, \hat{\mathbf{X}}) = E(||\mathbf{X} - \hat{\mathbf{X}}||^2) . \tag{6.3}$$

In practice, the expectation is replaced by spatial averaging over the actual squared error distortions for each block in a particular image. The most common way of assessing performance in image compression is to compute the *peak signal-to-noise ratio* or PSNR, $\text{PSNR} = 10 \log_{10} M^2/D$, where M is the maximum pixel amplitude level (usually $M = 255$). In general, it is recognized that the PSNR is far from an accurate indication of perceptual quality; however, it is generally helpful in comparing different versions of a particular compression scheme. It is less useful to compare entirely different compression schemes, since the perceptual effects of different types of degradation associated with different coding approaches can be impossible to assess from this performance measure. A variety of alternative distortion measures have also been considered with the hope of obtaining a perceptually more accurate assessment of dissimilarity between the input and reproduction vectors.

6.1.3 Nearest Neighbor Encoding

A VQ encoder can be viewed in a general way as a pattern classifier that assigns an input vector to one of a finite set of N possible classes. Any heuristic method of performing this assignment to suit a desired classification objective fits within the general definition and concept of VQ. However, in the prevailing use of VQ, the purpose of the encoder is to identify a region R_i of "similar" vectors to which the input vector belongs so that the decoder can specify a code vector y_i that is representative of the partition and provides a reasonable approximation to any input vector in R_i.

The notion of similarity and the goal of approximately reproducing the input vector lead naturally to a particular encoding rule. In *nearest neighbor encoding*, for a given input vector x, the encoder searches the codebook for that code vector which is "closest" to x in the sense of minimizing the distortion $d(x, y_j)$ over all vectors y_j in C. Such an encoder is said to satisfy the *nearest neighbor rule*. It is easy to show that for a given codebook the nearest neighbor encoder achieves optimal performance (for the given distortion measure) over all possible encoders.

6.1.4 Codebook Design

The most widely used method for VQ codebook design is known as the *generalized Lloyd algorithm* (GLA) or sometimes as the LBG algorithm, after the classic paper that studied this method for VQ design [2]. The design is based on a *training set* consisting of a large sample of M vectors experimentally generated from the signal source so that it is statistically representative of the vectors that will be encountered by the encoder in actual operation. The method iteratively improves the codebook starting from some initial codebook. Each iteration consists of two stages. First the training set is partitioned into N clusters, one for each code vector, by applying the nearest neighbor rule to each training vector and assigning it to the code vector that gives the least distortion. Next, a new codebook is formed by taking the centroid of each cluster as a code vector. The centroid for the ith cluster is that point in $\mathcal{R}^k$ which minimizes the average of the distortions $d(x_i, y_i)$ over all training vectors x_i that belong to that cluster. For the squared error distortion, the centroid is simply the average of the training vectors in the cluster. This procedure leads to a sequence of codebooks with a monotonically nonincreasing average distortion when computed over the training set.

The codebook so obtained is not generally globally optimal for the given source but satisfies certain local optimality properties, and in most applications it turns out to be quite effective. Experiments with different initial codebooks suggest that the resulting final codebooks give generally similar performance. Other design methods have been studied that are able to circumvent local optima and appear to offer the potential of finding a globally optimal codebook [3, 4].

An important issue in codebook design is the size of the training set M or, equivalently, the *training ratio* defined as M/N. The training ratio is the average number of training vectors per cluster that are used to generate a cluster centroid. A common rule of thumb is that a training ratio of 100 or larger is desirable, although often ratios as low as 10 are used. Actually, several factors influence the size of a training set needed for codebook design, including the statistical character of the signal source and the method of sampling the source to generate a training set.

First of all, the effectiveness of a training set of any given size depends on the degree to which the training vectors are statistically independent of one another. For example, a training set with a fixed number of 4 × 4 image blocks is better if fewer blocks per image are taken from more images (by subsampling from the set of available image blocks rather than using all available blocks from fewer images). By avoiding image blocks that are adjacent to one another, the effectiveness of the training set is enhanced.

If our objective were to minimize the average distortion incurred when the codebook is used to encode the training set itself, then increasing the training ratio would generally *decrease* the performance (i.e., it will increase the average distortion of the codebook). In fact, an extreme example of this is a training ratio of unity that gives zero distortion for any meaningful distortion measure. As the training ratio increases, the distortion measure will asymptotically approach the statistical average for the underlying signal source so that this method of assessing performance of a codebook is reliable only when an enormous training ratio is used.

Generally, it is more meaningful to assess the quality of a VQ coder by testing its performance on a data set, called the *test set*, that is *not* part of the training set but is still drawn from the same signal source. Note that a training set of satellite images and a test set of human face images would not satisfy this condition, since the training and test sets are not statistically representative of a common source. When assessing the performance of a VQ codebook from test data outside of the training set, a general rule is that the average distortion will decrease as the training set size is increased. Of course the size of the test set must also be reasonably large to obtain an average distortion value that is a reasonably accurate estimate of the statistical average for the source.

6.1.5 Performance and Complexity of Vector Quantization

VQ is often loosely described as the optimal way to quantize a vector. Specifically, consider any encoding system that operates in any way on the components of the input vector and generates a binary word (or a set of binary words) with a total of up to N possible distinct values, and an associated decoding system that reproduces an approximation to this vector from the binary data. Such an encoding method can never provide superior performance (and generally will give inferior performance) to a suitably designed VQ coding scheme with a codebook of size N [1]. The proof is simply to note that the given decoder can be used to generate a unique VQ codebook whose associated VQ decoder exactly replicates the given decoder operation. Then, by replacing the given encoder with a nearest neighbor encoder based on this codebook, the VQ will necessarily match or exceed the performance of the original coding scheme.

In general, in encoding a waveform or an image, the performance of VQ improves as the dimension increases. In fact, even if a sequence of samples are statistically independent, the average distortion per pixel decreases (slightly) as the dimension is increased. Asymptotically as the dimension approaches infinity, VQ theoretically can approach the ultimate distortion–rate performance bounds studied in information theory. Intuitively, it is clear that the benefit from increasing the dimension in coding a group of pixels in an image is the ability it provides to exploit more spatial correlation among the samples. In this chapter we use the term correlation in a general sense to include nonlinear as well as linear statistical interdependency among samples.

While VQ offers the best rate–distortion performance possible for coding a block of signal samples or parameters, it has possibly the worst complexity of any effective coding scheme. The complexity of performing a nearest neighbor search is proportional to the codebook size N and therefore grows exponentially with the dimension k for a fixed rate r. This is not a problem for moderate codebook sizes, but it is a sufficient obstacle to prevent direct VQ of an image block with sufficient size to adequately exploit spatial correlation. For example, suppose the rate of 0.25 bit per pixel (bpp) is specified. Then, for an 8×8 block, the codebook size is $N = 65536$, and for a 16×16 block, the size is $N = 10^{19}$. Thus a complexity barrier is quickly reached as the dimension grows for a given coding resolution.

The complexity problem does not eliminate VQ as a viable coding method, but it does generally limit the direct use of full search VQ to the role of a building block in a complete coding system. Thus other methods for eliminating redundancy, i.e., for exploiting the extensive spatial correlation in an image, are generally required prior to the quantization stage. When VQ is applied, the vectors usually represent a set of features extracted from an image rather than image pixels themselves. Alternatively, many coding systems can be viewed as a particular form of *structurally constrained* VQ that, in effect, performs VQ on large-dimension vectors by forcing constraints on the encoder and decoder mappings. These schemes may not resemble the paradigm of encoding via an exhaustive search through a single codebook.

6.1.6 Structured Vector Quantization Methods

Structured VQ methods can furnish superior tradeoffs between distortion–rate performance and complexity. The classic examples are *multistage VQ* (MSVQ) and *tree-structured VQ* (TSVQ), and we assume the reader is familiar with these basic methods (see, for example, [1]). Both MSVQ and TSVQ substantially reduce encoding complexity by encoding a vector in a sequential manner, where the outcome of prior search stages narrows down the set of possible code vectors that can be subsequently selected for the final reproduction. In TSVQ, the search complexity is reduced from N to roughly $\log N$ while the performance (as indicated by average distortion) is somewhat inferior to an unstructured VQ with a codebook of equivalent size. In MSVQ, both search complexity and storage requirements are reduced, but generally it incurs a greater performance penalty than does TSVQ.

A large family of structured vector quantizers, including MSVQ and TSVQ, can be modeled and studied as *product codes*. A product code vector quantizer is a VQ coding scheme where the b bits delivered to the decoder consist of a set of $s > 1$ indexes or binary words with lengths $\{b_1, b_2, \ldots, b_s\}$ bits with

$$\sum_{i=1}^{s} b_i = b = kr ,$$

(6.4)

and the decoder consists of (1) a set of *s feature codebooks* C_i, containing 2^{b_i} *feature code vectors* $\mathbf{c}_{i,n}$, $n = 1, 2, \ldots, 2^{b_i}$, and (2) a *synthesis map* $\mathbf{g}$ that synthesizes a reproduction $\hat{\mathbf{x}}$ of the input vector (to the encoder) $\mathbf{x}$ according to

$$\hat{\mathbf{x}} = \mathbf{g}(\hat{\mathbf{f}}_1, \hat{\mathbf{f}}_2, \ldots, \hat{\mathbf{f}}_s) ,$$

where $\hat{\mathbf{f}}_i$ is the feature code vector selected from codebook C_i by the index of size b_i bits. A typical product code encoder operates as follows. The source vector $\mathbf{x}$ is first decomposed into *features* $\mathbf{f}_1, \mathbf{f}_2, \ldots, \mathbf{f}_s$, where feature $\mathbf{f}_i$, $i = 1, 2, \ldots, s$, may be a scalar or a vector of dimension k_i. A set of operations $\mathbf{u}_i$ generate these features from the source vector according to

$$\mathbf{f}_i = \mathbf{u}_i(\mathbf{x}) .$$

(6.5)

The b bits available to encode a source vector are allocated among the s features, with b_i bits for feature $\mathbf{f}_i$. Feature $\mathbf{f}_i$ is encoded using codebook C_i and a suitable distortion measure to obtain $\hat{\mathbf{f}}_i$, the quantized version of $\mathbf{f}_i$.

In a *sequential search product code* [5], the s features are ordered and are extracted and quantized in the prescribed order. In the ith stage, feature $\mathbf{f}_i$ is extracted from $\mathbf{x}$ using a *feature extraction rule* or *feature analysis rule*:

$$\mathbf{f}_i = \mathbf{h}_i(\mathbf{x}, \hat{\mathbf{f}}_1, \hat{\mathbf{f}}_2, \ldots, \hat{\mathbf{f}}_{i-1}) .$$

(6.6)

The extracted feature $\mathbf{f}_i$ is quantized to $\hat{\mathbf{f}}_i$ by selecting a feature code vector $\mathbf{c}_{i,j}$ from the feature codebook C_i to minimize a *feature distortion measure* $d_i(\mathbf{f}_i, \mathbf{c}_{i,j})$, $i = 1, 2, \ldots, s$.

Two well-known examples of sequential search product codes are *shape–gain VQ* (SGVQ) and *mean–removed VQ* (MRVQ). Each has two feature codebooks, one a scalar codebook and the second a vector codebook with the dimension of the original vector. In SGVQ, the scalar shape codebook contains quantized gain values, typically representing the norm of the vector and the shape codebook represents normalized input vectors. In MRVQ, the scalar mean codebook represents the average of the vector components and the *residual* codebook represents the mean–removed input vectors.

Recently, a broader class of sequential search product codes, called *generalized product codes* was introduced, where each successive feature after the first may have multiple feature codebooks [6]. A specific codebook is selected based on the outcome of the previous feature quantization. This extension of the product code concept makes it possible to exploit the statistical interdependence between each pair of features $\mathbf{f}_i$ and $\mathbf{f}_{i+1}$ to more efficiently encode $\mathbf{f}_{i+1}$. Associated with every feature code vector, is a *codebook pointer* with a value equal to the index identifying the assigned codebook for feature $\mathbf{f}_{i+1}$.

6.1.7 Nonlinear Estimation with Vector Quantization

A VQ decoder may be viewed as a nonlinear estimator of the input $\mathbf{X}$ from a partial specification of $\mathbf{X}$ as given by the index identifying in which partition region $\mathbf{X}$ lies. In fact, when the code vectors are the centroids of the partition regions, the decoder in ordinary VQ is actually a nonlinear estimator of the input $\mathbf{X}$ given the transmitted index I. Although nonlinear estimation is in general a difficult problem and often quite intractable, when the observable data (on which the estimation is based) lie in a finite set of possible values, a table-lookup implementation of nonlinear estimation is possible and often quite practical.

Nonlinear estimation with quantized observables can be effectively exploited to circumvent complexity in VQ encoding. Rather than direct quantization of a large dimensional signal vector $\mathbf{Y}$, a variety of signal decompositions or data reduction methods can be used to obtain a reduced dimension feature vector $\mathbf{X}$ that partially characterizes the original high-dimensional vector. If $\mathbf{X}$ is quantized to $\hat{\mathbf{X}}$, then $\hat{\mathbf{X}}$ or the index I that specified it serves as the observable data for nonlinearly estimating $\mathbf{Y}$. This approach, called *nonlinear interpolative VQ* (NLIVQ) was introduced in [7] and extended in [8]. In NLIVQ, the decoder is estimating a signal vector $\mathbf{Y}$ from quantized observations of a different feature vector $\mathbf{X}$ that has somehow been extracted from $\mathbf{Y}$. Usually the encoder and decoder codebooks have different dimensions. In one version, the feature vector $\mathbf{X}$ is simply a subsampled version of the signal vector $\mathbf{Y}$ so that the decoder is generating an *optimal nonlinear interpolation* of the quantized observation $\hat{\mathbf{X}}$.

The simplest way to design an NLIVQ coding system is to first design an ordinary VQ codebook $\mathcal{C}$ for the feature vector $\mathbf{X}$ from a training set using the GLA algorithm. Then, the ith code vector $\mathbf{y}_i$ in the NLIVQ decoder codebook is found simply by computing the conditional expectation of $\mathbf{Y}$ given that $\mathbf{X}$ has been quantized with the ith code vector in $\mathcal{C}$. That is, $\mathbf{y}_i = E\{\mathbf{Y}|I = i\}$. This expectation is obtained from training data of $(\mathbf{X}, \mathbf{Y})$ pairs by simply taking the average of all training vectors $\mathbf{y}$ for which the corresponding $\mathbf{x}$ is quantized with index i.

The concept of NLIVQ can be modified in several ways. One way to improve performance is to jointly optimize the encoder and decoder codebooks [8]. One modification that can be useful is to consider not one but a set of feature vectors that are extracted from a given signal vector. In this case, suboptimal nonlinear decoders can generate a separate partial estimate of **Y** from each quantized feature and then combine these features in some way to obtain a final estimate of **Y**. One application of NLIVQ, described later in this chapter, is to improve the decoding of transform coded image blocks.

6.2 Vector Quantization with Memory

We have seen that the rate–distortion performance of a vector quantizer improves with an increase in the vector dimension and can approach information-theoretic bounds for sufficiently large dimensions. However, since the computational and storage complexity of VQ increases exponentially with an increase in vector dimension for a given rate, complexity constraints coupled with other issues such as encoding delay, limit the dimensionality of vectors that can be directly quantized.

We have seen previously that structured VQ provides a way to increase dimension while retaining manageable complexity at the expense of some performance loss over direct VQ. In this section, we describe a different approach to exploit correlation over a larger area. The main idea is to develop VQ systems that exploit intervector redundancy without an increase in vector dimension by remembering the previously quantized vectors and utilizing this information in encoding the current vector. Unlike ordinary memoryless VQ, where each vector is treated as an isolated entity, VQ with memory estimates either the current vector or some of its characteristics from previously quantized vectors. If sufficient intervector redundancy exists and the estimator is well-designed, incorporation of memory can provide increased performance compared to memoryless VQ, since correlation over a larger area is exploited.

Because the decoder has access to previously quantized vectors, it can emulate the estimation process at the encoder without any added side information. Thus the intervector redundancy exploitation is "free" in the sense that it does not contribute to the total bit rate. Although attractive in terms of rate–distortion performance, such "backward adaptation" if improperly designed may lack robustness. For example, channel errors can lead to a loss in synchronization between encoder and decoder and result in a catastrophic loss in performance. Also, occasionally the estimator may fail to provide a good estimate of the current vector and consequently it may be quantized poorly. Estimation of subsequent input vectors from this poor approximation is likely to be inadequate for modeling future input vectors. Thus the quantizer is forced off-track and may result in significantly lower performance than direct VQ.

The key to the design of a high performance vector quantizer with memory is an effective estimator of the current vector and its attributes. Since it is virtually impossible to develop analytically tractable models for joint pdfs of vectors, the design is often based on training data and may be heuristic. We shall consider in particular two VQ schemes that incorporate memory, *predictive vector quantization* (PVQ) and *finite state vector quantization* (FSVQ). PVQ can be viewed as a straightforward generalization of scalar DPCM to the vector case, wherein the intervector redundancy is exploited by a vector predictor, whereas FSVQ can be regarded as a *switched* VQ scheme that for each successive input vector, switches to one of a finite number of states based on the previously quantized vectors. In FSVQ, the intervector redundancy is exploited by a state transition function that performs the state selection.

6.2.1 Predictive Vector Quantization

Figure 6.1 is a schematic of a predictive vector quantizer [1]. A vector predictor in the PVQ encoder operates on previously quantized vectors to form a prediction

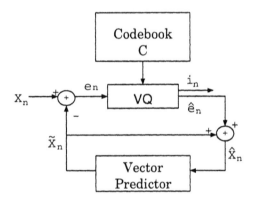

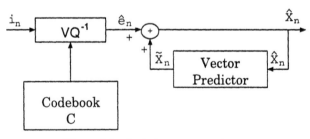

Figure 6.1   Predictive vector quantizer.

$\tilde{\mathbf{X}}_n = P(\hat{\mathbf{X}}_{n-1}, \hat{\mathbf{X}}_{n-2}, \ldots, \hat{\mathbf{X}}_{n-N})$ of the input vector $\mathbf{X}_n$. The residual vector $\mathbf{e}_n = \mathbf{X}_n - \tilde{\mathbf{X}}_n$ is vector quantized to $\hat{\mathbf{e}}_n$ by nearest neighbor search in a vector codebook $\mathcal{C}$. The corresponding code vector index i_n is transmitted to the decoder. The quantized vector $\hat{\mathbf{X}}_n$ is obtained by adding the quantized residual vector to the predicted vector. Thus $\hat{\mathbf{X}}_n = \tilde{\mathbf{X}}_n + \hat{\mathbf{e}}_n$. The PVQ decoder first generates $\hat{\mathbf{e}}_n$ through a table lookup. Next, the estimate $\tilde{\mathbf{X}}_n$ is formed and added to $\hat{\mathbf{e}}_n$ to obtain $\hat{\mathbf{X}}_n$.

PVQ was first studied by Cuperman and Gersho, who applied it to speech coding [9]. Since then it has been applied to image coding by several researchers, see in particular [10, 11]. For the case of image coding, a vector sequence is formed by scanning the blocks right to left, top to bottom. To ensure causality and limit complexity, a vector predictor for block $\mathbf{B}$ typically uses at most three vectors; namely, the vectors to the immediate left ($\mathbf{L}$), top ($\mathbf{U}$), and upper diagonal ($\mathbf{D}$) of the current vector, as shown in Fig. 6.2.

By drawing an analogy with DPCM, it is readily shown that the coding gain (SNR) of PVQ is the product of the prediction gain and the coding gain of the vector quantizer. Surprisingly, there is no theoretical proof that the performance of PVQ will exceed that of memoryless VQ for a given rate and dimension. The reason is that, although the prediction gain generally exceeds unity, the predictor tends to decorrelate the samples within a vector, resulting in a lower coding gain for $\mathbf{e}_n$ compared to direct quantization of $\mathbf{X}_n$. However, experiments have shown that in most cases, prediction gain more than offsets the loss in coding gain so that PVQ results in a net performance improvement over memoryless VQ.

Care should be exercised in selecting the dimension of the vector $\mathbf{X}_n$, since an increase in vector dimension forces the vector predictor to predict samples that are further away from previously quantized vectors; e.g., pixels in the bottom right-hand corner of block $\mathbf{B}$. Since correlation between samples decreases with an increase in distance, the predictor is unable to form a good estimate of distant samples and may result in a reduced prediction gain compared to that of a vector of lower dimensionality.

PVQ design requires the design of a vector predictor and a vector quantizer. Most design algorithms first design a predictor from the training vectors. Next,

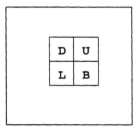

Figure 6.2   Predictor causality in PVQ.

a training sequence of residual vectors is formed by applying the original un-quantized vectors to the predictor and subtracting its output from the original vectors. Standard memoryless VQ design algorithms such as the GLA can be applied to this training sequence to obtain an initial codebook. The codebook can be improved iteratively by "closing the loop" and applying quantized vectors from the previous design as the predictor input. It is difficult to optimize the predictor for a given residual quantizer, and a previous attempt to jointly optimize the predictor and residual quantizer [12] has shown negligible increase in performance. Thus the predictor is generally held fixed, and only the residual quantizer is updated until convergence is obtained. Since the quantizer in PVQ is similar to memoryless VQ, we focus on different predictors for the remainder of this section.

For the case of linear prediction, a vector predictor is described via matrix multiplications; i.e., $\tilde{\mathbf{X}}_n = \sum_{i=1}^{N} \mathbf{A}_i \hat{\mathbf{X}}_{n-i}$ [9]. For the MSE distortion measure, the matrices $\mathbf{A}_i$ are solutions to the generalized multivariate Wiener–Hopf equations. The needed correlation values for these equations are generally estimated by averaging over the training data. Simplified linear vector predictors are sometimes defined by selecting a subset of pixels from adjacent image blocks to predict the current block or a subset of the current block.

Thus far we have discussed only linear predictors. Nonlinear prediction can in principle enhance performance by adding more flexibility. Recently, a nonparametric form of nonlinear predictor based on NLIVQ has been developed that estimates conditional expectations from a training sequence to obtain the minimum mean squared error (mmse) predictor [7]. It has been applied to code multispectral image sets, which are collections of images from a given spatial region and observed by sensors of different wavelengths. In this application, nonlinear prediction has been shown to improve performance over linear prediction [13]. Prediction also plays an important role in video coding; a detailed discussion of this is deferred to a later section.

6.2.2 Finite-State Vector Quantization

Figure 6.3 is a schematic of a finite-state vector quantizer [1]. Such a quantizer is specified by a next-state function f, a state space $\mathcal{S} = \{1, 2, \ldots, L\}$, an encoder mapping and a decoder mapping. The next-state function f depends on previously quantized vectors and the present state. Thus $S_{n+1} = f(\hat{\mathbf{X}}_n, S_n)$ where $S_n, S_{n+1} \in \mathcal{S}$. Each state i is associated with a state codebook $\mathcal{C}_i$, which is used to encode the current input vector $\mathbf{X}_n$ with a minimum distortion rule, and the code vector index is transmitted to the decoder.

Since the decoder has knowledge of the previously quantized vectors and the previous state, it can generate the current state S_n by applying the next-state

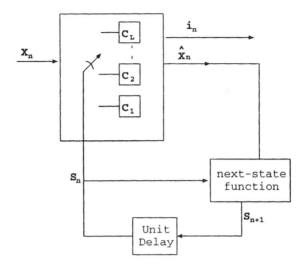

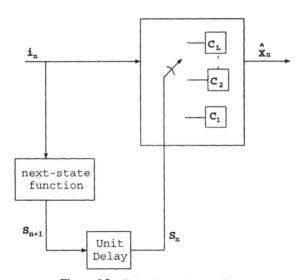

Figure 6.3   Finite-state vector quantizer.

function and access the corresponding codebook C_{S_n} without any side information. In essence, backward adaptation enables FSVQ to use a much larger codebook $C = \bigcup_{i=1}^{L} C_i$ without a corresponding increase in rate. The disadvantage is that when the input changes in an unpredictable way, the system cannot track it well and it may be trapped in a bad state. A simple solution is to periodically reset the FSVQ state to prevent it from staying trapped in a bad state.

FSVQ design requires the design of a state space, next-state function, and encoder and decoder mappings. Although no optimal solution to FSVQ design exists, a number of ad hoc design techniques are available. Here, we describe the popular "omniscient" design [14], thus named because the next-state function in the encoder during design operates on the unquantized vectors rather than the reproduced vectors. It is assumed that the decoder is "omniscient" and knows the state sequence thus generated. For the moment, assume that a state space along with next-state function has already been determined. The next-state function is applied to the training data to partition it into L subsequences, $\mathcal{T}_i, i \in \{1, 2, \ldots, L\}$, where L is the number of states by applying the following rule: If $f(\mathbf{X}_n, S_n) = S_{n+1}$, then $\mathbf{X}_{n+1} \in \mathcal{T}_{S_n}$. A separate codebook is designed for each subsequence and constitutes the state codebook for that particular state. Note that the state codebooks can be iteratively improved by "closing the loop" and applying the next-state function to the reproduced rather than the original vectors to obtain new training subsequences [1].

FSVQ was first applied to image coding in [15]. In this work, the next-state function examines the edge and shade characteristic of the quantized $\mathbf{D}$, $\mathbf{U}$, and $\mathbf{L}$ blocks to determine the best state for the current block $\mathbf{B}$. For example, a strong vertical edge in these previously quantized vectors indicates a high probability of a vertical edge in the current block. The next-state function should drive the FSVQ to a state whose associated state codebook has predominantly vertical edge code vectors.

A more recent FSVQ system employs a large codebook that is designed by standard VQ design algorithms. Based on the probability of transition between code vectors representing previously quantized blocks and code vectors in this large codebook, a small subset of the super codebook is used as the state codebook [16]. Thus intervector redundancy is employed to narrow the set of code vectors to those likely to be effective in coding the current vector, which leads to a reduction in rate compared to direct VQ with the large codebook.

6.3 Adaptive Vector Quantization

A vector quantizer is *adaptive* if the coding rule or codebook is changed in time to track changes in the short-term statistics of the input sequence of vectors. The adaptation is typically based on the current input vector or vectors within a small neighborhood of it. The motivation for such adaptation is twofold. First, since

input vectors such as those formed by dividing images into blocks exhibit significant variation, the tracking of input statistics through adaptation adds flexibility
and robustness to a quantizer and can result in improved performance. Second,
complexity constraints limit the dimension of vector that can be used. Adaptation
based on neighboring vectors can exploit intervector redundancy and thus exploit
correlation over a larger number of samples than in an individual vector.

Although PVQ and FSVQ studied in the previous section may be viewed as
adaptive, the term adaptive is usually associated with schemes where the codebook
changes gradually as the statistics of the signal, a sequence of input vectors,
gradually change. In contrast, FSVQ can make a major change in its codebook
with each new input vector. Also, adaptive VQ schemes are usually heuristically
designed from intuition and do not have the rigid structure of PVQ and FSVQ.

Adaptation can be either backward or forward [1]. *Backward adaptation* utilizes
information from previously quantized vectors. Since the decoder has access to
these vectors, no side information is needed for backward adaptation. *Forward
adaptation* on the other hand, extracts information from the current input vector or
even future vectors (via input buffering) and uses this information for adaptation.
Since forward adaptation is based on information not yet available at the decoder,
side information has to be transmitted for decoder tracking. Forward adaptation
is generally more effective since current and future information is utilized as
opposed to backward adaptation, which relies solely on the past. However, the
increase in rate due to side information must be justified by a sufficient increase in
performance. Next, we describe several examples of adaptive VQ restricting our
attention to those wherein a codebook is modified to track the input.

A simple example of a forward-adaptive VQ scheme, is *mean–adaptive VQ*
where a group of input vectors, say image blocks, is buffered before encoding
and the mean of the entire set of buffered pixels is computed, quantized, and
transmitted to the decoder. At the same time, the quantized mean is subtracted
from each component of the buffered vectors and the residual vectors are then
each coded with a fixed codebook. Unlike MRVQ, a mean value is quantized only
once for a group of vectors rather than once for each input vector.

In progressive code vector updating, the VQ codebook is slowly updated based
on the incoming vectors. In one such scheme [17], the distortion incurred in
quantizing each input vector is measured and, if it exceeds a certain threshold,
then that particular input vector is added to the codebook. To keep the codebook
size constant, the code vector that has not been used for the longest time is
removed from the codebook. Thus the codebook adapts itself by incorporating
within it those input vectors that it cannot code adequately and discarding those
code vectors that have been idle for a long time and therefore are less likely to be
effective in coding the current input sequence. In this forward adaptation scheme,
the side information required at the decoder to update the codebook consists of (1)

a flag informing the decoder that a code vector has been replaced and (2) the new code vector, which is the current input vector. Note that, since the decoder knows the code vector indexes for previously quantized vectors, it can extract the code vector with the longest idle time and its index need not be transmitted. In another variation, to limit the number of bits needed to specify the new code vector, a large secondary codebook is employed. The candidate code vector is quantized with this large secondary codebook and the corresponding code vector index transmitted to the decoder. Since the decoder has a copy of the secondary codebook, it can generate the new input vector by table lookup in this secondary codebook. Several other variations of progressive code vector updating exist but the philosophy of gradually modifying the codebook to track the changing statistical character of the input is the same in each such method.

In another form of adaptive VQ, adaptation is performed by designing and transmitting a new codebook rather than merely updating it one code vector at a time [18]. Although one may suspect that the bandwidth required to transmit a new codebook will be so excessive as to render such a method useless for data compression, it can be shown that if the codebook is replaced infrequently, the average bit rate for side information is low enough to justify the use of such a scheme. As in progressive code vector updating, the number of bits needed for specification of a new codebook can be constrained by allowing code vectors to take values only from a large, predesigned "universal" codebook [19]. Thus each new codebook is a subset of code vectors from this universal codebook. In such a case, the new codebook is specified to the decoder by transmitting the indexes of the selected code vectors from the universal codebook. Candidate code vector selection for each input image begins by using a number of blocks from the input image as "training data." If the training set size equals the desired codebook size, then the training data is quantized with the universal codebook and the corresponding code vectors form the new codebook. Otherwise, if the number of training vectors is large, then Lloyd iterations are first performed to design a codebook of desired size. The codebook thus obtained is quantized with the universal codebook to obtain the new codebook. In another variation of the scheme, the VQ encoder for an input image employs two codebooks: a relatively large fixed codebook and a small supplemental codebook. Only the supplemental codebook is updated for each image to reduce the side information needed for codebook specification.

Adaptive VQ design consists of designing an initial codebook and specifying the adaptation technique. The initial codebook is generally designed by applying a standard VQ design algorithm, such as the GLA, to a training set. If adaptive VQ employs a large secondary codebook, care should be taken that training data is sufficiently large and varied to ensure its robustness. The adaptation process is heuristic and based on an intuitive understanding of the input data. The examples described earlier in this section should provide the reader with a taste of adaptation techniques. Since the innovation and effectiveness of adaptive VQ as compared to direct VQ is rooted in its ability to extract information from an input sequence and its consequent utilization in effectively tracking input statistics, effective adaptation is critical to designing high performance adaptive VQ.

6.4 Vector Quantization in Transform and Subband Coding

An effective image coding technique is to decompose the image signal into a set of components that can be more efficiently quantized than the original signal. This class of coding schemes includes transform coding and subband coding, which are widely used in image compression. Vector quantization can be used with these signal decomposition techniques to boost the performance of the system.

In transform coding, the image is first partitioned into small-size blocks. A linear transformation (usually orthogonal) is applied to convert each image block into a block of transform coefficients, which are then quantized. The image is reconstructed by inverse transformation of the quantized transform coefficients. Parallel to transform coding is subband coding, including pyramids, whereby the image is decomposed into a set of spectral components by filtering and down-sampling. In fact, we may consider each transform coefficient in transform coding as a spectral subband. Readers are referred to other chapters in this book for more details of transform coding and subband coding.

The basic idea of vector quantization in the transform domain is a straightforward extension of its scalar counterpart: instead of scalar quantization, the transform coefficients are combined into groups, each group being a vector of some particular dimension. However, it is sometimes more instructive and more convenient for analytical purpose to formulate transform coding itself as a structured VQ technique rather than considering transform domain VQ as an improvement over conventional transform coding. In this more general point of view, an input vector consists of one or more image blocks, and a product code is generated by quantizing a set of low-dimensional feature vectors constructed from the transformed blocks. In the special case of conventional transform coding with scalar quantization, the dimension of the feature vectors is unity. The feature vectors are then quantized and combined to synthesize the output vector by the inverse transformation.

Compared with scalar quantization, vector quantization of the transform coefficients has several advantages. First, the Karhunen–Loeve transform (KLT), which is the optimal block transformation for decorrelating the signal, is hard to implement. Therefore, suboptimal but computationally efficient transformations like the discrete cosine transform (DCT) are commonly used to replace the KLT in image coding [20]. Since the transform coefficients are not completely decorrelated in these suboptimal transforms, VQ can be applied to exploit the remaining correlation. Second, for real images, linear transformations can exploit the linear dependency only in the first and second order statistics of the signal. In other words, the transform coefficients are at most statistically uncorrelated but not independent. The nonlinear dependency and the correlation in the higher order statistics remains to be exploited by VQ. Even if we assume that the transform coefficients are independent, identically distributed sources, VQ can still take advantage of the increased freedom in partitioning the vector space of the transformed image block with increased dimension. VQ can also be better tailored to fit the shape of the probability distribution of the transform coefficients than scalar quantization [21]. It can also give greater flexibility in bit allocation since fractional rates are possible.

In the light of its advantages, VQ has been extensively applied in the transform (usually DCT) domain. Recent research has been focused on (1) means to construct vectors from the transformed image such that the vectors can be efficiently quantized, (2) adaptive quantization and bit allocation for the vectors of transform coefficients, and (3) hybrid techniques to combine transform domain vector quantization with other coding methods to efficiently code the image. These issues are examined here.

To exploit the advantages of using VQ in the transform domain, the vectors should be defined so that there is a high degree of dependency among the elements of the vector. Basically, the vectors can be formed by grouping transform coefficients from different spectral bands of an individual image block (interband) or by grouping coefficients from the same spectral band across adjacent image blocks (intraband). In general, if the block size is small, we expect to find much correlation among transform coefficients of the same band across adjacent blocks, so there is more to gain in intraband VQ than interband VQ. On the other hand, if the block size is large, then interband VQ would be more favorable. Since current technology allows two-dimensional transformations of relatively large block sizes (e.g., 8×8, 16×16) to be computed in real time, interband vector quantization is often preferred to intraband vector quantization of the transform coefficients. Intraband transformed VQ can be found in some earlier works [22] in the literature.

In the basic form of the interband transformed VQ, a set of vectors are constructed for each image block by partitioning the corresponding block of transform coefficients, and each vector is coded by a distinct VQ tailored for the statistics of the transform coefficients that compose the vector. The bit rate for each VQ is usually determined by some bit allocation schemes. Various partitioning schemes and bit allocation schemes, either fixed or adaptive, have been used, some of which are described here. Usually the vectors are constructed in such a way that the elements in the same vector have similar variances [23], but other vector construction schemes have also been studied [24, 25] and good results have been obtained.

A special case of bit allocation is to assign zero bits to some of the vectors. This is equivalent to zonal sampling in which some of the transform coefficients are discarded. At the decoder, these missing coefficients are usually assigned zero amplitude; however, it is also possible to nonlinearly interpolate (i.e., estimate) their values from the other coded transform coefficients [7]. In a primitive form of this VQ scheme [26], a fixed zonal sampling pattern is used to construct a vector of low-order 4×4 DCT coefficients. This fixed zonal sampling method is effective for coding the regions of the image with low to medium detail where the gradient of the pixel intensity is moderate. In high-detail regions, however, DCT coefficients outside the sampling zone may have nonnegligible magnitudes. Therefore, VQ with fixed zonal sampling of DCT coefficients is often used as a building block in more complicated systems to code the low-detail regions, while the high-detail regions are coded by other techniques [26, 27]. Alternatively, a second stage can be adaptively applied to correct the error in high-detail regions that are incurred by abandoning the high-order DCT coefficients [28].

In addition to using a fixed partition and bit allocation scheme, the transform coefficients can be more efficiently quantized by adapting the partition and bit allocation to the local characteristic of the transformed block [24, 29–31]. In these schemes, each transformed block of the image is classified into one of several classes. The classification is based on the energy distribution of the transform coefficients, which varies with the type of scene contained in the image block such as shadings and different edge orientations. For each class of blocks, specific schemes for vector composition and bit allocation are used.

To reduce the computational complexity, structurally constrained VQ techniques are often used in the place of full search VQ to code the transform coefficients. In principle, any of the structurally constrained VQ techniques described elsewhere in this chapter could be applied. However, since the statistics of the transform coefficients of natural images usually have a small mean value, shape–gain VQ [32] is often preferred [23, 25, 27, 28, 33] as an alternative to full search VQ in the transform domain.

It is also possible to use VQ together with scalar quantization in a hybrid scheme to code the transform coefficients. A common example is to code the dc coefficients of the DCT with DPCM to exploit the interblock correlation while the ac coefficients are coded by VQ. In a more sophisticated application of VQ, the transform coefficients are adaptively scalar quantized, while the "quantizer selection vectors," or the side information that adapts the scalar quantizers, is vector quantized. This VQ scheme has been successfully implemented in a practical system to compress high definition television signals [34].

The philosophy of vector quantization in subband coding is similar to that in transform coding. Either interband VQ, intraband VQ, or a combination of both can be used to code the subbands. In earlier works, interband VQ has been used [35]. A more popular approach is to apply intraband VQ in a hybrid scheme to code the high-frequency bands at a high compression ratio while the low-frequency bands are coded by scalar quantization to provide the desired fidelity [36].

Intraband VQ has also been applied on multiresolution image pyramids [37], which is a special case of subband decomposition where each level of the pyramid corresponds to a particular spectral band of the image. In one proposal of pyramidal VQ [38], the pyramid is coded in a "closed loop" form with a multistage hierarchical VQ approach. The advantage of such a closed loop form is that the quantization error in a lower level of the pyramid will be accumulated in the higher levels, hence may be corrected with the coding of the higher levels. Recently, VQ has also been applied to code wavelet pyramids, and good results have been reported [39].

6.5 Vector Quantization in Interframe Video Coding

As described in detail in other chapters, video coding is typically performed by motion-compensated prediction or interpolation of the current frame from previously quantized frames. Here we briefly consider the role of VQ in interframe video coding.

A digital video sequence is generally formed by sampling a scene at discrete, equally spaced intervals in space and time and is specified by the pixel value at each sampling point. We will use the notation $F(\mathbf{p}, t)$, where $\mathbf{p} = (x, y)$ is the spatial coordinate vector and t is the time instant, to describe a spatio–temporal video signal. By fixing t at t_k, where k is the discrete time index, and varying the spatial coordinate vector $\mathbf{p}$, we get a snapshot at time t_k, referred to as the kth frame.

A simple interframe predictor for a video sequence is obtained by predicting the "background" area in the current frame by the spatially congruent area in the previous coded frame and setting the predictor to "zero" for the "foreground." Thus the estimate $\tilde{F}(\mathbf{p}, k)$ of the current frame using the prior coded frame $\hat{F}(\mathbf{p}, k-1)$ is

$$\tilde{F}(\mathbf{p}, k) = \begin{cases} \hat{F}(\mathbf{p}, k-1) & \mathbf{p} \in \text{background} \\ 0 & \mathbf{p} \in \text{foreground} . \end{cases} \tag{6.7}$$

This approach, called *conditional frame replenishment*, estimates only the background areas in the current frame [40]. In addition to the previous quantized frame, the predictor needs to know whether a pixel belongs to the background or not.

Transmitting this information for each pixel would be very inefficient. The need to efficiently specify which regions of the image are background versus foreground naturally suggests a block-based partition of the image. Thus the current frame is divided into spatially contiguous non-overlapping blocks, and a single membership decision is made for each block rather than each pixel. A block-based algorithm evaluates the prediction error for the two allowed cases for each block in the current frame and assigns it to the class that results in lower prediction error.

Block-based conditional replenishment can be viewed as a predictive vector coding scheme, where the successive blocks in time for a particular spatial location constitute a sequence of vectors. In particular, block conditional replenishment can be viewed as a form of adaptive VQ with a codebook that changes with each frame and that consists of two code vectors: C_0, which is the vector of all zeros, and C_1, which is the previously quantized block from the prior frame. Transmitting a binary-valued flag is equivalent to transmitting the index of the minimum distortion code vector. If it is 1, the block is replenished; otherwise it is not. Note that, since the decoder can generate the codebook for each block from the previously coded frame, it needs only a single bit index for its interframe prediction.

Performance of interframe prediction can be significantly enhanced with motion-based interframe prediction, commonly known as "motion-compensated prediction." Obviously, accurate motion estimation is the key to motion-compensated prediction. Note that although motion can be complex and nontranslational, for analytical tractability, objects are assumed to move parallel to the camera axis in straight lines. Experimental results have confirmed that such a model is adequate in most cases. With a block matching algorithm (BMA), it is assumed that all pixels in a block undergo uniform motion. Most contemporary video coders use block matching algorithms that, as in block-based conditional replenishment, can be viewed as a form of VQ.

The task of block matching algorithms is to find for each block the motion vector that minimizes some measure of the prediction error. Note that for a candidate motion vector $\mathbf{d} = (d_x, d_y)$, the predicted value of a pixel in block $\mathbf{B}$ is

$$\tilde{F}(\mathbf{p}, k) = \hat{F}(\mathbf{p} - \mathbf{d}, k - 1), \qquad \mathbf{p} = (x, y) \in \mathbf{B}. \tag{6.8}$$

If we assume a certain maximum displacement in both x and y directions, say $\mathbf{d}_{max}$, then the desired motion vector $\mathbf{d}^* = (d_x{}^*, d_y{}^*)$ should minimize the distortion

$$\mathbf{D} = \sum_{p \in \mathbf{B}} \| F(\mathbf{p}, k) - \hat{F}(\mathbf{p} - \mathbf{d}, k - 1) \|^2 \tag{6.9}$$

over all allowable motion vectors $\mathbf{d}$. The optimum motion vector can be obtained by searching for the nearest neighbor match to the input block $\mathbf{B}$, which we call the *target*, among all possible candidate prediction blocks and transmitted to the decoder. Thus the search for an optimal motion vector is equivalent to the operation of an adaptive VQ encoder, where the codebook consists of all candidate prediction blocks from the quantized prior frame. For a block **B** in frame k, the codebook is formed by applying a sliding window to a region centered around the current block location in the quantized previous frame $k - 1$. Since the decoder can generate the codebook from the previously coded frame, it needs to receive only the index (or equivalently, the motion vector) identifying the selected code vector in this codebook to form the prediction.

In the MPEG video coding algorithm [41], motion compensation is enhanced by including interpolation, wherein a block, the target, is estimated by averaging coded values from appropriately displaced blocks in two frames, one ahead of it in time and one behind it. For interpolation, each block needs two motion vectors. This scheme for generating the prediction (or more correctly, the *estimate*) of the current block can also be viewed as another form of VQ encoding. The equivalent codebook of candidate interpolation vectors is a structured codebook in which each code vector is an average of two code vectors, one from the codebook of candidate blocks from the future frame and the other from the codebook of candidate blocks from the previous frame. The search for an interpolation vector in the structured codebook that best matches a given target vector can be done in various ways. In an exhaustive search, every possible pair of candidates is tested. In a sequential search, one codebook is searched first and the result is used to modify the target vector for the second search, and then the other codebook is used to find the best match to the modified target vector. Finally, in an independent search, each codebook is searched independently to find the best match to the target vector. The indexes of the code vectors selected from the two codebooks specify the forward and backward motion vectors.

Once motion-compensated prediction or interpolation is performed, the remaining task of a video coder is to code the residual frame after the prediction is subtracted from the original frame. This coding task is equivalent to intraframe image coding with the key difference that the statistical character of a residual image is quite different from those of typical still images. In any case, the wealth of VQ-based image coding techniques is applicable to the coding of residual frames.

6.6 Variable Bit-Rate Vector Quantization

It is well-known that the amount of information in a scene varies with its context; for instance, edges carry more information than smooth areas. This fact motivates the use of variable rate coding schemes, where the number of bits consumed in coding a block of pixels depends on the information content of the scene contained in the block. In contrast to its fixed bit rate counterpart, variable rate coding takes advantage of the nonstationary property of image signals by allocating a varying number of bits to code different regions of the picture so that an efficient distortion–rate tradeoff may be achieved.

In general, a variable rate block coder can be modeled as the composition of a block quantizer and a variable bit rate lossless coder, as shown in Fig. 6.4. For each input block of pixels, **X**, the block quantizer selects a discrete symbol **I** from a finite alphabet $\mathcal{I}$ according to some encoding rule. The symbol **I** is then mapped by the lossless coder to a variable length binary code word, which is released to the channel and transmitted to the decoder. At the decoder, the binary code word is transformed into the reconstructed image block, usually by regenerating **I** as an intermediate step in the decoding process.

The symbol **I** could be a single index as in vector quantization or a composite symbol produced by a structured block quantization scheme. For example, the block quantizer can be a transform coder with scalar quantization for the transform coefficients, where **I** is the ordered set of indexes obtained by encoding the individual transform coefficients of the block. Another common example is an adaptive block coding scheme in which each image block is coded by a block coder selected from a finite set of candidate coders. In this case, **I** is composed of both the selected coder's identity and the actual index (or set of indexes) generated by the selected coder.

A simple way to construct a variable bit-rate system is to perform the quantization independent of the variable length lossless coding, so that the image blocks are encoded with the sole objective of minimizing the distortion, disregarding the bit rate that results after variable length coding. This approach is clearly suboptimal, since it does not consider overall system performance. For instance, it is possible that a better overall distortion–rate tradeoff will result from coding some particular image block with fewer bits and allowing a slightly higher distortion; the alternative in this case might require a large incremental expenditure of bits to achieve a very small reduction in distortion. The objective in the design of a variable bit-rate coder is not to minimize the distortion, but to obtain an efficient distortion–rate tradeoff. From another perspective, variable rate coding strives to

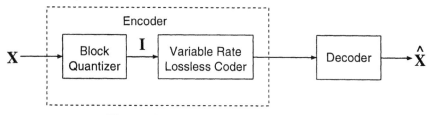

Figure 6.4 Model of a variable rate block coder.

minimize distortion for a given *average* bit rate. Allowing a slight distortion increase in one block might release a few extra bits for another block where the incremental bit allocation could provide a greater benefit by more substantially reducing distortion.

Let us now look at the problem in a more precise manner. There are three components in the variable rate block coder model: (1) the encoding rule, (2) the variable rate lossless coder, and (3) the decoder that defines the set of all possible reproduction image blocks or code vectors. Given the encoding rule, the variable length lossless coder is an entropy coder designed from the statistics of the indexes, where the average bit rate of the code words is lower bounded by the entropy of the indexes. Examples are Huffman coding, Lempel–Ziv coding, and arithmetic coding. For the decoder, optimality is governed by the same centroid condition as we have discussed earlier in this chapter for codebook design. Instead of optimal decoding by an unstructured table-lookup process as in VQ, some structural composition rules are applied to suboptimally (but with reduced complexity) reconstruct the image block from the decoded symbol $\mathbf{I}$. The decoding problem will be examined in more detail in the next section. Here, we may simply consider the decoder as a predefined mapping from binary code words to reproduction image blocks. Our emphasis in this section is on the design of the encoding rule.

Given the decoder and the entropy coder, the design of the encoding rule can be formulated as a constrained optimization problem similar to the classical bit allocation problem. The objective is to minimize the overall distortion subject to the constraint that the bit rate is less than or equal to a target bit rate, where the distortion and the bit rate can be either actually calculated from coding the particular input image or estimated, usually from a training set, as expected values for the possible input images. Notice the significance of the bit-rate constraint in the problem. Without the bit-rate constraint, the optimal strategy would be to code the image at the lowest possible distortion achievable by the decoder; i.e., the nearest neighbor encoding rule of VQ.

A Lagrangian formulation [42] can be applied to transform this constrained optimization problem into the more tractable unconstrained minimization problem:

$$\min\{D + \lambda R\} \tag{6.10}$$

for some $\lambda \geq 0$, where D is the distortion and R is the bit rate mentioned in the previous paragraph. Solving this for a particular value of λ gives a solution of the constrained problem for some particular rate constraint determined by the value of λ. The key to this Lagrangian formulation is in locating a λ such that the solution bit rate of the unconstrained problem (6.10) is equal to, or close to, the target bit rate. Some techniques for locating λ will be described later.

If the overall distortion is measured as the sum of the distortions of individual image blocks and these distortions are independent of each other, then (6.10) is equivalent to minimizing the Lagrangian separately for each block. Therefore, with a given decoder and entropy coder, the optimal encoding rule for an input block of pixels $\mathbf{X}$ with distortion measure $d(\cdot, \cdot)$ is

$$\min_{I \in \mathcal{I}} \{d(\mathbf{X}, \hat{\mathbf{X}}(I)) + \lambda r(I)\}, \tag{6.11}$$

where $\hat{\mathbf{X}}(I)$ and $r(I)$ are, respectively, the quantized image block and the length of the binary code word associated with the symbol I. This condition is analogous to the nearest neighbor condition of vector quantization. In fact, the optimal structure for variable bit rate block coding has been considered in the vector quantization literature as *entropy-constrained vector quantization* (ECVQ) [43].

The ECVQ encoder searches through a codebook for a code vector that minimizes the Lagrangian in (6.11) and transmits the variable length code word that corresponds to the index $\mathbf{I}$ of this "best matched" code vector. The vector is reconstructed by table lookup in the same way as in simple VQ. The Lagrangian can be considered as a biased distortion measure in which a penalty $\lambda r(\mathbf{I})$ is added to the actual distortion $d(\mathbf{X}, \hat{\mathbf{X}}(\mathbf{I}))$. If the length of a binary code word corresponding to a particular code vector is short, then the penalty for that code vector is small, otherwise the code vector is heavily penalized. Thus λ adjusts the sensitivity of the penalty to the code word length, and controls the average bit rate. As λ increases, the penalty increases, therefore, the average bit rate decreases. Similarly, the average bit rate increases if λ decreases.

An algorithm for the design of the ECVQ codebook from a training set was given in [43]. Similar to the GLA algorithm for VQ codebook design, this ECVQ codebook design algorithm employs a descent method that iteratively improves the codebook, starting from some initial configuration, until the distortion converges to a local minimum. An ad hoc technique is used to select a set of values of λ, and a codebook is designed for each value of λ. The codebook that results in an average bit rate closest to the target bit rate is employed. Once the codebook is chosen, actual operation requires the use of a modified nearest neighbor encoding rule that depends on the value of λ for that codebook and on the length of the binary code words that represent the code vectors. ECVQ has been applied to image coding with good results [44, 45].

Despite its optimality, ECVQ is not widely used because of its excessively high complexity. For ECVQ to be effective, the size of the codebook has to be at least an order of magnitude larger than the codebook of ordinary VQ of the same bit rate. Consequently, structured encoding (or decoding) methods have been developed.

A popular structured encoding method is the tree-structured vector quantization, which can be applied naturally to variable rate coding with the use of unbalanced trees. There are two approaches to generate the variable length code words with unbalanced TSVQ: (1) along the path traversed from the root node to a leaf, concatenating the intermediate code words that correspond to the selected branch at each level of the tree, or (2) entropy code the indexes that associate with the leaves. The latter is slightly more complicated, but generally results in a lower average bit rate. In unbalanced TSVQ, the encoding rule is embedded in the structure of the tree, which is designed with the objective of minimizing the expected distortion subject to the constraint on the expected bit rate.

An optimal pruning algorithm based on the work of Breiman, Friedman, Olshen, and Stone [46], called the *generalized BFOS* (GBFOS) algorithm, has been adopted by Chou, Lookabaugh and Gray to design unbalanced TSVQ [47]. The GBFOS algorithm searches for the value of λ as a slope on the lower convex hull of the finite set of solution points in the rate–distortion plane. The minimization of (6.10) is not performed explicitly, but recursively during the search for λ. Starting from an initial TSVQ of very high bit rate, the algorithm iteratively prunes off one node at a time to produce a series of unbalanced TSVQs of decreasing bit rate. At each iteration, the node that is selected and pruned is the one that will result in minimum rate of increase in distortion with respect to decrease in bit rate. Although each pruning operation is optimal in some sense, the overall design procedure does not produce trees with optimal performance. Also, only a limited set of rates are available without resorting to an awkward time sharing between two TSVQ structures. The unbalanced TSVQ thus designed is called *pruned tree–structured VQ* (PTSVQ) in the literature [1], and has been demonstrated to be effective for image coding [11].

In general, given a finite set of arbitrary block coders that operate at different bit rates, a variable rate coder can be constructed with an adaptive mechanism in which one of the block coders is selected to code each input image block. The following are some examples where a set of candidate block coders can be constructed:

1. *Transform coding.* Each candidate coder corresponds to a different scheme for allocating bits to code the transform coefficients.

2. *Hierarchical multistage coding.* An image block can be progressively coded in a hierarchy. Each level of the hierarchy represents a different candidate coder.

3. *Variable block size coding* [27, 48]. An image block can be coded as a single entity at a low bit rate or partitioned into smaller size blocks and coded at higher bit rates. Each candidate coder corresponds to a different block partitioning scheme.

Subject to the structural constraint defined by the set of candidate coders, the optimal strategy to code an image block is to first perform a trial encoding of the block with every candidate coder and then apply the biased nearest neighbor rule (6.11) to select the best coder based on the trial quantized image block and the code word output from each of the candidate coders [28]. This strategy has been applied to optimally adapt a variable rate block coding scheme to code images at a fixed overall bit rate, where the value of λ is estimated by an iterative algorithm that is essentially the same as the GBFOS algorithm, but originally developed in the context of bit allocation [49]. Starting from a heuristically chosen initial guess, the value of λ is iteratively updated in such a way that, in each iteration, the solution bit rate migrates toward the target bit rate and the rate of change in distortion with respect to bit rate is optimized. The algorithm terminates in a finite number of iterations when the solution bit rate is equal to, or close to, the desired target bit rate.

6.7 Enhanced Decoding

In spite of the theoretical advantages of VQ, in many applications the available computational power of an encoder requires the use of low-dimensional vectors, product code VQ schemes, or coding schemes that make little or no explicit use of VQ. Yet, even in such cases, it is possible to enhance the performance of the *decoder* with the aid of NLIVQ and related techniques without altering the operation of the encoder. In this section, we shall consider some of these VQ-based advanced decoding techniques that are capable of enhancing picture quality with given suboptimal encoders.

Consider a product code encoder that extracts a set of feature vectors from an input vector $\mathbf{X}$ and encodes the feature vectors into an ordered set of indexes, denoted by the compound index $\mathbf{I} = [I_1 I_2 \cdots I_K]$. The *optimal decoder* is one generates the minimum distortion reproduction of $\mathbf{X}$ given $\mathbf{I}$. For the well-known squared error distortion measure, the optimal decoder reproduces the input vector as the conditional expectation of $\mathbf{X}$ given $\mathbf{I}$. Since the set of possible values of $\mathbf{I}$ is finite, the optimal decoder could in principle be implemented by a table-lookup structure in which the compound index $\mathbf{I}$ has value i and addresses a codebook to fetch the code vector

$$\mathbf{C}(\mathbf{i}) = E[\mathbf{X} \,|\, \mathbf{I} = \mathbf{i}]. \tag{6.12}$$

This optimal decoding structure is shown in Fig. 6.5. Notice that the optimality of the structure in Fig. 6.5 is independent of the particular distortion measure being considered.

In the same way as centroids are computed for a given partition in ordinary VQ codebook design, the codebook of this optimal decoder can be computed from a training set of vectors $\{\mathbf{v}\}$ representing the source $\mathbf{X}$ by finding the average

$$\mathbf{C}(i) = \frac{1}{|\mathcal{R}_\mathbf{i}|} \sum_{v \in \mathcal{R}_\mathbf{i}} \mathbf{v}, \tag{6.13}$$

where $\mathcal{R}_\mathbf{i}$ is the subset of training vectors that are encoded with index $\mathbf{I} = \mathbf{i}$, and $|\cdot|$ denotes the number of vectors in the subset. However, the size of the codebook in

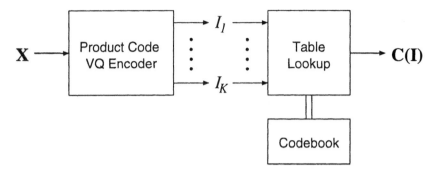

Figure 6.5   Optimal decoder for product code VQ.

such an optimal decoder grows exponentially with the bit rate; therefore, optimal decoding is feasible only at low rates. This formulation follows the approach of NLIVQ [7] except that here we consider a compound index **I** each of whose component indexes might be generated by a separate encoding module to represent a separate signal vector.

In NLIVQ, the original input vector **Y** is processed to extract a lower dimensional feature vector **X**, which is then encoded by VQ. The index so generated addresses a codebook at the decoder to look up the corresponding reproduction code vector, which is designed from a training set using (6.13) and has the same dimensionality as the input vector **Y**. This NLIVQ paradigm was first applied as an optimal nonlinear interpolation method in interpolative vector quantization of images, where the feature vector is obtained by decimating the input vector **Y** [50]. It can be easily shown that the reproduction code vectors at the NLIVQ decoder are indeed the optimally nonlinear interpolated version of the quantized feature vector.

In another application of NLIVQ [13], the input vector is composed of a set of cosituated image blocks in a sequence of highly correlated images. One of the image blocks in the set is taken as the feature vector. The decoder reproduces, from the index generated by quantization of the feature vector, the mmse estimate for the entire set of image blocks. Since NLIVQ is basically a low bit rate coding technique, it is often used as the first level in a multistage hierarchical scheme where the residual error is corrected by one or more subsequent coding stages.

Because of the excessive storage complexity required by the reproduction codebooks, optimal decoding is not practical at higher bit rates. Therefore structurally constrained vector decoding techniques that alleviate the codebook storage complexity by *synthesizing* the reproduction code vectors at the decoder have been studied. Structurally constrained vector decoding imposes constraints on the reproduction code vectors, hence results in generally higher distortion than optimal decoding. In fact we can view all conventional ways of decoding product codes as structurally constrained decoding, where the structure employed usually depends on the method used to extract the feature vectors from the input vector. However, the performance degradation can be minimized by carefully designing the decoder.

An example of such a compromise is the use of an additive vector decoding structure shown in Fig. 6.6 to decode enhanced quality pictures from bit streams produced by conventional transform encoding [51]. In this case, the transform coefficients are scalar quantized, and we can formulate a product code by considering each scalar encoded transform coefficient for an input image block as an element of the compound index **I**. Instead of the conventional way of decoding the indexes into the quantized transform coefficients followed by inverse transformation, each element I_k of the compound index addresses a distinct component codebook $\mathcal{C}_k$ to fetch a component code vector $\mathbf{C}_k(I_k)$. The component code vectors are defined in the spatial domain and have the same dimensionality as the input vector. The output image vector is then reconstructed as the vector sum

$$\hat{\mathbf{X}} = \sum_{k=1}^{K} \mathbf{C}_k(I_k). \tag{6.14}$$

It can be easily seen that the conventional decoder with inverse transformation is equivalent to this additive vector decoder if the ith code vector of the kth codebook is defined as $q_{ik}\mathbf{b}_k$, where q_{ik} is the reconstruction level of the kth transform coefficient associated with $I_k = i$, and $\mathbf{b}_k$ is the kth basis vector of the transformation.

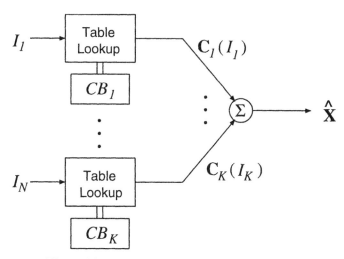

Figure 6.6   Additive vector decoder for product code VQ.

The key to enhanced decoding with the additive vector decoder is to properly design a set of component codebooks so that the output image blocks can be reconstructed with lower distortion than that achieved by the conventional method. For the mean squared distortion measure, an algorithm for designing a set of locally optimal codebooks has been developed in [51]. The main idea of this algorithm is that the distortion can be written as

$$d = E[\|\mathbf{X} - \sum_{k=1}^{K} \mathbf{C}_k(I_k)\|^2] = E[\|\mathbf{G}_j - \mathbf{C}_j(I_j)\|^2] \qquad (6.15)$$

for any $j \in \{1, 2, \ldots, K\}$, where

$$\mathbf{G}_j = \sum_{k=1, k \neq j}^{K} \mathbf{C}_k(I_k). \qquad (6.16)$$

Suppose all component codebooks except $\mathbf{C}_j$ are fixed. Then the code vectors of the component codebook $\mathbf{C}_j$ that minimizes (6.15) are given by the condition expectation

$$\mathbf{C}_j(i) = E[\mathbf{G}_j | I_j = i]. \qquad (6.17)$$

Notice that $\mathbf{G}_j$ and I_j are deterministic functions of the input image block $\mathbf{X}$; therefore, this conditional expectation can be estimated from a training set as an ensemble average similar to that in (6.13). Hence, starting from an initial configuration defined by the conventional decoder, we can iteratively apply (6.17) to improve one component codebook at a time, keeping all other codebooks fixed. Since the distortion is bounded from below and decreases monotonically with successive iterations, it converges to a local minimum.

Besides jointly decoding the intrablock indexes of a product code, enhanced decoding can also be achieved by exploiting the interblock correlation among the indexes. Consider now the simple case of a full search VQ. For each input vector, we can construct a compound index whose elements are the indexes encoded from the set of input vectors in its neighborhood including itself. Optimal decoding could also be applied to decode the output vector from the compound index by table lookup from a single codebook, but the codebook size is too large for any practical purpose.

A method to exploit the interblock correlation at the VQ decoder without excessive storage complexity is to use a block overlapping structure [52], which is called *lapped VQ*. In lapped VQ, the decoder tries to extract the maximum amount of information from each index by decoding it into a reproduction of higher dimensionality than the input vector. As illustrated in Fig. 6.7, each output vector extends beyond the area covered by the corresponding input vector into its neighborhood. The image is reconstructed as an overlapping patchwork of output vectors, where the pixel values in the lapped region are obtained by summing

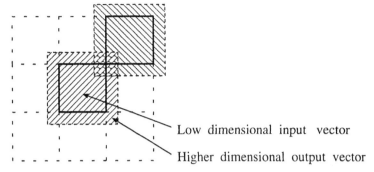

Low dimensional input vector

Higher dimensional output vector

Figure 6.7   Block overlapping at the decoder.

the corresponding elements of the overlapping code vectors. Intuitively, the inner elements of an output code vector are estimates of the pixels in the input vector, whereas the outer elements offer "predictions" of the neighborhood that attempt to correct for the estimation error after the inner region of the neighboring code vector has offered its best approximation. The decoder codebook of lapped VQ can be designed by an iterative algorithm similar to that described earlier for the additive vector decoder. Compared to conventional VQ decoding, lapped VQ not only reproduces images with higher SNR, but also exhibits less blockiness and less of a staircase effect along image edges.

The block overlapping technique can be applied with the additive vector decoding scheme we discussed earlier to improve product code decoding. In this lapped additive vector decoding scheme, the dimension of the component code vectors, and hence the dimension of the output image blocks, is higher than that of the input image blocks. The output image is constructed from the output image blocks by overlap and add as in lapped VQ. This lapped additive vector decoding method has been applied to decode enhanced quality pictures in transform coding, and more than 0.5 dB gain in SNR with improved perceptual quality has been reported [53].

6.8 Concluding Remarks

Since the first applications of vector quantization to image coding were reported in 1982 [54–56], there has been a tremendous growth in the use of VQ for image compression. Today, there is a great diversity of VQ-based techniques, and new methods continue to be introduced each year. Several commercial products for image and video coding have emerged, and it is generally recognized that VQ is a fundamental and generic tool that must be given serious consideration in the design of a new image or video compression scheme. In this chapter, we have covered some of the important techniques that are widely applicable and through the references, we offer pointers to the literature for further reading.

References

[1] A. Gersho and R. Gray, *Vector Quantization and Signal Compression*. Norwell, MA: Kluwer Academic Publishers, 1992.

[2] Y. Linde, A. Buzo, and R. M. Gray, "An algorithm for vector quantizer design." *IEEE Trans. Commun.*, vol. COM-28, pp. 84–95, Jan. 1980.

[3] K. Zeger, J. Vaisey, and A. Gersho, "Globally optimal vector quantizer design by stochastic relaxation." *IEEE Trans. Signal Processing*, vol. ASSP-40, pp. 310–322, Feb. 1992.

[4] K. Rose, E. Gurewitz, and G. C. Fox, "Vector quantization by deterministic annealing." *IEEE Trans. Inform. Theory*, vol. IT-38, pp. 1249–1258, 1992.

[5] W.-Y. Chan and A. Gersho, "High fidelity audio transform coding with vector quantization." In *Proc. Int. Conf. Acoust., Speech, Signal Processing*, April 1990, pp. 1109–1112.

[6] W.-Y. Chan and A. Gersho, "Generalized product code vector quantization: A family of efficient techniques for signal compression." *Digital Signal Processing*, vol. 4, pp. 95–126, April 1994.

[7] A. Gersho, "Optimal nonlinear interpolative vector quantization." *IEEE Trans. Commun.*, vol. COM-38, pp. 1285–1287, Sept. 1990.

[8] A. Gersho, "Optimal vector quantized nonlinear estimation." In *Proc. IEEE Int. Symp. Inform. Theory*, 1993, pp. 170.

[9] V. Cuperman and A. Gersho, "Vector predictive coding of speech at 16 kb/s." *IEEE Trans. Commun.*, vol. COM-33, pp. 685–696, July 1985.

[10] H.-M. Hang and J. W. Woods, "Predictive vector quantization of images." *IEEE Trans. Commun.*, vol. COM-33, pp. 1208–1219, Nov. 1985.

[11] E. A. Riskin, E. Daly, and R. M. Gray, "Pruned tree-structured vector quantization in image coding." In *Proc. Int. Conf. Acoust., Speech, Signal Processing*, Glasgow, Scotland, May 1989, pp. 1735–1738.

[12] P. C. Chang and R. M. Gray, "Gradient algorithms for designing predictive vector quantizers." *IEEE Trans. Acoust., Speech, Signal Processing*, vol. ASSP-34, pp. 679–690, Aug. 1986.

[13] S. Gupta and A. Gersho, "Feature predictive vector quantization of multispectral images." *IEEE Trans. Geoscience and Remote Sensing*, vol. 30, pp. 491–501, May 1992.

[14] J. Foster, R. M. Gray, and M. O. Dunham, "Finite-state vector quantization for waveform coding." *IEEE Trans. Inform. Theory*, vol. IT-31, pp. 348–359, May 1985.

[15] R. Aravind and A. Gersho, "Image compression based on vector quantization with finite memory." *Optical Eng.*, vol. 26, pp. 570–580, July 1987.

[16] N. M. Nasrabadi and Y. Feng, "A dynamic finite-state vector quantization scheme." In *Proc. Int. Conf. Acoust., Speech, Signal Processing*, April 1990, pp. 2261–2264.

[17] D. Paul, "A 500–800 bps adaptive vector quantization vocoder using a perceptually motivated distance measure." In *Conf. Record, IEEE Globecom'82*, 1982, pp. 1079–1082.

[18] M. Goldberg, P. R. Boucher, and S. Shlien, "Image compression using adaptive vector quantization." *IEEE Trans. Commun.*, pp. 180–187, Feb. 1986.

[19] S. Panchanathan and M. Goldberg, "Algorithms and architecture for image adaptive vector quantization." In *Proc. SPIE Conf. Visual Commun. Image Processing*, Cambridge, MA, Nov. 1988, vol. 1001, pp. 336–344.

[20] A. Jain, *Fundamentals of Digital Image Processing*. Englewood Cliffs, NJ: Prentice-Hall, 1989.

[21] T. D. Lookabaugh and R. M. Gray, "High-resolution quantization theory and the vector quantizer advantage." *IEEE Trans. Inform. Theory*, vol. IT-35, pp. 1020–1033, Sept. 1989.

[22] N. M. Nasrabadi and R. A. King, "Image coding using vector quantization: A review." *IEEE Trans. Commun.*, vol. COM-36, pp. 957–971, Aug. 1988.

[23] K. Aizawa, H. Harashima, and H. Miyakawa, "Adaptive discrete cosine transform coding with vector quantization for color images." In *Proc. Int. Conf. Acoust., Speech, Signal Processing*, April 1986, pp. 20.1.1–20.1.4.

[24] J. W. Kim and S. U. Lee, "Discrete cosine transform—Classified VQ technique for image coding." In *Proc. Int. Conf. Acoust., Speech, Signal Processing*, May 1989, pp. 1831–1834.

[25] Y. Du and J. Halfmann, "Comparison of coding performance of image transforms under vector quantization of optimized subbands." In *Proc. SPIE Int. Symp. Visual Commun. Image Processing*, Nov. 1989, pp. 1418–1429.

[26] B. Ramamurthi and A. Gersho, "Classified vector quantization of images." *IEEE Trans. Commun.*, vol. COM-34, pp. 1105–1115, Nov. 1986.

[27] D. J. Vaisey and A. Gersho, "Variable block-size image coding." In *Proc. Int. Conf. Acoust., Speech, Signal Processing*, April 1987, pp. 1051–1054.

[28] S.-W. Wu and A. Gersho, "Rate-constrained optimal block-adaptive coding for digital tape recording of HDTV." *IEEE Trans. Circuits Syst. Video Technol.*, vol. 1, pp. 100–112, March 1991.

[29] J. Marescq and C. Labit, "Vector quantization in transformed image coding." In *Proc. Int. Conf. Acoust., Speech, Signal Processing*, April 1986, pp. 4.5.1–4.5.4.

[30] M. Breeuwer, "Transform coding of images using directionally adaptive vector quantization." In *Proc. Int. Conf. Acoust., Speech, Signal Processing*, April 1988, pp. 788–791.

[31] Y. S. Ho and A. Gersho, "Classified transform coding of image using interpolative vector quantization." In *Proc. Int. Conf. Acoust., Speech, Signal Processing*, May 1989, pp. 1890–1893.

[32] M. J. Sabin and R. M. Gray, "Product code vector quantizers for waveform and voice coding." *IEEE Trans. Acoust. Speech, Signal Processing*, vol. ASSP-32, pp. 474–488, June 1984.

[33] T. Saito, H. Takeo, K. Aizawa, H. Harashima, and H. Miyakawa, "Adaptive discrete cosine transform image coding using gain/shape vector quantizers." In *Proc. Int. Conf. Acoust., Speech, Signal Processing*, April 1986, pp. 4.1.1–4.1.4.

[34] AT&T and Zenith Corp., *Digital Spectrum Compatible Technical Description*, Feb. 1991.

[35] P. H. Westerink, D. E. Boekee, J. Biemond, and J. Woods, "Subband coding of images using vector quantization." *IEEE Trans. Commun.*, vol. COM-36, pp. 713–719, June 1988.

[36] C. Podilchuk, N. Jayant, and P. Noll, "Sparse codebooks for the quantization of non-dominance sub-bands in image coding." In *Proc. Int. Conf. Acoust., Speech, Signal Processing*, April 1990, pp. 2101–2104.

[37] P. J. Burt and E. H. Adelson, "The Laplacian pyramid as a compact image code." *IEEE Trans. Commun.*, vol. COM-31, pp. 552–540, April 1983.

[38] Y. S. Ho and A. Gersho, "A pyramidal image coder using contour-based interpolative vector quantization." In *Proc. SPIE Int. Symp. Visual Commun. Image Processing*, Nov. 1989, vol. 1199, pp. 733–740.

[39] M. Antonini, M. Barlaud, P. Mathieu, and I. Daubechies, "Image coding using wavelet transform." *IEEE Trans. Image Processing*, vol. 1, pp. 205–220, April 1992.

[40] A. N. Netravali and B. G. Haskell, *Digital Pictures: Representation and Compression*. New York: Plenum Press, 1988.

[41] D. J. LeGall, "The MPEG video compression algorithm: A review." In *Proc. SPIE Conf. Image Processing Algorithms and Techniques*, Feb. 1991, pp. 444–457.

[42] H. Everett, "Generalized Lagrange multiplier method for solving problems of optimum allocation of resources." *Operations Res.*, vol. 11, pp. 399–417, 1963.

[43] P. A. Chou, T. Lookabaugh, and R. M. Gray, "Entropy-constrained vector quantization." *IEEE Trans. Acoust., Speech, Signal Processing*, vol. ASSP-37, pp. 31–42, Jan. 1989.

[44] P. Chou, "Application of entropy-constrained vector quantization to waveform coding of images." In *Proc. SPIE Int. Symp. Visual Commun. Image Processing*, Nov. 1989, pp. 970–978.

[45] Y. Kim and J. Modestino, "Adaptive entropy coded subband coding of images." *IEEE Trans. Image Processing*, vol. 1, pp. 31–48, Jan. 1992.

[46] L. Breiman, J. H. Friedman, R. A. Olshen, and C. J. Stone, *Classification and Regression Trees*. Belmont, CA: Wadsworth, 1984.

[47] P. A. Chou, T. Lookabaugh, and R. M. Gray, "Optimal pruning with applications to tree-structured source coding and modeling." *IEEE Trans. Inform. Theory*, pp. 299–315, March 1989.

[48] G. Sullivan and R. Baker, "Efficient quadtree coding of images and video." In *Proc. Int. Conf. Acoust., Speech, Signal Processing*, May 1991, pp. 2661–2664.

[49] Y. Shoham and A. Gersho, "Efficient bit allocation for an arbitrary set of quantizers." *IEEE Trans. Acoust., Speech, Signal Processing*, vol. ASSP-36, pp. 1445–1453, Sept. 1988.

[50] Y. S. Ho and A. Gersho, "A variable rate image coding scheme with vector quantization and clustering interpolation." In *Proc. Globecom'89*, 1989, pp. 25.5.1–25.5.1.

[51] S.-W. Wu and A. Gersho, "Improved decoder for transform coding with application to the JPEG baseline system." *IEEE Trans. Commun.*, pp. 251–254, Feb. 1992.

[52] S.-W. Wu and A. Gersho, "Lapped block decoding for vector quantization of images." In *Proc. SPIE Int. Symp. Visual Commun. Image Processing*, Nov. 1992.

[53] S.-W. Wu and A. Gersho, "Nonlinear interpolative decoding of standard transform coded images and video." In *SPIE Conf. Image Processing Algorithms and Techniques III*, Feb. 1992.

[54] A. Gersho and B. Ramamurthi, "Image coding using vector quantization." in *Proc. Int. Conf. Acoust., Speech, Signal Processing*, Paris, April 1982, vol. 1, pp. 428–431.

[55] R. L. Baker and R. M. Gray, "Image compression using non-adaptive spatial vector quantization." in *Conf. Record of the Sixteenth Asilomar Conf. on Circuits Syst. and Comput.*, Asilomar, CA, Oct. 1982.

[56] T. Murakami, K. Asai, and E. Yamazaki, "Vector quantizer of video signals." *Electron. Lett.*, vol. 7, pp. 1005–1006, Nov. 1982.

Chapter 7

Transform Coding

R. L. de Queiroz and K. R. Rao
Electrical Engineering Department
University of Texas at Arlington
Arlington, Texas

This chapter presents the role of discrete transforms, especially discrete cosine transform (DCT) and lapped orthogonal transform (LOT), in image coding. The application of these transforms both in still frame and image sequence coding is illustrated. Other functions such as quantization, motion estimation, human visual sensitivity, and variable length coding that are inherent to the overall coding scheme are also described.

7.1 Introduction

While various discrete transforms [1–8] such as Walsh–Hadamard, Haar, slant, discrete Fourier (DFT), DCT, discrete sine (DST), LOT, and discrete wavelet transform (DWT) have been investigated for application to image coding, only DCT has emerged as the most practical and efficient transform. The LOT [5, 6] and DWT (see [7, 8] and Chapter 8 in this book) have been extensively simulated in still frame and image sequence coding and have proven to be formidable competitors to the DCT. It is to be cautioned that transform by itself is only a part of the overall compression scheme, as the coding process may involve

Handbook of Visual Communications
223

preprocessing, quantizers, buffers, variable length coding, human visual sensitivity, motion compensation, multiplexers, network interfaces, etc. [9–13]. In hybrid schemes, this is supplemented with other techniques such as prediction, vector quantization, subband, etc. The objective of this chapter is to present the concepts of transform coding with some emphasis on DCT and LOT. Specific details regarding still frame image compression (JPEG, Joint Photographic Experts Group [14]) and image sequence coding (Recommendation H.261—Video Codec for Audiovisual Services at p×64 Kbit/sec) are presented to illustrate the practical applications and potential for additional interactive video services [15–19] The spectrum of applications range from very low bit-rate (say 40 Kbit/sec) coding to super high-definition (SHD) images. The spectrum includes different disciplines such as medical imaging, remote sensing, consumer electronics, printing and publishing, defense, television, sports, communications, and storage. It is hoped that the topics presented in this chapter will provide the background and incentive for exploring further in this constantly changing field. The reader is also encouraged to survey the literature to gain familiarity with the state of the art and be aware of the limitations that may be overcome eventually.

7.1.1 Linear Transforms over Images

The idea is to apply a linear transform over a sequence of N samples $x(n)$, in the spatial domain to obtain another sequence $y(n)$ $(n = 0, 1, \ldots, N - 1)$ in the transform domain. These transforms are invertible and obey a consistent system of linear equations which, in matrix notation, becomes

$$y = A_N x \qquad (7.1)$$

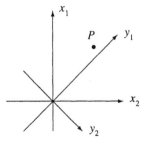

Figure 7.1 Illustration of the advantage of transforming two samples, using orthogonal transforms, when the pair of samples has an expected region of occurrence.

where $\mathbf{y}^T = [y(0), y(1), \ldots, y(N-1)]$, $\mathbf{x}^T = [x(0), x(1), \cdots, x(N-1)]$, and $\mathbf{A}_N = \{a_{ij}\}$, where $\{a_{ij}\}$ means a matrix with elements a_{ij} (in this case, i and j range from 0 through $N-1$). In a simple example ($N=2$) in Fig. 7.1 we have two samples (x_1 and x_2) extracted from a signal forming the pair $P = (x_1, x_2)$. Suppose P is likely to lie over a limited region. If $\mathbf{A}$ is an orthogonal matrix, the transform will correspond to a plane rotation, yielding a new pair of coordinates (y_1 and y_2). P can be expressed in both pair of axis, but choosing the (y_1, y_2) representation we can accurately code y_1, and perhaps discard y_2, without disturbing much the position of P.

The same principles apply to M-dimensional signals, leading to the main concept of transform coding (Fig. 7.2), which is to transform the signal and code more efficiently the samples that are more important, i.e., carry more energy of the input signal. At the decoder side, the samples are decoded and an inverse transform is carried out for recovering the signal.

To transform a two-dimensional (2-D) array we will use the so-called separable transforms, which reduce the problem as a succession of one-dimensional (1-D) transforms. First, each row of the image is transformed independently. The result is stored and over it is applied another 1-D transform, now, along the columns. Let a 2-D array of samples be $\mathbf{X} = \{x(n_1, n_2)\}$, for $0 \leq n_1 \leq N_1 - 1$, $0 \leq n_2 \leq N_2 - 1$. If $\mathbf{A}_{N_1}$ and $\mathbf{A}_{N_2}$ are the transform matrices applied to columns and rows respectively, then the transform array is

$$\mathbf{Y} = \mathbf{A}_{N_1} \mathbf{X} \mathbf{A}_{N_2}^T. \tag{7.2}$$

We will assume real orthogonal transforms; i.e., $\mathbf{A}_{N_1}^{-1} = \mathbf{A}_{N_1}^T$ and $\mathbf{A}_{N_2}^{-1} = \mathbf{A}_{N_2}^T$. Thus,

$$\mathbf{X} = \mathbf{A}_{N_1}^{-1} \mathbf{Y} (\mathbf{A}_{N_2}^{-1})^T = \mathbf{A}_{N_1}^T \mathbf{Y} \mathbf{A}_{N_2}. \tag{7.3}$$

This approach allows us to concentrate on the study of 1-D transform to perform the task of 2-D transforms.

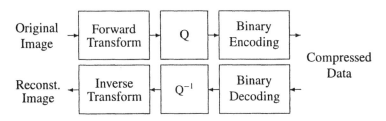

Figure 7.2   Basic concepts of transform coding.

7.2 Transforming the Signal

7.2.1 Block Transforms

In practice, orthogonal transforms are applied to consecutive blocks of M samples ($N = N_B M$, $N \gg M$), so that adaptive features based on block activity, detail, etc., can be introduced (Fig. 7.3). Also, there is no significant coding efficiency when the transform is applied to block length $M > 16$. Let the samples in any block be denoted as $\mathbf{x}_B^T = [x_0, x_1, \ldots, x_{M-1}]$; i.e., $x_i = x(kM + i)$ for any block index k. The corresponding transform vector is $\mathbf{y}_B^T = [y_0, y_1, \ldots, y_{M-1}]$. For a real unitary transform, $\mathbf{A}_M$, $\mathbf{A}_M^T = \mathbf{A}_M^{-1}$. The forward and inverse transforms are

$$\mathbf{y}_B = \mathbf{A}_M \mathbf{x}_B, \tag{7.4}$$

and

$$\mathbf{x}_B = \mathbf{A}_M^T \mathbf{y}_B. \tag{7.5}$$

The rows of $\mathbf{A}_M$, denoted $\mathbf{a}_n^T$ ($0 \leq n \leq M - 1$), are called the *basis vectors* because they form an orthogonal basis for the M-tuples over the real field. The transformed coefficients $[y_0, y_1, \ldots, y_{M-1}]$ represent the corresponding weights of vector $\mathbf{x}$ with respect to this basis. Then

$$\mathbf{A}_M = \begin{bmatrix} a_{00} & a_{01} & \cdots & a_{0,M-1} \\ a_{10} & a_{11} & \cdots & a_{1,M-1} \\ \vdots & \vdots & & \vdots \\ a_{M-1,0} a_{M-1,1} \cdots a_{M-1,M-1} \end{bmatrix} = \begin{bmatrix} \mathbf{a}_0^T \\ \mathbf{a}_1^T \\ \vdots \\ \mathbf{a}_{M-1}^T \end{bmatrix}. \tag{7.6}$$

As the block transform ignores the correlation between samples across the block boundaries, at low bit rates, an undesirable artifact called *block structure* arises in the reconstructed signal. Several techniques can reduce or eliminate this artifact (see the next section on lapped transforms).

It is well-known that the signal energy is preserved under an orthogonal transformation, assuming stationary signals; i.e.,

$$M\sigma_x^2 = \sum_{i=0}^{M-1} \sigma_i^2, \tag{7.7}$$

where σ_i^2 is the variance of y_i and σ_x^2 is the variance of the input samples.

Doubtless, the most popular block transform for image coding is the discrete cosine transform, which is orthogonal, has several desirable properties and can be implemented using fast algorithms. Indeed, the DCT is the transform used

$$\longleftarrow \text{------------} N = N_B M \text{ samples } \text{------------}\longrightarrow$$

$\leftarrow M \rightarrow$	$\leftarrow M \rightarrow$	$\leftarrow M \rightarrow$	$\leftarrow M \rightarrow$	$\leftarrow M \rightarrow$	$\leftarrow M \rightarrow$	$\leftarrow M \rightarrow$	$\leftarrow M \rightarrow$
0	1	2					$N_B - 1$

Figure 7.3   Sequence of N samples is divided into N_B blocks. Each block has M samples. 1-D transform is applied to each successive block.

in most image and video coding standards (see Chapter 11), and several VLSI implementations are now available for the DCT (see Chapter 15). Appendix 7.A presents a brief description of the DCT and its features.

7.2.2 Lapped Transforms

For lapped transforms [5, 6], the basis vectors can have length L, such that $L > M$, extending across traditional block boundaries. Thus, the transform matrix is no longer square and most of the equations for block transforms do not apply. We will concentrate our efforts on *orthogonal* lapped transforms and consider $L = N_o M$, where N_o is the overlap factor. As in the case of block transforms, we define the transform matrix as containing the orthonormal basis vectors as its rows. A lapped transform matrix $\mathbf{P}$ of dimensions $M \times L$ ($L = N_o M$) can be divided into square $M \times M$ submatrices $\mathbf{P}_i$ ($i = 0, 1, \ldots, N_o - 1$) as

$$\mathbf{P} = [\mathbf{P}_0 \ \mathbf{P}_1 \ \cdots \ \mathbf{P}_{N_o-1}]. \tag{7.8}$$

The orthogonality property is replaced by the perfect reconstruction (PR) property, defined by

$$\sum_{m=0}^{N_o-1-l} \mathbf{P}_m \mathbf{P}_{m+l}^T = \sum_{m=0}^{N_o-1-l} \mathbf{P}_{m+l} \mathbf{P}_m^T = \delta(l) \mathbf{I}_M \tag{7.9}$$

for $l = 0, 1, \ldots, N_o - 1$, where $\delta(l)$ is the Kronecker delta; i.e., $\delta(0) = 1$ and $\delta(l) = 0$ for $l \neq 0$.

As the transform matrix is no longer square, the division of the input signal into blocks is not helpful without considering all samples in the input signal. Following our notation, the input 1-D signal is composed of samples $x(n)$ for $0 \leq n \leq N - 1$. If we divide the signal into blocks, each of size M, we would have vectors $\mathbf{x}_k^T = [x(kM), x(kM + 1), \ldots, x(kM + M - 1)]$, for $0 \leq k \leq N_B - 1$. These blocks, by themselves, are useless for applying lapped transforms. The actual vector that is transformed by the matrix $\mathbf{P}$ has to have L samples and, at block number k, it is composed by the samples of $\mathbf{x}_k$ plus $L - M$ samples. These samples are chosen, picking $(L - M)/2$ samples at each side of the block $\mathbf{x}_k$, as shown in Fig. 7.4, for $N_o = 2$. However, the number of transform coefficients at

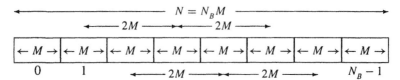

Figure 7.4 The signal samples are divided into N_B blocks, each of M samples. The lapped transform uses neighboring blocks samples, as in this example for $N_o = 2$, i.e., $L = 2M$, yielding an overlap of $(L - M)/2 = M/2$ samples on either side of a block.

each step is M, and in this respect, there is no change in the way we represent the transform-domain coefficients. To keep track of the block index, we will represent the transform vector at block k by

$$\mathbf{y}_k^T = [y(kM), y(kM + 1), \ldots, y(kM + M - 1)], \tag{7.10}$$

and the input vector of length L is denoted as $\mathbf{v}_k$, which is centered around the block $\mathbf{x}_k$ and is defined as

$$\mathbf{v}_n^T = \left[x\left(kM - (N_o - 1)\frac{M}{2} \right) \cdots x\left(kM + (N_o + 1)\frac{M}{2} - 1 \right) \right]. \tag{7.11}$$

Then, we have

$$\mathbf{y}_k = \mathbf{P}\mathbf{v}_k, \qquad 0 \le k \le N_B - 1. \tag{7.12}$$

The inverse transform is not direct as in the case of block transforms, i.e., with the knowledge of $\mathbf{y}_k$ we do not know the samples in the support region of $\mathbf{v}_k$, and neither in the support region of $\mathbf{x}_k$. We can reconstruct a vector $\hat{\mathbf{v}}_k$ from $\mathbf{y}_k$, as

$$\hat{\mathbf{v}}_k = \mathbf{P}^T \mathbf{y}_k \qquad 0 \le k \le N_B - 1, \tag{7.13}$$

where $\hat{\mathbf{v}}_k \ne \mathbf{v}_k$. To reconstruct the original sequence, it is necessary to accumulate the results of the vectors $\hat{\mathbf{v}}_k$, in a sense that a particular sample $x(n)$ will be reconstructed from the sum of the contributions it receives from all $\hat{\mathbf{v}}_k$, such that $x(n)$ was included in the region of support of the corresponding $\mathbf{v}_k$. This additional complication comes from the fact that $\mathbf{P}$ is not a square matrix. However, the whole analysis–synthesis system (applied to the entire input vector) is orthogonal, assuring the PR property using (7.13).

We can also describe the process using a sliding rectangular window applied over the samples of $x(n)$. As an M-sample block $\mathbf{y}_k$ is computed using $\mathbf{v}_k$, $\mathbf{y}_{k+1}$ is computed from $\mathbf{v}_{k+1}$, which is obtained by shifting the window to the right by M samples, as in Fig. 7.5.

As the reader may have noted, the region of support of all vectors $\mathbf{v}_k$ is greater than the region of support of the input vector. Hence, a special treatment has to

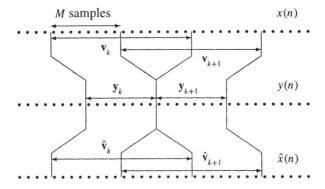

Figure 7.5   Illustration of a lapped transform with $N_o = 2$ applied to signal $x(n)$, yielding transform-domain signal $y(n)$. The input L-tuple as vector $\mathbf{v}_k$ is obtained by a sliding window advancing M samples, generating vectors $\mathbf{y}_k$. The sliding is also valid for the synthesis side.

be given to the transform at the borders. If we denote by $\mathbf{x}$ the input vector of $N = N_B M$ samples and by $\mathbf{y}$ the transform-domain vector, we can be consistent with our notation of transform matrices by defining a matrix $\mathbf{A}$ such that $\mathbf{y} = \mathbf{Ax}$ and $\hat{\mathbf{x}} = \mathbf{A}^T \mathbf{y}$. In this case, we have

$$
\mathbf{A}_N = \begin{bmatrix} \mathbf{P}_L & & & & & & \\ & \mathbf{P} & & & & & \\ & & \mathbf{P} & & & & \\ & & & \ddots & & & \\ & & & & \mathbf{P} & & \\ & & & & & \mathbf{P} & \\ & & & & & & \mathbf{P}_R \end{bmatrix}. \tag{7.14}
$$

$\mathbf{P}_L$ and $\mathbf{P}_R$ are special matrices applied to the borders of the signal and the displacement of the matrices $\mathbf{P}$ obeys the following

$$
\begin{bmatrix} \ddots & \ddots & & \ddots & \\ & \mathbf{P}_0 \mathbf{P}_1 \cdots \mathbf{P}_{N_o-1} & & \\ & & \mathbf{P}_0 \mathbf{P}_1 & \cdots & \mathbf{P}_{N_o-1} \\ & & & \ddots & \ddots & & \ddots \end{bmatrix}. \tag{7.15}
$$

$\mathbf{A}_N$ has N_B block rows, one for each transform operation over each vector $\mathbf{v}_k$ $(0 \le k \le N_B - 1)$.

Let the rows of $\mathbf{P}$ be denoted by vectors $\mathbf{p}_i^T$ ($0 \leq i \leq M - 1$), so that $\mathbf{P}^T = [\mathbf{p}_0 \mathbf{p}_1 \cdots \mathbf{p}_{M-1}]$. In an analogy to the block transform case, we have

$$y(kM + i) = \mathbf{p}_i^T \mathbf{v}_k. \tag{7.16}$$

The vectors $\mathbf{p}_i$ are the basis vectors of the lapped transform. They form an orthogonal basis for an M-dimensional subspace (there are only M vectors) of the L-tuples over the real field. As a remark, assuming infinite length signals, from the orthogonality of the basis vectors and from the PR property in (7.9), the energy is preserved, such that (7.7) is valid.

These formulations for lapped transforms are general, and if the transform satisfies the PR property equations in (7.9), they are independent of the contents of the matrix $\mathbf{P}$. For example, suppose a block transform, such as the DCT is chosen to transform the signal, but we want to use the direct algorithm for an $N_o = 4$ lapped transform. If $\mathbf{C}$ is the $M \times M$ DCT matrix, we can use

$$\mathbf{P} = [000\mathbf{C}000],$$

where $\mathbf{0}$ is an $M \times M/2$ null matrix. If we augment the length of the basis vectors, from the center to the borders, but maintaining PR, the formulas in this section would still be valid. The definition of $\mathbf{P}$ with a given N_o can accomodate any lapped transform whose length of the basis vectors lies between M and $N_o M$. For the case of block transforms we can set directly $N_o = 1$. This illustrates the fact that block transforms are a special case of lapped transforms [6, 20, 21]. The advantage of viewing block transforms as lapped transforms is clear as we know that lapped transforms, obeying (7.9), are paraunitary filter banks, if each basis function element is an element of the analysis or synthesis filters. Let $F_m(e^{j\omega})$ be the Fourier transform of the sequence composed by the elements of $\mathbf{p}_m$, and let the input signal be stationary with power spectral density (PSD) given by $S_{xx}(e^{j\omega})$. Then, the variance of each coefficient can be computed using the results for filter banks [6, 21, 22], as

$$\sigma_m^2 = \frac{1}{\pi} \int_0^\pi S_{xx}(e^{j\omega}) \, |F_m(e^{j\omega})|^2 \, d\omega. \tag{7.17}$$

This result is quite general and states the transform as a filtering process; i.e., subband processing.

We will focus our attention to the lapped orthogonal transform, which has symmetric basis vectors and fast algorithms based on the DCT. Appendix 7.B presents a brief description of the LOT, including flow graphs for fast implementations over finite length signals, which circumvent the boundary problems. The formulations here presented, however, are general and can be applied to any other lapped transform.

7.3 Performance of Transforms

In Fig. 7.1 we were considering the advantage of using transforms under a geo-metrical viewpoint. Now, we will consider the statistical version of this point of view. We want to compact variances in the frequency domain, in a way that fewer coefficients would have most part of the energy of the signal. A very popular measure of this compaction is provided by the *transform coding gain*, or G_{TC}. It is defined by the ratio between arithmetic and geometric means of the set of variances, and it is possible to show that this measure is also the gain of transform coding over PCM coding [13]. Then

$$G_{TC} = 10 \log_{10} \left[\frac{\frac{1}{M} \sum_{i=0}^{M-1} \sigma_i^2}{\left(\prod_{i=0}^{M-1} \sigma_i^2 \right)^{1/M}} \right], \qquad (7.18)$$

with σ_i^2 as the variance of the ith subband signal for $i = 0, 1, \ldots, M - 1$.

Another important issue concerning the choice of transform is the computa-tional complexity required for its implementation. This complexity can be mea-sured in operations per sample, where operations can be additions, multiplications, floating point operations, or any other measure of importance to the particular hardware where the implementation of the transform coder will take place.

Block transforms, such as the DCT, suffer an undesirable effect commonly present in low bit rate coding, called the *blocking effect* [3–5]. This is caused by artificial edges at the blocks' boundaries, having the appearance of bricks in the image, as shown in Fig. 7.6, using the improved Chen–Smith coder [23].

The LOT presents higher G_{TC} than the DCT and the blocking effect is largely reduced [5, 6]. On the other hand, LOT demands larger implementation complexity, establishing a tradeoff, which has to be considered before choosing the appropriate transform for a specific application.

7.4 Representation of a Transformed Image

The image samples are denoted as $x(n_1, n_2, k)$, where the three indices denote ver-tical, horizontal, and temporal resolutions. Considering still images and dropping the temporal index, we have $x(n_1, n_2)$ for $0 \le n_1 \le N_1 - 1$ and $0 \le n_2 \le N_2 - 1$, where $N_1 \times N_2$ are the dimensions of the image. Assume M divides both N_1 and N_2. A 2-D transform of each block results in $M \times M$ transform coefficients (see

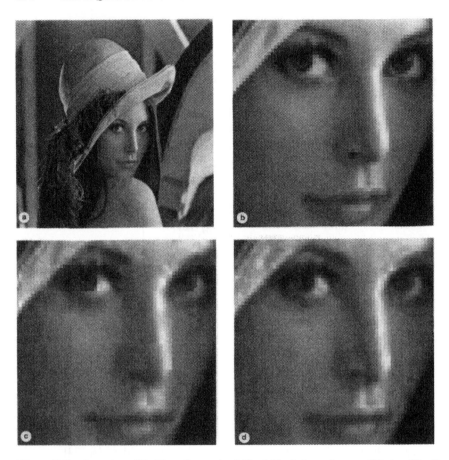

Figure 7.6   Illustration of blocking effects, using 256 × 256 pels image Lena, and blocks of 8 × 8 pels: (a) original image; (b) zoom of (a); (c) zoom of image coded at 0.5 bit/pel using DCT; (d) zoom of image coded at 0.5 bit/pel using LOT.

Fig. 7.7), which are combined to make the transform array $y(n_1, n_2)$. We adopt the following notation:

$$y(i, j; m, n) = y(mM + i, nM + j) \qquad (7.19)$$

for $0 \leq i, j \leq M - 1$, $0 \leq m \leq N_{B1} - 1$, and $0 \leq n \leq N_{B2} - 1$, where N_{B1} and N_{B2} are the number of blocks in vertical and horizontal directions, respectively. We also denote $N_B = N_{B1} N_{B2}$ as the total number of blocks in the image. The $y(0, 0; m, n)$ are called the dc coefficients, while the remaining $M^2 - 1$ elements of each block are called the ac coefficients. The position of the ac coefficients in a block is the set of all ij in the block except for the pair 00 and this set

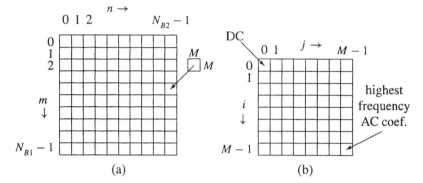

Figure 7.7 Illustration of the position of the transform coefficients $y(i, j; m, n)$. (a) An $N_1 \times N_2$ image is divided into $N_{B1} \times N_{B2}$ blocks, each of size $M \times M$. A 2-D transform is applied to each block resulting in M^2 transform coefficients. A transformed block is illustrated in (b), indicating the position of the dc coefficient and of the highest frequency ac coefficient.

will be represented as Ψ, i.e., Ψ represents the position of the ac coefficients. For video sequences, naturally, the blocks can have an extra temporal index as $y(i, j; m, n; k)$.

7.5 Quantizers and Entropy Coding

7.5.1 Quantizer Operation

The transforms discussed here involve numbers of long precision, and transform coefficients assume a wide range of values. To achieve compression, the coefficients are quantized to a finite number of levels. Assume each coefficient has a well-defined probability density function (PDF). For a well-chosen transform, as is the case of those discussed here, this PDF will be a concentrated symmetric function around the origin. The zero-mean Gausssian and Laplacian PDFs are usually chosen as models for this function. The DC coefficient in general has nonzero mean and is coded separately. The input to a quantizer is, thus, considered a random variable X with well-known statistics. The quantization–coding operation [12, 13] can be described in steps: (1) the input X is assigned to a quantization level number ℓ and the set of all possible values of ℓ is called the codebook; (2) ℓ is coded into binary numbers and transmitted to the receiver; (3) the receiver decodes the binary data and recovers the number ℓ; (4) with the number ℓ at hand, the receiver decodes a number $\hat{X}$, which is the quantized value of X. We can view this process as quantization (Q) and inverse quantization (Q^{-1}), noting that, connected back to back, they form a lossy operation.

$$X \xrightarrow{Q} \ell \xrightarrow{Q^{-1}} \hat{X}.$$

Let $\{t_{-q-1}, \ldots, t_{-2}, t_{-1}, t_1, t_2, \ldots, t_{q+1}\}$, where $t_{-q-1} = -\infty$, $t_{q+1} = \infty$ be a set of distinct increasing real numbers, so that $t_i < t_j$ if $i < j$. This set defines a partition of the range of all real numbers into $2q + 1$ nonoverlapping segments. To label these segments, let $1 \leq n \leq q$, then

$$I_{-n} \equiv (t_{-n-1}, t_{-n}], \qquad (7.20)$$
$$I_0 \equiv (t_{-1}, t_1), \qquad (7.21)$$
$$I_n \equiv [t_n, t_{n+1}). \qquad (7.22)$$

The assignment $x \to \ell$ is governed by: $\ell = n$ if $x \in I_n$. Here ℓ is assigned to a real number r_ℓ so that $\hat{x} = r_\ell$. Note that, in general, $r_\ell \in I_\ell$. Hence, the set of decision and reconstruction values completely specify the quantizer. We will consider two basic symmetric quantizers: the dead-zone quantizer and the nonuniform quantizer. For both, $t_n = -t_{-n}$ and $r_n = -r_{-n}$. Fig. 7.8 shows the relation between X and $\hat{X}$ for both quantizers.

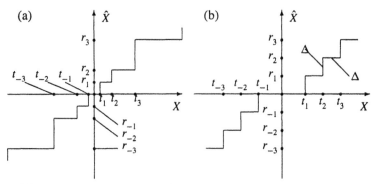

Figure 7.8 Basic quantizer types: (a) nonuniform; (b) deadzone. The decision and reconstruction values are shown and , therefore, the relation between original (X) and reconstructed ($\hat{X}$) signals.

7.5.2 Entropy Coding

As the number of occurrences of ℓ close to zero is greater than in other positions, we can use entropy coding to efficiently encode the output symbols of the quantizer. Let X have PDF $f(x)$, then, the probability of each quantizer output symbol is

$$p_i \equiv Pr[i \in I_k] = \int_{I_k} f(x)dx, \qquad (7.23)$$

and the entropy of the quantizer output is

$$H = -\sum_{i=-q}^{q} p_i \log_2 p_i. \qquad (7.24)$$

The form by which you assign a binary code to the quantizer level is crucial. There are several ways to do it, but we will consider only four possible cases. Assume X is to be quantized to one out of N_L levels, using a specific quantizer Q and that the entropy of the quantizer output is H for the specific PDF of X.

The first case is the trivial fixed length code assignment to $b = \log_2 N_L$ bits, where b is rounded to the smallest integer greater than $\log_2 N_L$.

The second possibility is to assign a variable length code to the output of Q, using the Huffman code. In this code, more frequent levels are assigned to shorter code words. The code generated is optimal in a sense that the average length is minimal for a given set of p_k. If a code word of n_k bits is assigned to level $\ell = k$, then it is possible to prove that [24]

$$H \le \sum_i p_i n_i \le H + 1. \qquad (7.25)$$

The design procedure of Huffman codes is left to [24, 25]. A disadvantage is that, as $n_i \ge 1$, the average length is always greater than unity. Thus, it is not as efficient if the source entropy is lower than 1 bit/symbol, without forming vectors. However, the code words can be instantaneously decoded. Another approach is the modified Huffman code, where a group of m less probable symbols is assigned to a new symbol, let us say, ELSE. The symbol ELSE is assigned to a code word, and another code word with $\log_2 m$ bits is appended to this code to distinguish each original symbol. This is useful to limit the maximum code word size.

The third method to encode the quantizer output levels is called *arithmetic coding*. In this, for a long enough sequence, the output bit rate is supposed to reach the entropy of a memoryless source. Unlike Huffman codes, this coding procedure can achieve rates below 1 bit/symbol, but it is not instantaneously decodable. Details on the implementation of arithmetic coders can be found in [26] and references therein.

The fourth method is the so-called run length coding. It is useful when the source emits long sequence of repetitive symbols. Thus, for a certain sequence, the code generated consists of two integers (m, n), meaning a run of m occurrences of the nth symbol.

7.5.3 Quantizer Design Issues

The dead-zone quantizer just has three parameters to design: the step size Δ, and the first and last decision levels, t_1 and t_q. Since we are assuming symmetric quantizers, t_{-1} and t_{-q} follow immediately, as well as the number of quantization levels. The uniform quantizer is achieved when $t_1 = \Delta/2$. Generally, q is set to a value such that $(q - \frac{1}{2})\Delta$ lies close to the maximum range of the coefficients, for a value of Δ that is expected to be most commonly used. Thus, in a uniform quantizer, the step size is the only parameter to be varied and it can control the rate and distortion of the coder, since larger Δ means more distortion and less entropy of the output symbols. The dead-zone is used most commonly to control buffers in entropy coding to match the bit rate produced by the coder and the bit rate the channel supports [27].

The nonuniform quantizer can be designed in a way to minimize the mean squared distortion produced by the quantizer, which is defined as

$$D_Q = \sum_{\ell=-q}^{q} \int_{l_\ell} (x - r_\ell)^2 f(x)\, dx. \qquad (7.26)$$

The well-known Lloyd–Max design method provides an iterative way to specify the decision and reconstruction values, which will minimize D_Q for a given PDF $f(x)$ [12, 13]. In this case, there is no concern with the rate by which the quantizer output symbols would be coded. Another method involves the restriction that the output symbols should be subject to a given entropy [28]. This is called the *entropy-constrained design method*. In this, the number of levels is not given as a parameter, but found from the design algorithm. The other parameters given are the entropy of the output set of symbols and the PDF is assumed. In both methods we can assume a unit-variance input PDF. Hence, X has to be scaled by its standard deviation to turn it into a unit-variance signal, before quantization.

7.6 Quantizer Selection

7.6.1 Uniform Quantizers

The coefficients $y(i, j; m, n)$ have to be quantized and encoded into a bit stream to be transmitted or stored. Assume the number of quantizer levels N_L is large enough such that between x and $x + dx$ there are $N(x)dx$ quantization levels.

We define $\lambda(x) = \lim_{N_L \to \infty} N(x)/N_L$. It can be shown that, for large N_L the distortion of the optimized quantizer is [12]

$$D_Q = \frac{1}{12} \frac{1}{N_L^2} \int_{-\infty}^{\infty} f(x)\lambda^{-2}(x)\,dx, \qquad (7.27)$$

and that the entropy of the output of the quantizer is bounded by [12]

$$H_Q \geq h(X) - \frac{1}{2}\log_2\{E\left[\frac{1}{(N_L\lambda(x))^2}\right]\}, \qquad (7.28)$$

where $h(X)$ is the differential entropy of the input random variable X and $E[\ ]$ denotes statistical expectation. For concave $\lambda(x)$, equality is achieved when $\lambda(x)$ is a constant, and thus, for a large number of quantizer levels, the entropy is minimized for a uniform quantizer. This hints a possible approach that can combine simplicity and performance for all coefficients: to use uniform quantizers with a large number of levels for all coefficients and to apply entropy coding to the output. In fact, this is the idea behind the JPEG standard, which we will discuss later.

7.6.2 Optimal Rate Allocation

Let us consider the problem of having ν random variables $\{X_1, X_2, \ldots, X_\nu\}$. Each X_i is supposed to have a PDF $f(x)$, variance σ_i^2 and zero mean. Each one is quantized using a quantizer q_i allowing a distortion D_i and achieving a mean bit rate R_i. Hence we can construct the following sets: $\{\sigma_1\sigma_2\ldots\sigma_\nu\}, \mathbf{q} = \{q_1q_2\ldots q_\nu\}$, $\{D_1 D_2 \ldots D_\nu\}$, and $\{R_1 R_2 \ldots R_\nu\}$. Each quantizer belongs to a set of admissible quantizers $\mathbf{Q} = \{Q_0, Q_1, \ldots, Q_{N_Q}\}$, where each Q_i is a unit-variance normalized quantizer, allowing a distortion d_i and achieving a bit rate ρ_i, when quantizing a random variable with PDF $f(x)$. The rate-allocation problem is the one to find

$$D_{\min}(R) = \min_{\mathbf{q}} D = \min_{\mathbf{q}} \frac{1}{\nu}\sum_{i=1}^{\nu} D_i \qquad (7.29)$$

subject to

$$R = \frac{1}{\nu}\sum_{i=1}^{\nu} R_i \leq B, \qquad (7.30)$$

where B is the bit rate specified in bits/sample. The quantizers are assumed normalized, such that, once the quantizer i is selected for X_n $(q_n = Q_i)$, then $D_n = \sigma_n^2 d_i$.

The globally optimal choice of quantizers can be found by testing all $(N_q + 1)^\nu$ combinations allowed by $\mathbf{q}$. For each one, the pair (R, D) is computed and plotted in the R–D plane, generating a plot of all possible pairs (R, D). Once a bit rate is specified, we look up in the plot for the point with minimum D and as D is specified, we can look for the point with minimum R. A full search is generally impossible[1] and using dynamic programming, we can track only the points (R, D) lying in the lower convex hull of the whole set of plots [29]. Computation is drastically reduced to reasonable values, but in our cases, it can still be a very expensive and slow routine.

If, for Q_1 through Q_{N_Q}, we have

$$d_i \approx \epsilon 2^{-2\rho_i}, \qquad (7.31)$$

where ϵ is a constant, then the optimal rate allocation is given by [13]

$$R_i = R + \frac{1}{2} \log_2 \frac{\sigma_i^2}{\sigma_{gm}^2}, \qquad (7.32)$$

with σ_{gm}^2 as the geometric mean of the variances $\{\sigma_1^2, \sigma_2^2, \ldots, \sigma_v^2\}$. This rate allocation is impractical because it can generate negative numbers. The constraint of nonnegative numbers leads to [13]

$$R_i = \max(0, \Theta), \qquad (7.33)$$

$$\Theta = \xi + \log_2 \sigma_i, \qquad (7.34)$$

for $i = 0, 1, \ldots, v$, where ξ is a parameter incorporing all fixed terms of (7.32), plus an adjustment factor. For ε as an arbitrary small quantity, ξ is adjusted by iterative methods until

$$0 \leq B - \frac{1}{v} \sum_{i=1}^{v} R_i \leq \varepsilon. \qquad (7.35)$$

Another important practical concern is the availability of quantizers with the bit rates specified by (7.33). Thus, Θ has to be quantized to the nearest value of ρ_i ($i = 0, 1, \ldots, N_Q$). Also it is assumed that the PDF used to optimize a quantizer Q_i (achieving a rate ρ_i with distortion d_i) is the same as the PDF of the coefficient to which this quantizer will be assigned, which is not necessarily true in practical cases. In general, the random variables used in the rate-allocation process are the block coefficients $y(i, j)$; i.e., coefficients $y(i, j; m, n)$ for any block position (m, n).

7.7 Human Visual Sensitivity Weighting

The human eye has discriminative sensitivity to different spatial frequencies. The human visual sensitivity (HVS) weighting array is meant to devise the relative

---

[1] In the cases in consideration here, v can be in excess of 500 and N_Q can be as high as 20. Therefore, full search would force us to plot many more points than the number of particles inside the universe.

importance of each transform coefficient for the reconstructed image, in the subjective viewpoint of a human observer. Although spatial frequency is not the only important issue when devising the HVS model, it gives a simple and useful way to incorporate a subjective performance criteria into the transform coding process. The conversion between the 1-D continuous model and the 2-D weighting array can be done for the DCT or for the LOT. For the DCT, the method used in [30] is based on sampling a 2-D isotropic model, while for the LOT the method used in [22] is based on evaluation of the energy of the Fourier transform of the HVS model into each coefficient, in a method that can be used for most subband-transform methods. The differences between the HVS arrays for LOT and DCT for the same 1-D model are smaller than the differences among the arrays found using the same transform but different HVS models. Hence, we can assume approximately the same HVS array for both DCT and LOT, and we can use the sampling method from [30].

If the 1-D model is given by $w(f)$ where f is radial frequency in cycles per degree of the visual angle subtended, then the HVS array composed by elements η_{ij} for $(i, j) \in \{\Psi \cup (0, 0)\}$ are found by

$$\eta_{ij} = \frac{w(f_{ij})}{\alpha^2(i)\alpha^2(j)}, \tag{7.36}$$

where $\alpha(0) = 1/\sqrt{2}, \alpha(n) = 1 \ (n > 0)$,

$$f_{ij} = \frac{f_s\sqrt{i^2 + j^2}}{2M}, \tag{7.37}$$

and f_s is a sampling frequency parameter. f_s can be varied according to the ratio of the distance of the viewer to the screen width and according to the number of pixels displayed per line. Furthermore, the model is relative, such that the maximum value in the array can be set to unity. Some models for $w(f)$ are [27–32]

$$w(f) = 2.46(0.1 + 0.25f)e^{-0.25f} \tag{7.38}$$

$$w(f) = 2.6(0.192 + 0.114f)\,e^{(-0.114f)^{1.1}} \tag{7.39}$$

$$w(f) = (0.2 + 0.45f)e^{-0.18f} \tag{7.40}$$

$$w(f) = (0.31 + 0.69f)e^{-0.29f}. \tag{7.41}$$

The higher the weights, the more important the coefficients are, subjectively. To weight our decision of which coefficients are more important, we can change the quantizer allocation in two different ways. For uniform quantizers we can define different step sizes Δ_{ij} for each coefficient $y(i, j; m, n)$ in the block and set $\Delta_{ij} = \Delta/\eta_{ij}$. Thus, more distortion will occur to less important coefficients. Where rate allocation is used, we can use $\sigma_{ij}\eta_{ij}$ $(ij \in \Psi)$ as input to the rate-allocation process instead of solely the standard deviations σ_{ij}. In this way, the rate-allocation process will save bits from less important coefficients to give to subjectively more important ones.

7.8 Transform Coders: Zonal Sampling

The zonal sampling strategy is based on the selection of some transformed coefficients for transmission, discarding the remaining. In principle, the transformed block is divided into regions and each one is subject to a particular encoding fidelity criterion as illustrated in Fig. 7.9a. As the input signal has its spectral flatness measure different than unity, we can expect compaction of the signal energy in few coefficients as measured by (7.18). A first method to achieve signal compression is to allocate quantizers with different number of levels to each region, i.e., with different numbers of bits assigned to each coefficient in a particular region. Regions with less concentration of energy receive few bits, while regions with greater concentration of energy receive more bits. The bits are distributed according to the energy distribution of the transform and rate allocation follows the $\log_2$ rule discussed previously. See [10] for details on the energy distribution using the DCT. An example of bit-allocation is shown in Fig. 7.9b.

7.8.1 Chen–Smith Adaptive Coder

Chen and Smith [36] devised a technique to distribute different bit allocations to different blocks of the signal, according to the ac energy of the blocks. The brief description of the algorithm follows:

- Transform the image using blocks of $M \times M$ pels.

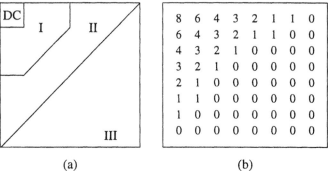

(a)   (b)

Figure 7.9 Zonal sampling method: (a) transformed block is divided into regions of importance according to energy distribution of the transform coefficients; (b) a bit-allocation example yielding an average of 1 bit/pel.

- Quantize and code separately the coefficients $y(0, 0; m, n)$ (the dc coefficients) using uniform quantizers.
- Compute the ac energy of each block as

$$E(m, n) = \sum_{(i,j) \in \Psi} y^2(i, j; m, n). \tag{7.42}$$

Sort the energies, and classify the blocks (in sorted order) into N_C equally populated classes. Thus, there will be N_B / N_C blocks in each class. Construct the class map $C(m, n)$ with the classification of each block. $C(m, n) = k$ if block (m, n) belongs to the class k ($k = 1, 2, \ldots, N_C$).

- For all blocks belonging to the same class, compute the average variance of the transform coefficients and then, their standard deviations. Construct N_C standard deviation maps with the standard deviation of the coefficients, found from

$$\sigma_k^2(i, j) = \sum_{m=0}^{N_{B1}-1} \sum_{n=0}^{N_{B2}-1} \delta(C(m, n) - k) \, y_{ij}^2(m, n) \qquad (i, j) \in \Psi, \tag{7.43}$$

where δ is the Kronecker delta function and $1 \leq k \leq N_C$.

- Merge all N_C ac standard deviation maps and decide the bit allocation. Following our previous section, from all $\sigma_k^2(i, j)$ we shall find an equivalent set of rates $R_k(i, j)$ to minimize the distortion for a given budget R. We also pose the constraint, in this case, that each $R_k(i, j)$ is an integer lying between 0 and $B_{\max}$, where $B_{\max}$ is the number of bits of the maximum quantizer available. Create N_C bit-allocation maps with a one-to-one correspondence with the elements of the standard deviation maps.

- Reestimate the standard deviations using the bit-allocation maps:

$$\hat{\sigma}_k(i, j) = c \, 2^{R_k(i,j)-1} \qquad 1 \leq k \leq N_C \qquad (i, j) \in \Psi, \tag{7.44}$$

where c is a normalization factor. It was suggested to choose c as the maximum $\sigma_k(i, j)$ for which $R_k(i, j) = 1$, for avoiding excessive clipping.

- Send the class map, the normalization coefficient c, and bit-allocation maps as side information. For example, if we chose $N_C = 8$ and $B_{\max} = 7$ we can encode the maps with 3 bits/sample, while the coefficient can be efficiently quantized with 16 or 32 bits.

- Quantize, encode, and send all the coefficients, using the reestimated variances. A coefficient $y(i, j; m, n)$ (block (m, n)), which belongs to class k ($C(m, n) = k$), is scaled (divided by $\hat{\sigma}_k(i, j)$), applied to a quantizer with $2^{R_k(i,j)}$ levels, and encoded with $R_k(m, n)$ bits. If $R_k(i, j) = 0$, the particular coefficient is not transmitted.

The receiver may first decode the side information and the dc coefficients. Given the class map, the bit-allocation maps, and the normalization factor c, the decoder can reconstruct the standard deviations used to scale the quantizers as in (7.41). With the maps reconstructed and with the knowledge of the transmission order, the decoder can exactly determine the position of the incoming coefficient, the class of its block, how many bits were assigned to it, and the variance used for quantization. Therefore, the receiver can decode the coefficients, apply an inverse transform, and obtain the reconstructed image.

In its original form, the Chen–Smith coder does not have a good overall performance. Coders such as the JPEG baseline system [14] yield compressed images with much better quality . The improved Chen–Smith (ICS) coder [23] results from a collection of improvements on the original algorithm in almost all parts, but following the zonal sampling strategy and using ac energy classification. Greater efficiency is achieved using entropy coding, several fractional bit-rate quantizers, and an efficient method to provide the coefficient variances to the receiver. The resulting coder easily surpasses the JPEG in all images tested and is far better than the original Chen–Smith coder [36].

7.9 Joint Picture Experts Group Baseline System

The Joint Picture Experts Group, (JPEG) is a combined committee of the International Telegraph and Telephone Consultative Committee (CCITT) and International Standards Organization (ISO) in charge of defining a compression standard for still images. JPEG received several contributions from institutions and individuals from around the world. The committee finally agreed about an algorithm that is efficient but not excessively complex. This algorithm has a property that raises the JPEG coder above all other image compression algorithms: it is becoming a de facto standard. Everyday, more and more companies launch new JPEG products in the market, including extremely fast DCT VLSI chips, JPEG chips, image compression boards for personal computers, graphics programs admitting files in the JPEG format, motion video using JPEG, etc.

There are three versions of the JPEG coder:

1. *Baseline system.* This is the basic part that is, in most cases, the only JPEG implementation available. This part is mandatory for all DCT-based decoders.

2. *Extended system.* This is a superset of the baseline system with optional parts of the standard. It includes arithmetic coding, input numbers with precision greater than 8 bits, and progressive encoding, allowing progressive image transmission (PIT), i.e., progressive decoding.

3. *Independent function—lossless compression.* This part is not DCT based.

We do not intend to explain completely the operation of the JPEG coder, but the concepts behind its most important parts. Certainly, the reader would

Figure 7.10   JPEG baseline system basic flow-graph for the encoder section. Decoder parts follow the same path in reversed order, with inverse operations.

not be able to implement a fully compatible JPEG coder with the information contained here. Implementation details and complete information may be found in [14].

The basic structure of the JPEG baseline encoder is shown in Fig. 7.10. The original image is translated from its original format to a common reference, then DCT is applied to 8×8 blocks and the transformed block is quantized using uniform quantizers. The dc coefficient is submitted to a DPCM-style prediction error encoding procedure, and the prediction errors substitute for the actual coefficients in the quantized and encoded array. Thus, each block is submitted to binary encoding, which is divided into generation of a composite symbols and Huffman coding. The decoder operation is trivial, given the encoder, by inverting the order and the operations. We will proceed with a brief conceptual description of the basic parts of baseline system.

7.9.1 Input–Output Formatting

The image to be compressed is stored under any predefined graphic format. There is a very large variety of graphics formats, such as GIF, TIFF, PGM, BMP, among several others, and several relevant parameters such as number of lines and columns, color bases (RGB, CMYK, etc.), pixel representation, pallettes, and storing order. Therefore, it is the function of any implementation of JPEG to convert the data from the target formats to three 2-D arrays of pixels. These arrays, as we defined, have N_1 and N_2 as multiples of 8, which can be done by replicating the last pixel in each line and the last row, until the necessary number of lines and columns is reached. The image is translated to the CCIR 601 components [37]: Y (luminance), Cr and Cb (chrominance). Each chrominance image is spatially decimated by a factor 2:1 (format 4:2:2) and then each component is encoded–decoded independently. For a monochrome image, the chrominance components (all zeros) are discarded. All basic components of the system work similarly for all components. Therefore, we will base our presentation on a single monochrome image, the component Y, although specific tables will significantly vary for chrominance components. The reader is, thus, referred to [14] for a complete description of the JPEG.

7.9.2 Scaling and Quantization

The input image samples $x(n_1, n_2)$ are supposed to be represented by 8 bits, therefore, having values ranging between 0 and 255. A level shift is applied to these samples subtracting 128 of each one, so that they range between -128 and 127. Following our notation, an 8×8 DCT is applied to these samples resulting in the coefficients $y(i, j; m, n)$. Using the orthonormal DCT each coefficient would range from -1023 through 1023. Instead of the unique quantizer step Δ in Section 7.5.3, we use a set of $M^2 = 64$ steps Δ_{ij} ($ij \in \{\Psi, 00\}$), similar to those used in Section 7.7, but the HVS weights η_{ij} do not necessarily follow any of the models discussed there. Each $y(i, j; m, n)$ is quantized using a uniform quantizer of step $s_{ij} = \Delta_{ij}S$, where S is a scaling factor, or quality factor, to control the actual step size and, therefore, the bit rate achieved by the JPEG coder, as discussed in Section 7.5.3. The default[2] quantization table for luminance is

$$\{\Delta_{ij}\} = \begin{bmatrix} 16 & 11 & 10 & 16 & 24 & 40 & 51 & 61 \\ 12 & 12 & 14 & 19 & 26 & 58 & 60 & 55 \\ 14 & 13 & 16 & 24 & 40 & 57 & 69 & 56 \\ 14 & 17 & 22 & 29 & 51 & 87 & 80 & 62 \\ 18 & 22 & 37 & 56 & 68 & 109 & 103 & 77 \\ 24 & 35 & 55 & 64 & 81 & 104 & 113 & 92 \\ 49 & 64 & 78 & 87 & 103 & 121 & 120 & 101 \\ 72 & 92 & 95 & 98 & 112 & 100 & 103 & 99 \end{bmatrix}.$$

The quantization procedure is reduced to

$$\ell(i, j, m, n) = \text{round}\left(\frac{y(i, j; m, n)}{s_{ij}}\right),$$

where round(u) is the integer nearest to u and ℓ is the quantizer output level. The inverse quantization (dequantization) is given by

$$\hat{y}(i, j; m, n) = \ell(i, j, m, n)s_{ij}.$$

The dc coefficients are subject to an extra processing, the prediction error encoding, given by

$$\ell(0, 0, m, n) \leftarrow \ell(0, 0, m, n) - \ell(0, 0, m, n - 1).$$

The substitution can be done in place, deleting the old level number. Here, $\ell(0, 0, m, n)$ is assumed zero outside the image bounds.

Similarly, at the decoder, the actual dc quantized level is reconstructed by

$$\ell(0, 0, m, n) \leftarrow \ell(0, 0, m, n) + \ell(0, 0, m, n - 1).$$

---

[2]This is an example provided by JPEG. Other user-defined tables can be used as well.

After quantization, all processing is lossless, i.e., perfectly invertible without any distortion; thus, the description of the encoding procedure is sufficient to determine both encoding and decoding operations.

7.9.3 Composite Symbols

Since, in JPEG baseline system, each block is encoded independently, let us examine one particular block and drop the indexes (m, n) for convenience; $\ell(i, j)$ represents $\ell(i, j, m, n)$, for a given block position pair mn. The set of $\{\ell(i, j)\}$ is converted into a 1-D sequence by scanning the block in the following order:

		0	1	2	3	4	5	6	7
					j	$\rightarrow$			
0		0	1	5	6	14	15	27	28
1		2	4	7	13	16	26	29	42
2		3	8	12	17	25	30	41	43
3	i	9	11	18	24	31	40	44	53
4	$\downarrow$	10	19	23	32	39	45	52	54
5		20	22	33	38	46	51	55	60
6		21	34	37	47	50	56	59	61
7		35	36	48	49	57	58	62	63

i.e., in a zigzag path starting from the dc coefficient and ending with the highest frequency ac coefficient. The dc is coded separately and all the AC coefficients along the zigzag scan are represented as follows: the length of the run of zeros Zr between the last nonzero element and actual $\ell(i, j)$, plus $\ell(i, j)$ itself. If there is a run of more than 15 zero elements, a special procedure takes place. After the last nonzero element, an end-of-block (EOB) mark is inserted. Each nonzero element is assigned to a category (K) and to an offset (Z_{offset}) using the following rule

$$ K = m \quad \Leftrightarrow \quad \ell(i, j) \in \{\text{m integer}|[-2^m + 1, -2^{m-1}] \cup [2^{m-1}, 2^m - 1]\}, $$

and N_i is an m-bit number to represent $\ell(i, j)$ inside the segment defined by $K = m$ (1 bit for sign and $m - 1$ for magnitude). For example, if $\ell(i, j) = -38$, then $K = 6$ because -38 belongs to the set $\{-63, \ldots, -32, 32, \ldots, 63\}$. Z_{offset} is given by the sign (negative: sign bit $= 1$) plus the offset in the segment $\{32, \ldots, 63\}$ $(38 - 32 = 6 \rightarrow 00110)$. Therefore, each nonzero number occurrence is described by the set of numbers

$$ \ell(i, j) > 0 \rightarrow Zr/K/Z_{\text{offset}} \equiv \text{composite symbol}. $$

The mapping into category and offset is also valid for the dc coefficient, and each block is structured as $K_{\text{dc}} / Z_{\text{offset,DC}} /$ sequence of composite symbols / EOB.

7.9.4 Huffman Coding

First, the level numbers are converted into composite symbols, then each composite symbol is encoded using modified Huffman codes, upper limited to a maximum of 16 bits per code word. The symbols to be Huffman encoded are the zero-run Zr and the category K for each composite symbol. Both are encoded in the same code word. The dc has a special Huffman table for only its category representation. The encoder can design the Huffman tables and include them into the data as side information. JPEG provides examples of *good* Huffman tables, which were chosen after several tests. In these, the number of categories is limited to 10. Table 7.1 shows the Huffman code example for the DC difference categories and for the ac combined symbols (Zr, K), for $0 \leq Zr \leq 5$. Since there are 10 categories, runs up to 16 zero elements, and an EOB mark, the table would have 161 entries, we have limited its size and we refer the reader to [14] for the complete table.

7.9.5 JPEG Compression Results

Using the example tables, the JPEG baseline system is quite efficient for most images, considering its moderately low complexity. The only parts that are time consuming are the DCT implementation and, to some extent, the Huffman decoding procedure. Image Lena compressed to 1 bit/pel and 0.5 bit/pel using the JPEG baseline system is shown in Fig. 7.11.

7.10 Interframe Image Coding

The previous sections dealt with compression techniques applied to still frames. For an image sequence such as videophone, videoconference, television (standard TV: NTSC, PAL, SECAM), high definition TV (HDTV) [38], or super high definition (SHD) TV (either for storage or transmission), additional reduction in bit rate can be achieved by taking advantage of the temporal correlation. The basic premise is that, if there is no significant change between adjacent frames or fields, the bit rate required to code the change can be very small. The image sequence can be made adaptive by coding on a block-by-block basis; i.e., by testing if there is a significant change between corresponding blocks of successive frames or fields. This process can be improved if the motion of a pel or a group of pels (block) can be estimated within a frame interval. Interframe prediction aided by motion estimation on a block-by-block basis has been efficiently developed for image sequences resulting in VLSI chips, chip sets, codecs, etc., dedicated to the various applications cited here. An added advantage is that, when motion estimation fails (fast motion, scene change, etc.), intraframe coding methods can be applied; i.e., adaptive block coding under various modes (intra, inter, with or without motion

Table 7.1   Example of Huffman codes.

			DC differences		
Category	Code word	Category	Code word	Category	Code word
0	00	4	101	8	111110
1	010	5	110	9	1111110
2	011	6	1110	10	11111110
3	100	7	11110	11	111111110

			AC—up to $Zr = 5$		
Zr	K	Code word	Zr	K	Code word
0	1	00	3	1	111010
0	2	01	3	2	111110111
0	3	100	3	3	111111110101
0	4	1011	3	4	1111111110001111
0	5	11010	3	5	1111111110010000
0	6	1111000	3	6	1111111110010001
0	7	11111000	3	7	1111111110010010
0	8	1111110110	3	8	1111111110010011
0	9	1111111110000010	3	9	1111111110010100
0	10	1111111110000011	3	10	1111111110010101
1	1	1100	4	1	111011
1	2	11011	4	2	1111111000
1	3	1111001	4	3	1111111110010110
1	4	111110110	4	4	1111111110010111
1	5	11111110110	4	5	1111111110011000
1	6	1111111110000100	4	6	1111111110011001
1	7	1111111110000101	4	7	1111111110011010
1	8	1111111110000110	4	8	1111111110011011
1	9	1111111110000111	4	9	1111111110011100
1	10	1111111110001000	4	10	1111111110011101
2	1	11100	5	1	1111010
2	2	11111001	5	2	11111110111
2	3	1111110111	5	3	1111111110011110
2	4	111111110100	5	4	1111111110011111
2	5	1111111110001001	5	5	1111111110100000
2	6	1111111110001010	5	6	1111111110100001
2	7	1111111110001011	5	7	1111111110100010
2	8	1111111110001100	5	8	1111111110100011
2	9	1111111110001101	5	9	1111111110100100
2	10	1111111110001110	5	10	1111111110100101

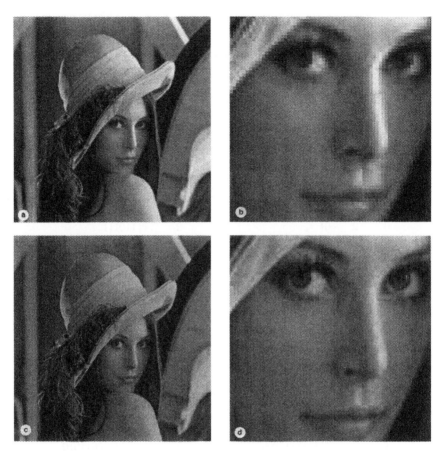

Figure 7.11   Image Lena compressed using JPEG baseline system: (a) 256 × 256 pels image compressed to 1 bit/pel; (b) zoom of (a); (c) 512 × 512 pels image compressed to 0.5 bit/pel; (d) zoom of (c).

compensation, etc.) achieves large data compression as it takes advantage of both spatial and temporal correlations. The penalty here is the increase in the codec complexity. Also, additional bits indicating the coding mode need to be transmitted to the receiver so that the decoder can process them accordingly. This technique, called *interframe motion-compensated hybrid coding,* is the basis for various standards that are either being recommended or adopted. The motion compensation (MC) technique has also been extended to reconstruction of skipped frames at the receiver when frames are coded at a reduced rate in the encoder. As the temporal interpolation is along the motion trajectory, the overall quality of the reconstructed frames can be further improved. Such a scheme has been adopted in MPEG1 [16], MPEG2 [39], MPEG++ [40] (proposed by the Advanced Television Research Consortium to the FCC for terrestrial transmission of HDTV), etc. To illustrate these concepts, the video source coder in generalized form as per Recommendation H.261 [17–19], video codec for audiovisual services at p×64 Kbit/sec (CCITT Study Group XV—Recommendation H Series) is shown in Fig. 7.12. The corresponding decoder is shown in Fig. 7.13. Some of the functions of both the coder and decoder are optional. H.261 does not specify how the various functions are to be implemented nor the pre- and postprocessing required for the codec. H.261 is part of the H series that relate to frame structure, terminals, telephone systems, terminal equipment, etc. Block diagram of the entire codec is shown in Fig. 7.14.

To accommodate both 525 lines (NTSC) and 625 lines (PAL, SECAM) systems, the coder can accept the video signal in two formats (Fig. 7.15): common intermediate format (CIF) and quarter CIF (QCIF). Both are noninterlaced (progressive). The temporal reference (TR) is indicated by a 5-bit number (32 possible values). The TR is formed by incrementing its value in the previously transmitted picture header by 1 plus the number of non-transmitted pictures (at 29.97 Hz) since the last transmitted one.

Video multiplex is arranged in a hierarchical structure with four layers as (1) picture; (2) group of blocks (GOB) (Fig. 7.16); (3) macroblock (MB) (Fig. 7.17); (4) block. The division into these layers is to introduce appropriate control and

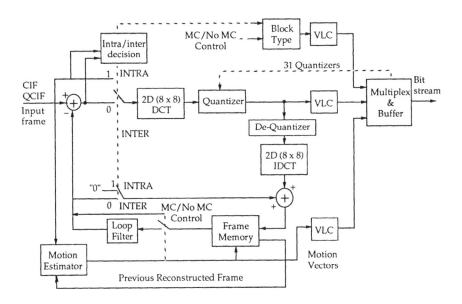

Figure 7.12   H.261 source coder.

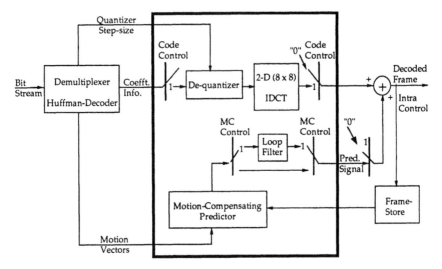

Figure 7.13   H.261 source decoder.

overhead bits (for various modes and for regulating the bit rate) and also to prevent error propagation.

The interframe or intraframe coding decision is made on a macroblock (MB) basis. An MB consists of four Y blocks, one Cb block, and one Cr block (each block is of size 8×8). One motion vector (MV) per MB based on a 16×16 Y block, whose range is up to ± 15 pels or lines/frame, is allowed. This MV is appropriately scaled for the color difference blocks (Cb and Cr). A 2-D spatial loop filter that operates on the predicted 8×8 blocks is optional at the encoder. This is a separable filter of coefficients $\{1/4, 1/2, 1/4\}$ applied to both horizontal

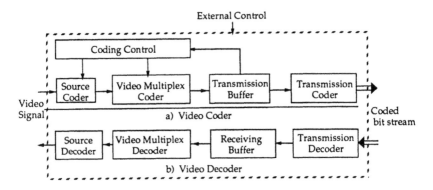

Figure 7.14   H.261 outline block diagram of the video codec.

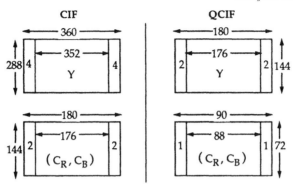

Figure 7.15 Noninterlaced picture resolutions for common intermediate format (CIF) and quarter CIF (QCIF). Aspect ratio is 4:3 and temporal resolutions are 30, 15, 10, and 7.5 pictures/sec.

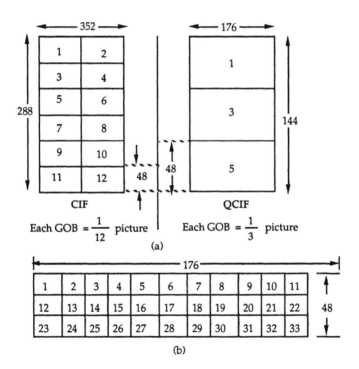

Figure 7.16 (a) H.261 arrangement of GOBs in a picture. (b) H.261 arrangement of macroblocks in a group of blocks (GOB).

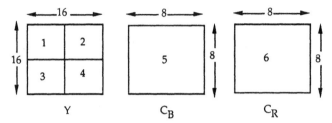

Figure 7.17   H.261 arrangement of blocks in a macroblock (MB).

and vertical edges. The weights are modified at the block edges. The loop filter reduces the high-frequency artifacts caused by MC and the quantization noise in the feedback loop. The filter is switched on or off for all six blocks in an MB. The control mechanism for the various modes such as intraframe or interframe with or without MC, loop filter on or off, etc., as stated earlier is left to the designer. The VLC that indicates to the receiver the actual state of the MB coding is specified in Table 7.2. At this stage it is appropriate to digress from the H.261 coder and describe briefly the motion compensation method.

7.10.1 Motion Estimation and Compensation

Interframe prediction can be improved if the motion of individual pels or group of pels can be reasonably estimated. This is called motion-compensated *prediction* or *prediction aided by motion estimation* (ME). Pel-recursive algorithms (PRAs) that address pel-by-pel motion, and block matching algorithms (BMAs) [41–43] have been extensively developed. In practice PRAs have been seldom used, as they are not only computationally intensive, but also require large overhead. On

Table 7.2   H.261 VLC table for MTYPE (o means that the item is present in the MB).

	MQUANT	MVD	CBP	TCOEFF	VLC
INTRA				o	0001
INTRA	o			o	0000 001
INTER			o	o	1
INTER	o		o	o	00001
INTER+MC		o			0000 0000 1
INTER+MC		o	o	o	0000 0001
INTER+MC	o	o	o	o	0000 0000 01
INTER+MC+LF		o			001
INTER+MC+LF		o	o	o	01
INTER+MC+LF	o	o	o	o	0000 01

the other hand, VLSI chips that can implement BMAs in real time even for HDTV have been developed [44]. Also, by cascading these chips, both the block size and the motion vector range can be extended.

7.10.2 Block Matching Algorithm

BMA (Fig. 7.18) assumes that all pels within a block such as an 8×8 or a 4×4 block have uniform motion within a frame interval. This assumption improves as the block size gets smaller at the cost of increased overhead in terms of coding the MV for each block. The MV range is limited by the window in the previous frame in which the search for the ME of the block in the present frame is carried out.

Motion estimation is based on the maximum correlation of (or minimum distortion) of an $M_1 \times M_2$ block $\mathcal{B}\ell_i$ in the present frame with the $M_1 \times M_2$ block $\mathcal{B}\ell_{i-1}$ within the search window of the previous frame. The distortion can be expressed as

$$\frac{1}{M_1 M_2} \sum_{l_1=1}^{M_1} \sum_{l_2=1}^{M_2} f \left(x(n_1 + l_1, n_2 + l_2, i) \right.$$
$$\left. -x(n_1 + l_1 + m_1, n_2 + l_2 + m_2, i - 1) \right), \quad (7.45)$$

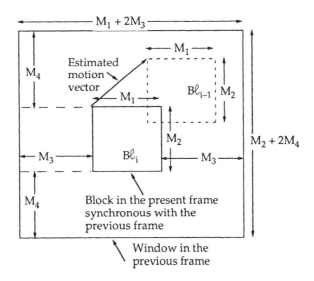

Figure 7.18   Block matching algorithm. Motion estimation of an $(M_1 \times M_2)$ block $\mathcal{B}\ell_i$ in the present frame within an $(M_1 + 2M_3) \times (M_2 + 2M_4)$ window in the previous frame. The MV range here $\pm M_3$ pels/frame and $\pm M_4$ lines/frame. The $(M_1 \times M_2)$ block $\mathcal{B}\ell_i$ in the present frame has the best match with the $(M_1 \times M_2)$ block in the previous frame within the search window.

where (m_1, m_2) is the motion vector. While various distortion criteria such as the mean squared error (Mse), mean absolute error (Mae), etc., have been investigated, simulation results using various test sequences indicate that MAE is as good as any criterion for estimating the motion. For MAE, (7.45) reduces to

$$\frac{1}{M_1 M_2} \sum_{l_1=1}^{M_1} \sum_{l_2=1}^{M_2} \left| x(n_1 + l_1, n_2 + l_2, i) \right.$$
$$\left. -x(n_1 + l_1 + m_1, n_2 + l_2 + m_2, i - 1) \right|. \quad (7.46)$$

While motion estimation for each color component Y, Cb, and Cr is ideal, ME based on Y, applied to Cb and Cr (appropriately scaled) is adequate. Brute search implies computing the distortion for every pel and line displacement within the search window. This method requires computation of $(2M_4 + 1)(2M_3 + 1)$ distortions followed by the identification of the least distortion. Various algorithms that reduce this number of distortions (although not optimal) have been developed and some of these have been implemented in practice. There are a number of variations to the basic interframe hybrid coding aided by MC. For example, in MPEG2, ME based on 16×16 or 8×8 block is suggested. Also, frames are interpolated based on forward MV (motion vector from previous frame) or backward MV (motion vector from future frame) or both. Other techniques such as hierarchical BMA for ME have been proposed [45].

A 2-D (8×8) DCT is applied to all the six blocks in an MB (whatever the mode may be; Table 7.2), and the DCT coefficients are scanned in a zigzag fashion. The dc coefficient of an intrablock is uniformly quantized to 8 bits with no dead zone and all the other coefficients (intrablock ac coefficients and all interblock DCT coefficients) are quantized using one of the 31 nearly uniform quantizers (step size value; 2, 4, 6, . . . , 62 quantizer selection indicated by overhead) and the dead zone.

7.10.3 Macroblock Addressing

If all the quantized coefficients in an MB are zero, then it is called a *skipped MB*. To indicate the skipped MB, a VLC that denotes the difference between the absolute addresses of the MB and the last transmitted MB within GOB is developed (see Table 7.3). For a transmitted MB, another VLC, which denotes whether individual (8×8) blocks in the MB are coded or not, is developed. This code describes the coded block pattern (CBP) (see Fig. 7.19 and Table 7.4).

Table 7.3 H.261 VLC table for MBA.

MBA	Code	MBA	Code	MBA	Code
1	1	12	0000 1001	23	0000 0100 010
2	011	13	0000 1000	24	0000 0100 001
3	010	14	0000 0111	25	0000 0100 000
4	0011	15	0000 0110	26	0000 0011 111
5	0010	16	0000 0101 11	27	0000 0011 110
6	0001 1	17	0000 0101 10	28	0000 0011 101
7	0001 0	18	0000 0101 01	29	0000 0011 100
8	0000 111	19	0000 0101 00	30	0000 0011 011
9	0000 110	20	0000 0100 11	31	0000 0011 010
10	0000 1011	21	0000 0100 10	32	0000 0011 001
11	0000 1010	22	0000 0100 011	33	0000 0011 000

Note: Code is position of MB within a GOB. For the first transmitted MB, in a GOB, use the absolute address. For subsequent MBs, MBA is the difference between the absolute address and the last transmitted block. *Key:* MBA Stuffing 0000 0001 111; VLC for start code 0000 0000 0000 0001.

Table 7.4 H.261 VLC table for CBP.

CBP	Code	CBP	Code	CBP	Code	CBP	Code
60	111	35	0001 1100	62	0100 0	38	0000
4	1101	13	0001 1011	24	0011 11	29	0000
8	1100	49	0001 1010	36	0011 10	45	0000
16	1011	21	0001 1001	3	0011 01	53	0000
32	1010	41	0001 1000	63	0011 00	57	0000
12	1001 1	14	0001 0111	5	0010 111	30	0000
48	1001 0	50	0001 0110	9	0010 110	46	0000
20	1000 1	22	0001 0101	17	0010 101	54	0000
40	1000 0	42	0001 0100	33	0010 100	58	0000
28	0111 1	15	0001 0011	6	0010 011	31	0000 0011 1
44	0111 0	51	0001 0010	10	0010 010	47	0000 0011 0
52	0110 1	23	0001 0001	18	0010 001	55	0000 0010 1
56	0110 0	43	0001 0000	34	0010 000	59	0000 0010 0
1	0101 1	25	0000 1111	7	0001 1111	27	0000 0001 1
61	0101 0	37	0000 1110	11	0001 1110	39	0000 0001 0
2	0100 1	26	0000 1101	19	0001 1101		

Note: The CBP entries mean blocks (in MB) in which at least one coefficient is transmitted. Pattern number $= 32P_1 + 16P_2 + 8P_3 + 4P_4 + 2P_5 + P_6$. $P_i = 1 \rightarrow$ block coded; $P_i = 0 \rightarrow$ block not coded.

$$(Y)\ \begin{array}{|c|c|}\hline 1 & 2 \\ \hline 3 & 4 \\ \hline \end{array} + (C_r)\ \boxed{5} + (C_b)\ \boxed{6} \qquad \text{(block number)}$$

Similar to Table 7.1 for JPEG, 2-D VLC for the most frequent combination of run-lengths of zero coefficients (RUN) preceding a nonzero coefficient (LEVEL) is developed. This coding is based on quantization and zigzag scan. The remaining combinations are encoded with a 20-bit word consisting of 6 bits ESCAPE, 6 bits RUN, and 8 bits LEVEL. Table 7.5 lists the common modules in JPEG, H.261, and MPEG1 (Motion Picture Experts Group for digital storage media) [46]. The overriding criteria is to provide an incentive to develop VLSI chips as many of the functions are common to the three applications and also to provide compatibility between the corresponding systems.

Some highlights for the H.261 algorithm are described here to give the reader a task of transform coding. Various other aspects, such as error detection and

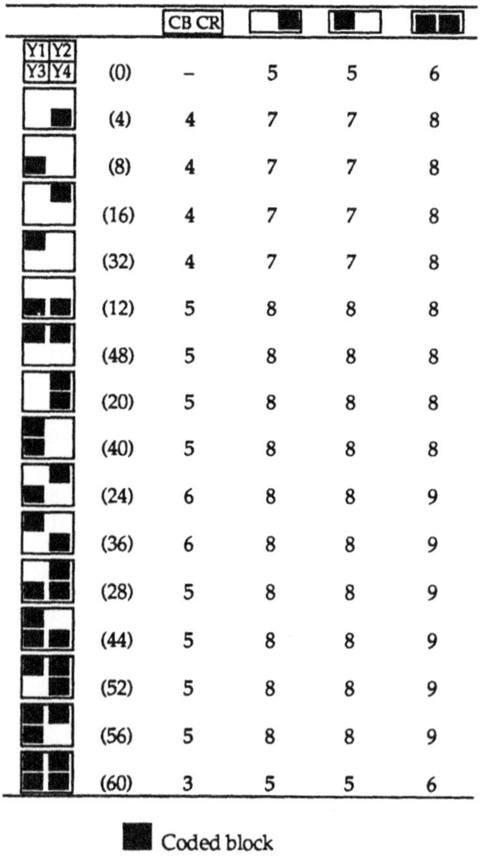

Y1 Y2 / Y3 Y4	CB CR	■	■	■ ■
(0)	–	5	5	6
(4)	4	7	7	8
(8)	4	7	7	8
(16)	4	7	7	8
(32)	4	7	7	8
(12)	5	8	8	8
(48)	5	8	8	8
(20)	5	8	8	8
(40)	5	8	8	8
(24)	6	8	8	9
(36)	6	8	8	9
(28)	5	8	8	9
(44)	5	8	8	9
(52)	5	8	8	9
(56)	5	8	8	9
(60)	3	5	5	6

■ Coded block

Figure 7.19   63 coded block patterns.

Table 7.5   Common modes used in JPEG, H.261, and MPEG1.

	JPEG (Baseline)	H.261	MPEG1
8 × 8 DCT	Yes	Yes	Yes
Zigzag scan	Yes	Yes	Yes
2-D run-length coding	Yes	Yes	Yes
Quantizer	Default tables	32 step sizes	Intra: similar to JPEG Inter: similar to H.261
Motion estimation		Forward	Forward/backward
Full block matching	No	Integer pel accuracy	Half-pel accuracy
Rate buffer control	No	Yes	Yes
VLC: TCOEFF	Downloadable	Fixed	Superset of H.261
MBA, CBP	No	Fixed	Same as H.261
MV, MTYPE	No	Fixed	Superset of H.261

correction, multiplexer, buffer, channel coder, network interface, and corresponding decoding functions at the receiver, form an integral part of the codec. It is intuitive that replacing the DCT by LOT can reduce the blocking artifacts. This replacement requires a number of other modifications to the basic coding algorithm. Other techniques include groups of transform coefficients (either intra or inter) that can be coded as vectors. Such a scheme called *vector quantization* (VQ) has been proposed, investigated, and implemented. The VQ can be applied to prediction errors in the case of DPCM, transform coefficients or signals in a particular subband. This is the topic of the following section.

7.11 Vector Quantization

It is well known from Shannon's rate–distortion theory that a better performance can be achieved by coding data as vectors rather than as scalars. The VQ [12] process is quite straightforward and involves the following (Fig. 7.20): (1) design a codebook consisting of code vectors or patterns representative of the input vectors to be coded; (2) compare the input vector with members of the codebook and find one with the best match or the least distortion; (3) send a code index or address for the corresponding code vector from (2); (4) retrieve this code vector from the replica of the codebook stored at the receiver as the output vector.

Based on a large training sequence, a locally optimal algorithm has been developed by Linde, Buzo, and Gray, popularly known as the *LBG algorithm* [47], that has been applied to speech and image coding. There are a number of variations

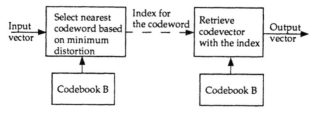

Figure 7.20   Basic vector quantization scheme.

and enhancements to the basic LBG algorithm and as well as several modifications of the VQ process. VQ can be applied to samples of any signal, prediction errors in the case of DPCM, prediction errors of transform coefficients, samples in a subband, groups of transform coefficients, i.e., VQ can be combined with DPCM, transforms, subband, and other redundancy schemes to achieve additional compression. Also, adaptivity can be incorporated in the VQ or any of these hybrid techniques [48]. The reader is referred to Chapter 6, in this book, for details in the role of VQ in image coding.

7.12 Conclusions

A brief description of the transformed image coding is presented. Transform by itself plays a minor role in the overall coding scheme. The combination of various techniques coupled with adaptive features is the key for efficient bit-rate reduction. Adaptation, however, introduces additional hardware and one has to compromise between quality and complexity (hence cost), which are governed by the consumer acceptance. By integrating various services, such as JPEG, H.261, MPEG1, and MPEG2 into a single system, that system can be made versatile leading to an expanded market.

Appendix 7.A: Discrete Cosine Transform

We do not intend to cover all aspects of the DCT, but to give the minimum amount of information necessary to help the reader to follow our presentation of block transforms. A complete reference on the DCT is [4], where most of topics of interest related to the DCT are covered.

The close relation with the discrete Fourier transform and the availability of fast algorithms make the DCT one of the most popular choices for image processing. Also the DCT is shown to be an asymptotically optimal block transform for highly correlated signals.

As we saw in Section 7.2.1, the block transform is completely specified by the $M \times M$ matrix $\mathbf{A}_M$, with the forward and inverse transform operations presented in (7.4) and (7.5) respectively. This matrix has elements a_{ij}, and for the DCT they are described as

$$a_{ij} = \alpha(1)\sqrt{\frac{2}{M}} \cos\left[\left(j + \frac{1}{2}\right)\frac{i\pi}{M}\right] \tag{7.47}$$

for $0 \leq i, j \leq M - 1$, where $\alpha(0) = 1/\sqrt{2}$ and $\alpha(1) = 1$ for $i > 0$.

Actually, there are several classifications for the DCT, according to the arguments of the cosine term. The definitions just shown correspond to the so-called DCT type II, while the inverse transform, given by $\mathbf{A}_M^T$ is the so-called the DCT type III. Also, there are several possible ways to implement this transform through fast algorithms. In [4] it is possible to find several flow graphs for implementing the DCT for several values of M. Also, integer implementations and direct 2-D transforms are particularly interesting methods, which can increase the speed of practical transform operations. The methods based on the implementation of the DCT through the implementation of the DCT type IV and through the DFT (using split-radix fast DFT) can virtually achieve the minimum complexity for 1D implementation. The total number of additions (n_{add}) and multiplications (n_{mult}), for DCT of length M, are

$$n_{\text{add}} = \frac{M}{2} \log_2 M + 1 \quad , \quad n_{\text{mult}} = \frac{3M}{2} \log_2 M - M + 1.$$

As a final remark, assuming AR(1) signal models with autocorrelation function as $r_x(n) = 0.95^{|n|}$, the G_{TC} for the DCT with $M = 8$ is 8.83 dB.

Appendix 7.B: Lapped Orthogonal Transform

The LOT was first developed using optimization routines, yielding a transform that did not possess fast algorithms. The factorized form of the LOT, allowing fast implementation and better performance than the one developed in [49], was introduced in [5]. Since then, other similar factorizations based on the same principles have been developed.

The LOT has an overlap factor $N_o = 2$, meaning that the length of the basis vectors is twice the block size. Therefore, there is an overlap of $M/2$ samples at each side of the block, as discussed in Section 7.2.2. We will refer to one only type of fast LOT [6], which can be used for $M \leq 16$. Basically, there are two techniques to find the transform: by maximizing the transform coding gain and by using QR factorization, which is the decomposition of a square matrix into the product of an orthogonal matrix and an upper triangular one. Both lead to the same structure, which still does not completely allow a fast implementation. An approximation of an orthogonal matrix by a reduced number of plane rotations is, then, applied. However, we will not discuss the design method (which is exhaustively presented in [6]), reserving ourselves to present directly the flow graph of the LOT in Fig. 7.21 for $M = 8$.

For the borders, if the input signal $x(n)$ has length N and N_B blocks of size M, one can create a signal with $N + M$ samples by including $M/2$ samples in each extremity of $x(n)$. These samples may be symmetric reflections of the $M/2$ boundary samples of $x(n)$, in each border. Thus, it would be possible to transform $N_B + 1$ blocks and the first should be discarded. For the inverse transform, one can create one extra block in each extremity (in the transform domain), perform the inverse transform over $N_B + 2$ blocks, and discard the first two blocks obtained. To find the extra transform-domain blocks, one may copy the values of the coefficients of the neighbor block, which are inside bounds, but inverting the sign of the odd coefficients; i.e., coefficients 1,3,5,... have their sign inverted and those numbered 0,2,4... are not altered. This will assure perfect reconstruction and orthogonality of the process.

For size-8 LOT, the rotation angles $\{\theta_1, \theta_2, \theta_3\}$ can be either $\{0.13, 0.16, 0.13\}$ using the maximum G_{TC} design or $\{0.145, 0.17, 0.13\}$ using the QR factorization.

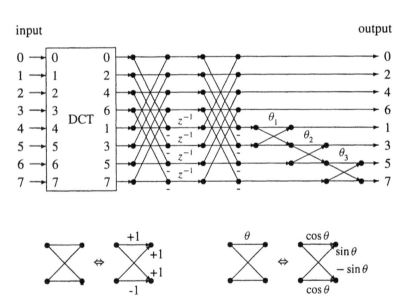

Figure 7.21   Flow graph for the 1-D LOT with $M = 8$. All resulting coefficients have to be multiplied by 1/2 for orthonormality. Input and output coefficient order are indicated. Forward transform is achieved by following the flow graph from left to the right and inverse transform is achieved following the paths from right to the left, replacing the DCT by its inverse, and moving delays to the upper $M/2$ branches.

The difference in G_{TC} is as small as 0.05 dB, but the QR approach yields basis functions with better filtering characteristics.

Finally, assuming AR(1) signal models with adjacent correlation coefficient 0.95, the G_{TC} for the size-8 LOT is 9.20 dB using the QR factorization and approximation by plane rotations. This increase in the gain, compared to the DCT, comes at the expense of the extra computations, which are evident from the flow graph presented for the LOT.

References

[1] N. Ahmed and K. R. Rao, *Orthogonal Transforms for Digital Signal Processing*. New York: Springer-Verlag 1975.

[2] D. F. Elliott and K. R. Rao, *Fast Transforms: Algorithms, Analyses, and Applications*. New York: Academic Press, 1982.

[3] K. R. Rao Ed., *Discrete Transforms and their Applications*. New York: Van Nostrand Reinhold, 1985.

[4] K. R. Rao and P. Yip, *Discrete Cosine Transform: Algorithms, Advantages, Applications*. San Diego, CA: Academic Press, 1990.

[5] H. S. Malvar and D. H. Staelin, "The LOT: Transform coding without blocking effects." *IEEE Trans. Acoust., Speech, Signal Processing*, vol. ASSP-37, pp. 553–559, April 1989.

[6] H. S. Malvar, *Signal Processing with Lapped Transforms*. Norwood, MA: Artech House, 1992.

[7] J. W. Woods, Ed., *Subband Coding of Images*. Hingham, MA: Kluwer Academic, 1991.

[8] J. Shapiro, "Embedded image coding using zerotrees of wavelet coefficients." *IEEE Trans. Signal Processing*, vol. 41, pp. 3445–3462, Dec. 1993.

[9] A. N. Netravali and B. G. Haskell, *Digital Pictures: Representation and Compression*. New York: Plenum Press, 1988.

[10] R. J. Clarke, *Transform Coding of Images*. Orlando, FL: Academic Press, 1985.

[11] M. Rabbani and P. W. Jones, *Digital Image Compression Techniques*. Bellingham, WA: SPIE Optical Engineering Press, 1991.

[12] A. Gersho and R. M. Gray, *Vector Quantization and Signal Compression*. Hingham, MA: Kluwer Academic, 1992.

[13] N. S. Jayant and P. Noll, *Digital Coding of Waveforms*. Englewood Cliffs, NJ: Prentice-Hall, 1984.

[14] Final Text for ISO/IEC DIS 10918-1, *Information Technology—Digital Compression and Coding of Continuous Tone Still Images—Part 1: Requirements and Guidelines*, Jan. 14, 1992; *Part 2. Compliance Testing—CD 10918-2*, 12/16/91.

[15] CCITT Recommendation H.261, *Video Codec for Audiovisual Services at $p \times 64$ Kbit/sec*, COM-XV-R37-E, Aug. 1990.

[16] ISO 11172, *Coding of Moving Pictures and Associated Audio for Digital Storage Media at up to 1.5 Mbps*, ISO/IEC JTC, 29N071, Dec. 1991.

[17] Special Issue on Video Coding for 10 Mbit/s, *Signal Processing: Image Commun.*, vol. 5, Feb. 1993.

[18] Special Issue on 64 Kbit/sec Coding, Part I, *Signal Processing: Image Commun.*, vol. 2, Dec. 1990.

[19] M. L. Liou, "Overview of the $p \times 64$ Kbit/sec video coding standard," *Commun. ACM*, vol. 34, pp. 60–63, April 1991.

[20] M. Vetterli and D. Le Gall, "Perfect reconstruction filter banks: some properties and factorizations." *IEEE Trans. Acoust., Speech, Signal Processing*, vol. ASSP-37, pp. 1057–1071, July 1989.

[21] P. P. Vaidyanathan, *Multirate Systems and Filter Banks*. Englewood Cliffs, NJ: Prentice-Hall, 1993.

[22] R. L. de Queiroz and K. R. Rao, "HVS weighted progressive transmission of images using the LOT." *J. Electron. Imaging*, vol. 1, pp. 328–338, July 1992.

[23] E. M. Rubino, H. S. Malvar, and R. L. de Queiroz, "Improved Chen-Smith image coder." In *Proc. IEEE Int. Symp. Circuits Syst.*, Chicago, May 1993, pp. 267–270.

[24] R. G. Gallagher, *Information Theory and Reliable Communications*. New York: J. Wiley and Sons, 1968.

[25] D. Huffman, "A method for the construction of minimum redundancy codes." *Proc. IRE*, vol. 40, pp. 1098–1101, 1952.

[26] P. G. Howard and J. S. Vitter, "Practical implementations of arithmetic coding." In *Image and Text Compression*, ed. J. A. Storer. Hingham, MA: Kluwer Academic, 1992.

[27] W. H. Chen and W. K. Pratt, "Scene adaptive coder." *IEEE Trans. Commun.*, vol. COM-32, pp. 225–232, March 1984.

[28] N. Farvardin and J. W. Modestino, "Optimal quantizer performance for a class of non-Gaussian memoryless sources." *IEEE Trans. Inform. Theory*, vol. IT-30, pp. 485–497, May 1984.

[29] P. H. Westerink, "Sub-band coding of images," Ph.D. thesis, Delft Techninical University The Netherlands, 1989.

[30] B. Chitprasert and K. R. Rao, "Human visual weighted progressive image transmission." *IEEE Trans. Commun.*, vol. COM-38, pp. 1040–1044, July 1990.

[31] K. H. Tzou, T. R. Hsing, and J. G. Dunham, "Applications of physiological human visual system model to image compression." *Proc. SPIE*, vol. 504, pp. 419–424, 1984.

[32] H. Lohscheller, "A subjectively adapted image communication system." *IEEE Trans. Commun.*, vol. COM-32, pp. 1316–1322, Dec. 1984.

[33] N. B. Nill, "A visual model weighted cosine transform for image compression and quality assessment." *IEEE Trans. Commun.*, vol. COM-33, pp. 551–557, June 1985.

[34] J. L. Mannos and D. J. Sakrison, "The effect of visual fidelity criterion on the encoding of images." *IEEE Trans. Inform. Theory*, vol. IT-20, pp. 525–536, July 1974.

[35] K. N. Ngan, K. S. Leong, and H. Singh, "Cosine transform coding incorporating human visual system model." Presented at *SPIE Fiber'86*, Cambridge, MA, Sept. 1986, pp. 165–171.

[36] W. H. Chen and C. H. Smith, "Adaptive coding of monochrome and color images." *IEEE Trans. Commun.*, vol. COM-25, pp. 1285–1292, Nov. 1977.

[37] CCIR Recommendation 601, *Encoding Parameters of Digital Television for Studios*. CCIR Tech. Report Int. Telecommun. Union, vol. XI—Part 1, Plenary Assembly, Geneva, Switzerland, 1982.

[38] International Workshop on HDTV'93, Ottawa, Canada, Oct. 1993.

[39] *Information Technology—Generic Coding of Moving Pictures and Associated Audio*, jointly prepared by SC29/WGII (MPEG) and Experts Group for ATM Video Coding in the ITU-T SG 15. Recommendation H.262 ISO/IEC 13818-2, Committee Draft.

[40] K. Joseph *et al*, "MPEG++: A robust compression and transparent system for digital HDTV." *Signal Processing: Image Commun.*, vol. 4, pp. 307–323, 1992.

[41] M. I. Sezan and R. L. Lagendijk, *Motion Analysis and Image Sequence Processing*. Hingham, MA: Kluwer Academic, 1993.

[42] B. Liu and A. Zaccarin, "New fast algorithms for the estimation of block motion vectors." *IEEE Trans. Circuits Syst. Video Technol.*, vol. 3, pp. 148–157, April 1993.

[43] R. W. Young and N. G. Kingsbury, "Frequency domain estimation using a complex lapped transform," *IEEE Trans. Image Processing*, vol. 2, pp. 2–17, Jan. 1993.

[44] T. Nishitani, P. H. Ang, and F. Calthoor, *VLSI Video/Image Signal Processing.* Hingham, MA: Kluwer Academic, 1993.

[45] H. Hölzlwimmer, A. V. Brandt, and W. Tengler, "A 64 Kbit/sec motion compensated transform coder using vector quantization with scene adaptive codebook." In *Proc. Int. Conf. Commun.,* Seattle, WA, June 1987, pp. 151–156.

[46] K. M. Yang and S. Singhal, "Design of a multi-function video decoder based on a motion compensated predictive-interpolative coder." In *Proc. SPIE, Visual Commun. Image Processing,* Lausanne, Switzerland, Oct. 1990, vol. 1360, pp. 1530–1539.

[47] Y. Linde, A. Buzo, and R. M. Gray, "An algorithm for vector quantizer design." *IEEE Trans. Commun.,* vol. COM-28, pp. 84–95, Jan. 1980.

[48] N. M. Nasrabadi and R. A. King, "Image coding using vector quantization: A review." *IEEE Trans. Commun.,* vol. COM-36, pp. 957–971, Aug. 1988.

[49] P. Cassereau, "A new class of optimal unitary transforms for image processing." Master thesis, Massachusetts Institute of Technology, Cambridge, May 1985.

[50] Special Issue on 64 Kbit/sec Coding, Part II, *Signal Processing: Image Commun.,* vol. 3, Feb. 1991.

Chapter 8

Subband and Wavelet Filters for High-Definition Video Compression*

T. Naveen[†]
Video and Networking Division
Tektronix, Inc.
Beaverton, Oregon

and

J. W. Woods
Center for Image Processing Research
Rensselaer Polytechnic Institute
Troy, New York

We review the structure of some of the 1-D subband *finite impulse response* (FIR) filters found in the literature. We develop a bit allocation algorithm that takes into account the impulse response of the filters being used in subband synthesis. Using this filter based bit allocation algorithm, we compare the subband compression capabilities of nine filter sets at different bit rates. Linear phase even-tap and

---

*Based on "A Filter Based Bit Allocation Scheme for Subband Compression of HDTV" by J. W. Woods and T. Naveen which appeared in *IEEE Trnas. Image Processing,* vol 1, pp. 436–440, July 1992. © 1992 IEEE.

[†]T. Naveen was at Rensselaer when this work was performed.

Handbook of Visual Communications
265

odd-tap quadrature mirror filters (QMFs), and nonlinear phase orthogonal and linear phase biorthogonal perfect reconstruction filters were considered for the comparison. Using DPCM and PCM in subband compression of high-definition (HD) video, we have found that QMFs have an edge over the rest, in subjective as well as objective (mean squared error) comparisons.

8.1 Introduction

In recent years, subband coding has gained attention as a powerful method for compressing both still images and video, including HDTV [1, 2, 3, 4, 5, 6]. Subband image coding is based on the decomposition of the input image into relatively narrow subbands where each of these subbands is then decimated and encoded with a coder and bit rate accurately matched to the statistics of that particular subband. At the receiver, these subbands are interpolated, filtered, and added back together to give the reconstructed image frame (cf. Fig. 8.1).

Numerous 1-D subband filter sets are available in the literature: even-tap and odd-tap QMFs, orthogonal and biorthogonal perfect reconstruction filter sets, IIR filters, etc. The goal of this chapter is to compare the compression capabilities of these filter sets. One approach to make this comparison is to assume a wide-sense stationary (WSS) source and compute the information-theoretic overall rate–distortion functions of the subband coding system with various filter sets. Tabatabai [7] analyzed the effects of channel and quantization errors in the reconstructed signal. Fischer [8] gives a method to compute such a rate–distortion function for an orthogonal perfect reconstruction filter bank, assuming a colored WSS Gaussian source. This approach is not directly applicable to a general biorthogonal perfect reconstruction filter bank or a generalized Gaussian source. Another approach would be to perform realizable coding of the subbands and

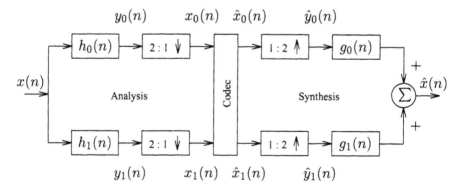

Figure 8.1   General 1-D subband analysis and synthesis systems.

compare the operational rate–distortion performances of various filter banks. References [4, 9] compared subband filters by quantizing the subbands of a standard image generated by various filters, using a single fixed set of uniform quantizers and measuring the entropy of the quantizer outputs. Note that these references were not making optimal quantization of the subbands. Westerink [3] compared even-tap QMFs by encoding a standard image, using optimal quantization of uniform-sized subbands. In this chapter, we make a similar comparison between QMFs and perfect reconstruction filter sets.

A major component of subband coding schemes is the allocation of the bits for encoding the various subbands. Some of the work reported in the subband image coding literature involves making decisions on the bit allocations perceptually [2, 4]. In contrast, in references [1, 3, 5, 6], M equi-sized subbands were generated by QMFs and encoded on an objective (mse) basis. For sufficiently bandlimiting filters, it has been found that the overall distortion D in the reconstructed image can be written as the sum of the distortion (reconstruction error variance) σ_{rk}^2 in each subband channel [1], and so the optimum bit allocation $\{r_k\}$ can be found such that:

$$D = \min \sum_k \sigma_{rk}^2 \quad \text{and} \quad \frac{1}{M} \sum_k r_k \leq R, \quad r_k \geq 0, \quad (8.1)$$

where the subscripts are indexes of the subbands. The assumption that the overall distortion can be written as the sum of separate distortions is valid for QMFs that are half-band symmetric filters, whose low-pass and high-pass filters are mirror images of each other. On the other hand, the low-pass and high-pass perfect reconstruction filters are not mirror images of each other when they are linear phase even-order FIR filters [4, 10]. The frequency response of such a filter set is shown in Fig. 8.2. When using such nonmirrored filters, the quantization noise in various subbands will not be equally weighted in the reconstruction, even if the subbands are of same size and bandwidth. Thus, to make a valid bit allocation, in [11] we have introduced scaling factors w_k in (8.1) as

$$D = \min \sum_k w_k \sigma_{rk}^2 \quad \text{such that} \quad \frac{1}{M} \sum_k r_k \leq R, \quad r_k \geq 0, \quad (8.2)$$

where w_k takes into consideration the filter set being used in the reconstruction (synthesis) of an image from its subbands. The weighting factor w_k represents the energy contribution of the kth channel to the overall reconstruction error when a unit variance white noise is entered to that channel in the synthesis filter bank.

In this chapter, we use a simple additive white noise representation for the quantizers. A more accurate model for the quantizer at low bit rates is a gain-plus-additive white noise representation [13]. A paper by Uzun and Haddad [14] uses this model to arrive at the weighting factors. An equation similar to (8.2) has been derived in [15], where the terms w_k are introduced to model the human visual system rather than, as here, the effects of finite length filters.

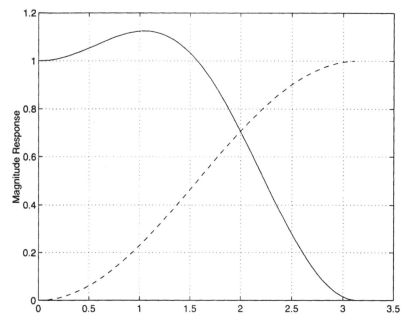

Figure 8.2   The frequency response of LeGall and Tabatabai's 5 × 3 subband filter set [12].

In Section 8.2, we review the structure and the design philosophy of some of the finite impulse response 1-D subband filters found in the literature, outlining their relationship to the wavelet theory. We compare the interband decorrelating capabilities of some of these filter sets for an AR(1) WSS source. Next, we present a method to arrive at the weighting factors for any given set of filters and use these weights in an optimal bit allocation. We then use this allocation method to compare the performances of subband image coding filters. In Section 8.3, we review the input and output power spectral densities for linear shift invariant systems and extend the result to an interpolator. In Section 8.4, we present a method to compute the scaling factors w_k of (8.2). In Section 8.5, we provide an iterative algorithm to obtain the bit allocation. Section 8.6 gives the details of our encoding procedure. In Section 8.7, we present the simulations results, where we compare the performance of the various filters both objectively and subjectively on a test video. Section 8.8 concludes this chapter.

8.2 Review of Subband Filter Sets

In Fig. 8.1, neglecting the coding errors and transmission losses, we can write

$$\hat{X}(\omega) = \frac{1}{2}\left[G_0(\omega)H_0(\omega) + G_1(\omega)H_1(\omega)\right]X(\omega)$$
$$+ \frac{1}{2}\left[G_0(\omega)H_0(\omega+\pi) + G_1(\omega)H_1(\omega+\pi)\right]X(\omega+\pi), \quad (8.3)$$

or equivalently in z domain,

$$\hat{X}(z) = \frac{1}{2}\begin{bmatrix}G_0(z)\\G_1(z)\end{bmatrix}^T\begin{bmatrix}H_0(z)H_0(-z)\\H_1(z)H_1(-z)\end{bmatrix}\begin{bmatrix}X(z)\\X(-z)\end{bmatrix}. \quad (8.4)$$

The second term in (8.3) is due to aliasing, which can be made to disappear (necessary and sufficient) by setting

$$G_0(\omega)H_0(\omega+\pi) + G_1(\omega)H_1(\omega+\pi) = 0. \quad (8.5)$$

The necessary and sufficient solution to (8.5) in z domain is

$$\begin{bmatrix}G_0(z)\\G_1(z)\end{bmatrix} = C(z)\begin{bmatrix}H_1(-z)\\-H_0(-z)\end{bmatrix} \quad (8.6)$$

for some $C(z)$, which is usually taken to be a constant, c. In the Fourier domain, this is equivalent to

$$G_0(\omega) = c \cdot H_1(\omega+\pi),$$
$$G_1(\omega) = -c \cdot H_0(\omega+\pi), \quad (8.7)$$

and in spatial (time) domain,

$$g_0(n) = c \cdot (-1)^n h_1(n),$$
$$g_1(n) = -c \cdot (-1)^n h_0(n). \quad (8.8)$$

Upon cancellation of the aliased component in the output, the overall transfer function is given by

$$T(\omega) = \frac{\hat{X}(\omega)}{X(\omega)} = \frac{c}{2}\left[H_0(\omega)H_1(\omega+\pi) - H_0(\omega+\pi)H_1(\omega)\right],$$
$$T(z) = \frac{\hat{X}(z)}{X(z)} = \frac{c}{2}\left[H_0(z)H_1(-z) - H_0(-z)H_1(z)\right]. \quad (8.9)$$

Ideally the filter bank output should be a delayed replica of the input,

$$T(\omega) = e^{-j\omega D}. \quad (8.10)$$

The necessary and sufficient condition for this is [16]:

$$\frac{c}{2}\left[H_0(z)H_1(-z) - H_0(-z)H_1(z)\right] = \text{const} \cdot z^{-2l-1}, \qquad l \in \Im, \qquad (8.11)$$

where $\Im$ denotes the set of integers[1].

We denote the lengths of filters h_i and g_i by L_{h_i} and L_{g_i} respectively, for $i = 0, 1$. In the following, we bring together some of the considerations given in the literature for the design of subband filters. As we shall explain later, some of these considerations are conflicting. For image coding applications, these criteria need not be satisfied exactly and approximations are sufficient.

1. Easy to implement (computationally efficient). This could be achieved through one or more of the following: symmetric filter coefficients, short length filters, multiplierless implementation of the convolution, and fast transform equivalent of the convolution.

2. The basis functions generated by h_0 are orthogonal. That is, the impulse response of filter h_0 and its shifted versions (by even shifts) form an orthogonal set. This ensures that there is no redundancy in the transform coefficients.

3. For the same reason, the basis functions generated by h_0 and h_1 should be orthogonal.

4. Perfect reconstruction in the absence of coding and channel errors.

5. The aliased components in the subbands should be small. This is achieved by making the frequency response of h_0 to be as close as possible to that of an ideal half-band filter.

6. The filters should have linear phase, which is important in image compression.

7. The overall transfer function $T(\omega)$ should be maximally flat at zero frequency. This is important since the energy in images is concentrated near zero frequency, and it is undesirable to introduce large distortions there.

8. $T(\omega)$ should have maximal slope through the $\omega = \frac{\pi}{2}$ point.

9. The coding gain should be maximized. This would involve signal adaptive design of the filters [17].

10. The filters should be such that the energy of the signal is concentrated in a single subband (as much as possible). This criterion is relevant when one of the subbands is not transmitted because of bit allocations, or when one low priority subband is lost in the channel [18].

11. Step response of the filters should have small overshoots. Otherwise, ringing artifacts occur in the encoded image.

12. There should be regularity. That is, iterated synthesis applied to a sequence consisting of only one nonzero entry should look reasonably nice, even after

---

[1] We use $\Im$ to represent the set of integers instead of the more commonly used representation Z, so as not to confuse with the Z-transformation.

several iterations. This feature is needed when one subband is made zero while encoding with low bit rates. The definition of regularity is given in Section 8.2.2

13. Ideally, the subband signals should be uncorrelated. That is, referring to Fig. 8.1, for a zero-mean WSS input $x(n)$,

$$E\{x_i(k)x_j(l)\} = \sigma_i^2 \delta_{ij}\delta_{kl} \quad \forall k,l \quad \text{and} \quad i,j \in \{0,1\}. \tag{8.12}$$

This would let us encode various subbands and their components independently.

We say a subband filter set is *orthogonal* if criteria 2 and 3 are satisfied by it. We note here that some of these considerations are conflicting; e.g., 5 and 11. Though one cannot achieve zero overshoot and ideal frequency responses simultaneously, one can use a cost function for the optimization, which is a combination of both the step and frequency responses of the filters. Numerous approaches have been reported in the literature to design filters h_i and g_i, $i = 0, 1$, that satisfy exactly or approximately (8.8) and (8.10), in addition to some of the other considerations given here. In this section, we discuss some of these approaches from a signal processing viewpoint. We start with the quadrature mirror filters, followed by perfect reconstruction filters. We then outline the relationship between perfect reconstruction filter banks and wavelet theory. We finally look into the decorrelating capabilities of various subband filter sets.

8.2.1 Linear Phase Quadrature Mirror Filters

The QMF meets (8.10) approximately by selecting the high-pass filter according to

$$h_1(n) = (-1)^n h_0(n),$$
$$\text{or} \quad H_1(\omega) = H_0(\omega + \pi). \tag{8.13}$$

We can see that the low-pass and high-pass filters are mirror images of each other about the frequency $\frac{\pi}{2}$ and that their lengths are identical, $L_{h_0} = L_{h_1} = L$. The transfer function then becomes

$$T(\omega) = \frac{c}{2}\left[H_0^2(\omega) - H_0^2(\omega + \pi)\right]. \tag{8.14}$$

We should note here that $T(\omega)$ cannot equal the ideal response (8.10) for these linear phase QMFs, except in a trivial two-tap filter $h_0 = [0.5, 0.5]$, and for infinite length filters [10]. It can be shown that when the L is odd, for linear phase QMFs, $T(\frac{\pi}{2}) = 0$. This is a serious amplitude distortion and is not acceptable for any coding applications. For this reason, from here on, we take the length of the filter L to be even for QMFs.

Even-Tap QMFs   In even-tap QMF design, in addition to (8.8) and (8.13), we constrain $h_0(n)$ to be symmetric, linear phase:

$$h_0(l) = h_0(L - 1 - l), \qquad l = 0, 1, ..., L - 1,$$
$$H_0(\omega) = |H(\omega)| e^{-j\omega\left(\frac{L-1}{2}\right)}, \tag{8.15}$$

for some $H(\omega)$. Then the phase response of the overall system becomes linear, with arg $\{T(\omega)\} = -\omega(L - 1)$, and

$$T(\omega) = \frac{c}{2}\left[|H(\omega)|^2 + |H(\omega + \pi)|^2\right]e^{-j\omega(L-1)}. \tag{8.16}$$

Reference [19] optimally designs $h_0(n)$ to approximate the ideal frequency response. The filter coefficients of QMFs 12B, 16B, and 32C from [19] are given in Appendix 8.A. The impulse response of filter h_0 and its shifted versions (by even shifts) form an approximately orthogonal set; that is,

$$\langle h_0(n - 2l), h_0(n - 2k)\rangle \approx 0.5\delta_{kl}, \quad (k, l) \in \Im^2, \tag{8.17}$$

with $\langle \cdot, \cdot \rangle$ denoting the inner product[2]. The approximation is very good (see Table 8.1).

In addition, the $h_0(n)$ and $h_1(n)$ are orthogonal with respect to even shifts:

$$\langle h_1(n - 2l), h_0(n - 2k)\rangle = 0, \qquad (k, l) \in \Im^2. \tag{8.18}$$

Odd-Tap QMFs   In odd-tap QMFs, we still have the length of filters L to be even. However, in addition to satisfying (8.8) and (8.13), we force a zero coefficient in the symmetric (linear phase) h_0 [20]:

$$h_0(L - 1) = 0,$$
$$h_0(l) = h_0(L - 2 - l), \quad l = 0, 1, ..., L - 2,$$
$$H_0(\omega) = |H(\omega)| e^{-j\omega\left(\frac{L}{2}-1\right)}. \tag{8.19}$$

Similar to the even-tap QMF case, the phase response of the overall system becomes linear, with arg $\{T(\omega)\} = -\omega(L - 2)$, and

$$T(\omega) = \frac{c}{2}\left[|H(\omega)|^2 + |H(\omega + \pi)|^2\right]e^{-j\omega(L-2)}. \tag{8.20}$$

Adelson, Simoncelli, and Hingorani designed five-tap, seven-tap, and nine-tap QMF filter sets in [20]. In this chapter, these filter sets are referred to as AS 5 × 5, AS 7 × 7, and AS 9 × 9, and their impulse responses are listed in Appendix 8.A. The design algorithm in this reference attempts to make the impulse response of the filter h_0 and its even translates form an approximately orthogonal set (see

---

[2]That is, $\langle a(n), b(n)\rangle \overset{\Delta}{=} \sum_{n=-\infty}^{\infty} a^*(n)b(n)$, where * denotes complex conjugation.

Table 8.1   The inner products of various QMFs.

Filter set	$\langle h_i(n), h_i(n)\rangle$	$\max_{(k,l)\in\Im^2, k\neq l} \lvert\langle h_i(n-2l), h_i(n-2k)\rangle\rvert$
QMF 12B	0.49997848	0.00076142
QMF 16B	0.50002730	0.00036911
QMF 32C	0.50003153	0.00012938
QMF AS 5×5	0.49999312	0.00289552
QMF AS 7×7	0.49999520	0.00536129
QMF AS 9×9	0.50001121	0.00039800

Table 8.1). We see that the orthogonality of the basis functions improves with the length of the filters. The orthogonality of the odd-tap QMFs is better than that of even-tap QMFs of comparable length. In addition to orthogonality, the design of AS 9×9 achieves a maximal slope in its frequency response at $\frac{\pi}{2}$. The $h_0(n)$ and $h_1(n)$ are orthogonal with respect to even shifts:

$$\langle h_1(n-2l), h_0(n-2k)\rangle = 0, \qquad (k, l) \in \Im^2. \tag{8.21}$$

8.2.2 Perfect Reconstruction Filters

In the QMF solution to satisfy (8.10), we restricted the analysis filters (and hence the synthesis filters) to be symmetric (linear phase), mirror images of each other, and of the same filter length. By removing some of these restrictions, we have a lot more freedom in the design of filters, leading to perfect reconstruction (PR). We outline some of these PR filters here. The relationship between wavelet theory and perfect reconstruction subband filtering has received a lot of attention [16, 21, 22, 23, 24].

Orthogonal (or Paraunitary) Filter Banks   In this family of filters, the impulse responses of the filters $h_i(n)$ are orthogonal to their even translates,

$$\langle h_i(n-2l), h_i(n-2k)\rangle = \delta_{kl}, \qquad k, l \in \Im, \qquad i = 0, 1; \tag{8.22}$$

and the filters $h_0(n)$ and $h_1(n)$ are orthogonal to each other at even translates,

$$\langle h_0(n-2l), h_1(n-2k)\rangle = 0, \qquad k, l \in \Im. \tag{8.23}$$

To satisfy (8.22), the length of h_0 has to be even. To satisfy both (8.22) and (8.23), the necessary and sufficient condition is [16]

$$H_1(z) = z^{-2k-1} H_0(-z^{-1}), \tag{8.24}$$

for some $k \in \Im$. This means that the filters h_0 and h_1 have equal length, $L_{h_0} = L_{h_1} = L$. If we take $k = \frac{L-2}{2}$, (8.24) is equivalent to

$$h_1(n) = (-1)^{n+1} h_0(L - n - 1). \tag{8.25}$$

Taking (8.8) and (8.25) together, h_0 describes all the four filters involved in the orthogonal perfect reconstruction filters in the subband analysis and synthesis banks. These orthogonal PR filter sets are called *PR-QMFs* [25]. We note here that the finite length orthogonal perfect reconstruction filters are not linear phase [23], except in a trivial two-tap filter set with $h_0 = [0.5, 0.5]$.

Expanding the filters in a polyphase notation,

$$H_i(z) = H_{i0}(z^2) + z^{-1} H_{i1}(z^2), \tag{8.26}$$

a polyphase matrix for the filter set is defined as

$$H_p(z^2) = \begin{bmatrix} H_{00}(z^2) & H_{01}(z^2) \\ H_{10}(z^2) & H_{11}(z^2) \end{bmatrix}. \tag{8.27}$$

For orthogonal filter banks,

$$[H_p(z^{-1})]^T \cdot H_p(z) = I, \tag{8.28}$$

or $H_p(z)$ is *paraunitary*. For this reason, orthogonal filter banks are also called *paraunitary* filter banks [26].

Vaidyanathan and Hoang in [27] designed paraunitary filters using a lattice structure for the filter banks. Two of these filters, designated as *HV 12B* and *HV 16B*, are listed in Appendix 8.A. This reference did not consider regularity of the filters.

Regularity of a filter is defined as follows. Define a piecewise constant function as

$$f^{(i)}(x) = 2^{i/2} \cdot h_0^{(i)}(n), \tag{8.29}$$

where n is selected so that $2^{-i} n \leq x < 2^{-i}(n + 1)$, and where $h_0^{(i)}$ is the inverse Z-transform of

$$H_0^{(i)}(z) = \prod_{l=0}^{i-1} H_0(z^{2^l}). \tag{8.30}$$

Then the filter h_0 is termed *regular* if the sequence of functions $f^{(i)}(x)$ converges pointwise to a continuous function. It is necessary for the low-pass filter $H_0(z)$ to have a zero at half the sampling frequency or $H_0(e^{j\pi}) = 0$ [16]. This would also mean that the corresponding high-pass filter $H_1(z)$ has a zero at the dc frequency. As mentioned earlier, iterated synthesis by a regular filter applied to a sequence consisting of only one nonzero entry looks smooth. Thus we can expect smooth reconstructions while reconstructing data in a two-band system from a single subband. This property is useful when one subband is made zero while encoding the data with low bit rates, or while zooming in on an image using these filters.

Reference [23] designed paraunitary filters satisfying the regularity condition. Some of these filters are also listed in [22]. Two of these filters, referred to as W 6×6 and W 8×8 are listed in Appendix 8.A.

Biorthogonal Filter Banks   All perfect reconstruction filters, whether or not orthogonal, are termed *biorthogonal*. This is due to the fact that for every perfect reconstruction filter set,

$$\langle \tilde{g}_0(n - 2k), h_1(n - 2l) \rangle = 0,$$
$$\langle \tilde{g}_1(n - 2k), h_0(n - 2l) \rangle = 0,$$
$$\langle \tilde{g}_i(n - 2k), h_i(n) \rangle = \delta_k, \qquad k, l \in \mathfrak{I}, \qquad i = 0, 1, \qquad (8.31)$$

where $\tilde{h}$ denotes the time inversion,

$$\tilde{h}(n) \stackrel{\Delta}{=} h(L - 1 - n). \tag{8.32}$$

Since the orthogonality conditions (8.22) and (8.23) are no longer necessary for biorthogonal filter banks, there is more freedom to design these filters. Linear phase perfect reconstruction real FIR filter banks using filters $h_0(n)$ and $h_1(n)$ have one of the following forms [16]:

1. Both filters are symmetric and have odd lengths, differing by an odd multiple of 2.

2. One filter is symmetric and the other is antisymmetric; both lengths are even and equal to or differ by an even multiple of 2.

3. One filter is of odd length, the other is even; both have all their zeros on the unit circle. Either both filters are symmetric or one is symmetric and the other is antisymmetric.

Most of the biorthogonal filter sets available in the literature satisfy either condition (1) or (2). A very short length filter set given by LeGall and Tabatabai in [12], which we refer to as G 5×3, is listed in Appendix 8.A. This filter set has very simple coefficients and can be implemented with only shifts and adds. Reference [10] designed biorthogonal filter sets using lattice structured filters for conditions (1) and (2). These filters have lengths of 64×64 and 23×25. Reference [16] gives filters with lengths of 18×18, 24×24, and 20×20, satisfying (2). Using a combination of step-response error and frequency-response error as the objective function, Kronander in [4] designed a biorthogonal filter bank using the lattice structure given in [10]. This filter set, K 9 × 7, is listed in Appendix 8.A. Reference [28] designed biorthogonal filter banks using a structure for $H_0(z)$ and $G_0(z)$:

$$H_0(z) = \frac{1}{2}\left[1 + zA(z^2)\right],$$
$$G_0(z) = 2z^{-1} + \left[1 - zA(z^2)\right]B(z^2), \tag{8.33}$$

where $A(z^2)$ and $B(z^2)$ are derived from Lagrange half-band filters. These biorthogonal filter sets are reported to result in considerably lower aliasing in the low-band signal compared to other filter sets [28].

Similar to the orthogonal filter bank case, we can impose the regularity constraint on the low-pass filters h_0 and g_0. Many examples of biorthogonal filter sets having the regularity property are listed in [24], some of which are cross-listed in [21]. One of these filter sets, W 5 × 7 in this chapter and listed in Appendix 8.A, is very close to being orthogonal. The five-tap filter in this set corresponds to the Gaussian filter of [29] with parameter a in that paper equal to 0.6. The approximate orthogonality of this filter set can be seen in Table 8.2.

8.2.3 Relationship of the Perfect Reconstruction Filter Banks to the Wavelet Theory

The orthogonal and biorthogonal PR filter banks, having the regularity property, are related to the wavelet theory. A wavelet transform splits the original space in two, then splits one of the resulting half spaces in two, etc.

Orthogonal Wavelets   The impulse responses of orthogonal regular subband filter banks are related to the discrete wavelet transformation. The time reversed versions of two filter impulse responses $h_0(n)$ and $h_1(n)$, i.e. $\tilde{h}_0(n)$ and $\tilde{h}_1(n)$, together with their even translates, form an orthonormal basis for $l^2(\Im)$. That is, the set $\{\tilde{h}_0(n - 2l), \tilde{h}_1(n - 2k); l, k \in \Im\}$ forms an orthonormal basis for $l^2(\Im)$.

Table 8.2   The inner products for W 5 × 7.

$\langle h_0(n - 2l), h_0(n - 2k) \rangle$		$\langle h_1(n - 2l), h_1(n - 2k) \rangle$		$\langle h_0(n - 2l), h_1(n - 2k) \rangle$
$l = k$	$\max_{(k,l)\in\Im^2, k\neq l} \lvert\langle\cdot,\cdot\rangle\rvert$	$l = k$	$\max_{(k,l)\in\Im^2, k\neq l} \lvert\langle\cdot,\cdot\rangle\rvert$	$\max_{(k,l)\in\Im^2} \lvert\langle\cdot,\cdot\rangle\rvert$
0.49000001	0.00250000	0.51053578	0.00271684	0.00607143

Call V_{-1} the *space* $l^2(\Im)$, and call V_0 the *subspace* of V_{-1} spanned by $\tilde{h}_0(n)$ and its even translates. Call W_0 the *subspace* of V_{-1} spanned by $\tilde{h}_1(n)$ and its even translates. Then W_0 is the orthogonal complement of V_0 in V_{-1}:

$$V_{-1} = V_0 \oplus W_0,$$
$$V_0 \perp W_0,$$
$$V_0 \subset V_{-1}. \tag{8.34}$$

Filtering followed by subsampling of a discrete signal $x(n)$, as in Fig. 8.1, gives us $x_0(n)$ and $x_1(n)$, which are the representations of the signal $x(n)$ with the orthonormal basis $\{\tilde{h}_0(n-2l), \tilde{h}_1(n-2k); l, k \in \Im\}$. Up-sampling $x_0(n)$, followed by filtering with $\tilde{h}_0(n)$ gives the projection of $x(n)$ onto V_0. Similarly up-sampling $x_1(n)$, followed by filtering with $\tilde{h}_1(n)$ gives the projection of $x(n)$ onto W_0. Thus the subband analysis structure gives a way to decompose a space V_{-1} into V_0 and W_0.

With a cascaded structure of Fig. 8.3, we can decompose V_0 into V_1 and W_1 and so on. Then we have

$$V_j \subset V_{j-1} \qquad j = 0, 1, \ldots$$
$$V_{j-1} = V_j \oplus W_j \qquad j = 0, 1, \ldots \tag{8.35}$$

which leads to

$$\cdots V_2 \subset V_1 \subset V_0 \subset V_{-1}$$
$$V_{-1} = W_0 \oplus W_1 \oplus W_2 \oplus \cdots. \tag{8.36}$$

The representation of a signal $x(n)$ in terms of the orthonormal basis functions of W_j is the discrete wavelet transform, which is conveniently achieved using the cascaded structure of Fig. 8.3.

From the discrete bases in $l^2(\Im)$ space, we can obtain a continuous function $\psi \in L^2(\Re)$ from which a family $\{2^{-k/2}\psi(2^k - l), k, l \in \Im\}$ of orthonormal basis

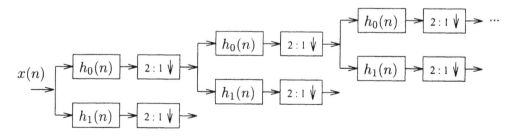

Figure 8.3   Discrete wavelet transform on sequences.

for $L^2(\Re)$ can be obtained [23]. Notice that the members of this family are the shifted and scaled versions of the prototype (or *mother*) wavelet ψ. Define the scaled Fourier series $H_0(\xi)$ of $h_0(n)$ as

$$H_0(\xi) = \sqrt{2} \sum_n h_0(n) \exp\{-jn\xi\}. \tag{8.37}$$

Define $\phi(x)$ to be the inverse Fourier transform:

$$\phi(x) = \text{FT}^{-1}\{(2\pi)^{-1/2} \prod_{k=1}^{\infty} H_0(2^{-k}\xi)\}, \tag{8.38}$$

where the Fourier transform is defined to be

$$\text{FT}\{f\} = (2\pi)^{-1/2} \int f(x) \exp\{-j\xi x\}dx. \tag{8.39}$$

Then the mother wavelet can be obtained as

$$\psi(x) = \sqrt{2} \cdot \sum_n h_1(n) \cdot \phi(2x - n), \tag{8.40}$$

where $h_1(n)$ is given by (8.25). From this $\psi(x)$, we can obtain the family of orthonormal basis functions for $L^2(\Re)$:

$$\psi_{kl}(x) = 2^{-k/2}\psi(2^k - l), \qquad k, l \in \Im. \tag{8.41}$$

Then the wavelet representation of any $f \in L^2(\Re)$ can be written as

$$f = \sum_{k,l \in \Im} \langle f, \psi_{kl} \rangle \psi_{kl}. \tag{8.42}$$

Biorthogonal Wavelets   In the biorthogonal wavelet representation of a signal, we have two hierarchies of approximation spaces,

$$\begin{aligned} \cdots V_2 \subset V_1 \subset V_0 \subset V_{-1} \\ \cdots \tilde{V}_2 \subset \tilde{V}_1 \subset \tilde{V}_0 \subset \tilde{V}_{-1}. \end{aligned} \tag{8.43}$$

Let W_j be a complement to V_j in V_{j-1}, and let $\tilde{W}_j$ be a complement to $\tilde{V}_j$ in $\tilde{V}_{j-1}$. In this general biorthogonal representation, W_j is not an orthogonal complement of V_j but the following holds:

$$\begin{aligned} W_j \perp \tilde{V}_j, \\ \tilde{W}_j \perp V_j, \end{aligned} \tag{8.44}$$

indicating biorthogonality.

Similar to the orthogonal wavelet case, a biorthogonal wavelet representation of a continuous function $f(x)$ can be achieved from regular biorthogonal subband filters [16, 24]. Define scaled Fourier series $H_0(\xi)$ and $\tilde{G}_0(\xi)$ of $h_0(n)$ and $\tilde{g}_0(n)$, respectively, as

$$H_0(\xi) = \sqrt{2} \sum_n h_0(n) \exp\{-jn\xi\},$$

$$\tilde{G}_0(\xi) = \sqrt{2} \sum_n \tilde{g}_0(n) \exp\{-jn\xi\}. \tag{8.45}$$

Define $\phi(x)$ and $\tilde{\phi}(x)$ to be the inverse Fourier transforms:

$$\phi(x) = \mathrm{FT}^{-1}\{(2\pi)^{-1/2} \prod_{k=1}^{\infty} H_0(2^{-k}\xi)\},$$

$$\tilde{\phi}(x) = \mathrm{FT}^{-1}\{(2\pi)^{-1/2} \prod_{k=1}^{\infty} \tilde{G}_0(2^{-k}\xi)\}. \tag{8.46}$$

Then define two parent wavelets for $L^2(\Re)$:

$$\psi(x) = \sqrt{2} \cdot \sum_n h_1(n) \cdot \phi(2x - n),$$

$$\tilde{\psi}(x) = \sqrt{2} \cdot \sum_n \tilde{g}_1(n) \cdot \tilde{\phi}(2x - n). \tag{8.47}$$

Then we have

$$\left\langle \tilde{\phi}(x - l), \phi(x - k) \right\rangle = \delta_{kl},$$

$$\left\langle \tilde{\psi}(x - l), \psi(x - k) \right\rangle = \delta_{kl},$$

$$\left\langle \tilde{\phi}(x - l), \psi(x - k) \right\rangle = 0,$$

$$\left\langle \tilde{\psi}(x - l), \phi(x - k) \right\rangle = 0. \tag{8.48}$$

Finally, define two dual bases for $L^2(\Re)$:

$$\psi_{kl}(x) = 2^{-k/2}\psi(2^k - l), \quad k, l \in \Im.$$

$$\tilde{\psi}_{kl}(x) = 2^{-k/2}\tilde{\psi}(2^k - l), \quad k, l \in \Im. \tag{8.49}$$

Then the biorthogonal wavelet representation of any $f \in L^2(\Re)$ can be written as

$$f = \sum_{k,l \in \Im} \left\langle f, \tilde{\psi}_{kl} \right\rangle \psi_{kl} = \sum_{k,l \in \Im} \left\langle f, \psi_{kl} \right\rangle \tilde{\psi}_{kl}. \tag{8.50}$$

8.2.4 Special Designs of Subband Filters

As mentioned earlier, Kronander in [4] designed a linear phase biorthogonal filter using a combination of step-response error and frequency-response error as the objective function. Using the same objective function he also designed a seven-tap filter set without the perfect reconstruction property. This filter set K 7×7 listed in Appendix 8.A, has low-pass and high-pass filters of equal lengths.

References [17, 18] design a paraunitary filter bank that optimizes the coding gain for a given input signal. Assuming a constant quantizer performance factor [13], the coding gain over PCM of a two-band subband scheme with an orthogonal filter bank is given by

$$G_{\text{SBC}} = \frac{\frac{1}{2}(\sigma_{x_0}^2 + \sigma_{x_1}^2)}{(\sigma_{x_0}^2 \sigma_{x_1}^2)^{\frac{1}{2}}}, \tag{8.51}$$

where $\sigma_{x_0}^2$ and $\sigma_{x_1}^2$ are the variances of subband signals $x_0(n)$ and $x_1(n)$ respectively in Fig. 8.1. Maximizing this gain involves finding the filter set that minimizes the variance of one of the two subbands. When the spectrum of the input signal $x(n)$ is nonincreasing and has components beyond $\omega = \frac{\pi}{2}$, which is true for many natural images, the design goal would be to approximate ideal half-band filters [17].

Multiplierless PR filter sets are given in [25]. These filter sets have integer coefficients. Then multiplication by any filter coefficient $h(n)$ can be performed using binary shifts and adds, making the filter implementation multiplierless.

8.2.5 Orthogonality of Filters and Stochastic Orthogonality

So far, we were concerned with the orthogonality of the filter sets, i.e. (8.22), (8.23), and (8.31). In this section, we deal with another kind of orthogonality, the *stochastic orthogonality* of the subband signals for a WSS source. Assuming a zero mean source, stochastic orthogonality and stochastic uncorrelatedness are the same [30]. Ideally, for coding and other applications, all the samples of different subbands that do not coincide in location are independent random variables. It is well-known that for Gaussian random variables, uncorrelatedness is equivalent to independence. In this section, we will be interested in the uncorrelatedness of these random variables, which for a non-Gaussian random variable is easier to verify than independence. The ideal uncorrelatedness requirement for the random variables in Fig. 8.1 is

$$E\{x_i(k)x_j(l)\} = \sigma_i^2 \delta_{ij} \delta_{kl} \qquad \forall k, l \quad \text{and} \quad i, j \in \{0, 1\}. \tag{8.52}$$

This would mean that intraband as well as interband samples are uncorrelated. Let $x(n)$ have a finite even length of M and define a vector X as

$$X \stackrel{\Delta}{=} [x(0)\, x(1) \cdots x(M-1)]^T. \tag{8.53}$$

Define a vector $\Theta(n)$ as the signal obtained by interleaving the samples of low-pass and high-pass subbands:

$$\Theta \triangleq \left[x_0(0)\, x_1(0)\, x_0(1)\, x_1(1) \cdots x_0(\frac{M}{2})\, x_1(\frac{M}{2}) \right]^T . \tag{8.54}$$

Similarly for the encoded signals,

$$\hat{X} \triangleq \left[\hat{x}(0)\, \hat{x}(1) \cdots \hat{x}(M-1) \right]^T \tag{8.55}$$

and

$$\hat{\Theta} \triangleq \left[\hat{x}_0(0)\, \hat{x}_1(0)\, \hat{x}_0(1)\, \hat{x}_1(1) \cdots \hat{x}_0(\frac{M}{2})\, \hat{x}_1(\frac{M}{2}) \right]^T . \tag{8.56}$$

Then the two-channel subband analysis and synthesis processes can be described as the transformations

$$\Theta = AX,$$
$$\hat{X} = B\hat{\Theta}, \tag{8.57}$$

where A and B are $M \times M$ matrices with

$$A = \begin{bmatrix} h_0(L-1) & h_0(L-2) & h_0(L-3) & h_0(L-4)\cdots & 0 & 0 \\ h_1(L-1) & h_1(L-2) & h_1(L-3) & h_1(L-4)\cdots & 0 & 0 \\ 0 & 0 & h_0(L-1) & h_0(L-2)\cdots & 0 & 0 \\ 0 & 0 & h_1(L-1) & h_1(L-2)\cdots & 0 & 0 \\ & & & \vdots & & \\ h_0(L-3) & h_0(L-4) & h_0(L-5) & h_0(L-6)\cdots h_0(L-1) & h_0(L-2) \\ h_1(L-3) & h_1(L-4) & h_1(L-5) & h_1(L-6)\cdots h_1(L-1) & h_1(L-2) \end{bmatrix} \tag{8.58}$$

and

$$B = \begin{bmatrix} g_1(0) & g_0(0) & 0 & 0 & \cdots g_1(2) & g_0(2) \\ g_1(1) & g_0(1) & 0 & 0 & \cdots g_1(3) & g_0(3) \\ g_1(2) & g_0(2) & g_1(0) & g_0(0)\cdots g_1(4) & g_0(4) \\ & & & \vdots & \\ 0 & 0 & 0 & 0 & \cdots g_1(1) & g_0(1) \\ 0 & 0 & 0 & 0 & \cdots g_1(0) & g_0(0) \end{bmatrix} . \tag{8.59}$$

In obtaining A and B we have assumed equal length filters, which can be obtained by padding the filter coefficients with zeroes when needed. Then the perfect reconstruction requirement is

$$BA = I_{M \times M}, \tag{8.60}$$

which is satisfied by all biorthogonal filter banks. Note that in general A is not invertible.

From (8.57), the mean squared error of the reconstructed signal can be written as

$$E\left\{(X - \hat{X})^T(X - \hat{X})\right\} = E\left\{(\Theta - \hat{\Theta})^T B^T B(\Theta - \hat{\Theta})\right\}. \qquad (8.61)$$

The paraunitary (orthogonal PR) filter banks satisfy (8.8) and (8.25). For these filter sets A is invertible and unitary:

$$B = A^{-1} = A^T. \qquad (8.62)$$

For paraunitary filter banks (8.61) becomes

$$E\left\{(X - \hat{X})^T(X - \hat{X})\right\} = E\left\{(\Theta - \hat{\Theta})^T(\Theta - \hat{\Theta})\right\}, \qquad (8.63)$$

which says that the mean square quantization error of the paraunitary subband coefficients equals the mean square of the reconstructed signal. The relationship between mean squared error of subband coefficients and mean squared error of the reconstructed signal for a general subband filter set is considered in the following sections.

The transformation A is in general *not* decorrelating, except when the source correlation is given by

$$R_{xx} = A\Lambda A^T \qquad (8.64)$$

for some diagonal matrix Λ. Intuitively the statement that A is not in general decorrelating is appealing, when considering an ideal half-band filter bank. For a colored source with a power spectral density (PSD) of $P_{xx}(\omega)$, an ideal filter bank would generate a low-pass subband with a PSD

$$P_{x_0 x_0}(\omega) = P_{xx}\left(\frac{\omega}{2}\right), \qquad \omega \in [-\pi, \pi], \qquad (8.65)$$

which is a constant with respect to ω only if $P_{xx}(\omega)$ is a constant for $\omega \in [-\frac{\pi}{2}, \frac{\pi}{2}]$.

A weaker requirement compared to (8.52) would be interband uncorrelatedness:

$$E\{x_i(k)x_j(k)\} = \sigma_i^2 \delta_{ij} \qquad \forall k \quad \text{and} \quad i, j \in \{0, 1\}, \qquad (8.66)$$

which is satisfied by an ideal half-band subband filter bank and approximately satisfied by other filter banks. Thus the cross-correlation of the two subband signals can be used as a measure of their performance on a given input:

$$E\left\{x_0(n)x_1(n)\right\} = E\left\{(h_0 * x)(2n) \cdot (h_1 * x)(2n)\right\}, \qquad (8.67)$$

which after some trivial algebra for a WSS source becomes

$$E\left\{x_0(n)x_1(n)\right\} = \sum_l \left[\sum_k h_0(k)h_1(k+l)\right] R_{xx}(l) \qquad \forall n. \qquad (8.68)$$

Define a *subband filter set quality function* Q_{01} as

$$Q_{01}(l) \triangleq \langle h_0(k), h_1(k+l)\rangle. \qquad (8.69)$$

Then (8.68) can be written as

$$E\left\{x_0(n)x_1(n)\right\} = \sum_l Q_{01}(l)R_{xx}(l) \qquad \forall n, \qquad (8.70)$$

which should be zero to achieve uncorrelatedness across the subbands. For ideal half-band subband filter bank, $Q_{01}(l) = 0$ for all l, and consequently it produces uncross-correlated subbands. We should note here that for long length QMFs, $Q_{01}(l) \approx 0$, and they give best decorrelation between the subbands, as we see later.

The autocorrelation function $R_{xx}(n)$ is known to be an even function. Then, the necessary and sufficient condition to make (8.70) equal zero for any R_{xx} is to make $Q_{01}(l)$ an odd function,

$$Q_{01}(l) = -Q_{01}(-l) \qquad (8.71)$$

for all l. This says that

$$\sum_k h_0(k)\left[h_1(k+l) + h_1(k-l)\right] = 0 \qquad \forall l, \qquad (8.72)$$

which, using the notation introduced in (8.32), becomes

$$(h_0 * \tilde{h}_1)(L-1-l) + (h_0 * \tilde{h}_1)(L-1+l) = 0 \qquad \forall l. \qquad (8.73)$$

For orthogonal filter sets we have $\sum_k h_0(k)h_1(k+l) = 0$ for even l and (8.68) reduces to

$$E\left\{x_0(n)x_1(n)\right\} = \sum_l \left[\sum_k h_0(k)h_1(k+2l+1)\right] R_{xx}(2l+1) \qquad \forall n. \quad (8.74)$$

Then to achieve uncross-correlated subbands, (8.71) should be satisfied for odd values of l.

We computed the cross-correlation given in (8.68) for a few subband filter sets for a *first-order autoregressive* (AR(1)) source for which

$$R_{xx}(l) = \rho^{|l|}, \quad l \in \Im. \qquad (8.75)$$

Table 8.3 lists these numbers when $\rho = 0.95$. Even-tap QMFs give the best decorrelation between the subbands for this source. So the even-tap QMFs are more suitable while encoding various subbands independently.

8.2.6 Summary Review of Subband Filter Sets

As evidenced by the previous review, quite a large number of subband filter sets are available in the literature. We have listed some of them in Appendix 8.A. From our study, we have found that all the two-band FIR subband filter sets follow (8.8). A Venn diagram of various FIR subband filter sets available in literature is shown in Fig. 8.4. We note here that long linear phase QMFs are approximately orthogonal and approximately PR. Due to resource limitations, only a few of the filter sets that are representative of the various families of filters are compared in this chapter. Specifically, we compare QMF 12B, QMF 16B, QMF 32C, AS 9 × 9, W 6 × 6, W 8 × 8, G 5 × 3, K 7 × 7, and K 9 × 7 subband filter sets.

In the following sections, we present our method to compute the scaling factors w_k of (8.2) and compare the subjective and objective performances of these filter sets. We start by looking at the output power spectral density of an interpolator.

8.3 Power Spectral Densities

In a subband coding system, we encode various sample rate signals and then interpolate, filter, and add them back together to obtain the output signal. In this section, we review the input and output power spectral densities for LSI systems and extend the results to an interpolator. Assuming that the quantization noise is white, we use these results in Section 8.4 to estimate the contribution of the noise in the final output, giving rise to the scaling factors, w_k [11]. Chapter 3 in a book

Table 8.3   Measure of cross-correlation between the subbands for an AR(1) source with $\rho = 0.95$.

Filter set	$E\left\{x_0(n)x_1(n)\right\}$	Filter set	$E\left\{x_0(n)x_1(n)\right\}$
QMF 12B	-2.551263e-17	W 6 × 6	5.562071e-03
QMF 16B	-9.214871e-18	W 8 × 8	-3.057945e-03
QMF 32C	-1.450650e-17	G 5 × 3	5.781445e-03
AS 5 × 5	1.106756e-02	K 9 × 7	-3.411520e-03
AS 7 × 7	-8.294532e-03	W 5 × 7	9.200726e-03
AS 9 × 9	-7.900085e-04	K 7 × 7	5.444855e-03

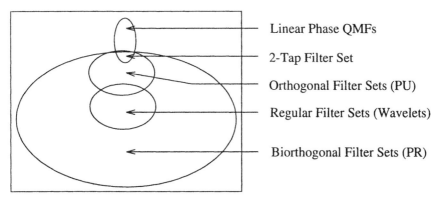

Figure 8.4   A Venn diagram of various two-band FIR subband filter sets.

by Akansu and Haddad [25] gives an analysis of the noise in the reconstructed signal using a polyphase structure for the subband synthesis system.

The relationship between the input and output power spectral densities for an LSI system (Fig. 8.5) is given by

$$P_{yy}(\omega) = |H(\omega)|^2 \, P_{xx}(\omega). \tag{8.76}$$

An interpolator is a linear shift variant system (Fig. 8.6). A twofold interpolator inserts a zero between adjacent samples of the input sequence $x(n)$. For the

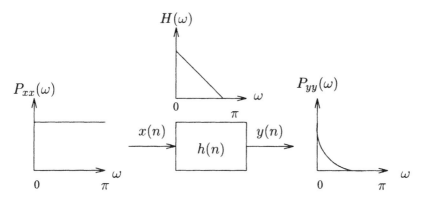

Figure 8.5   An LSI system with an input random signal and its input and output power spectral densities.

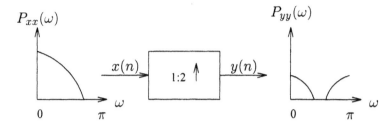

Figure 8.6 A twofold interpolator and its input and output average power spectral densities.

analysis, let us assume that every odd sample of the output $y(n)$ is zero. The autocorrelation function of the random variable $y(n)$, defined as

$$R_{yy}(n + m, n) \triangleq E\{y(n + m)y(n)\}, \tag{8.77}$$

is then zero whenever either m or n is odd. The input and output autocorrelations of the interpolator are related as

$$R_{yy}(n + m, n) = \begin{cases} R_{xx}\left(\frac{m}{2}\right), & m \text{ and } n \text{ even} \\ 0, & \text{otherwise.} \end{cases} \tag{8.78}$$

We can see that the output of the interpolator is wide-sense cyclostationary with a period of 2; i.e., for each integer k,

$$R_{yy}(n + m + 2k, n + 2k) = R_{yy}(n + m, n),$$
$$\mu_y(n + 2k) = \mu_y(n), \tag{8.79}$$

where μ is the statistical mean. To characterize this signal by a power spectral density, we compute the *time-average autocorrelation function* [31] over a single period, as

$$\bar{R}_{yy}(m) \triangleq \frac{1}{2} \sum_{n=0,1} R_{yy}(n + m, n)$$
$$= \begin{cases} \frac{1}{2} R_{xx}\left(\frac{m}{2}\right), & m \text{ even} \\ 0, & m \text{ odd.} \end{cases} \tag{8.80}$$

The Fourier transform of $\bar{R}_{yy}(m)$ then yields the *average power spectral density* of the cyclostationary stochastic process $y(n)$:

$$P_{yy}(\omega) = \begin{cases} \frac{1}{2} P_{xx}(2\omega), & 0 \leq \omega \leq \frac{\pi}{2} \\ \frac{1}{2} P_{xx}(2\pi - 2\omega), & \frac{\pi}{2} \leq \omega \leq \pi. \end{cases} \tag{8.81}$$

Thus the increase in sampling rate has an effect on the average power spectral density that is the same as on the Fourier transform of a deterministic signal; i.e., *spectral imaging.* In addition there is a gain factor of one half. In the Z-transform domain, (8.81) can be written as

$$P_{yy}(z) = \frac{1}{2} P_{xx}\left(z^2\right).$$

(8.82)

In the next section, we look at a subband synthesis system and obtain the filter based weighting factors to be used in (8.2).

8.4 Noise in a Subband Synthesis System

Assuming that the quantization noise introduced by the encoder of a particular subband signal is white, with a variance of σ_x^2, we can estimate the contribution of this noise to the final output. A branch in a typical subband synthesis system is given in Fig. 8.7. Since $x(n)$ is white, from (8.81) we can write

$$P_{xx}(\omega) = \sigma_x^2,$$

$$P_{y_1 y_1}(\omega) = \frac{\sigma_x^2}{2}, \qquad 0 \le |\omega| \le \pi.$$

(8.83)

Using (8.76) for the LSI case,

$$P_{y_2 y_2}(\omega) = \frac{\sigma_x^2}{2} \left|G_1(\omega)\right|^2.$$

(8.84)

The random variable y_2 is colored. Due to the interpolation, the power spectral density of y_3 can be written as

$$P_{y_3 y_3}(\omega) = \begin{cases} \frac{\sigma_x^2}{4} \left|G_1(2\omega)\right|^2, & 0 \le |\omega| \le \frac{\pi}{2} \\ \frac{\sigma_x^2}{4} \left|G_1(2\pi - 2\omega)\right|^2, & \frac{\pi}{2} \le |\omega| \le \pi. \end{cases}$$

(8.85)

The final filtering stage makes the power spectral density of the output y_4:

$$P_{y_4 y_4}(\omega) = \begin{cases} \frac{\sigma_x^2}{4} \left|G_1(2\omega)G_2(\omega)\right|^2, & 0 \le |\omega| \le \frac{\pi}{2} \\ \frac{\sigma_x^2}{4} \left|G_1(2\pi - 2\omega)G_2(\omega)\right|^2, & \frac{\pi}{2} \le |\omega| \le \pi. \end{cases}$$

(8.86)

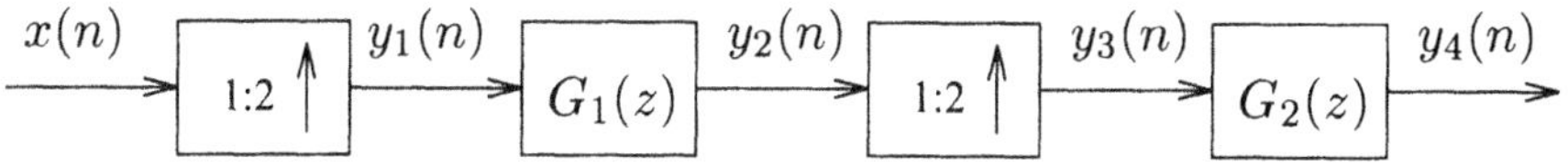

Figure 8.7   A channel of the subband synthesis system.

In a subband coding situation, we are more interested in the variance of the signals y_2 and y_4 than the entire power spectrum. Using Parseval's Theorem,

$$\sigma_{y_2}^2 = \bar{R}_{y_2 y_2}(0) = \frac{1}{2\pi} \int_{-\pi}^{\pi} P_{y_2 y_2}(\omega) d\omega = \frac{\sigma_x^2}{4\pi} \int_{-\pi}^{\pi} |G_1(\omega)|^2 d\omega$$
$$= \frac{\sigma_x^2}{2} \sum_{n=-\infty}^{\infty} |g_1(n)|^2 . \tag{8.87}$$

Similarly the variance of the signal y_4 can be written as

$$\sigma_{y_4}^2 = \int_{-\pi}^{\pi} P_{y_4 y_4}(\omega) = \frac{\sigma_x^2}{4} \sum_{n=-\infty}^{\infty} |(g_2 * \xi)(n)|^2 , \quad \text{where}$$

$$\xi(n) \triangleq \begin{cases} g_1(n/2), & n \text{ even} \\ 0, & n \text{ odd.} \end{cases} \tag{8.88}$$

Using this procedure, the scaling factor for the variances in the case of one or more levels of interpolation can be easily found. For example, when only one level of interpolation and filtering is involved, the scaling factor of the variance for a particular subband is

$$w = \frac{1}{2} \sum_n |g_1(n)|^2 , \tag{8.89}$$

where g_1 is the filter the subband encounters in the reconstruction. Similarly, for two levels of interpolation and filtering,

$$w = \frac{1}{4} \sum_n |(g_2 * \xi)(n)|^2 , \tag{8.90}$$

with $\xi(n)$ as defined in (8.88). The scaling factors found for the 1-D case can be used in a separable fashion for the 2-D subband coding as $w_k = w_k^h w_k^v$, where w_k, w_k^h, and w_k^v refer to the scaling factors for the kth 2-D subband, its horizontal, and its vertical weighting factors, respectively.

8.5 Bit Allocation Algorithm

In the case of M subbands of equal bandwidth, each subband has been subsampled by $\sqrt{M}$ in each dimension. Assuming error-free transmission, and the use of PCM (or DPCM) coding of individual subbands, and assuming a *constant quantizer performance factor* ϵ_* [13], (8.2) can be written as[3]

$$\min D = \sum_{k=1}^{M} \epsilon_*^2 2^{-2r_k} \sigma_k^2 w_k \qquad \text{such that} \qquad \frac{1}{M} \sum_{k=1}^{M} r_k \le R, \quad r_k \ge 0, \quad (8.91)$$

where the terms w_k are the weighting factors found using the method given in the previous section, σ_k^2 is the variance of the subband k if PCM is used, and the variance of the prediction error if DPCM is used to code the subband. Using the Lagrangian multipliers to minimize the reconstruction error variance with the overall bit rate constraint given in (8.91), we get the approximately optimal bit assignment:

$$r_{k,opt} = R + \frac{1}{2} \log_2 \frac{\sigma_k^2 w_k}{\left[\prod_{l=1}^{M} \sigma_l^2 w_l \right]^{\frac{1}{M}}} \qquad \forall k. \qquad (8.92)$$

To avoid negative bit allocations, we have used a quickly converging iterative algorithm, as in [15]. In the next section we describe the encoder used in our simulations.

8.6 Description of the Encoder

We have used a *frame-by-frame* coding technique to encode the video, using the separable procedure given in [1] to analyze each frame of the video into 16 subbands (see. Fig. 8.8) prior to encoding and synthesize them after decoding. DPCM was used on the lowest frequency subband and PCM on the higher frequency subbands. Uniform threshold quantizers (255 levels) were used for the DPCM, and uniform threshold quantizers (255 levels) with a central dead zone for the PCM. The quantizers were designed assuming a generalized Gaussian pdf [32] for quantizer inputs. The pdf estimations were made based on the subbands of the first frame, using the Kolmogorov–Smirnov test [3]. The bit allocations were made using the first frame's statistics for each filter set and were then used for the rest of the sequence. Noninteger values of r_k, resulting from the bit allocation algorithm, were not a problem since we were using uniform threshold quantizers and entropy constrained encoding, where we could achieve noninteger bit rates very easily.

We have used *arithmetic coding* [33] to achieve the entropy rates for encoding the indices output by the quantizers in PCM and DPCM. The required histograms were obtained from the generalized Gaussian model used for the quantizer. To provide protection from uncorrected channel errors, the input to the arithmetic encoder is grouped into blocks. An *end of block* marker of length 5 bits is used to separate

---

[3]Strictly speaking, the equation $D(r) = \epsilon_*^2 2^{-2r} \sigma^2$ is an asymptotic result, which is valid only at higher rates.

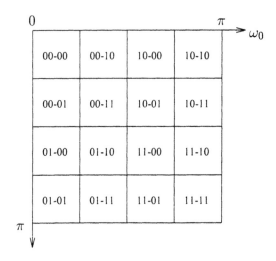

Figure 8.8   Sixteen subband partition in 2-D frequency domain.

the blocks. The subband $LL - LL$ requires the highest priority, hence, only two lines of this subband are blocked together for arithmetic encoding purpose. For this subband, a bit error would corrupt the data on two lines only. Similarly, in the other three low frequency subbands ($LL - LH$, $LL - HL$, $LL - HH$), four lines of the subbands are blocked together, whereas the entire subbands were treated as a block for the higher frequency bands. For these bands, one could use various prediction and interpolation strategies to fill the corrupted blocks, or just fill them with zeros, producing a softer image with a mild increase in aliasing error. In our simulations, no channel errors were introduced. In the next section, we provide the results of video coding using the bit allocation procedure given previously with each of the various subband filters being compared.

8.7 Results

We have compared the compression performance of nine different subband filter sets at various bit rates using the filter specific bit allocation scheme discussed previously. The filters consisted of three even-tap QMFs (Johnston's 12B, 16B, and 32C) [19], one odd-tap QMF (AS 9 × 9), two biorthogonal perfect reconstruction filters (Kronander's 9 × 7 and LeGall and Tabatabai's 5 × 3 filter sets, K 9 × 7 and G 5 × 3, respectively, in this chapter) [4], a subjectively optimized filter (Kronander's 7 × 7 filter set, here K 7 × 7) [4], and two orthogonal wavelet filters

(6×6 and 8×8 filter sets) [22]. The impulse responses of these filters are listed in Appendix 8.A.

The scaling factors for various subbands while using the filter sets under study are given in Table 8.4. For the filter comparison, we have used the *MIT sequence*, a 1600×960 progressively scanned gray-level sequence of rate 30 frames/sec and of length 30 frames. The simulations were made at different bit rates for each filter set.

There is a significant improvement in the subband compression capabilities of the perfect reconstruction filter sets with the new filter-based (weighted) bit allocation given by (8.2) compared to the normally used (unweighted) bit allocation given by (8.1). As an example, the average peak-to-peak signal-to-noise ratios (PSNRs) of the reconstructed sequences generated using the G 5×3 filter set with and without a filter based bit allocation are shown in Fig. 8.9. In the PSNR computations for this chapter, we have excluded a border of 16 pixels wide on each side of the image, so that the boundary treatment used in the filtering does not affect the filter comparison. The rates reported in this chapter include the header information necessary for the arithmetic decoder and the quantizer descriptions. We see that for this filter set, the PSNR has improved by about 1.7 dB with the filter-based bit allocation scheme with respect to the unweighted bit allocation, at higher bit rates.

Table 8.4   Scaling factors of subbands for the filter sets under study.

Subband	G 5×3	K 7×7	K 9×7	QMFs, W 6×6 and W 8×8
LL-LL	0.4727	0.9696	0.8224	1.00
LL-LH	0.5747	0.9527	0.7961	1.00
LL-HL	0.5747	0.9527	0.7961	1.00
LL-HH	0.6988	0.9358	0.7708	1.00
LH-LL	1.6248	1.1038	1.7188	1.00
LH-LH	0.6338	0.9704	0.9372	1.00
LH-HL	1.9756	1.0782	1.6640	1.00
LH-HH	0.7706	0.9479	0.9074	1.00
HL-LL	1.6248	1.1038	1.7188	1.00
HL-LH	1.9756	1.0782	1.6640	1.00
HL-HL	0.6338	0.9704	0.9372	1.00
HL-HH	0.7706	0.9479	0.9074	1.00
HH-LL	5.5851	1.2424	3.5925	1.00
HH-LH	2.1786	1.0922	1.9589	1.00
HH-HL	2.1786	1.0922	1.9589	1.00
HH-HH	0.8499	0.9602	1.0681	1.00

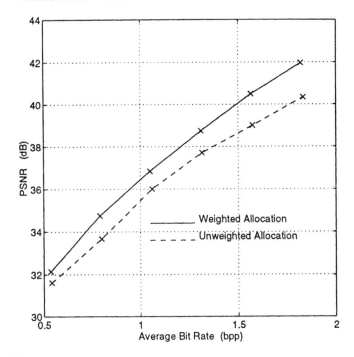

Figure 8.9   The PSNR performance of the G 5 × 3 filter set with and without the filter-based bit allocation (actual rates, PSNR figure excludes image borders).

The resulting average PSNR of the reconstructed sequences using the procedure described for different filter sets are compared in Fig. 8.10. In Fig. 8.10, the filter identification information is given in the same order as their performance at a bit rate of 1.8 bpp. We can see from Fig. 8.10 that all the filters give almost the same PSNR performance. Upon closer observation, we can say that, in general, the even-tap filters (QMFs and wavelet filters) performed better than the odd-tap filters (biorthogonal perfect reconstruction filters and the K 7 × 7 filter). The QMF 16B filter (followed closely by QMF 32C) gave the best quantitative performance at all the bit rates we have studied. From these results, we can see that for compression of images, orthogonality of a filter set is an important property. We can conjecture that the QMFs performed well because of their superior decorrelating capability on AR(1) sources as evidenced by Table 8.3.

To make a subjective evaluation of the compressed video, we recorded a 640 × 480 window out of the full size 1600 × 960 pixel image on a U-matic NTSC VCR. The subjects were positioned at a distance of 3–4 times the picture height while evaluating the video. The AS 9 × 9 filter set was not included in the subjective evaluations. The following informally summarizes the viewers comments.

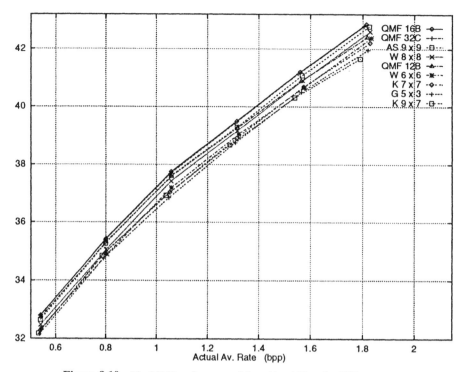

Figure 8.10   The PSNR performance of the subband filters for *MIT* sequence.

At a rate of 0.75 bpp, the sequences produced by the QMFs, orthogonal wavelet filter sets and K 7 × 7 filter sets were grouped together as similar sequences, best at this bit rate. The K 9 × 7 filter set produced a bad sequence, where as the G 5 × 3 filter set produced an intermediate sequence.

At 1.0 bpp, even though the sequences now had very little background noise, they were still not rated well for HDTV. The QMF 32C produced the best sequence at this rate, followed by QMF 12B, QMF 16B, and the orthogonal wavelet filter sets, which were all alike. These were followed by K 7 × 7, G 5 × 3, and K 9 × 7 in that order.

At 1.25 bpp, the sequences were thought acceptable for HDTV purposes. At this bit rate, QMF 12B and the orthogonal wavelet filters produced the best sequences. They were rated to be almost identical to the original sequence. The filters QMF 16B, QMF 32C, and the K 7 × 7 produced next best results. These were followed by the G 5 × 3 and K 9 × 7 filter sets.

At the higher rates of 1.5 and 1.75 bpp, all the sequences with one exception were almost the same as the original sequence, the exception being the sequence produced by K 9 × 7 filter set at 1.5 bpp.

8.8 Conclusions

In this chapter, we have reviewed the structure of some subband filter sets and their relationship to the wavelet theory. We have seen that all the two-band FIR subband filter sets in the literature fall into a certain group in Fig. 8.4. We have found that the QMFs are very close to being orthogonal. We have compared the decorrelation capabilities of these filters on an AR(1) source and found that even-tap QMFs performed well. This superiority of even-tap QMFs is a direct consequence of their close-to-ideal frequency response.

We have introduced a filter based bit allocation procedure for subband compression. A significant improvement in the subband compression capabilities of the perfect reconstruction filter sets has been observed with the new filter based bit allocation compared to the normally used bit allocation. In our study we have seen that, by allocating bits based on the mean squared error criterion, the longer filters gave the best quantitative performance. However, the orthogonal wavelet filters, which are just 6×6 and 8×8, were rated subjectively equivalent to the QMFs at all the bit rates we have studied.

In our studies we have seen that the background noise was the most objectionable distortion, and a bit rate of 1.25 bpp was needed to reduce it to an acceptable level. Within the realm of intraframe coding, we could allocate bits to various subbands on a perceptual basis or combine this with a spatially adaptive scheme to reduce this figure.

The method to compute the information-theoretic rate–distortion functions of orthogonal subband coding systems given in [8] needs to be generalized for biorthogonal subband coding systems. Similar analysis for Laplacian or a generalized Gaussian source would be applicable for image coding applications.

Appendix 8.A Subband Filter Coefficients

The subband analysis–synthesis banks shown in Fig. 8.11. use the impulse responses of Tables 8.5–8.9. These would have to be multiplied by $\sqrt{2}$ to obtain the filters for Fig. 8.1.

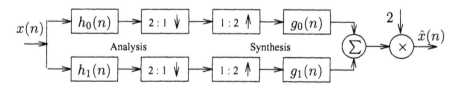

Figure 8.11   General 1-D subband analysis and synthesis systems.

Table 8.5   Impulse responses of even-tap QMFs.

QMF 12B	QMF 16B	QMF 32C
−0.006 443 977	0.002 898 163	0.000 691 057 9
0.027 455 390	−0.009 972 252	−0.001 403 793 0
−0.007 581 640	−0.001 920 936	−0.001 268 303 0
−0.091 382 500	0.035 968 530	0.004 234 195 0
0.098 085 220	−0.016 118 690	0.001 414 246 0
0.480 796 200	−0.095 302 340	−0.009 458 318 0
	0.106 798 700	−0.000 130 385 9
	0.477 346 900	0.017 981 450 0
		−0.004 187 483 0
		−0.031 238 620 0
		0.014 568 440 0
		0.052 947 450 0
		−0.039 348 780 0
		−0.099 802 430 0
		0.128 557 900 0
		0.466 405 300 0

Note: Impulse responses are low-pass filter coefficients from beginning to center.

Table 8.6   Impulse responses of odd-tap QMFs.

AS 5 × 5	AS 7 × 7	AS 9 × 9
−0.053 81	0.005 25	0.019 95
0.250 00	−0.051 78	−0.042 71
0.607 62	0.255 25	−0.052 24
	0.603 55	0.292 71
		0.564 58

Note: Impulse responses are low-pass filter coefficients from beginning to center.

Table 8.7   Impulse responses of orthonormal PR filters.

HV 12B	HV 16B	W 6 × 6	W 8 × 8
0.139 923	0.085 802 8	0.235 233 603 892 023 422	0.162 901 714 025 722 359
0.433 225	0.333 460 0	0.570 558 457 915 656 351	0.505 472 857 545 974 078
0.483 920	0.502 875 0	0.325 182 500 262 768 672	0.446 100 069 123 479 579
0.110 218	0.270 339 0	−0.095 467 207 783 983 659	−0.019 787 513 117 921 420
−0.181 087	−0.122 716 0	−0.060 416 104 155 179 252	−0.132 253 583 684 454 048
−0.062 621	−0.162 335 0	0.024 908 749 865 818 829	0.021 808 150 237 399 213
0.087 273	0.055 371 3		0.023 251 800 535 572 058
0.019 887	0.086 587 1		−0.007 493 494 665 131 923
−0.039 030	−0.034 900 5		
0.000 751	−0.039 024 4		
0.012 109	0.021 711 7		
−0.003 911	0.012 625 6		
	−0.010 706 2		
	−0.001 556 0		
	0.003 317 4		
	−0.000 853 6		

Note: Impulse responses are low-pass filter coefficients of the analysis bank.

Table 8.8   Impulse responses of biorthogonal PR filters.

G 5 × 3		K 9 × 7		W 5 × 7	
H0	H1	H0	H1	H0	H1
−0.125	−0.25	0.036 869 684 595 88	0.059 310 992 403 00	−0.05	3.0/280.0
0.250	0.50	0.002 902 050 784 03	0.004 668 429 195 73	0.25	−3.0/56.0
0.750		−0.164 740 146 432 83	−0.309 310 992 403 00	0.60	−73.0/280.0
		0.307 180 052 709 85	0.490 663 141 608 54		17.0/28.0
		0.635 576 716 686 14			

Note: Impulse responses are low-pass and high-pass filter coefficients from beginning to center of the analysis bank.

Table 8.9   Impulse response of the K 7 × 7 filter set.

H0	H1
0.012 661 127 811 13	0.022 570 044 695 91
−0.064 988 174 046 60	−0.039 642 948 574 68
0.237 338 872 188 87	−0.272 570 044 695 91
0.629 976 348 093 19	0.579 285 897 149 37

Note: Impulse responses are low-pass and high-pass filter coefficients from beginning to center of the analysis bank.

Acknowledgments

This work was supported in part by the National Science Foundation under grant number NCR-9003754 and in part by Philips Laboratories, North American Philips Corporation. The authors wish to thank Anil Murching and Gary Lilienfield for proofreading the manuscript, and Y. H. Kim for providing source code for parts of our simulations.

References

[1] J. W. Woods and S. D. O'Neil, "Sub-band coding of images," *IEEE Trans. Acoust., Speech, Signal Processing*, vol. ASSP-34, pp. 1278–1288, Oct. 1986.

[2] H. Gharavi and A. Tabatabai, "Sub-band coding of monochrome and color images," *IEEE Trans. Circuits Syst.*, vol. CAS-35, pp. 207–214, Feb. 1988.

[3] P. H. Westerink, "Subband coding of images," Ph.D. thesis, Delft University of Technol., Delft, The Netherlands, 1989.

[4] T. Kronander, "Some aspects of perception based image coding," Ph.D. thesis, Linkoping University, Linkoping, Sweden, 1989.

[5] N. Tanabe and N. Farvardin, "Subband image coding using entropy-coded quantization over noisy channels," *J. Selected areas Commun.*, vol. 10, pp. 926–943, June 1992.

[6] Y. H. Kim and J. W. Modestino, "Adaptive entropy coded subband coding of images," *IEEE Trans. Image Processing*, vol. 1, pp. 31–48, Jan. 1992.

[7] A. Tabatabai, "Optimum analysis/synthesis filter bank structures with application to sub-band coding systems," in *Proc. IEEE Int. Symp. Circuits Syst.*, 1988, pp. 823–826.

[8] T. R. Fischer, "On the rate-distortion efficiency of subband coding," *IEEE Trans. Inform. Theory*, vol. 38, pp. 407–413, Mar. 1992.

[9] K. Irie and R. Kishimoto, "A study on perfect reconstructive subband coding," *IEEE Trans. Circuits Syst. Video Technol.*, vol. 1, pp. 42–48, Mar. 1991.

[10] T. Q. Nguyen and P. P. Vaidyanathan, "Two-channel perfect-reconstruction FIR QMF structures which yield linear-phase analysis and synthesis filters," *IEEE Trans. Acoust., Speech, Signal Processing*, vol. ASSP-37, pp. 676–690, May 1989.

[11] J. W. Woods and T. Naveen, "A filter based bit allocation scheme for subband compression of HDTV," *IEEE Trans. Image Processing*, vol. 1, pp. 436–440, July 1992.

[12] D. Le Gall and A. Tabatabai, "Sub-band coding of digital images using symmetric short kernel filters and arithmetic coding techniques," in *Proc. IEEE Int. Conf. Acoust., Speech, Signal Processing*, New York, NY, Apr. 1988, pp. 761–764.

[13] N. S. Jayant and P. Noll, *Digital Coding of Waveforms*. Englewood Cliffs, NJ: Prentice-Hall, 1984.

[14] N. Uzun and R. A. Haddad, "Modeling and analysis of quantization errors in two channel subband filter structures," in *Proc. SPIE Conf. Visual Commun. Image Processing*, Nov. 1992.

[15] B. Mahesh and W. A. Pearlman, "Image coding on a hexagonal pyramid with noise spectrum shaping," in *Proc. SPIE Conf. Visual Commun. Image Processing IV*, Nov. 1989, vol. 1199, pp. 764–774.

[16] M. Vetterli and C. Herley, "Wavelets and filter banks: Theory and design," *IEEE Trans. Signal Processing*, pp. 2207–2232, Sept. 1992.

[17] P. Desarte, B. Macq, and D. T. M. Slock, "Signal-adapted multiresolution transform for image coding," *IEEE Trans. Inform. Theory*, vol. 38, pp. 897–904, Mar. 1992.

[18] D. Taubman and A. Zakhor, "A multi-start algorithm for signal adaptive subband systems," in *Proc. IEEE Int. Conf. Acoust., Speech, Signal Processing*, 1992, vol. III, pp. 213–216.

[19] J. D. Johnston, "A filter family designed for use in quadrature mirror filter banks," in *Proc. IEEE Int. Conf. Acoust., Speech, Signal Processing*, Apr. 1980, pp. 291–294.

[20] E. H. Adelson, E. Simoncelli, and R. Hingorani, "Orthogonal pyramid transform for image coding," in *Proc. SPIE Conf. Visual Commun. Image Processing*, 1987, pp. 50–58.

[21] M. Antonini, M. Barlaud, P. Mathieu, and I. Daubechies, "Image coding using wavelet transform," *IEEE Trans. Image Processing*, vol. 1, pp. 205–220, Apr. 1990.

[22] A. N. Akansu, R. A. Haddad, and H. Caglar, "The binomial QMF-wavelet transform for multiresolution signal decomposition," *IEEE Trans. Signal Processing*, 1991.

[23] I. Daubechies, "Orthonormal bases of compactly supported wavelets," *Commun. Pure Appl. Math.*, vol. XLI, pp. 909–996, 1988.

[24] A. Cohen, I. Daubechies, and J. C. Feauveau, "Biorthogonal bases of compactly supported wavelets," *Commun. Pure Appl. Math.*, vol. 45, pp. 485–560, June 1992.

[25] A. N. Akansu and R. A. Haddad, *Multiresolution Signal Decomposition*. San Diego, CA: Academic Press Inc., 1992.

[26] P. P. Vaidyanathan, "Quadrature mirror filter banks and perfect-reconstruction techniques," *IEEE ASSP Mag.*, pp. 4–20, July 1987.

[27] P. P. Vaidyanathan and P.-Q. Hoang, "Lattice structures for optimal design and robust implementation of two-channel perfect reconstruction QMF banks," *IEEE Trans. Acoust., Speech, Signal Processing*, vol. ASSP-36, pp. 81–94, Jan. 1988.

[28] C. W. Kim and R. Ansari, "FIR/IIR exact reconstruction filter banks with applications to subband coding of images," in *Proc. IEEE 34th MidWest Symp. Circuits Syst.*, Montery, C.A., May 1991, pp. 227–230.

[29] P. J. Burt and E. H. Adelson, "The Laplacian pyramid as a compact image code," *IEEE Trans. Commun.*, vol. COM-31, pp. 532–540, Apr. 1983.

[30] H. Stark and J. W. Woods, *Probability, Random Processes, and Estimation Theory for Engineers*. Englewood Cliffs, NJ: Prentice-Hall, 1986.

[31] J. G. Proakis, *Digital Communications*. New York, NY: McGraw-Hill, 1989.

[32] N. Farvardin and J. W. Modestino, "Optimum quantizer performance for a class of non-Gaussian memoryless sources," *IEEE Trans. Inform. Theory*, vol. 30, pp. 485–497, May 1984.

[33] C. B. Jones, "An efficient coding system for long source sequences," *IEEE Trans. Inform. Theory*, vol. 27, pp. 280–291, May 1981.

Chapter 9

Hierarchical Coding*

F. Bosveld, R. L. Lagendijk, and J. Biemond
Delft University of Technology
Department of Electrical Engineering
Information Theory Group
Delft, The Netherlands

Hierarchical coding of video is of interest especially due to emerging broadband networks and over-the-air digital transmission links. Hierarchical coding techniques can be used within the process of data compression (source encoding) as well as within the process of channel adaptation (channel encoding). Hierarchical source coding supports compatibility between video systems with different resolution displays. Hierarchical channel coding facilitates the robust transmission of the video sequence through communication networks where errors are likely. When bursts of errors occur, an improved error resilience warrants the continuity of the service with gracefully decreased video components. In this chapter, we investigate the concept of hierarchical source coding as well as the subsequent hierarchical channel coding from both the theoretical and practical points of view.

9.1 Introduction

Data compression of video is becoming increasingly important because of the recent advance of broadband networks and over-the-air digital transmission links.

---

*This work was supported in part by the EC-RACE-2026 (DART) project and NATO grant 5-2-05/CRG 900834.

These new transmission media facilitate and stimulate the introduction of new services ranging from videotelephony and multimedia communications to the distribution of high-definition TV (HDTV). Scalable open architecture video compression schemes are of particular interest since these will increase interworking, flexibility, error resilience, and compatibility while simultaneously reducing bandwidth requirements and equipment costs. The envisioned video encoders should be able to operate independent of the various decoder display formats, which can vary from high-resolution domestic HDTV sets to low-resolution portable receivers. Scalability in bit rate, spatial and temporal resolution (frame rate) is therefore pursued as this allows the encoded video signal to be reconstructed at a variety of bit rates and resolutions. The open architectures of the encoders should guarantee that currently defined and future video standards can be encoded by extending the encoders with some (basic) building blocks. Further, the envisioned encoders have to be flexible with respect to the selected transmission channel. Channel adaptation modules should be exchangeable so that several existing and emerging transmission channels can be accommodated.

Generally, the basis of an open architecture system is formed by hierarchical coding techniques, which are used within the process of data compression (source encoding) as well as within the process of channel adaptation (channel encoding). Hierarchical *source* coding schemes decompose a high-resolution video signal into several *subsignals*, which are of varying importance with respect to the quality of the reconstructed signal (see Fig. 9.1). Reconstruction of a limited number of subsignals yields a signal with a lower resolution than the original signal. In the subsequent encoding stage *layers* are constructed that contain parts of the encoded video signal. The lowest resolution video signal is reconstructed by decoding the information of the first layer of the hierarchy. Higher resolution

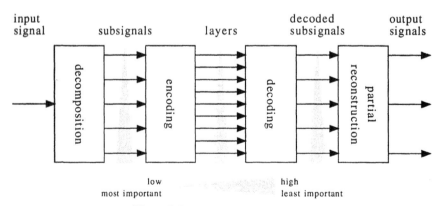

Figure 9.1   Hierarchical source coding.

video signals are reconstructed by decoding the additional information contained in higher layers. Typically, layers contain encoded subsignal information as well as some refinement information about lower layers. Motion and synchronization information is generally also included. Hierarchical source coding schemes are frequently used to achieve compatibility between several video standards [1–5] and are therefore also called *compatible* schemes.

Hierarchical *channel* coders adapt an encoded video signal to a specific channel in such a way that error resilience is improved. When channel errors occur, the reconstructed signal should degrade gracefully, which means that the subjective quality of the reconstructed signal decreases gradually as error rates increase. This can be achieved by decomposing the signal considered into several subsignals, which are subsequently protected using forward error correcting codes of differing strength. Lower layers are more heavily protected and are therefore more likely to be used error free in the reconstruction. Another way to achieve graceful degradation is to transmit the most important layers using the most reliable paths of the transmission channel. This is done, for example, in the channel adaptations for emerging ATM channels, for recording tapes, and for terrestrial broadcasting [4, 6, 7]. Frequently, hierarchical source and channel coding techniques are used jointly, which usually implies that a single decomposition is used for both the source and the channel coding.

In this chapter, we investigate the concept of hierarchical source coding as well as the subsequent hierarchical channel coding from both a theoretical and practical points of view. Section 9.2 starts out with an overview of some applications and transmission channels and the constraints they impose on the encoding of the layers. Next, in a concise overview of the literature, basic techniques, and problems of compatible coding are discussed, such as the basic decomposition methods and the various strategies to encode the subsignals. The problems refer to the difficult incorporation of motion information and to the multistage quantization of the subsignals. Multistage quantization, i.e., the layered quantization of a signal, is thoroughly investigated in Section 9.3, where we confine ourselves to an *intraframe compatible subband coding scheme*. This scheme encodes an HDTV signal at 140 Mbit/sec using three layers such that two compatible video signals are supported. Section 9.4 presents the channel adaptation of this coding scheme to ATM-based networks. It is assumed that for the lowest layer, a constant bit rate contract is negotiated while for the higher two layers variable bit rate contracts are obtained. Simulation results are subsequently discussed in Section 9.5, where it will be shown that compatibility is achieved at the cost of a small performance loss and that the compatible scheme has a high ATM cell loss resilience as opposed to a similar noncompatible scheme.

9.2 Compatible Coding

9.2.1 Application and Channel Constraints

In general, every application involving video compression benefits from a compatible concept in which the video encoder and decoder are highly independent of each other with respect to resolution and bit rate [8, 9]. Connections between different resolution transmitters and receivers at various bit rates become possible using the most appropriate or least expensive channel. In the following, we briefly discuss four potential compatible applications that use different transmission or storage channels.

1. *Compatible HDTV distribution* (terrestrial broadcasting). This application is probably the most challenging research subject of the past decade. Here, the aim is to encode an HDTV signal hierarchically in an RF spectrum of 6–8 MHz bandwidth. Using advanced modulation techniques like QAM and QPSK, this implies that the bit rate of the HDTV signal has to be compressed to 24–32 Mbit/sec. A compatible TV signal should be reconstructable at approximately 7 Mbit/sec [4, 10]. A single transmitter will serve many kinds of receivers ranging from high-resolution domestic HDTV sets to low-resolution portable LCD receivers. Roof or central antennas are used by high-resolution receivers that can display either the HDTV signal or multiple TV signals. Handheld low-resolution receivers with small antennas can visualize the compatible TV signal and are low cost, thanks to limited memory and power needs. Graceful degradation is of utmost importance, since the CNR is different for each individual receiver depending on the size of the antenna and distance to the transmitter.

2. *Image and video archive systems* (optical storage). Image and video archive applications, like the photo-CD and CD-I, are becoming increasingly popular. Currently, with the photo-CD application, photos are stored three times at resolutions varying from 768×512 pixels to 3072×2048 pixels. In this way, detailed magnification of user-selected areas is facilitated using the random-access property of the medium. The use of hierarchical compression techniques to store these photos has two distinct advantages. First, the available disk space is used more efficiently since no information is stored twice. Second, fast visual search procedures are facilitated by using very low-resolution images as browsing pictures. The full-screen display of a photo involves an intermediate-resolution image while the highest resolution images can be used for magnification.

3. *Multimedia communications* (ATM-based networks). ATM-based networks will facilitate the transmission of a variety of signals (audio, video, and data) that may have completely different bit stream characteristics (CBR, VBR, and bursty). Therefore, multimedia communications will become feasible using desktop workstations where people communicate via on-screen windows. Such visual

communication will inevitably involve size adjustments of the communication windows, which mean a change in required resolution from a scalable coding point of view. Compatible schemes can easily support this feature by flexibly receiving more or fewer layers.

4. *Digital video recording* (magnetic tapes). Hierarchical schemes can be used in the digital recording of video signals as well. Since magnetic recording channels are not very reliable, hierarchical channel coding techniques are necessary to improve the image quality. Hierarchical source coding techniques can be used for the multiresolution playback and recording and also facilitate certain trick modes like fast forward and slow motion. For example, a fast-forward mode would be achieved by reading only the lowest resolution data from the fast-moving tape and show this on-screen.

Depending on the application and transmission or storage channel used, certain requirements on the coding of the layers may show up. Without loss of generality, consider, for example, a two-layer compatible HDTV/TV distribution scheme for ATM-based networks. Since HDTV is the vending service here, it must be encoded with minimal impairment. In practice, this means that the two encoded layers together use the maximum channel bandwidth of 140 Mbit/sec. In contrast, the quality of the received TV signal is of secondary interest and could be left unconstrained. In general, this yields a VBR stream for the lowest layer as is illustrated in Fig. 9.2a.

Alternatively, it might be that the transmission scheme guarantees that the lowest layer is encoded at 140/4 = 35 Mbit/sec to facilitate the simultaneous reception and display of four independent TV signals. This yields two CBR layers as illustrated in Fig. 9.2b. A similar reasoning holds for the aforementioned image and video archive system using optical storage media, where multiple browsing images have to be displayed on a single screen. Transmission channels can also impose constraints on the coding stages of the layers. This is the case, for example, if layers need to be transmitted in CBR channels. With the compatible recording of

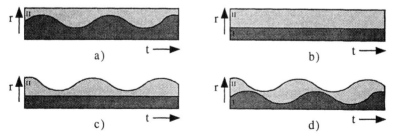

Figure 9.2   Bit stream characteristics of (I) lowest layer and (II) highest layer.

video, there is an equivalent constraint as layers need to be recorded on particular subsets of tracks which are served by a single head.

Other application or channel constraints may impose different requirements on the coding stages. Two additional transmission characteristics are shown in Figs. 9.2c and 9.2d, where the high-resolution image is encoded in a VBR stream. The transmission scheme that is discussed in Sections 9.3 and 9.4 encodes the layers according to Fig. 9.2c.

9.2.2 Basic Techniques and Problems

The first coding schemes employing the concept of compatibility were proposed by Chen, Tzou, and Fleisher; LeGall, Gaggioni, and Chen; and Bellisio and Tzou [1, 11–13]. In these schemes, a progressive-scan HDTV signal was *intraframe* encoded in the 140 Mbit/sec channel of the emerging ATM-based B-ISDN network while the compatible progressive-scan TV signal was encoded in 45 Mbit/sec. Together, these papers introduced the three basic methods to create the required subsignals; namely, the Laplacian pyramid, the subband or wavelet decomposition, and the DCT transform. Since these decompositions still serve as the basis of all proposed scalable schemes, we describe them briefly for a two-layer compatible HDTV/TV scheme.

1. *Laplacian pyramid* ([11, 14]). The Laplacian pyramid generates two sub-signals of which the low-resolution subsignal is equal to the compatible low-resolution signal. It is created by applying a 2:1 decimation (half-band filtering and subsequent 2:1 subsampling) of the high-resolution video signal in the vertical and horizontal directions. Subsequently, the obtained low-resolution subsignal is vertically and horizontally 1:2 interpolated (1:2 up-sampling followed by half-band filtering) and subtracted from the original high-resolution signal. The difference signal obtained, called the *Laplacian signal*, is the second subsignal. An advantage of the Laplacian pyramid is the freedom of choice of the low-pass filter. Since any low-pass filter can be used to generate the low-resolution signal, the smoothness of this signal is easily controlled as is the amount of aliasing.

2. *Subband decomposition* ([12, 15]). The subsignals are formed by subbands that are generated by decomposing the high-resolution video signal into several subbands. The subbands containing the low-resolution frequencies ($[-\frac{\pi}{2}, \frac{\pi}{2}] \times [-\frac{\pi}{2}, \frac{\pi}{2}]$) are required to reconstruct the low-resolution video signal. The required subband filters are more constrained than the pyramid filters since they have to cancel out the aliasing terms in the transfer function of the complete filter bank. Although the quality of the low-resolution signal is not as easily controlled as it is with the Laplacian pyramid, it is sufficiently high when using regular subband filters. Additionally frequency roll-off can be used to smooth the reconstructed low-resolution signal [16].

3. *DCT transform* ([1, 17]). In this method, the high-resolution signal is divided into $U \times V$ subblocks, each of which is transformed into $U \times V$ DCT coefficients. Subsignals are composed of all DCT coefficients with equal indices. The low-resolution signal can be reconstructed by a scaled inverse DCT transformation of the $\frac{U}{2} \times \frac{V}{2}$ "low-frequency" coefficients of all subblocks. The applied low-pass filter to obtain the low-resolution signal is implicitly specified in the DCT transform and is therefore not flexible at all. The quality of the low-resolution signal is reasonable but can be improved by changing either the transform [18] or by using some specific "high-frequency" coefficients in the inverse transform [10].

In Europe, the research into compatible coding schemes originally focused on the hierarchy of video standards of Table 9.1.

This hierarchy consists of three progressive-scan video signals (HDP, EDP, and VT) and two interlaced signals (HDI-EDI). Pecot, Tourtier, and Thomas [3], Vandendorpe and Delogne [19], and Bosveld, Lagendijk, and Biemond [2] proposed intraframe hierarchical schemes which support the HDP–EDP–VT hierarchy. The interlaced HDI and EDI have to be deinterlaced before encoding and are interlaced again after decoding. *Intrafield* coding schemes have been proposed by Breeuwer and de With [5] and, more recently, by Guillemot [20]. These schemes support the HDI–EDI compatibility and use different subband-based filters for the decomposition of the odd and even fields to guarantee equidistant lines in the EDI signal. A spatio–temporal subband decomposition is proposed in [21] that supports the full hierarchy of Table 9.1.

To reduce the required bit rate of the HDTV signal to approximately 30 Mbit/sec, several schemes are proposed that employ motion compensation. Schemes based on a spatio–temporal pyramid and on subband–wavelet decompositions have been proposed by Uz, Vetterli, and LeGall [4], Woods and Nareen [22] and Zafar, Zhang, and Jabbari [23], respectively. These schemes predict the images of various resolution using motion information that is estimated from the lowest resolution signal and refined at higher resolution levels. A different approach is discussed in [10, 24–26] for subband- and DCT-based systems. In these papers, a high-accuracy motion

Table 9.1   Possible hierarchy of video services.

Video service	Scanning parameters	Sampling parameters		Bit rate
		Luminance Y	Chrominance UV	Mbit/sec
HDP	1250/50/1:1	144 MHz/1920	72 MHz/960	2304
HDI	1250/50/2:1	72 MHz/1920	36 MHz/960	1152
EDP	625/50/1:1	36 MHz/960	18 MHz/480	576
EDI	625/50/2:1	18 MHz/960	9 MHz/480	288
VT	312.5/25/1:1	4.5 MHz/480	2.25 MHz/240	72

field is determined from the high-resolution signal. The transmitted compatible signals are predicted using scaled-down motion vectors. Although this increases the complexity because of the existence of multiple prediction loops, it avoids drift in the decoding loops for the compatible signals in the receivers. In-band motion compensation, i.e., the prediction of subbands or DCT coefficients, is used in the hierarchical coding schemes of [26–28] In the latter two, it is shown that subbands have to be interpolated before motion compensation. This is caused by the aliasing that makes the relation between subbands of shifted frames ambiguous.

Various compatible schemes using motion information have been studied within the second phase of the ISO-Motion Picture Expert Group (MPEG-2) [29]. Their aim is to standardize a scalable codec that can compress a CCIR 601 video signal into 4–9 Mbit/sec. Several schemes have been proposed that are compatible with the previously defined MPEG-1 standard for encoding SIF signals [30, 31]. This implies that existing MPEG-1 decoders can reconstruct the compatible low-resolution SIF signal. Other proposed schemes extend the MPEG-1 syntax to accommodate scalability of resolution and bit rate [32–36].

The schemes presented by Ohm [37, 38] and Taubman, and Zakhor [39] are new developments in hierarchical coding schemes. They present velocity-based spatio–temporal subband schemes that temporally decompose the signal by a subband decomposition along the motion trajectory. This facilitates a high data compression and avoids the blurring of moving objects in the compatible signals.

In [40] three basic coding strategies for the encoding of the subsignals are identified; namely, the distributed coding strategy, the error feedback strategy, and the selection strategy. For the two-layer compatible HDTV/TV scheme where the HDTV signal is encoded for 140 Mbit/sec ATM transmission these strategies are briefly discussed:

1. *Distributed coding strategy*. With this strategy, the available bit rate for the high-resolution signal is *divided* between the layers *prior* to the actual encoding. The lowest layer consists of the subsignals necessary to reconstruct the low-resolution signal while the highest layer is composed of the other subsignals. Therefore, the bit streams of the layers will be CBR (see Fig. 9.2b). To obtain a high-quality, high-resolution signal, the bit rate of the lowest layer has to be approximately 70 Mbit/sec.

2. *Error feedback coding strategy*. This strategy starts with encoding the lowest layer at a certain bit rate. All *quantization errors* made in this quantization stage are *fed back* to the quantization stage of the second layer, where they are encoded together with the remaining subsignals. This prevents quantization errors from propagating into the reconstruction of the high-resolution signal at the receiver. The bit streams of the layers are CBR (see Fig. 9.2b). The advantage of this approach is that the bit rate of the lowest layer can be chosen freely under 70 Mbit/sec while maintaining a high-quality, high-resolution signal.

3. *Selection strategy.* Here, all subsignals are encoded simultaneously at 140 Mbit/sec, which guarantees the best coding performance for the high-resolution signal. The lowest layer is subsequently created by *selecting* the appropriate encoded subsignals. The bit stream of the lowest layer is the sum of the bit rates of selected subsignals and will in general be VBR (see Fig. 9.2a).

With the error feedback strategy, the most important subsignals are encoded in a multistage way; coarse approximations are sent in the lowest layer that are subsequently refined by bits carried in the higher layers. In general, these multistage coding methods will affect the coding efficiency of the high-resolution signal since they perform less well than similar single-stage coding methods. However, proper design of the involved multistage quantizers will minimize the performance loss [16, 39, 41]. We discuss the performance of the multistage quantizers extensively in the following section.

9.3 Intraframe Hierarchical Source Coding

In this section we introduce a compatible intraframe subband coder to exemplify the error feedback coding strategy and to facilitate the discussion of multistage quantizers. The scheme is capable of encoding progressive-scan video signals into three separate layers in such a way that two compatible lowerresolution video signals are provided. It is assumed that the bit rate available for each of the layers is constrained by either channel properties or negotiated network contracts. After discussing the structure of the scheme in Section 9.3.1, four candidate strategies for the multistage encoding of the subbands are evaluated in Section 9.3.2; namely, concatenated coding, conditional entropy coding, conditional quantization, and embedded quantization. Subsequently, Section 9.3.3 discusses the distribution of the available bits among the subbands of the layers. The encoding of the color components is finally discussed in Section 9.3.4.

9.3.1 Structure of Compatible Intraframe Scheme

The basic structure of the encoder and decoder is pictorially illustrated in Fig. 9.3. It is reconstructed using all 31 subbands. The high-resolution input signal is decomposed into 31 subbands (i.e., subsignals) using the QMF16B filter [42]. The two compatible video signals, denoted L^1 and L^2, can be reconstructed using subbands 1–7 and 1–19, respectively. The high-resolution signal, denoted L^3, of the first layer, denoted by l^1, carries the encoded subbands 1–7. These subbands are encoded under supervision of a private forward bit allocation algorithm ($BA1$), which optimally distributes the available bits among the subbands by selecting quantizers from a set of admissible quantizers. The baseband, containing low-pass

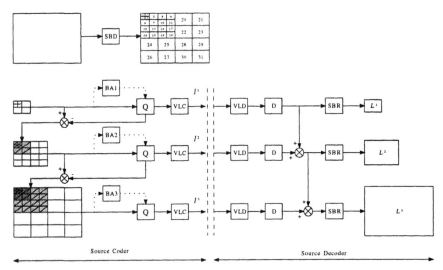

Figure 9.3   Compatible intraframe subband encoder and decoder.

information, is DPCM encoded while the other six subbands are PCM encoded. The code words of these subbands are VLC encoded using arithmetic codes before transmission. As explained in Section 9.2, the error feedback strategy dictates that the quantization errors of this layer have to be fed back to the next layer to prevent the propagation of these errors into the reconstructions of the L^2 and L^3 signals at the receiving side.

The second layer is, therefore, composed of the following encoded subbands: (1) subbands 8–19 and (2) subbands 1–7, which contain the quantization errors of layer l^1. Like layer l^1, a forward bit allocation algorithm $BA2$ distributes the available bits among the subbands that are all PCM encoded. Note that the quantizers used to encode subbands 1–7 should take into account that the signal characteristics are different from those of subbands 8–19. Similarly, the third layer is encoded under supervision of bit allocation $BA3$ and consists of the encoded fed back quantization errors of layer l^2 and subbands 20–31. Subbands 8–19 now contain quantization errors resulting from the quantization stage of layer l^2, while subbands 1–7 contain errors that result from the two quantization stages of layers l^2 and l^1. The decoder reconstructs the L^n signals ($1 \le n \le 3$) by receiving and decoding the appropriate layers l^n, which are subsequently summed before using them in subband reconstructions.

Comparing the encoder with a similar subband-based noncompatible scheme (Fig. 9.4) shows that the basic difference is that the most important subbands (1–19) are encoded in a multistage way; a coarse approximation is sent to a lower layer that is refined by data carried in the higher layers. More specifically, consider the coding

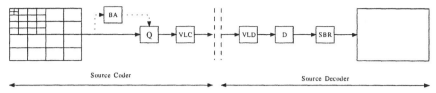

Figure 9.4 Noncompatible intraframe subband encoder and decoder.

process of subband 1. In the noncompatible scheme this subband is DPCM encoded at a high rate as it contains the most important low frequencies. In the compatible scheme this subband is first DPCM encoded at a medium rate and is successively refined by multiple PCM stages. This is done to guarantee the independence between the encoder and the various resolution decoders. Also the coding processes of subbands 2–19 differ for the two schemes. In the noncompatible scheme they are PCM encoded using a single quantizer as opposed to the compatible scheme where they are multistage quantized using two or three successive quantizers.

Therefore, it can be concluded that the compatible scheme is competitive with the noncompatible scheme, with respect to the quality of the reconstructed signal L^1, only if the multistage encoding methods do not introduce any significant performance loss.

9.3.2 Multistage Coding

Multistage quantization was first proposed by Wang and Goldberg [43] in a non-compatible coding scheme. Here, DCT coefficients were multistage PCM encoded to increase the coding efficiency. Theoretical analyses were done by Equitz and Cover [44], who showed that the successive approximation of signals can be done without performance loss only if the individual solutions of the rate distortion problems can be written as a Markov chain. Barnes and Frost [45] researched the performance of multistage scalar and vector quantizers to reduce the size of code-books. More recently, several multistage coding strategies have been proposed for use in compatible coding schemes by Shapiro [46], Taubman and Zakhor [39], Naveen *et al.* [47, 48], and Bosveld, Lagendijk, and Biemond [41].

In this section, four multistage PCM coding strategies are discussed; namely, concatenated coding, conditional entropy coding, conditional quantization, and embedded quantization ([41, 47]). After some notational remarks (see also [45]), each strategy is described and evaluated. Subsequently, multistage DPCM coding is analyzed. For reasons of simplicity, multistage coding of a single input signal (for example a subband) is investigated for two-layer systems. All results obtained are compared with the performance of single-stage scalar quantization using uniform threshold quantizers (UTQs).

Notational Conventions   Let X be the random input signal of $\mathbf{R}$ with probability density function (pdf) $p_X(x)$. A particular realization of X is denoted by x, and its quantized version by y. The applied quantizer can be defined by (1) the finite set of representation levels or code vectors $A = \{y_1, y_2, \ldots, y_K\}$ of $\mathbf{R}$; (2) the partition $P = \{S_1, S_2, \ldots, S_K\}$ of $\mathbf{R}$ where cells S_i satisfy $\cup_{i=1}^{K} S_i = \mathbf{R}$ and $S_i \cap S_j = \emptyset$ for $i \neq j$; and (3) the mapping $Q : \mathbf{R} \rightarrow A$; which defines the relationship between the code vectors and the partition as $Q(x) = y_i$ if and only if $x \in S_i$. With K we denote the number of code vectors and cells.

The structure of a typical two-stage quantizer is pictorially illustrated in Fig. 9.5. In general, an N-stage quantizer consists of a sequence of N quantizers (A^n, Q^n, P^n), $1 \leq n \leq N$, where quantizer Q^1 quantizes the original signal x^1. Quantizers Q^n quantize the resulting quantization error signals $x^{n+1} = x^n - Q^n(x^n)$, $1 \leq n \leq N - 1$. The number of code vectors of Q^n is denoted with K^n. Signal x^1 is related to the quantization errors as follows [45]:

$$x^1 = \sum_{n=1}^{N} Q^n(x^n) + x^{n+1}. \tag{9.1}$$

In practice, all quantizers Q^n are constructed using an encoder mapping $\mathcal{E}^n : \mathbf{R} \rightarrow J^n$ and a decoder mapping $\mathcal{D}^n : J^n \rightarrow A^n$, where J^n is the index set $\{1, 2, \ldots, K^n\}$. The indices to be transmitted at stage n are denoted j^n, the corresponding random variable J^n. The output bit rate r^n of the VLC coder of stage n is assumed to be equal to the entropy $H(\cdot)$ of the indices:

$$r^n = H(J^n) = -\sum_{i=1}^{K^n} P(J^n = i) \log_2 P(J^n = i), \tag{9.2}$$

with $P(J^n = i)$ the probability of index i. The distortion after stage n is denoted d^n and is equal to the variance of x^{n+1}:

$$d^n = \sigma_{x^{n+1}}^2. \tag{9.3}$$

In the sequel, uniform threshold quantizers are frequently used since these quantizers are known to be among the best scalar entropy-constrained quantizers in the mse sense [49]. UTQs can be described by their step size δ and the number of representation levels or code vectors K. For an odd K, the cells S_i are given by the interval $[u_{i-1}, u_i)$ where

$$u_i = (i - \frac{1}{2} - \frac{K-1}{2}) \, \delta, \qquad 1 \leq i \leq K - 1, \tag{9.4}$$

with $u_0 = -u_K = -\infty$. The representation levels y_i are given by

$$y_i = \frac{\int_{S_i} x p(x) dx}{P(y_i)}, \qquad 1 \leq i \leq K, \tag{9.5}$$

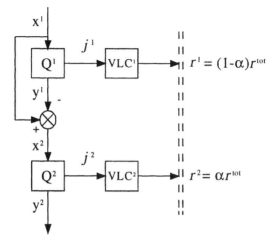

Figure 9.5   Concatenated coding strategy.

with probabilities $P(y_i)$ defined as

$$P(y_i) = \int_{S_i} p(x)dx, \qquad 1 \leq i \leq K. \tag{9.6}$$

In practice, with symmetric pdf models of the subband data, K is always chosen to be odd. Otherwise, the entropy rate would be lower bounded by 1 because of the decision level at 0 [49]. The mse distortion D of the UTQ is defined as

$$D = \sum_{i=1}^{K} \int_{S_i} (x - y_i)^2 p(x)dx, \tag{9.7}$$

and the output rate R is equal to the entropy of the representation levels,

$$R = -\sum_{i=1}^{K} P(y_i) \log_2 P(y_i). \tag{9.8}$$

Concatenated Coding   The most straightforward way to encode the subbands in a multistage way is to concatenate multiple scalar quantizers. This structure was already shown in Fig. 9.5. The input signals of the two concatenated quantizers Q^1 and Q^2 are denoted by x^1 and x^2, respectively. In this two-stage quantizer, x^2 represents the quantization errors of quantizer Q^1 and is therefore equal to

$$x^2 = x^1 - Q^1(x^1) = x^1 - y^1. \tag{9.9}$$

The indices j^1 and j^2 are losslessly encoded using VLC coders before transmission. The performance of this two-stage quantizer is evaluated by varying the individual bit rate of each quantizer while keeping the total rate constant. In Fig. 9.5 this is specified by two parameters:

- The total bit rate r^{tot};
- A parameter α, $0 \le \alpha \le 1$ which splits the available bit rate r^{tot} between the two quantizers; namely, $r^1 = (1 - \alpha)r^{tot}$ and $r^2 = \alpha r^{tot}$.

Observe that if $\alpha = 0$ or $\alpha = 1$ the two-stage quantizer degenerates to a single-stage quantizer. The distortion after two stages is denoted by $d^{tot} = d^2$.

Given $p_{X^1}(x)$ and bit rate r^1, we are able to optimally encode x^1 using a UTQ (see (9.4)–(9.8)). For subband data, $p_{X^1}(x)$ can reasonably be modeled by a generalized Gaussian (GG) function, which is given for unit variance by

$$p_{X^1}(x) = \frac{bc}{2\Gamma(\frac{1}{c})} \exp\left(-|bx|^c\right), \quad \text{with} \tag{9.10}$$

$$b = \sqrt{\Gamma\left(\frac{3}{c}\right) / \Gamma\left(\frac{1}{c}\right)}. \tag{9.11}$$

For typical subband signals the shape parameter c is approximately equal to 0.75. This model with $c = 0.75$ is used in the rest of this section and is illustrated in Fig. 9.6a. The shape parameter c is equal to 2.0 or 1.0 when the input signal is Gaussian or Laplacian distributed, respectively.

The shape of the pdf of the quantization error signal x^2, i.e., $p_{X^2}(x)$, is not so straightforward since it depends on the pdf $p_{X^1}(x)$ and the applied quantizer Q^1. However, if these are known, pdf $p_{X^2}(x)$ is analytically given by [17]

$$p_{X^2}(x) = \sum_{i=1}^{K^1} p_{X^1|J^1=i}(x) P(J^1 = i). \tag{9.12}$$

The terms $p_{X^1|J^1}(x)$ are called the *conditional quantization error pdfs* and describe the quantization errors resulting from the cells S_i^1. Fig. 9.6b shows these conditional pdfs for a five-level UTQ, while in Fig. 9.6c the result of the weighted summation of (9.12) is shown. Consequently, an optimal UTQ can be designed to quantize X^2 using $p_{X^2}(x)$ as given in (9.12) and bit rate r^2.

An experimentally determined performance of the two-stage quantizer is shown in Fig. 9.7 for $r^{tot} = 5$ bpp and various values of α. The performance is expressed using the SNR measure, which is defined as

$$\text{SNR} = 10 \log_{10}\left(\frac{\sigma_{x^1}^2}{d^{tot}}\right), \tag{9.13}$$

with $\sigma_{x^1}^2 = 1$.

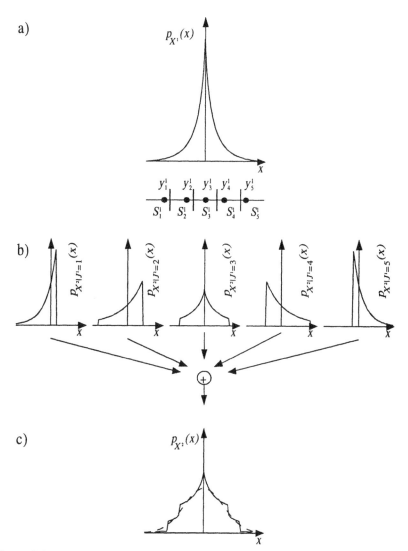

Figure 9.6   Construction of quantization error pdf: (a) $p_{X^1}(x)$; (b) $p_{X^1|J^1=i}(x)$, (c) $p_{X^2}(x)$ and mapped pdf (dashed line). Note that all pdfs shown have unit variance.

Also shown in Fig. 9.7 is the performance of the single-stage quantization using a UTQ at 5 bpp (dashed line). It can be seen that the two-stage quantizer performs less well than the single-stage quantizer. For $r^1 < 3$ bpp (i.e., $\alpha > 0.4$) the performance degradation is approximately 1.0 dB, while for $r^1 > 3$ bpp (i.e.

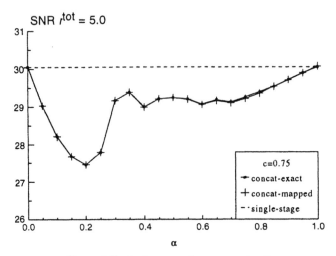

Figure 9.7   Performance of concatenated coding.

$\alpha < 0.4$) the degradation is 2.5 dB maximally. Similar results are obtained for a wide range of c values.

In practice, this kind of concatenated coding causes an expansion of the number of quantizers that should be designed and stored. Namely, for each predesigned admissible quantizer at the first stage, a separate set of admissible quantizers for the second stage has to be designed and stored. This expansion can be reduced by replacing each possible $p_{X^2}(x)$ with a predefined and limited set of pdfs. This approximation process is called pdf *mapping* and uses the Kolomov–Smirnov fitting method. The best fitted pdf is also shown in Fig. 9.6c (dashed line). As can be seen from Fig. 9.7, the resulting performance loss is almost negligible as the performance curves using the exact pdf and the mapped pdf nearly coincide.

Conditional Entropy Coding   Concatenated coding implies that the input signal x^1 is encoded using quantizers Q^1 and Q^2, which are optimized for the pdf of the different input signals at a certain rate. Symbolically this can be written as

$$p_{X^1}(x), r^1 \rightarrow Q^1, \tag{9.14}$$
$$p_{X^1}(x), Q^1, r^2 \rightarrow p_{X^2}(x), r^2 \rightarrow Q^2. \tag{9.15}$$

Once the quantizers have been designed, x^1 is encoded using two independent quantizers, each of which is followed by its own VLC. Knowledge about the selected indices j^1 of quantizer Q^1 is not used in the second stage. Use of this knowledge can, however, improve the coding efficiency of the second stage and thus the coding efficiency of the two-stage quantizer as a whole. In Section 9.3.2

it is shown how this knowledge can be used in the quantization of the second stage (conditional quantization) while in this section the conditional entropy coding strategy is introduced, which focuses on the VLC encoding of j^2 (Fig. 9.8). Note that this implies that the quantizers Q^1 and Q^2 are designed in the same way as is done with concatenated coding.

With the concatenated coding strategy, the VLC coders encode the indices j^n into a bit stream with a bit rate equal to the entropy of j^n, denoted by $H(j^n)$. Knowledge of j^1 in the encoding of j^2 reduces the entropy of the signal to be transmitted since

$$H(j^2|j^1) \;=\; \sum_i H(j^2|j^1 = i)P(j^1 = i) \;\; < \;\; H(j^2), \qquad (9.16)$$

if and only if j^2 and j^1 are mutually dependent. This implies that $H(j^2|j^1 = i)$ and $H(j^2|j^1 = j)$ should differ for some $i \neq j$; i.e.,

$$[lll]H(j^2|j^1 = i) \neq H(j^2|j^1 = j), \qquad (9.17)$$
$$\Rightarrow P(j^2 = l|j^1 = i) \neq P(j^2 = l|j^1 = j),$$
$$\Leftrightarrow \int_{S_l^1} p_{X^2|J^1=i}(x) \neq \int_{S_l^1} p_{X^2|J^1=j}(x), \qquad \text{for some } i, j, i \neq j.$$

Dependency between j^2 and j^1 exists, therefore, if the representation levels y_l^2 of Q^2 have different probabilities caused by different conditional quantization error pdfs. For typical GG distributions and UTQs this is generally true (see Fig. 9.6).

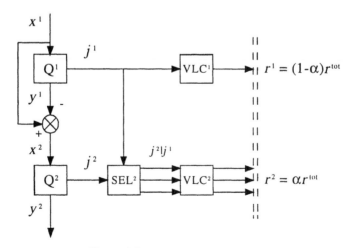

Figure 9.8 Conditional entropy coding.

Implementing a VLC coder to encode $j^2|j^1$ involves the design and storage of K^1 VLC codes. For UTQs, K^1 can be as high as 2048 levels which yields complex arithmetic codes. The complexity can be limited by conditioning j^2 on partitions of the index set J^1. These partitions are called *classes* and are denoted by C_v^1. The conditional entropy measure becomes then

$$H(j^2|j^1) = \sum_{v=1}^{{}^{C}N^1} H(j^2|j^1 \in C_v^1) \, P(j^1 \in C_v^1), \tag{9.18}$$

where ${}^{C}N^1$ denotes the number of index classes. It has been observed that the performance loss due to this grouping is negligible if indices j^1 associated with similarly shaped pdfs are grouped together. A very good tradeoff between complexity and performance is achieved by the following three-way grouping:

$$C^1 : J^1 \rightarrow \begin{cases} C_1^1 : \text{if } j_i^1 < \frac{K^1-1}{2} \text{i.e.} y_i^1 < 0, \\ C_2^1 : \text{if } j_i^1 = \frac{K^1-1}{2} \text{i.e.} y_i^1 = 0, \\ C_3^1 : \text{if } j_i^1 > \frac{K^1-1}{2} \text{i.e.} y_i^1 > 0. \end{cases} \tag{9.19}$$

The three conditional quantization errors pdfs $p_{X^2|J^1 \in C_v^1}(x)$ arising from this grouping are shown in Fig. 9.9. The complexity of this grouping can be reduced even further while maintaining the same performance. Because of the symmetry in pdfs of $X^2|J^1 \in C_1^1$ and $X^2|J^1 \in C_3^1$, it is possible to combine these classes by reversing the sign of signal $X^2|J^1 \in C_3^1$. Subsequently, both conditional signals have the same pdf and can thus be jointly encoded.

The performance of the two-stage quantizer using the conditional entropy coder is shown in Fig. 9.10. Note that the output bit rate of the second quantizer is still αr^{tot} so that the gain over concatenated coding shows up in a smaller distortion d^2 and d^{tot}. As can be seen, for $r^1 < 3$ bpp ($\alpha > 0.4$, $r^{\text{tot}} = 5$ bpp) the coding performance increases by approximately 0.5 dB compared to concatenated coding. Above $r^1 > 3$ bpp ($\alpha < 0.4$) the improvement is negligible. The performance obtained if j^2 is conditioned on the individual indices j^1 instead of on the three classes is also shown in Fig. 9.10. As can be seen, the performance difference is very small.

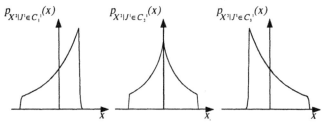

Figure 9.9   Pdfs of conditional quantization errors $X^2|J^1 \in C_v^1$.

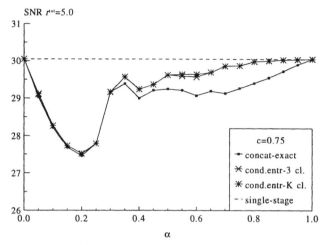

Figure 9.10   Performance of conditional entropy coding.

Conditional Quantization   Knowledge about j^1 can also be used in the design of the second quantization stage. The dissimilar shapes of the pdfs of the $X^2|J^1$ signals (see Fig. 9.6) suggest a strategy to encode them separately with optimally designed quantizers. With this conditional quantization strategy (Fig. 9.11) the

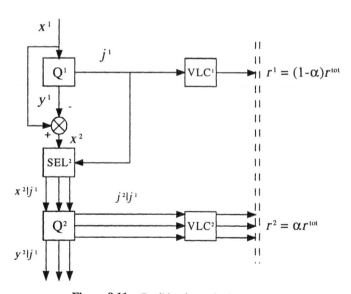

Figure 9.11   Conditional quantization strategy.

second quantization stage consists of several quantizers that are designed for a certain $p_{X^2|J^1}(x)$ at a certain rate r_j^2. Together with the optimally designed UTQ Q^1 this can be symbolically written as

$$p_{X^1}(x), r^1 \rightarrow Q^1, \tag{9.20}$$

$$p_{X^2|J^1=i}(x), r_j^2 \rightarrow Q_j^2, \tag{9.21}$$

or if a grouping is used to reduce the complexity

$$p_{X^1}(x), r^1 \rightarrow Q^1, \tag{9.22}$$

$$p_{X^2|J^1 \in C_v^1}(x), r_j^2 \rightarrow Q_j^2.$$

The pdfs $p_{X^2|J^1 \in C_v^1}(x)$ will in general not be distributed according to the GG distribution (see Fig. 9.6). At low bit rates r^1 they are usually strongly asymmetric and nonuniform while at high bit rates they are essentially uniformly distributed. In practice, for each $p_{X^2|J^1 \in C_v^1}(x)$ special quantizers have to be designed with various output bit rates, which greatly increase the complexity. A pdf mapping of all $p_{X^2|J^1 \in C_v^1}(x)$ to a limited set of pdfs is therefore necessary to manage the complexity.

A more fundamental problem is raised by the term r_j^2 in (9.21); namely, how should the available bit rate r^2 be decomposed into the components r_j^2 such that signal X^2 is optimally encoded? Fortunately, this is similar to the bit allocation problem in SBC applications, for which efficient algorithms are available. This allocation procedure (to be described in Section 9.3.3) minimizes the overall distortion d^2 at total bit rate r^2 by selecting quantizers Q_j^2 from a set of optimally predesigned quantizers to encode each signal $X^2|J^1 \in C_v^1$.

The following experiment illustrates the process of conditional quantization. For high bit rates of r^1 ($r^1 > 4$ bpp; i.e., $\alpha < 0.25$ and $r^2 < 1$ bpp), all $X^2|J^1$ signals are more or less uniformly distributed. Optimal uniform quantizers are selected by the allocation algorithm from a set with bit rates ranging from 1 bpp to 8 bpp. Figure 9.12 shows that for these bit rates a major performance gain is obtained. Because $r^2 < 1$ bpp and the scalar quantization applied, some signals $X^2|J^1$ are mapped to zero ($r_j^2 = 0$) while other signals are more accurately quantized ($r_j^2 \geq 1$). For lower rates of r^1 signals $X^2|J^1$ are no longer uniformly distributed and the previously sketched strategy performs worse than the concatenated coding strategy.

Embedded Quantization   In the previous paragraph, the potential coding performance of conditional quantization was demonstrated. Essentially it is based on the possibility to refine the individual $X^2|J^1$ signals at the second quantization stage. The design methodology of Q^1 and the different Q_j^2 cause the complexity to

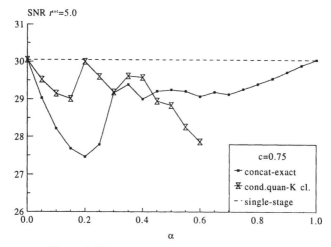

Figure 9.12 Performance of conditional quantization.

grow in a tree-structured way, which makes this method very unattractive. In this section the embedded coding strategy is discussed. It is shown that this strategy, using a class of embedded uniform threshold quantizers, outperforms the other strategies but has reasonable complexity.

The embedded quantization approach is illustrated in Fig. 9.13. At first sight this configuration looks quite different from the previous multistage schemes (Figs. 9.5, 9.8, and 9.11). However, it will be shown that if Q^1 and Q^2 are properly designed the scheme is equivalent to the conditional quantization scheme.

In Fig. 9.13 the original signal x^1 is quantized by Q^1 as well as by Q^2 with outputs j^1 and j^2, respectively. Note that Q^2 quantizes now x^1 instead of y^1. Quantizers Q^1 and Q^2 have been designed in such a way that the cells $S_i^1, 1 \leq i \leq K^1$ and $S_j^2, 1 \leq j \leq K^2$ are related as follows:

$$S_i^1 = \cup_{j=n_i}^{m_i} S_j^2, \qquad n_{i+1} > m_i \geq n_i > m_{i-1}, \qquad (9.23)$$

where n_i and m_i are indices of Q^2. Figure 9.14 illustrates this relation, showing that each cell of Q^2 is entirely contained in one cell of Q^1. Therefore, the probabilities of indices j^2 can be conditioned on the probabilities of indices j^1:

$$P(j^2 = j) = P(j^2 = j, j^1 = i) \qquad (9.24)$$
$$= P(j^2 = j | j^1 = i) \, P(j^1 = i),$$

which yields

$$H(j^2) = H(j^2 | j^1) + H(j^1). \qquad (9.25)$$

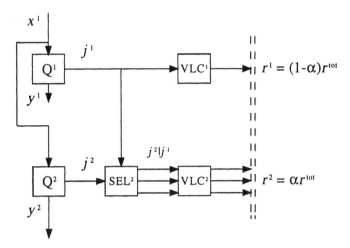

Figure 9.13   Embedded quantization.

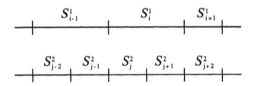

Figure 9.14   Embedded quantizers.

Therefore, we can transmit the information of j^2 by using a refinement signal $j^2|j^1$ that refines the already transmitted j^1 signal. It is easily seen that this scheme is equivalent to the scheme depicted in Fig. 9.11 in the sense that each conditional quantization error $X^2|J^1$ is quantized with a second quantizer Q_j^2 equivalent to a certain part of a larger quantizer Q^2 with more representation levels.

In the following experiment a class of embedded uniform threshold quantizers is used. This class consists of a set of quantizers Q^m, $1 \leq m \leq M$ with various output rates. Quantizer Q^M has the highest rate, denoted R^M, and is a UTQ. The quantizers with lower rates Q^m, $1 \leq n < M$ are derived from this quantizer using a pruning algorithm. The pruning algorithm first combines those cells S_i^M and S_{i+1}^M that decrease the output rate most for a given distortion increase. When a certain desired rate is reached the resulting quantizer is stored as Q^{M-1}. Other quantizers Q^m, $1 \leq m < M$ are found in a similar recursive pruning way. Observe that Q^m, $m < M$ will not in general be a UTQ but an approximation. The performance of embedded quantization is shown in Figs. 9.15 and 9.16 where the performances are illustrated

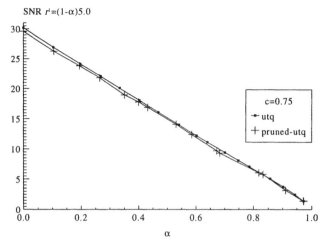

Figure 9.15 Performance of Q^1 for embedded quantization.

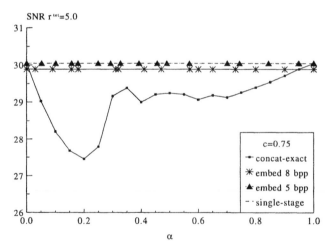

Figure 9.16 Performance of embedded quantization.

of Q^1 and Q^2, respectively. The performances of the pruned quantizers Q^1 are slightly less than the performance of a UTQ. The performance of Q^2 depends on the rate R^M for which Q^M has been optimized (Fig. 9.16). If $R^M = 5$ bpp then the performance coincides with the single-stage performance (dashed line). If $R^M = 8$ bpp then Q^2 is a pruned quantizer itself and will thus perform less well than the single-stage quantization (starred line).

Multistage Predictive Coding in Conclusion   To summarize the previous results, we conclude that the coding performance of the two-stage quantizer increases when going from the concatenated coding strategy to the conditional entropy coding strategy to the conditional quantization strategy. Unfortunately, the complexity of these methods also gets higher for each strategy and thus makes these strategies increasingly unattractive. The best coding performance results from the embedded coding strategy, where a refinement of indices instead of signal values takes place. In this method the complexity is moderate and comparable to the complexity of conditional entropy coding.

The following should however be noted. With the embedded quantization strategy, Q^1 is no longer an optimal UTQ. Still, the overall performance of the two-stage quantizer is better than the observed performances of the other strategies where Q^1 was a UTQ. Barnes and Frost [45] show that this is generally the case for all multistage quantizers and propose therefore a design methodology where all quantizers Q^n are designed simultaneously. This is done by minimizing the quantization error of the N-stage quantizer in an iterative loop where one particular quantizer Q^m, $1 \leq m \leq N$ is modified with the other quantizers Q^n, $1 \leq n \neq m \leq N$ in place. The drawback of these alternatively designed multistage quantizers is, however, that their encoding tree becomes entangled. This means that a minimal distortion approximation in stage $1 \leq n < N$ as was done in previous paragraphs does not lead to the minimal quantization error at stage N. Instead, a full search of the encoding tree is necessary to guarantee that the quantization error in stage N is minimized. This makes it difficult to incorporate these alternatively designed multistage quantizers in the compatible coder of Fig. 9.3.

Multistage Differential Predictive Coding   To conclude this section, the process of multistage DPCM coding is analyzed. Fig. 9.17 shows a two-stage DPCM coder using the concatenated coding strategy. The first coding stage consists of a DPCM encoder that removes the redundancy from input signal x^1 by subtracting a prediction signal $P^1(y^1)$. The difference signal e^1 can therefore be expressed as

$$e^1 \;=\; x^1 - P^1(y^1). \tag{9.26}$$

After quantization, the signal y^1 is reconstructed by adding the quantized version of e^1 to the prediction $P^1(y^1)$:

$$y^1 \;=\; P^1(y^1) + Q^1(e^1). \tag{9.27}$$

Note that the prediction process P^1 uses only previously quantized pixels of y^1. The second stage of the two-stage DPCM encoder consists of a PCM coder that quantizes the signal x^2 which is, as before, the difference between x^1 and y^1:

$$x^2 \;=\; x^1 - y^1. \tag{9.28}$$

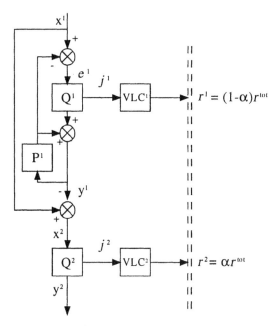

Figure 9.17 Concatenated coding using DPCM.

However, using (9.26) and (9.27) to substitute x^1 and y^1 it can be shown that x^2 is equal to the quantization error made in the first quantization stage [17]. Namely,

$$
\begin{aligned}
x^2 &= x^1 - y^1, \\
&= e^1 + P^1(y^1) - P^1(y^1) - Q^1(e^1), \\
&= e^1 - Q^1(e^1).
\end{aligned}
\tag{9.29}
$$

This result has three important consequences. First, we can optimize the two-stage DPCM coder if the pdf of e^1 is known. The pdf of x^2 follows then from (9.12) as for the multistage PCM coding case. Second, it is possible to use any of the previously discussed multistage quantizers in the multistage DPCM coder, since these are all based solely on the knowledge of the pdfs of x^1 and x^2. Third, the results for multistage PCM encoding and multistage DPCM encoding will be very similar. For multistage DPCM coders, the performance can be decreased only by the reduced accuracy of the prediction $P^1(y^1)$ for increasing α. Namely, for these values of α, y^1 is more coarsely quantized which yields a less accurate prediction.

Figure 9.18 shows the obtained performances for multistage DPCM coding when using the concatenated coding strategy and the embedded coding strategy. Experimental results show that, for regular low-frequency subband data, the pdf of e^1 can reasonably well be modeled with a GG pdf with shape parameter $c = 0.75$.

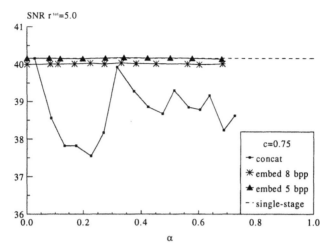

Figure 9.18   Performance of multistage DPCM.

As can be seen, the performance curves of the concatenated coding strategy and the embedded coding strategy are situated as expected. When using the embedded UTQ family with $R^M = 5$, it should be observed that the performance loss caused by the less accurate prediction is almost negligible. It can therefore be concluded that, as with multistage PCM, multistage DPCM yields negligible performance losses if embedded quantizers are used.

9.3.3 Bit Allocation Algorithm

The bit allocation algorithms $BA1$–$BA3$ in Fig. 9.3 can be used to allocate either the available bits among the subbands of layer or to allocate the allowed distortion of a certain layer. The algorithm is a combination of the bit allocations procedures proposed in [15, 50]. First, the algorithm determines the activity class of each 8 × 8 block in all subbands based on a variance measure. For the DPCM encoded subband the variance estimate is taken from the prediction error. Based on the average measure of the four activity classes in all subbands, the allocation algorithm selects a quantizer from the set of admissible ones for each activity class. It thereby uses performance tables that indicate the expected resulting bit rate and distortion per class for each admissible quantizer after quantization and VLC coding. These performance tables are based on the assumed (normalized) pdf of the subband data. In our implementation, the tables contain 20 quantizers entries for eight possible pdf functions. Before evaluation the tables are scaled to match the actual activity class variances. The final assignment is such that the distortion is minimized for a certain bit rate or that the bit rate is minimized for a certain distortion. More precisely, if $r_{j,k}^n$ and $d_{j,k}^n$ denote the bit rate and distortion of the kth activity class of the jth subband at layer l^n, then the distortion

$$d^n = \sum_{j=1}^{{}^sN^n} \sum_{k=1}^{{}^cN^n} \frac{{}^cM_{j,k}^n}{{}^sM_j^n} d_{j,k}^i \tag{9.30}$$

is minimized under the condition that

$$r^n = \sum_{j=1}^{{}^sN^n} \sum_{k=1}^{{}^cN^n} \frac{{}^cM_{j,k}^n}{{}^lM^n} r_{j,k}^i \tag{9.31}$$

is constant or vice versa. Here, ${}^sN^n$ denotes the number of subbands at layer l^n and ${}^cN^n$ the number of activity classes in each subband (${}^cN^n = 4$). The number of pixels in the jth subband of layer l^n is denoted by ${}^sM_j^n$. Further, ${}^cM_{j,k}^n$ and ${}^lM^n$ signify the number of pixels in activity class k in subband j and the total number of pixels in layer l^n, respectively. The overall bit rate R^n for signal L^n is given by

$$\begin{aligned} R^1 &= r^1, \\ R^2 &= r^2 + \frac{1}{4}r^1, \\ R^3 &= r^3 + \frac{1}{4}r^2 + \frac{1}{16}r^1, \end{aligned} \tag{9.32}$$

while the distortion of signal L^n is given by

$$D^n = d^n. \tag{9.33}$$

After the bit allocation, each activity class is multistage encoded and transmitted. Also the classification tables are transmitted for the correct decoding of the subband data at the receiving side. Classification tables made for a particular subband have to be used in all subsequent quantization stages, since blocks belonging to a particular activity class have their own quantization history.

9.3.4 Color Component Encoding

The color difference signals are encoded in a way similar to the luminance signal. The U and V components are separately decomposed into 31 subbands using the QMF16B filter [42]. The subbands containing the low-pass information are DPCM encoded while the other subbands are PCM encoded. Layers are created in a way similar to the luminance signal by employing the error feedback strategy and multistage quantization.

The available bit rate of each layer is divided between the luminance and chrominance signals on an 80 : 20% ratio *prior* to the actual bit allocations. The bit allocation for the chrominance signals selects quantizers for both the U and V subbands thereby minimizing the joint MSE of these signals.

9.4 Hierarchical Channel Coding for Asynchronous Transfer Mode

In this section we discuss the adaptation of the hierarchical subband coder introduced in the previous section to ATM-based networks. These multigigabit networks are based on optical fiber technology and are expected to become the all-purpose communication networks of the future. Information is transported using small packets, called *cells*, according to the standardized Asynchronous Transfer Mode (ATM) protocol [51]. Thus various services with highly diverse bit stream characteristics can be accommodated such as, for example, HDTV distribution, multimedia communications, videotelephony, and all kinds of archive-retrieval services.

In the following, several user-related aspects of ATM-based networks are discussed. The impact of data transmission via cells on video compression schemes is analyzed, and the need for hierarchical channel coding techniques is explained. Subsequently, ATM adaptation of the hierarchical encoder of Section 9.3 is described; namely, synchronization issues and the shaping of the produced bit streams for each of the three layers.

9.4.1 Video Coding for Asynchronous Transfer Mode-Based Networks

ATM-based networks transport the information through the network using cells composed of a header part and a payload part. The header consists of 5 bytes and contains primarily routing information. The payload is composed of 48 bytes of which a few bytes are used for standardized ATM adaptation layer (AAL) protocols [52]. A service communicates by means of private virtual channels (VCs) which are used to deliver the cells correctly. This mechanism enables the transmission of bit streams with highly diverse characteristics such as constant bit rate (CBR) streams, variable bit rate (VBR) streams, and bursty streams.

Video communications can benefit from ATM transmission since CBR transmission is no longer obligatory. Therefore, constant-quality VBR sche-mes become feasible as opposed to the existing variable-quality CBR schemes. These VBR schemes try to maintain a certain locally or globally defined quality level and transmit only as many bits as required. This property is exploited by the network, which fills the transmission bandwidth of a fiber by multiplexing various VCs of independent VBR coders. Statistical multiplexing allows the network resources to be used efficiently. To regulate this multiplexing procedure or, more generally, to manage the proper allocation of network resources, the network and encoder negotiate a Quality of Service (QoS) contract at the moment of the call setup. This contract specifies the expected bit-stream characteristics of the encoder in terms of burstiness, mean and peak bit rate as well as the desired quality of the VC such as cell loss and bit error rates, cell insertion rate, and maximum delay time. For video codecs this implies that, on one hand, they have to monitor and control their produced bit stream, but that, on the other hand, they can rely on certain negotiated channel characteristics.

Cell losses are inevitable in ATM networks, caused mainly by buffer overflow in network nodes. In particular, during congestion periods, the cell loss rate (CLR) can become extremely high. Since retransmission of cells is impractical in a video transmission environment because of the high bit rates, video compression schemes a priori need to take into account cell losses. Several hierarchical channel coding schemes that take special precautions to decrease the effects of missing or erroneous cells have been proposed [53, 54]. These schemes achieve graceful degradation under relatively high cell loss rates by decomposing the video signal into two layers. The lowest, i.e., most important, layer is transmitted in high-priority cells of a VC, while the second layer is transmitted in the remaining low-priority cells. When network buffers overflow, high-priority cells are served first, keeping the high-priority cell loss rate at an acceptable level. All low-priority cells are, however, discarded. The priority information is contained in the header by means of the cell loss priority (CLP) indicator.

Although the CLP indicator is suitable to be used for hierarchical *channel coding* applications, it is, however, less suitable for hierarchical *source coding* applications. Namely, in video transmission, the decoders (and not the network) decide which layers are used in the reconstruction and only those layers are to be received via the network. The CLP indicator is not suitable for this purpose since the indicator cannot be used for user-defined channel selection inside the network. Further, only two subchannels can be accommodated. This is clearly not sufficient for advanced compatible systems.

An alternative approach is to use multiple virtual channels to accommodate hierarchical channel coding, such as in the compatible scheme proposed in [55]. Each layer is transmitted in a single VC so that receivers can receive the various layers separately. Thus several VC connections exist between the decoder and the encoder for a single service[1]. The use of an additional layer in the reconstruction now implies the setup and reception of an extra VC. In the approach sketched earlier, hierarchical channel coding is incorporated by negotiating different cell loss and bit error rates for the different VCs. The lowest layer uses the most reliable VC (for example, $CLR < 10^{-8}$) while the highest layers are transmitted using less reliable VCs ($CLR < 10^{-3}$).

---

[1]It should be noted that the use of virtual channels is not limited to video. Multiple audio channels, subtitles, and teletext information will probably be transmitted using additional VCs.

9.4.2 Asynchronous Transfer Mode Adaptation of Compatible Subband Scheme

Figure 9.19 shows the adaptation of the compatible intraframe subband encoder discussed in Section 9.3 to an ATM-based network. The encoding stages of all layers are extended by a multiplexer (MUX), a buffer (BUF), and an ATM adaptation layer (AAL) module. The multiplexer interleaves the arithmetically encoded indices of the quantizers with the required side information and inserts start codes with synchronization capabilities. The resulting bit stream is subsequently buffered before packetization and transmission by the ATM adaptation layer module.

In the remainder of this section we assume that the negotiated QoS contracts contain bit stream characteristics similar to those given in Table 9.2. As can be seen, the lowest layer is transmitted as a CBR stream of 6 Mbit/sec while the two higher layers are transmitted as VBR streams. The idea behind this choice is that the continuity of the service is guaranteed by the CBR layer as the network can easily maintain such a relatively low bit rate contract for the lowest layer. The constant quality can be achieved by relatively cheap QoS contract for the VBR layers carrying the higher resolution information. Note that although the cumulative bit rate characteristics of the layers are also shown in Table 9.2, these will presumably not be part of the negotiated contracts. Further it is assumed that the mean bit rate values are evaluated over $\frac{1}{50}$ sec, which is equal to the time between two successive frames.

The hierarchical source coder uses the negotiated mean bit rate values as target values for the forward bit allocations $BA1$–$BA3$. Bit allocation $BA1$ distributes the available bits among the 7×4 activity classes by selecting the appropriate quantizers as discussed in Section 9.3.2. Bit allocations $BA2$ and $BA3$, on the contrary, allocate the allowed mse distortions among the 19×4 and 31×4 activity

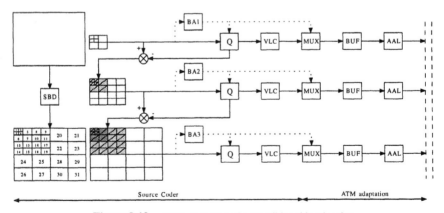

Figure 9.19   ATM adaptation of compatible subband coder.

Table 9.2   Negotiated bit stream characteristics (in Mbit/sec).

Stage	Stream	AAL	r^n		R^n	
n	type	class	Mean	Peak	Mean	Peak
1	CBR	1	6	6	6	6
2	VBR	2	11	29	17	35
3	VBR	2	49	101	70	140

classes and consider the negotiated mean bit rates as maximum mean bit rates to be spent on the encoding of a single frame. When the maximum bit rate is not sufficient to warrant the desired mse quality level, the allocation algorithms $BA2$ and $BA3$ revert to allocating the maximum bit rate with a minimized distortion. The shaping of the resulting bit streams into a CBR stream and two VBR streams (see Table 9.2) is discussed in Section 9.4.2.

Synchronization and Side Information   Figure 9.20 shows the syntax of the bit streams produced by the three multiplexers. As can be seen, the bit streams are composed of start codes (SOF-Y, SOF-UV, SOB, SOL), side information (SI) and VLC data. A start code indicates the start of an encoded object, which may be a frame containing luminance information (SOF-Y), a frame containing chrominance information (SOF-UV), subbands (SOB), or some consecutive scan lines of a subband called a *slice* (SOL). A start code consists of a synchronization word of 32 bits followed by 8 bits that indicate the type and sequence number of the encoded object. Decoders use these start codes to synchronize the decoding of the three separate VCs and recover synchronization after cell loss in a certain VC.

The transmitted side information in the frame object contains the variances of the subband activity classes. Decoders require these variances to perform the same bit/distortion allocation as the encoder and to recover the used set of quantizers and VLC tables. Also contained are the mean value of the luminance component or chrominance components. Further, the frame object consists of several subband

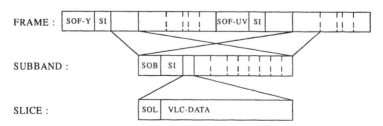

Figure 9.20   Syntax of the layer bit streams after multiplexing.

objects that contain the encoded subbands that are not entirely mapped to zero (i.e., $r_i^n > 0$).

Subband objects contain side information (if any) and several slice objects. The side information contains the activity classification table of the subband that is transmitted in the first subband object of a given subband. This table is necessary to decode the different blocks of the subband data with the appropriate tables. For the DPCM encoded (low-frequency) subband of layer l^1 the side information also carries the prediction coefficients used by the encoder.

Slices are constructed by several consecutive lines of a subband and transmitted in a slice object. The number of lines in a slice, or equivalently the number of slices in a subband, is dynamically determined by the multiplexer depending on the bit rates of the assigned quantizers of the subband activity classes. This is done in such a way that each slice contains approximately 2500 bits. Consequently, the overhead of slice start codes is less than 1.5% of the bit rate used. Accurately quantized subbands may be transmitted in as much as 50 slices, while coarsely quantized subbands are carried in a single slice.

Packetization and Bit-Stream Shaping   The bit streams produced by the multiplexers are temporarily buffered and next packetized by the AAL modules. These modules map the encoder-specific bit streams into the payloads of ATM packets according to a particular protocol. Two AAL protocols, denoted AAL I and II, are currently under standardization and are meant to adapt connection-oriented CBR sources and VBR sources to ATM-based networks, respectively [52]. The AAL modules use some bytes of the payload to facilitate the detection of lost cells (both AAL I and II), forward error correction (AAL II), and to indicate the start, continuation, or end of a message (AAL II). As the lowest layer is CBR it is packetized using the protocol of AAL I, while the other two (VBR) layers are packetized using AAL II.

In order not to violate the negotiated contracts, the bit streams of all layers need to be controlled or *shaped* according to the negotiated parameters. This shaping is done using a buffer between the multiplexer and the packetizer and by modifying the order in which the subbands are encoded. As an example, consider Fig. 9.21a. Here, the bit streams of layers l^1 and l^3 are shown when the encoded subbands are sequentially multiplexed in the order indicated in Fig. 9.19, and the output of the multiplexer is packetized instantaneously (i.e., no buffering). Clearly, the bit rates fluctuate heavily in time and violate the negotiated peak bit rates (dotted line) because of the accurate encoding of the low-resolution subbands. The mean bit rates, however, conform to the negotiated contract since the forward bit allocations have taken care of that.

Bit-stream shaping is now used to reduce the peak bit rates. A buffer between multiplexer and AAL is directly filled with the output of the multiplexer and is emptied at the negotiated peak bit rate as long as the buffer is not empty. For layer

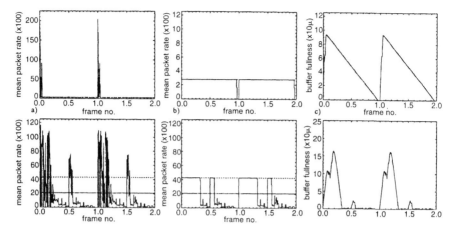

Figure 9.21 Bit streams of layer l^1 (upper row) and layer l^3 (lower row): (a) unshaped bit streams, (b) shaped bit streams, (c) buffer fullness.

l^1 this means that the output bit stream has indeed become CBR, while for layers l^2 and l^3 the peak bit rate of the VBR stream can never exceed the negotiated value. If the buffers of these layers underflow, the packetization is temporarily stopped. Packetization is started again when the buffer fills up again. The AAL modules interpret these events as the end of a message (EOM) and the start of a message (SOM). Figure 9.21b shows the resulting shaped bit streams while Fig. 9.21c shows the fullness of the buffers. The resulting streams now conform to the QoS contracts.

The shape of the VBR stream can be modified even more by using some information from the forward bit allocation procedures. Immediately after the bit allocation procedure, the expected bit rates of all subbands are known. The shape of the output bit stream can therefore be predicted directly after the bit allocation and can be influenced by changing the order of encoding of the subbands.

9.5 Experimental Results

In this section we discuss three experiments that address several aspects of hierarchical coding. The first experiment elaborates on the problem of multistage quantization. In addition to Section 9.3.2, the performance of three-stage quantizers is considered as these are highly relevant to advanced compatible coding schemes. The second experiment compares the HDTV coding performance of three particular compression schemes; namely, (1) the noncompatible subband coder of Fig. 9.4, (2) the compatible subband coder using concatenated coding, and (3) the compatible subband coder using embedded coding. For schemes (2) and (3) also the quality of the compatible signals is compared. The ATM cell loss resilience of the three schemes is subsequently discussed in the third experiment. Encoded bit streams are exposed to cell loss rates varying from 10^{-8} to 10^{-3}.

9.5.1 Three-Stage Quantization

Section 9.3.2 described four methods of multistage quantization; namely, concatenated coding, conditional entropy coding, conditional quantization, and embedded quantization. Using two-stage quantizers for the sake of simplicity and for illustrational purposes, it was concluded that embedded quantization has the best performance with moderate complexity. In the following we confirm that this conclusion holds for three-stage quantizers as well. These three-stage quantizers are highly relevant for the compatible coding scheme since they quantize some of the most important subbands (see Fig. 9.3). The performance of a three-stage quantizer is visualized using a rate–distortion (R–D) plot in which the cumulative bit rate and resulting distortion after each stage, i.e., the points $(\sum_{i=1}^{n} r^i, d^n)$ for $1 \leq n \leq 3$, are connected. This is illustrated in Fig. 9.22a for an imaginary three-stage quantizer with $(r^1, r^2, r^3) = (2.4, 1.0, 1.7)$ bpp. As a reference, the optimal R–D curve of single-stage quantization using a UTQ is also shown. The input signal x^1 is assumed to be distributed according to a generalized Gaussian pdf with shape parameter $c = 0.75$.

The results of concatenated coding using the predesigned quantizers of the compatible coding scheme are shown in Fig. 9.22b. Each R–D curve represents a three-stage quantizer which is composed of three selected quantizers (Q^1, Q^2, Q^3) with bit rates (r^1, r^2, r^3) such that $r^{tot} = r^1 + r^2 + r^3 = 5$ bpp. The endpoints of these curves $(r^{tot} = 5)$ deviate significantly from the optimal reference curve as well as the intermediate R–D points (i.e., the coding performances after the second coding stage). This confirms the conclusion of Section 9.3.2 that concatenated coding causes a significant performance loss.

Figure 9.22c shows the R–D curves of the three-stage quantizers of Fig. 9.22b when conditional entropy coding is used. Due to the conditional entropy coding of indices j^2 and j^3, the total bit rate of the three quantizers (Q^1, Q^2, Q^3) will be less than 5 bpp; i.e., $r^{tot} = r^1 + r^2 + r^3 < 5$ bpp. This can be observed since the R–D curves are closer to the optimal reference curve than those of the concatenated coding. Thus it is confirmed that conditional entropy coding outperforms concatenated coding.

The performance of the three-stage quantizer using embedded coding is shown in Fig. 9.22d. The coding performance of the pruned UTQ at 5 bpp is shown and connected to the performances of the other pruned UTQs with bit rates lower than 5 bpp. Any of these pruned quantizers may be selected in the first and second coding stage without influencing the overall performance of the three-stage quantizer.

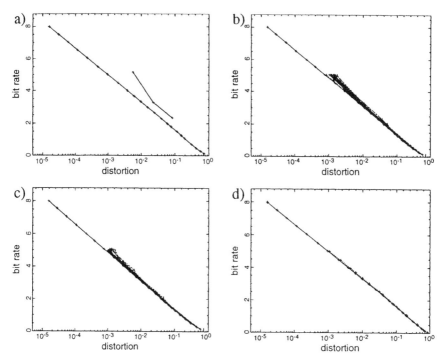

Figure 9.22 Performance of three-stage quantizer: (a) example, (b) concatenated coding, (c) conditional entropy coding, (d) embedded coding.

Since this curve lies at a small distance from the optimal reference curve, it is confirmed that embedded quantization is best suited for multistage quantization. Similar results have been obtained for a wide range of bit rates and c values.

9.5.2 Performance Comparison

Next, we evaluate the coding performance of three particular schemes; namely, the non-compatible subband coder of Fig. 9.4, the compatible subband coder using concatenated coding, and the compatible subband coder using embedded coding. The results are obtained using the "Mobile" sequence. This progressive-scan sequence consists of 40 frames (720×576 pixels) and has a frame rate is 50 Hz[2].

Figure 9.23 shows the obtained SNRs of the HDTV signal for each system when encoding the HDTV signal at 70 Mbit/sec. The compatible signals are encoded by the compatible schemes at 6 Mbit/sec and at 17 Mbit/sec as tabulated in Table 9.2.

---

[2]Since this sequence has not HDTV dimensions, the actual applied bit rates are down-scaled versions of the ones tabulated in Table 9.2.

The noncompatible scheme outperforms the other schemes by 0.1 dB for the embedded quantization method and by 0.6 dB for the concatenated coding method. These performance losses are due to the tradeoffs in achieving compatibility. The SNRs of the compatible signals are also calculated (Fig. 9.24). The coding

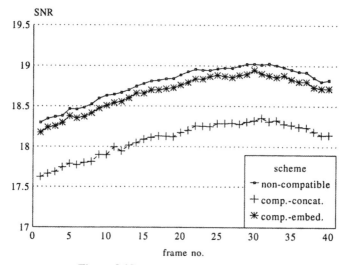

Figure 9.23   HDTV coding performances.

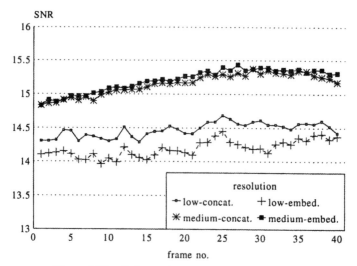

Figure 9.24   Coding performances of compatible signals.

performance refers only to the quantization process since no original signals exist. It can be observed that the quality of compatible signals generated by the concatenated coding scheme is higher than the quality of embedded quantization scheme. This could, however, be expected since the quantizers used at the first stage in concatenated coding are optimal UTQs as opposed to that of the suboptimal pruned quantizers used in the embedded quantization method.

9.5.3 Asynchronous Transfer Mode Cell Loss Resilience

Finally, we illustrate the ATM cell loss resilience for the noncompatible scheme and the compatible scheme using concatenated coding and embedded coding. None of the three coding schemes use error concealment methods or forward error correcting codes (FEC) to reduce the effects of cell losses. The FECs of the VBR AAL modules are not used either. Single cell losses are simulated for each scheme according to the method described in [29] while decoding the encoded data set four times. For the noncompatible scheme, cell loss rates of 10^{-6} and 10^{-3} are applied (Fig. 9.25) while the three-layer compatible schemes are exposed to the following two CLR combinations (l^1, l^2, l^3): $10^{-8}, 10^{-6}, 10^{-6}$ and $10^{-8}, 10^{-3}, 10^{-3}$ (Figs 9.26 and 9.27).

As can be noted, the HDTV reconstruction quality of the noncompatible scheme is severely affected by ATM cell losses for a CLR of 10^{-3}. The performance for the compatible scheme using concatenated coding, on the contrary, is hardly affected even for CLRs of 10^{-3} for the higher layers. The performance for the

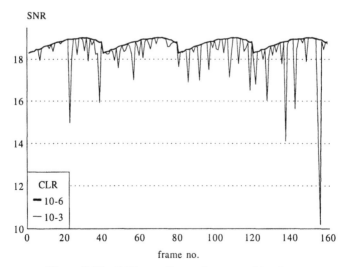

Figure 9.25 Cell loss resilience of noncompatible scheme.

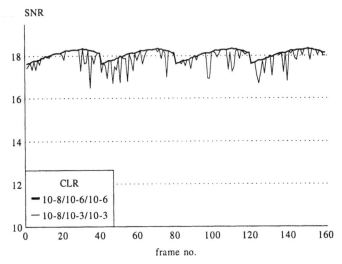

Figure 9.26   Cell loss resilience of compatible scheme using concatenated coding.

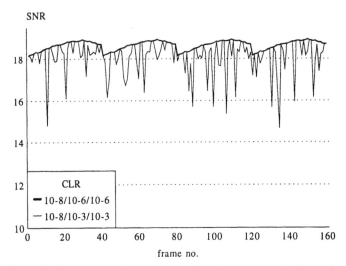

Figure 9.27   Cell loss resilience of compatible scheme using embedded coding.

compatible scheme using embedded coding is moderately affected for CLRs of 10^{-3}. This is cell loss in the second layer automatically implies the loss of corresponding information in the third layer since part of the transmitted code word is lost.

Subjective evaluations of the reconstructed sequences confirm these numerical results as errors in the compatible schemes are hardly visible as opposed to the annoying artifacts caused by cell losses in the noncompatible scheme.

9.6 Discussion

Hierarchical source and channel coding techniques facilitate the introduction of scalable open architecture video compression schemes. With these schemes, the video encoder and decoder are highly independent of each other, as decoders may reconstruct the video signal at a variety of bit rates and spatial and temporal resolutions.

In this chapter we have described how hierarchical source coding techniques decompose the video signal into a hierarchy of subsignals that are subsequently encoded and transmitted in multiple layers. Different applications and transmission channels may impose different constraints on these encoding stages. An intraframe compatible subband coder has been introduced to facilitate the discussion about multistage quantization methods: concatenated coding, conditional entropy coding, conditional quantization, and embedded quantization. Next, the scheme has been adapted to ATM-based networks to ensure constant quality and cell loss resilience. Experimental results showed that compatibility is achieved at the cost of a small performance loss and that the compatible scheme has, as expected, a high ATM cell loss resilience as opposed to a similar noncompatible scheme.

Further, it is observed that embedded quantization is superior to concatenated quantization. The coding performance loss of the compatible scheme for the HDTV signal when using embedded quantization is only 0.1 dB. Concatenated coding causes a performance loss of approximately 0.6 dB. However, with respect to the cell loss resilience it shows that concatenated coding outperforms embedded coding. This is because for embedded coding, cell loss in the second layer automatically implies the loss of corresponding information in the third layer since part of the transmitted code word is lost.

References

[1] K. H. Tzou and T. C. Chen,"Compatible HDTV coding for broadband ISDN."In *GLOBECOM'88*, Florida, 1988, pp. 743–749.

[2] F. Bosveld, R. L. Lagendijk, and J. Biemond,"A refinement system for hierarchical video coding."In *SPIE Visual Commun. Image Processing Conf.*, Laussanne, Switzerland, Oct. 1990, pp. 575–586.

[3] M. Pecot, P. J. Tourtier, and Y. Thomas,"Compatible coding of television images, part 2 compatible system."*Signal Processing: Image Commun.*, vol. 2, pp. 249–268, 1990.

[4] K. M. Uz, M. Vetterli, and D. J. LeGall,"Interpolative multiresolution coding of advanced television with compatible subchannels."*IEEE Trans. Circuits Syst. Video Technol.*, vol. 1, pp. 86–99, 1991.

[5] M. Breeuwer and P. H. N. de With,"Source coding of HDTV with compatibility to TV."In *SPIE Visual Commun. Image Processing Conf.*, Lausanne, Switzerland, 1990, pp. 765–776.

[6] D. S. Lee and K. H. Tzou,"Hierarchical DCT coding of HDTV for ATM networks."In *ICASSP'90*, Albuquerque, NM, April 1990.

[7] P. H. N. de With, A. M. A. Rijckaert, J. Kaaden, and H. W. Keesen,"Digital consumer HDTV recording based on motion-compensated DCT coding of video signals."*Signal Processing: Image Commun.*, vol. 4, pp. 401–420, Aug. 1992.

[8] V. M. Bove and A. B. Lippman,"Scalable open-architecture television."*SMPTE J.*, pp. 2–5, Jan. 1992.

[9] J. C. Ellershaw, M. J. Biggar, and A. W. Johnson,"Interworking of video codecs with different spatial resolutions and aspect ratios."*Picture Coding Symp.*, Cambridge, MA, 1990.

[10] G. Schamel,"Graceful degradation and scalability in digital coding for terrestrial transmission."*Int. Workshop HDTV'92*, Kawasaki, Japan, Nov. 1992.

[11] T. C. Chen, K. H. Tzou, and P. E. Fleisher,"A hierarchical HDTV coding system using a DPCM-PCM approach."In *SPIE Visual Commun. Image Processing Conf.*, Boston, 1988, pp. 804–811.

[12] D. J. LeGall, H. Gaggioni, and C. T. Chen,"Transmission of HDTV signals under 140 Mbits/sec using a sub-band decomposition and discrete cosine transform coding."In *Signal Processing of HDTV*. North-Holland, The Netherlands, Elsevier Science Publishers B.V., 1988.

[13] J. A. Bellisio and K. H. Tzou,"HDTV and the emerging ISDN network."In *SPIE Visual Commun. Image Processing Conf.*, Boston, 1988, pp. 772–786.

[14] P. J. Burt and E. Adelson,"The Laplacian pyramid as a compact image code."*IEEE Trans. Commun.*, vol. COM-31, 1983.

[15] J. W. Woods and S. D. O'Neil,"Subband coding of images."*IEEE Trans. Acoust., Speech, Signal Processing*, vol. ASSP-34, pp. 1278–1288, 1986.

[16] J. W. Woods and T. Naveen,"Motion compensated multiresolution transmission of HD video using multistage quantizers."In *ICASSP'93*, Minneapolis, April 1993, pp. 582–585.

[17] N. S. Jayant and P. Noll,*Digital Coding of Waveforms: Principles and Applications to Speech and Video*.Englewood Cliffs, NJ: Prentice-Hall, 1984.

[18] L. Vandendorpe,"Hierarchical coding of digital moving picture."Ph.D. thesis, University of Louvain, 1992.

[19] L. Vandendorpe and P. Delogne,"Hierarchical encoding of HDTV by transform coefficients block splitting."In *SPIE Image Processing Algorithms Techniques*, 1990, pp. 343–354.

[20] C. Guillemont,"Compatible HDTV/TV hierarchical scheme for secondary distribution of TV and HDTV signals."In *SPIE Visual Commun. Image Processing Conf.*, Boston, Nov. 1992, pp. 1475–1483.

[21] F. Bosveld, R. L. Lagendijk, and J. Biemond,"Compatible spatio–temporal subband encoding of HDTV."*Signal Processing*, vol. 28, Sept. 1992.

[22] J. W. Woods and T. Naveen,"Motion compensated multiresolution transmission of HDTV."In *Proc. GLOBECOM 91*, New York, Dec. 1991.

[23] S. Zafar, Y. Q. Zhang, and B. Jabbari,"Multiscale video representation using multiresolution motion compensation and wavelet decomposition."*IEEE J. Selected Areas Commun.*, vol. 11, pp. 24–35, 1993.

[24] L. Vandendorpe and P. delogne,"Hierarchical transform and subband coding of HDTV."*Fourth Int. Workshop HDTV*, Turin, Italy, Sept. 1991.

[25] HIVITS-RACE; Exhibition day paper,"An HDTV/TV compatible subband coding algorithm," May 1992.

[26] J. F. Vial, M. Pecot, P. J. Tourtier, and Y. Thomas,"In-band interpolation applied to motion-compensated subband coding."*Fourth Int. Workshop HDTV*, Turin, Italy, Sept. 1991.

[27] H. Gharavi,"Multilayer subband-based video coding."*IEEE Trans. Commun.*, vol. COM-39, pp. 1288–1291, 1991.

[28] F. Boucherok and J. F. Vial,"Compatible multi-resolution coding scheme."*Int. Workshop HDTV'92*, Kawasaki, Japan, Nov. 1992.

[29] ISO-IEC JTC1/SC2/WG11,*Test Model 2 for MPEG 2*, MPEG 92/245 edition, July 1992.

[30] R. ter Horst, A. Koster, K. Rijkse, E. Fert, G. Nocture, and L. Tranchard,"MUPCOS: A multi-purpose coding scheme."*Signal Processing: Image Commun.*, vol. 5, pp. 57–89, 1993.

[31] G. Morrison and I. Parke,"COSMIC: A compatible scheme for moving image coding."*Signal Processing: Image Commun.*, vol. 5, pp. 91–103, 1993.

[32] T. Chiang and D. Anastassiou,"Two-layer coding of interlaced HDTV for graceful degradation."*Int. Workshop HDTV'92*, Kawasaki, Japan, Nov. 1992.

[33] M. R. Civanlar and A. Puri,"Scalable video coding in frequency domain."In *SPIE Visual Commun. Image Processing Conf.*, Boston, Nov. 1992, pp. 1124–1134.

[34] Y. Yu and D. Anastassiou,"Interlaced video coding with field based multiresolution representation."*Signal Processing: Image Commun.*, vol. 5, pp. 185–198, 1993.

[35] C. Gonzales and E. Viscito,"Flexibly scalable digital video coding."*Signal Processing: Image Commun.*, vol. 5, pp. 5–20, 1993.

[36] T. Hanamura, W. Kameyama, and H. Tominaga,"Hierarchical coding scheme of video signal with scalability and compatibility."*Signal Processing: Image Commun.*, vol. 5, pp. 159–184, 1993.

[37] J. R. Ohm,"Temporal domain sub-band video coding with motion compensation."In *ICASSP'92*, San Francisco, March 1992, pp. 229–232.

[38] J. R. Ohm,"Layered vq and sbc techniques for packet video applications."*Workshop Packet Video'93*, Berlin, 1993.

[39] D. Taubman and A. Zakhor,"Multi-rate 3-d subband coding of video."Submitted to *IEEE Trans. Image Processing*, vol. 3, pp. 572–588, 1994.

[40] F. Bosveld, R. L. Lagendijk, and J. Biemond,"Hierarchical coding of HDTV."*Signal Processing: Image Commun.*, vol. 4, pp. 195–225, 1992.

[41] F. Bosveld, R. L. Lagendijk, and J. Biemond,"Compatible HDTV transmission using conditional entropy coding."In *ICASSP'93*, Minneapolis, April 1993, pp. 674–677.

[42] J. D. Johnston,"A filter family designed for use in quadrature mirror filter banks."In *ICASSP'80*, 1980, pp. 291–294.

[43] L. Wang and M. Goldberg,"Progressive image transmission by transform coefficient residual quantization."*IEEE Trans. Commun.*, vol. COM-36, pp. 75–87, Jan. 1988.

[44] W. H. R. Equitz and T. M. Cover,"Successive refinement of information."*IEEE Trans. Inform. Theory*, vol. 37, pp. 269–275, 1991.

[45] C. F. Barnes and R. L. Frost,"Residual vector quantizers with jointly optimized code books."*Advances Electron. and Electron Phys.*, vol. 84, pp. 1–59, 1992.

[46] J. M. Shapiro,"Application of the embedded wavelet hierarchical image coder to very low bit rate image coding."In *ICASSP'93*, Minneapolis, April 1993, pp. 558–561.

[47] T. Naveen, F. Bosveld, J. W. Woods, and R. L. Lagendijk,"Multiresolution transmission of video using multistage conditional quantization."*Picture Coding Symp.*, Lausanne, Switzerland, March 1993.

[48] T. Naveen, F. Bosveld, J. W. Woods, and R. L. Lagendijk,"Rate constrained multiresolution transmission of video."Submitted to *IEEE Trans. Circuits Syst. Video Technol.*, 1993.

[49] N. Farvardin and J. W. Modestino,"Optimum quantizer performance for a class of non-Gaussian memoryless sources."*IEEE Trans. Inform. Theory*, vol. 30, pp. 485–497, 1984.

[50] P. H. Westerink, J. Biemond, and D. E. Boekee,"An optimal bit allocation algorithm for sub-band coding."In *ICASSP'88*, New York, April 1988, pp. 757–760.

[51] CCITT SG XVIII,*Draft Recommendation I.121: Broadband Aspects of ISDN*, June 1990.

[52] CCITT SG XVIII,*Draft Recommendation I.363: B-ISDN ATM Adaptation Layer (AAL) Specification*, 1992.

[53] M. Ghanbari,"A motion vector replenishment video codec for ATM-networks."*Signal Processing: Image Commun.*, vol. 3, pp. 143–156, 1991.

[54] G. Morrison and D. Beaumont,"A two-layer video coding for ATM-networks."*Signal Processing: Image Commun.*, vol. 3, pp. 179–196, 1991.

[55] F. Bosveld, R. L. Lagendijk, and J. Biemond,"Compatible HDTV distribution using fixed distortion subband coding."*Int. Workshop HDTV'92*, Kawasaki, Japan, Nov. 1992.

Chapter 10

Model-Based Coding

K. Aizawa
Department of Electrical Engineering
University of Tokyo
Tokyo, Japan

10.1 Introduction

Historically, progress in image coding techniques has been made by incorporating results from other fields such as information theory. Most of the existing coding methods such as predictive coding, transform coding, and vector quantization belong to information-theory based methods, in which image signals are considered random signals and compressed by exploiting their stochastic properties.

Apart from these information-theoretic coding methods, research on new approaches to image coding that are related to both image analysis and computer graphics has recently intensified. An essential difference between conventional coding methods and these new approaches is the image model they assume. Contrary to conventional coding methods that efficiently represent 2-D waveforms of image signals, these new approaches represent image signals using structural image models that take into account some sense of the 3-D properties of the scene. A major advantage of this new coding method is that it describes image content in a structural way. Its application areas are naturally different from those of waveform coding.

Although model-based coding is still in its infancy, a number of proposals have been made. In this chapter, the approaches to model-based coding are first overviewed and categorized from the point of view of the image source models. Next, 3-D model-based coding method is described, and examples are illustrated. A hybrid method is described that makes use of 3-D model-based coding and waveform coding. The current status of system implementations and applications of model-based coding are also mentioned. Finally, the major remaining problems are discussed.

10.2 Model-Based Approaches to Image Coding

To efficiently encode image signals it is necessary to select an image model and apply it to representation of image signals. In the past, image compression has been regarded largely as a problem for information theory, and image signals have been modeled by using stochastic properties; that is, by exploiting the spatial and temporal correlation in either the spatial domain or the transform domain. For example, waveform coding such as transform coding has been successful in representing image signals as waveforms.

On the other hand, from an image analysis point of view, images can be considered as having structural features such as contours and regions. These image features have been exploited to encode images at very low bit rates while retaining enough visible structures in the reconstruction so as to maintain a certain level of quality. Notably, in the last few years, this structural model-based approach has adopted 3-D structural models of the scene. In this case, the coder and decoder have an object model; the coder analyzes input images and the decoder generates output images using the model (see Fig. 10.1).

Although the details of these structural model-based approaches vary widely, they can be classified into a few categories based on the complexity of the models they use: (1) 2-D feature based coding, (2) 3-D feature based coding, and (3) 3-D model-based coding. Methods classified into 2 and 3 are often called *model-based coding*. Each approach is briefly described and summarized in Table 10.1.

10.2.1 Two Dimensional Feature-Based Coding

These coding methods exploit visibly important 2-D features such as edges, contours, and regions. Coding methods such as contour-based coding and region-based coding are examples. The former method extracts contours, encodes shapes and intensities of contours, and reconstructs an image from them [2, 3]. The latter method segments images into homogeneous regions and encodes their shapes and intensities [4, 3]. Unlike works [5, 61] of the early 1980s that have encoded only

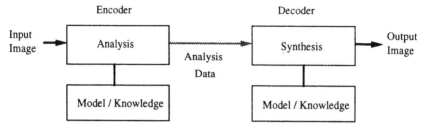

Figure 10.1   General description of a model-based coding system.

Table 10.1   Coding method categories. According to a computer vision framework [53], this classification roughly corresponds to the hierarchy of the image analysis, that is, primal sketch, $2\frac{1}{2}$-D sketch, and 3-D model, respectively.

Coding techniques	Image source model, exploited properties
Waveform coding	Statistical properties such as spatiotemporal correlation. 2-D motion model for motion compensation
2-D feature-based coding	2-D features such as contours, edges, regions. 2-D motion model
3-D feature-based coding	Planar/curved patches model or global surface model. 3-D motion model
3-D model-based coding	Parameterized detailed 3-D model of the objects. 3-D motion and action parameters

contour information and output binary images, recent methods aim to reconstruct images with natural intensity levels.

In the case of image sequences, moving regions that are detected as changing areas between two successive frames are modeled and coded as arbitrarily shaped 2-D objects translating two dimensionally [7].

10.2.2 Three-Dimensional Feature-Based Coding

Information such as surface structure and motion information estimated from image sequences are utilized in image coding. Surfaces of the object are modeled by general geometric models such as planes or smooth surfaces.

Several different approaches have been proposed. Hötter *et al.* [8, 9] and Diehl [10] have proposed a method utilizing a segmented surface model, in which changing regions caused by object motion are detected and modeled by planar patches or parabolic patches. Ostermann *et al.* [9, 11], Morikawa and Harashima [12], and Koch [13] proposed a coding method utilizing global surface models, in which a smooth surface model of the scene is estimated from an image sequence. These methods have also been discussed in the context of motion compensation. They have been applied along with motion compensation and interpolation to improve the performance of conventional waveform coding methods.

10.2.3 Three-Dimensional Model-Based Coding

The 3-D model-based coding utilizes detailed parameterized object models. Obtaining such a detailed model from general scenes is extremely difficult. However, when the object to be coded is restricted to specific classes, special knowledge obtained from a 3-D model of the object can be used in the coding system. For instance, in the case of videophone image transmission, a 3-D face model is sufficient for describing scene objects since most of the images consist of a moving head and shoulders. This assumption frees us from the requirement of reconstructing a 3-D model from 2-D images. From this point of view, Aizawa, Harashima and Saito [14–16], and Welsh *et al.* [17–19], independently have proposed model-based coding schemes that utilize parameterized 3-D models of a person's face. Prior to their proposals, *semantic coding* had been proposed by Forchheimer, Fahlander, and Kronander [10], but the reconstruction image of their original proposal was too synthetic, though its concept was very similar to the model-based coding. Up to now, most of the contributions to 3-D model-based coding have focused on human facial images and the parameterized facial models are usually given in advance.

Because 3-D model-based coding uses parameterized models and is more graphics oriented, it is expected to have broader applications than the conventional waveform coding methods. Some interesting applications have been proposed. *Speech-driven image animation*, in which images are synthesized only by speech parameters, and *virtual space teleconferencing* [21, 22] are two particular examples. We will discuss application aspects in the following section.

To improve the fidelity of the reconstructed images, there have been proposals of 3-D model-based/waveform hybrid coding, in which waveform coding is used to compensate errors that occur in the model-based coding process. Several methods have been combined with facial model-based coding, including MC/DCT [23, 24], vector quantization [25, 26], and contour coding [27].

The automatic modeling and analysis pose big problems to this approach. It is fair to say that automatic modeling has not been reported so far. Some automatic motion tracking (analysis) has been done for restricted conditions such that the model is made in advance and the initial position of the face is known. A real-time coding and decoding experimental system [28] is presented, where it is assumed that the model and the initial position of the face are given, and the face motion is tracked by using facial feature points that were detected by simple threshold logic. Choi *et al.* [29, 30] and Li, Rovivainen, and Forchheimer [31] reported direct estimation method of head and facial movements.

Since coding schemes related to the 2-D feature-based approach and 3-D feature-based approach are described in other chapters, this chapter focuses on summarizing research about 3-D model-based coding method.

10.3 A General Description of Three-Dimensional Model-Based Coding

Contributions to 3-D model-based coding up to now have concentrated on facial image analysis and synthesis, since the facial images are very important for broad applications including image communication. Hence, in this section, the research on modeling, analysis, and synthesis of facial images, has been done in the context of model-based coding, is reviewed. The proposal by Aizawa *et al.* [14] is also described as an example for 3-D model-based coding scheme.

10.3.1 Three-Dimensional Model-Based Coding for a Person's Face

The diagram of the system proposed by Aizawa *et al.* [14] is shown in Fig. 10.2. This model-based coding system consists of three main components: a 3-D facial model, an encoder, and a decoder.

The encoder separates the object from the background, estimates the motion of the person's face, analyzes the facial expressions, and then transmits the necessary analysis parameters. The encoder will add new depth information and initially unseen portions of the object into the model by updating and correcting it if required. The decoder synthesizes and generates the output images by using a 3-D facial model and the received analysis parameters. A very low-rate transmission can be potentially achieved by the use of this model-based image coding scheme because the encoder sends only the analysis parameters.

Analysis and synthesis of human facial images play a very important role for this model-based coding system. In the following sections, synthesis and analysis methods of human facial images are overviewed from the point of view of their applications to model-based coding.

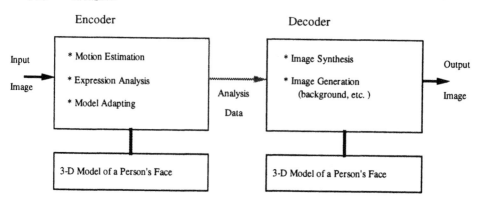

Figure 10.2   General description of a model-based analysis synthesis image coding system for a person's face.

10.3.2 Modeling and Synthesis

Face modeling   Modeling an object is the most important part in model-based coding because both analysis and synthesis methods strongly depend on the model they use.

There have been studies in the computer graphics field that handle human face animation. For example, Parke developed a parameterized model of a human face with geometrical details like eyes, mouth, etc. [32]. However, such work has produced results that lack surface details and reality because only wire frame models and shading techniques were used to reconstruct a human face. For image communication purposes, a person's face must be modeled in sufficient detail.

To develop an accurate model, an original intensity facial image is used through the texture mapping technique [14, 16, 17]. A 3-D wire frame generic face model (see Fig. 10.3) that approximately represents a human face is utilized. In the modeling process, the 3-D wire frame generic face model is adjusted to fit a frontal face image of the object person, and the frontal face image is texture mapped on the adjusted wire frame.

Even this modeling process is hard to fully automate. An interactive system for building a facial model has been developed at the University of Tokyo, which can be used not only for model-based coding but also for experimental studies of facial expressions by psychologists [33]. This system is also being used and improved at the University of of Illinois at Urbana–Champaign, Carnegie Mellon University, and other places.

Once a 3-D facial model has been obtained it can be easily manipulated; e.g., rotated in any direction. The rotated images still appear natural, although the depth information of the created 3-D facial model is only a rough approximation.

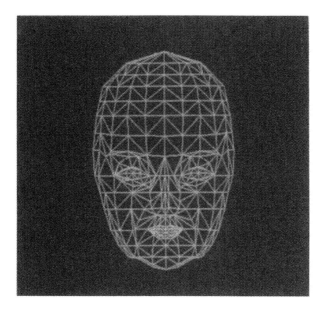

Figure 10.3   A 3-D wire frame generic face model.

Using a 3-D wire frame model and texture of an original image becomes a typical approach to modeling faces in model-based coding. However, because this process uses only a frontal image and does not take into account the real depth information of the person's face, the 3-D shape of the face can be very different from that of the real face, although the difference is rarely noticeable as long as the face model does not rotate very much from the initial position.

To improve the geometrical accuracy of the model, various approaches have been proposed. A side view of a face is used as well as a frontal view. The 3-D generic wire frame model is adjusted to both frontal and side views of a person [34, 35]. Stereo range data is used for adjusting the 3-D generic facial model [25]. The depth information of the frontal-adjusted model is updated and corrected by using an image sequence [14, 30].

Typically, the wire frame facial models are composed of a surface triangularized by 100–500 triangle elements. The resolution of triangularization is sometimes not sufficient to describe the details of the face when extreme facial expressions are synthesized on the 3-D face model. The larger the number of triangle elements is, the better the quality of the synthesis image is, although the complexity of modeling and analysis will grow. There have been attempts to use 3-D range data to model a face [36]. The range data were obtained by using a 3-D scanner (Cyberware) that can acquire both range and color data with sufficiently high density.

As for the human body model, a stick figure model has been proposed for describing and coding the human body motion [37].

Synthesis of facial movements   Texture mapping of original facial images onto a 3-D wire frame model gives rise to natural looking images. In addition, to synthesize naturally animated images, synthesis of facial movements (expressions) plays a very important role. When the 3-D wire frame model is available, facial movements can be reproduced in a number of ways. There is hierarchy of levels for the parameterization for synthesizing facial actions.

1. *Texture reproduction level.* Reproduce facial movements by updating the texture; for example, through clip-and-paste methods [14], template methods [17], or model-based/wave-form combined approaches [24].

2. *Node control level.* Synthesize facial movements by controlling the nodes of the wire frame model of the face. Interpolate of both intensities and node positions of 3-D model templates [29].

3. *Shape control level.* Synthesize facial movements by controlling the shape of the wire frame model of the face. For example, shape parameterization [14, 20] by making use of a facial action coding system (FACS) [39], or heuristic control of shapes of facial components such as eyes, mouth [40].

4. *Muscle control level.* Parameterize and build muscle models in the wire frame model and synthesize facial actions by controlling the muscles [1, 41].

5. *Abstract level.* Control the preceding parameters according to more abstract parameters such as description of emotions.

In computer graphics, so far, the majority have adopted the shape control method (e.g., [32]). These researchers have generally attempted to implement their own descriptions, though they start with some general description such as FACS. A muscle model approach has been applied [1], too. Recently, a facial model that has models of three skin layers and muscles has been investigated [41].

In model-based coding proposals, the shape control method and the texture reproduction method have been commonly utilized. For example, one proposal of the texture reproduction approach uses templates for facial components (eyes, mouth), which are stored in advance, and images are updated using these templates [17]. There are a variety of shape control approaches, from heuristic control to more refined control based on facial anatomy such as FACS based control.

Another approach is an interpolation method in terms of the facial shape as well as facial texture, in which a number of stored templates of 3-D facial models are interpolated into different facial expressions [29].

To improve the reconstruction fidelity, texture updating is required. Some approaches were proposed that also encode the differential signal between input images and synthesized images using the DCT [23], VQ [26], and contour coding [27].

10.3.3 Analysis of Facial Images

Analysis problems are much more difficult than synthesis problems. At present no system can work automatically both in modeling objects and in analyzing the sequence, even for simple head-and-shoulder images.

The analysis methods strongly depend on the chosen model and the synthetic output images. In addition, analysis implies several additional problems such as segmentation of objects, estimation of global motion, and estimation of local motion. In the case of facial images, global motion and local motion correspond to head motion and facial expressions, respectively. Motion analyses of facial images have been reported under some restrictive situations, such as that the model is made in advance and the initial position of the face is known. One feature of those facial motion analysis methods for model-based coding is that they attempt to use the information of the 3-D model when analyzing the image. Approaches to analyzing facial motion images follow.

1. *Detection of face and facial features.* Face-region detection is by active contour model [17]. Segmentation of moving head-and-shoulder shapes use shape constraints, temporal contour correction, and contour smoothness [42]. Extraction of facial feature components (eyes, mouth) uses a deformable template [43].

2. *Head motion (global motion) tracking.* Motion estimation of head motion is tracked by marks plotted on the face [14]. Global motion estimation of the head uses a displacement vector field obtained by the block-matching technique [44].

3. *Facial expression (local motion) analysis.* Feature points are extracted and tracked by a simple thresholding method [28] or by using an active contour model [45]. Facial expressions (action units) are estimated by using feature point displacements [46]. Direct estimation of head motion and facial expressions (action units) is by using the spatiotemporal gradient [30]. Facial muscle movement is estimated based on optical flow [47]. Estimation of facial muscle movement based on physical and anatomical models [48, 41].

10.4 An Example of a Three-Dimensional Model-Based Coding for a Person's Face

To illustrate an example of modeling and synthesis of 3-D model-based coding for a person's face, the proposal by Aizawa *et al.* [14] is described. The proposal by Choi and Takebe [30] is described for the analysis method.

10.4.1 Modeling a Person's Face Using a Generic Face Model

To develop a sufficiently detailed model for a person's face, a 3-D wire frame generic face model (see Fig. 10.3) and a original facial image are utilized. A person's face is modeled as follows:

1. The 3-D wire frame generic face model is scaled and adjusted so that it can fit a frontal face image of the object person.

2. The original frontal face image is texture mapped on the adjusted wire frame model (see Fig. 10.4).

When the 3-D face model is obtained, it can easily be manipulated. For example, Fig. 10.5 shows a rotated view of the 3-D model. Although the depth is not necessarily a good approximation, the rotated image still appears natural.

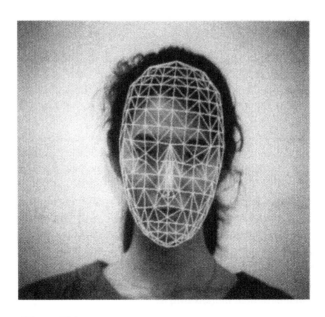

Figure 10.4   A frontal face image and adjusted wire frame model.

Figure 10.5   Images synthesized by rotating 3-D model.

10.4.2 Synthesis of Facial Expressions

The representation of facial expressions plays a very important role in 3-D model-based coding system. Two different methods for synthesizing facial expressions are now described. One technique uses a clip-and-paste method and the other a facial structure deformation method.

The *clip-and-paste method* is performed by analyzing 3-D head motion parameters, extracting specific expressive regions from an input image, transmitting them, and then placing them in the corresponding region of the synthesized image. The regions that must be clipped are identified on the 3-D wire frame model which moves in unison with the person's head (see Fig. 10.6). Figure 10.6 shows two frames out of the input image and synthesized image sequences. In this case, the 3-D movement of the head is analyzed by the least mean square method using six white points that are initially plotted on the subject's face. The depth information of these points is first roughly estimated using 3-D facial model and then successively updated in conjunction with the estimation of the head motion parameters.

The *facial structure deformation method* simulates facial expressions by deforming the 3-D facial model, which can be done in various ways. the facial action coding system [39] is adopted to describe facial actions. FACS itself was originally used in psychological studies and employs a somewhat more qualitative representation. FACS describes a set of minimal basic actions (action units or AUs) performable on a human face. There are 44 possible AUs that are designed to be closely connected with the anatomy of the human face. Any facial actions can be synthesized by parameterizing AUs into deformation rules on the 3-D model. The deformation rules control the location of control nodes of the wire frame facial model. Currently 34 AUs out of 44 are parameterized in our system.

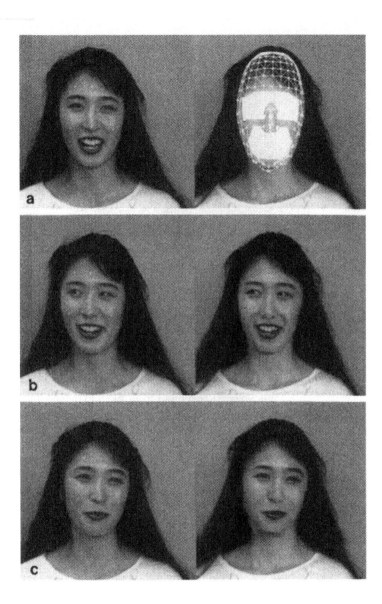

Figure 10.6 Clip-and-paste method: (a) eye and mouth area as clipped regions; (b) and (c) are two frames showing images by clip-and-paste method. The input images are on the left side and the synthesized images are on the right side. The hair and shoulders are modeled by 2-D plane and those of the first frame are translated two dimensionally.

Figure 10.7 shows examples of the synthesized images. Four faces in each pictures are synthesized by using the same combination of the AUs parameters.

As for bit transmission rates for these two methods when a frame rate is approximately 10 frames/sec, the clip-and-paste method is expected to require between 1 and 10 kbits/sec and the facial structure deformation methods less than 1 kbits/sec.

10.4.3 Analysis of Facial Motion

Facial motion analysis based on FACS by Choi and Takebe [30] is described. In their scheme, facial motion is separated into global motion (head motion) and local motion (facial expressions). Head motion is treated as a rigid body motion, and facial expressions are modeled as the AUs. Head motion and facial expressions are estimated hierarchically by using two consecutive frames; head motion is estimated, head motion is compensated, and facial expressions (AUs) are estimated. These steps are iterated to improve precision. Both head motion and facial expression are estimated by direct estimation [49, 50] that requires no a priori knowledge of point correspondences. For example, when analyzing facial expressions, the linear combination of AU parameters is estimated such that the frame difference between the new frame and the head-motion compensated frame is minimized. It is assumed that the initial location of the head is known.

The results of the analysis experiment which uses the CCITT standard image sequence *Clare*, is shown in Fig. 10.8. The image synthesized by using the analysis parameters is a fair quality reproduction, even though it only makes use of the texture information of the first frame.

10.5 Model-Based/Waveform Hybrid Coding

Model-based coding still has many problems, such as how to deal with unmodeled objects and how to improve the quality of the synthesized image, how to deal with the analysis errors. One way to practically solve these difficulties for the time being is to incorporate waveform coding into model-based coding. If some waveform coding methodologies are utilized together with model-based coding in a hybrid way, it can compensate model-based coding; that is, it can encode unmodeled objects and compensate the unsatisfactory reconstructed image caused by analysis and synthesis errors. Some hybrid type approaches have been proposed that employ this paradigm.

Pearson [51] and Nakaya, Aizawa, and Harashima [23, 24] discussed this hybrid coder from a general perspective. The general description of the model-based/waveform hybrid scheme is illustrated in Fig. 10.9a. On the encoder side,

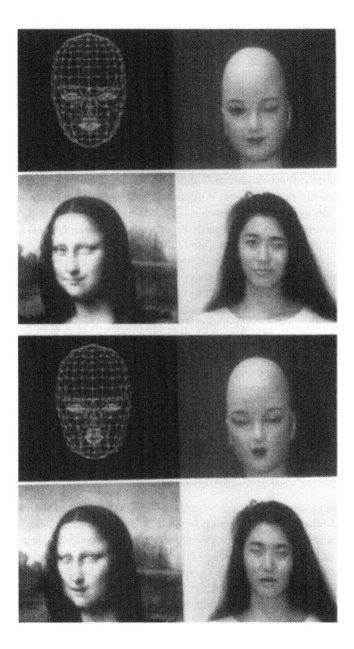

Figure 10.7 Images synthesized by the structure deformation method. The four different facial models are driven by the same AU parameters.

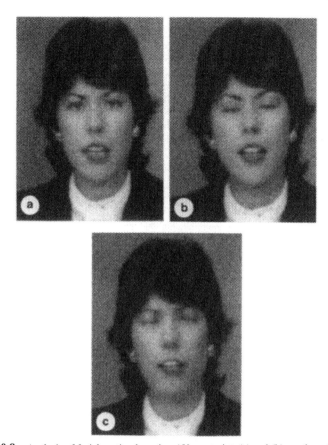

Figure 10.8   Analysis of facial motion by using AU constraint. (a) and (b) are the original image frames (1, 2) and (c) is a image synthesized by using the texture of frame 1 and analysis parameters corresponding to the original image frame 2.

image synthesis is performed using the analysis parameters extracted in the analysis part. The difference signals between synthesized images and input images are encoded by the waveform coder. The information extracted in the analysis part is also utilized to control the waveform coder. This control mechanism makes it possible, for example, to avoid the transmission of unnecessary information.

The configuration of the hybrid coding scheme for motion image sequence is shown in Fig. 10.9b. A local decoder is included in the transmitter, which allows the use of previous frames. This general description is similar to the conventional MC/DCT coder structure if the analysis part and synthesis part is replaced by motion detection and motion compensation, respectively.

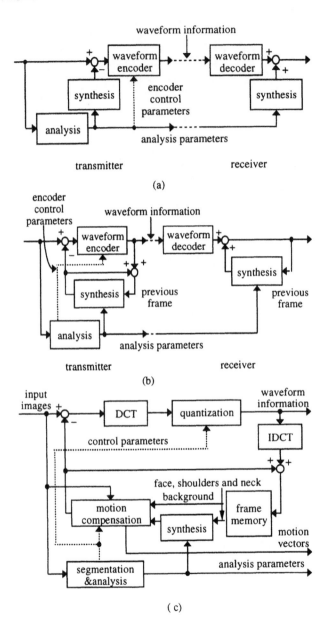

Figure 10.9 Model-based/waveform hybrid coding: (a) general description of model-based/waveform hybrid coding; (b) model-based/waveform hybrid coding of motion image sequences; (c) model-based/MC-DCT hybrid coder.

So far MC-DCT [23], vector quantization [26], and contour coding [27] have been combined with model-based coders. The clip-and-paste method of the previous section is also considered a hybrid coder. In addition, various 2-D and 3-D feature-based coding methods that utilize a general shape model and transmit the model failure information are also considered hybrid coders according to this general description of the hybrid coder. A bit-allocation scheme, which assigns more bits to facial area, combined with waveform coder has also been attempted [52].

The diagram of the simulated model-based/MC-DCT coder [23] is described in Fig. 10.9c. In the experiments, the hybrid coder was compared to the conventional MC-DCT (RM8) at both 64 kbits/sec and 16 kbits/sec. The comparisons are shown in Fig. 10.10. In the hybrid coder of this experiment, model-based coding is applied to the face region and MC-DCT compensates the errors of the blocks outside the face. As shown in Fig. 10.10, the hybrid coder has significant visible improvements. At a rate of 64 kbits/sec, the artifacts around the head in the background are avoided because the hybrid coder applies model-based coding to a human face, that is, the major moving object; thus the hybrid coder needs much less transmission bits for the face, and MC-DCT can allocate more bits for coding the regions outside of the face. At very low bit transmission rates such as 16 kbits/sec, where the MC-DCT does not work well at all, the improvement by hybrid coder is very clear.

10.6 Applications and Implementations

There already exist some implementations and applications of 3-D model-based coding. Because 3-D model-based coding uses the 3-D properties of the objects, it has much broader application than conventional waveform coding methods. The following lists some of the existing work and ideas.

1. *Virtual space teleconference* [53, 21, 22]. The idea of virtual space teleconferencing is a kind of advanced model-based coding, in which various 3-D CG data bases are used together with the 3-D model-based coding scheme. The concept is illustrated in Fig. 10.11. The communication partners are coded by 3-D model-based coding and displayed together with various 3-D CG data. It will provide an advanced communication interface with realistic sensations.

2. *Structured video* [54, 55, 56]. Because model-based coding describes images in highly structured way, it can potentially provide new creative and communicative environments. By using the 3-D properties of the scene, it is possible to create new scenes from preexisting material. (Fig. 10.7 is a simple example.) Video modeling will provide a way to handle and edit video materials and compose new scenes employing common computer graphics technology. It will also be able to provide video indexing for video database applications.

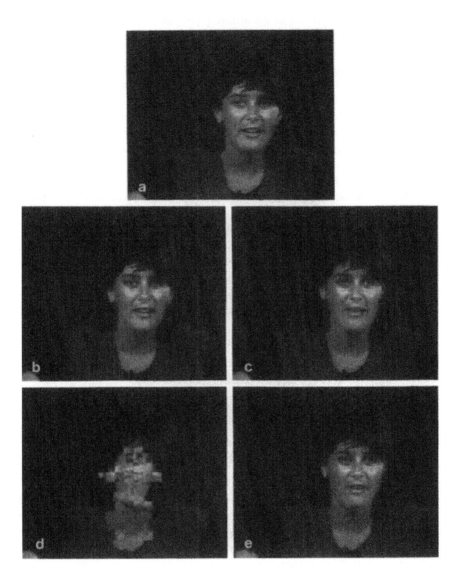

Figure 10.10   Comparison of model-based/MC-DCT coding and standard MC-DCT coding (RM8): (a) original image of 82nd frame; (b) and (c) are decoded images of 82nd frame at 64 kbits/sec by MC-DCT and model-based/MC-DCT hybrid methods, respectively; (d) and (e) are decoded images of 82nd frame at 16 kbits/sec by MC-DCT and model-based/MC-DCT hybrid methods, respectively.

Figure 10.11   Illustration of the concept of virtual space teleconference.

3. *Speech- or text-driven facial animation system for an advanced human–machine interface* [57, 58, 59, 18]. One of the applications outside image communications is enhancement of current prerecorded message systems and voice-activated databases. By using text data or voice to drive a 3-D facial model, a *talking head* substantially improves the interface between the human user and the message system. In the same context, the synthesized talking head can be utilized in the human–computer interface. In comparison to the image communication application that require high complexity to analyze video, this kind of applications requires much less computation. A prototype real-time working system has been developed.

4. *Real-time implementation of a model-based coding system* [28]. A prototype real-time system has been developed. The image motion analysis used in the system is rather simple. A threshold operation is applied to input image, and feature points that are considered to represent basic shapes of facial components (eyes, mouth) are detected. It is assumed that the initial modeling is done beforehand and the initial position of the face is known. The experimental system that analyzes and encodes motion images in real-time works at 9600 bits/sec [28].

5. *Synthesis of facial expressions for psychological studies* [33]. Since image synthesis systems used for model-based coding are able to manipulate a facial model and generate variety of different facial expressions, attempts have been made to apply the facial synthesis method to psychological studies so that judgemental experiments can use the facial image controlled by parameters as stimuli [33].

In addition, there have been attempts to convert 2-D images into stereo images by using a 3-D facial model so that one can watch stereo images at the display by just receiving information for 2-D images [60].

10.7 Remaining Problems for Three-Dimensional Model-Based Coding

Model-based coding is in its infancy, and many problems remain to be solved. The following are, in our opinion, some of the difficult issues in model-based coding.

10.7.1 Modeling Objects

Modeling objects is the most important issue in model-based coding. The complexity of analysis and synthesis depends on the model adopted.

Currently well-approximated 3-D models that are obtained from a priori knowledge are used by most researchers. A difficult problem is how to deal with unknown objects. It seems necessary that modeling should be done hierarchically in terms of approximation degree, so that unknown object can be roughly modeled and a priori 3-D models are further applied to known objects. Hybrid coding of model-based coding and waveform coding is considered a kind of bilevel hierarchical model.

Another problem is caused by the complexity of the well-approximated model. If a model has finer details, it is more realistic for synthesis but more complex for analysis. It seems unwise to use a complex model for every object. Some objects such as faces should be modeled by using a well-approximated complex model, while the other objects had better be coded based on roughly approximated models. Model hierarchy will be also helpful for this problem.

10.7.2 Evaluation Problem

For conventional waveform coding techniques a common evaluation measure is the mean squared error. Though it has often been claimed that it is not always a good criterion, it has been a force that guides the progress of waveform coding in the right way. What is a good quality criterion for model-based coding? There has been no discussion on this point. Though Pearson [6] deals with differences between an original image and an image synthesized by texture-mapped models, squared error is still used for the quality measure. One of the attractive points of model-based coding is that it is independent of the squared error measure. There is no guidance at the present time on how to quantitatively evaluate the quality of model-based coding systems.

10.7.3 Promising Applications

Currently, research is being done on model-based coding without seriously considering what its promising applications might be. Invoking this question is necessary for deciding the future direction of model-based coding. At the beginning, model-based coding was intended for image communication applications. However, when we think of its high asymmetry (analysis is far more difficult than synthesis) and the growing bandwidth in future communication systems, image compression for real-time telecommunication will not be a good application for the model-based coding.

Instead, one way-communication applications may be important areas in which database applications, broadcasting-type communication applications, and machine-interface applications are included. The major advantage of model-based coding is not in compression, but in describing scenes in a structural way in codes that can be easily operated and edited. Thus, model-based coding can be applied to creating new image sequences by modeling and analyzing stored old image sequences. Such manipulations of image content may be the most important application of model-based coding.

Acknowledgements

The author would like to thank David Wuertele of the University of Tokyo for his helpful suggestions.

References

[1] S. M. Platt and N. I. Badler, "Animating facial expressions." *Comput. Graphics*, vol. 13, pp. 245–242, Aug. 1981.

[2] S. Carlsson, "Sketch based coding of gray level images." *Signal Processing*, vol. 15, pp. 57–83, 1988.

[3] M. Kunt, A. Ikonomopoulos, and M. Kocher, "Second generation image coding techniques." *Proc. IEEE*, vol. 73, pp. 795–812, 1985.

[4] M. Gilge, T. Englehardt, and R. Mehlan, "Coding of arbitrarily shaped segments based on a generalized orthogonal transform." *Signal Processing: Image Commun.*, vol. 1, pp. 153–180, 1989.

[5] A. Matsunaga and Y. Yasuda, "Video transmission over low bit rate channel." *Picture Coding Symp.*, Davis, CA, 1983.

[6] D. E. Pearson, "Texture mapping in model-based image coding." *Signal Processing: Image Commun.*, vol. 2, pp. 377–396, 1990.

[7] M. Hötter, "Object-oriented analysis-synthesis coding based on moving two-dimensional objects." *Signal Processing: Image Commun.*, vol. 2, pp. 409–428, Dec. 1990.

[8] M. Hötter and J. Ostermann, " Analysis synthesis coding based on planar rigid moving objects." *Int. Workshop on 64 kbps Coding of Moving Video*, Hannover, Germany, 1988.

[9] H. G. Musmann, M. Hötter, and J. Ostermann, "Object-oriented analysis-synthesis coding of moving images." *Signal Processing: Image Commun.*, vol. 1, pp. 117–138, Oct. 1989.

[10] N. Diehl, "Object-oriented motion estimation and segmentation in image sequences." *Signal Processing: Image Commun.*, vol. 3, pp. 23–56, 1991.

[11] J. Ostermann, "Modelling of 3-D moving objects for an analysis–synthesis coder." in *Proc. SPIE, Sensing and Reconstruction of Three-Dimensional Objects and Scenes*, vol. 1260, pp. 240–249, 1990.

[12] H. Morikawa and H. Harashima, "Three-D structure extraction coding of image sequences." *J. Visual Commun. Image Representation*, vol. 2, pp. 332–344, 1991.

[13] R. Koch, "Dynamic 3-D scene analysis through synthesis feedback control." *IEEE Trans. Pattern Anal. Machine Intell.*, vol. PAMI-15, pp. 556-568, 1993.

[14] K. Aizawa, H. Harashima, and T. Saito, "Model-based analysis synthesis image coding for a person's face." *Signal Processing: Image Commun.*, vol. 1, pp. 139–152, 1989.

[15] K. Aizawa, H. Harashima, and T. Saito, "Model-based synthetic image coding system." *Picture Coding Symp.*, Stockholm, 1987.

[16] K. Aizawa, T. Saito, and H. Harashima, "Construction of a three-dimensional personal face model for knowledge-based image data compression." In *Nat. Conf. Record IEICEJ*, Musashino, Japan, Sept. 1986, pp. 1–221 [in Japanese].

[17] W. J. Welsh, "Model-based coding of videophone images." *Electron. and Commun. Eng. J.*, vol. 3, pp. 29–36, Feb. 1991.

[18] W. J. Welsh, A. D. Simons, A. D. Hutchinson, and R. A. Searby, "Synthetic face generation for enhancing a user interface." In *Proc. Image Commun.*, France, 1990, pp. 177–182.

[19] W. J. Welsh, "Model-based coding of moving images at very low bit rate." *Picture Coding Symp.*, Stockholm, 1987.

[20] R. Forchheimer and T. Kronander, "Image coding-from waveforms to animation." *IEEE Trans. Acoust. Speech, Signal Processing*, vol. ASSP-37, pp. 2008–2023, Dec. 1989.

[21] F. Kishino and K. Yamashita, "Communication with realistic sensation applied to a teleconferencing system." *IEICE Tech. Report*, IE89-35, 1989 [in Japanese].

[22] G. Xu, H. Agawa and Y. Nagashima, F. Kishino, and Y.Kobayashi, "Three-dimensional face modeling for virtual space teleconferencing systems." *IEICE*, vol. E73, 1990.

[23] Y. Nakaya and H. Harashima, "Model-based/waveform hybrid coding for low-rate transmission of facial images." *IEICE Trans. Commun.*, vol. E75-B, pp. 377–384, 1992.

[24] Y. Nakaya, K. Aizawa, and H. Harashima, "Texture updating methods in model-based coding of facial images." *Picture Coding Symp.*, Boston, 1990.

[25] T. Fukuhara, K. Asai, and T. Murakami, "Model-based image coding using stereoscopic images and hierarchical structuring of new 3-D wire-frame model." *Picture Coding Symp.*, Tokyo, 1991.

[26] T. Fukuhara, K. Asai, and T. Murakami, "Hierarchical division of 3-D wire-frame model and vector quantization in a model-based coding of facial images." *Picture Coding Symp.*, Boston, 1990.

[27] T. Minami, I. So, T. Mizuno, and O. Nakamura, "Knowledge-based coding of facial images." *Picture Coding Symp.*, Boston, 1990.

[28] M. Kaneko, A. Koike, and Y. Hatori, "Real-time analysis and synthesis of moving facial images applied to model-based image coding." *Picture Coding Symp.*, Tokyo, 1991.

[29] C. S. Choi, T. Okazaki, H. Harahsima, and T. Takebe, "A system of analyzing and synthesizing facial images." *IEEE Int. Symp. Circuit Syst.*, Singapore, 1991, pp. 2665–2668.

[30] C. S. Choi and T. Takebe, "Analysis and synthesis of facial expressions in knowledge-based coding system for facial image sequence." In *ICASSP'91*, Toronto, 1991.

[31] H. Li, P. Rovivainen, and R. Forchheimer, "Three-D motion estimation in model-based facial image coding." *IEEE Trans. Patt. Anal. Machine Intell.*, vol. PAMI-15, pp. 545–555, 1993.

[32] F. I. Parke, "Parameterized models for facial animation." *IEEE Comput. Graphics Applications*, vol. 12, pp. 61–68, Nov. 1982.

[33] H. Yamada, H. Chiba, K. Tsuda, and K. Maiya, "New approach to the research on the facial expressions: Model-based facial action synthesizing computer (MASC) system." *Int. J. Psychology*, vol. 27, pp. 47, 1992.

[34] T. Akimoto and Y. Suenaga, "Three-D facial model creation using generic model and front and side views of face." *IEICE Trans. Inform. & Syst.*, vol. 75-D, pp. 191–197, 1992.

[35] L. Tang, M. Pouyat, K. Aizawa, and T. S. Huang, "Accuracy of modelling human face using a generic model." *Picture Coding Symp.*, Lausanne, Switzerland, 1993.

[36] K. Waters and D. Terzopoulos, "Modeling and animating faces using scanned data." *J. Visualization Comput. Animation*, vol. 2, pp. 129–131, 1991.

[37] T. Kimoto and Y. Yasuda, "Hierarchical representation of the motion of a walker and motion reconstruction for model-based image coding." *Opt. Eng.*, vol. 20, pp. 888–903, 1991.

[38] R. Forchheimer, O. Fahlander, and T. Kronander, "Low bit-rate coding through animation." *Picture Coding Symp.*, Davis, CA, March 1983, pp. 113–114.

[39] P. Ekman and W. V. Friesen, "Facial action coding system." *Consulting Psychologists Press*, 1977.

[40] M. Kaneko, A. Koike, and Y. Hatori, "Coding of facial image sequence based on a 3-D model of the head and motion detection." *J. Visual Commun. Image Representation*, vol. 2, pp. 39–54, March 1991.

[41] D. Terzopoulous and K. Waters, "Analysis and synthesis of facial image sequence using physical and anatomical models." *IEEE Trans. Patt. Anal. Machine Intell.*, vol. PAMI-15, pp. 569–579, 1993.

[42] Buck, "Segmentation of moving head-and-shoulder shapes." *Picture Coding Symp.*, Boston, March 1990.

[43] A. L. Yullie, P. W. Hallinan, and D. S. Cohen, "Feature extraction from faces using deformable templates." *Int. J. Comput. Vision*, vol. 8, pp. 99–111, 1992.

[44] A. Koike, M. Kaneko, and Y. Hatori, "Model-based image coding with 3-D motion estimation and shape change detection." *Picture Coding Symp.*, Boston, 1990.

[45] T. S. Huang, S. Reddy, and K. Aizawa, "Human facial motion modeling, analysis and synthesis for video compression." In *Proc. Visual Commun. Image Processing SPIE*, Boston, 1991, pp. 234–241.

[46] C. S. Choi, K. Aizawa, H. Harashima, and T. Takebe, "Analysis and synthesis of facial expressions in model-based image coding." *Picture Coding Symp.*, Boston, 1990.

[47] K. Mase, "An application of optical flow—Extraction of facial expressions." In *Proc. MVA*, Tokyo, 1990.

[48] J. Fischl, B. Miller, and J. Robinson, "Parameter tracking in a muscle-based analysis/synthesis coding system." *Picture Coding Symp.*, Lausanne, Switzerland, 1993.

[49] T. S. Huang and R. Y. Tsai, "Image sequence analysis: Motion estimation." In *Image Sequence Analysis*, ed. T. S. Huang. Springer-Verlag, 1981.

[50] M. Yamamoto, P. Boulanger, J.-A. Beraldin, M. Rioux, and J. Domey, "Direct estimation of deformable motion parameters from range image sequence." In *Third Int. Conf. Comput. Vision*, Osaka, 1991, pp. 460–464.

[51] D. Pearson, "Model-based image coding." In *Proc. GLOBECOM'89*, Nov. 1989, Dallas, 16.1, pp. 554–558.

[52] H. Ueno, K. Dachiku, K. Ozeki, and F. Sugiyama, "A study on facial region detection in the standard video coding method." *Int. Workshop on 64 kbit/sec Coding of Moving Video*, Sept. 1990, 5-2.

[53] H. Harashima and F. Kishino, "Intelligent image coding and communications with realistic sensation—Recent trends." *IEICE*, vol. E74, pp. 1582–1592, June 1991.

[54] H. Holtzman, "3-D video modeling." In *CHI Conf. on Human Factors in Computing Systems*, CA, May 1992.

[55] H. D. Lin and D. G. Messerschmitt, "Video composition methods and their semantics." In *ICASSP'91*, 1991, pp. 2833–2836.

[56] H. Morikawa, E. Kondo, and H. Harashima, "Structural description of moving pictures for coding." *Picture Coding Symp.*, Tokyo, 1991.

[57] M. Kaneko, A. Koike, and Y. Hatori, "Synthesis of moving facial images with mouth shape controlled by text information." *IEIECJ Tech. Report*, IE89-4, 1989 [in Japanese].

[58] S. Morishima, K. Aizawa, and H. Harashima, "An intelligent facial image coding driven by speech and phoneme." In *ICASSP'89*, Glasgow, 1989.

[59] S. Morishima, "A human-machine interface using media conversion and model-based coding schemes." In *CG International*. Springer-Verlarg, 1992.

[60] M. Tanimoto and S. Nakashima, "Basic experiment of 2d–3d conversion for a new 3d visual communication." *Picture Coding Symp.*, Boston, 1990.

[61] D. E. Pearson and J. A. Robinson, "Visual communication at very low bit rates." *Proc. IEEE*, vol. 73, pp. 795–812, 1985.

Chapter 11

Image and Video Coding Standards

R. Aravind, G. L. Cash, H.-M. Hang, B. G. Haskell, and A. Puri
Visual Communications Research Department
AT&T Bell Laboratories
Holmdel, New Jersey

Most image or video applications involving transmission or storage require some form of data compression to reduce the otherwise inordinate demand on bandwidth and storage. Compatibility among different applications and manufacturers is very desirable and often essential. This chapter describes several standard compression algorithms developed in recent years.

11.1 Introduction

The International Organization for Standardization (ISO) Joint Photographic Experts Group (JPEG) has developed an algorithm for coding single-frame color images. It is based on the discrete cosine transform (DCT), and it also has extensions for progressive coding. Starting from an original red, green, blue (RGB) picture of 24 bits per picture element (pel or pixel), the JPEG algorithms give good

An altered version of this material has been published in *AT&T Technical Journal*, Jan–Feb. 1993. Copyright © 1993, AT&T; all rights reserved; reprinted with permission. H.-M. Hang is currently with the Dept. of Electronics Eng., National Chiao Tung University, Hsinchu, Taiwan, R.O.C.

image quality at compression factors of 10 to 20; i.e., bit rates between 1 and 2 bits per pixel.

The International Telegraph and Telephone Consultative Committee (CCITT) Study Group 15 (SG15) and its experts group on video telephony has finalized a set of coding standards, known informally as the P×64 standard, for sending videophone or videoconference pictures on integrated services digital network (ISDN) facilities. The standard is applicable over a bandwidth range from 56 kilobits per second (kb/sec) to 2 megabits per second (Mb/sec). It relies not only on the DCT, but also on motion-compensated prediction to compress data generated by the moving imagery.

The ISO Moving Picture Experts Group (MPEG) has developed both audio and video compression algorithms that can compress entertainment or educational video for storage and transmission on various digital media, including compact disc, remote video databases, movies on demand [1], cable television (CATV), and fiber to the home. Requirements are for implementation of normal play, fast forward/reverse, random access, normal reverse, and simple very-large-scale integration (VLSI). The MPEG algorithm utilizes all the P×64 methodology, as well as some new techniques, most notably conditional motion-compensated interpolation.

11.2 JPEG Still–Color Image Coding

The need for an international standard for continuous-tone still image compression resulted, in 1986, in the formation of JPEG. This group was chartered by ISO and the CCITT to develop a general-purpose standard suitable for as many applications as possible. After thorough evaluation and subjective testing of a number of proposed image-compression algorithms, the group agreed, in 1988, on a DCT-based technique. From 1988 to 1990, the JPEG committee refined several methods incorporating the DCT for lossy compression. A lossless method was also defined. The committee's work has been published in two parts: *Part 1. Requirements and Guidelines* [2] describes the JPEG compression and decompression method; *Part 2. Compliance Testing* [3] describes tests to verify whether a coder–decoder (codec) has implemented the JPEG algorithms correctly.

To appreciate the need for image compression, consider the storage and transmission requirements of an uncompressed image. A typical digital color image has 512×480 pixels. At 3 bytes per pixel (one each for the red, green, and blue components), such an image requires 737,280 bytes of storage space. To transmit the uncompressed image over a 64 kb/sec channel takes about 1.5 min. The JPEG algorithms offer "excellent" quality for most images compressed to about 1.0 bit/pixel. This 24:1 compression ratio reduces the required storage of the 512×480 color image to 30,720 bytes, and its transmission time to about 3.8 sec. Applications for image compression may be found in desktop publishing, education, real estate, and security, to name a few.

In the next section, we give an overview of the JPEG algorithms. In subsequent sections, we present some operating parameters and definitions, and describe each of the JPEG operating modes in more detail.

11.2.1 Overview of the JPEG Algorithms

The JPEG committee could not satisfy the requirements of every still image compression application with one algorithm. As a result, the committee proposed four different modes of operation:

1. *Sequential DCT-based.* Figure 11.1 presents a simplified diagram of a sequential DCT codec. In this mode, 8×8 blocks of the input image are formatted for compression by scanning the image left to right and top to bottom. A block consists of 64 samples of one component that make up the image. Each block of samples is transformed to a block of coefficients by the forward discrete cosine transform (FDCT). The coefficients are then quantized and entropy coded.

2. *Progressive DCT-based.* This mode offers a means of producing a quick "rough" decoded image when the medium separating the coder and decoder has a low bandwidth. The method is similar to the sequential DCT-based algorithm, but the quantized coefficients are partially encoded in multiple scans.

3. *Lossless.* In this mode, the decoder renders an exact reproduction of the digital input image. The differences between input samples and predicted values, where the predicted values are combinations of one to three neighboring samples, are entropy coded.

4. *Hierarchical.* This mode is used to code an input image as a sequence of increasingly higher resolution frames. The first frame is a reduced resolution version of the original. Subsequent frames are coded higher resolution differential frames.

The color space conversion process in Fig. 11.1 is not a part of the standard. In fact, JPEG is color space independent. As a first step in the compression process, many image-compression schemes take advantage of the human visual system's low sensitivity to high-frequency chrominance information [4] by reducing the chrominance resolution. Many images (usually RGB) are typically converted to a luminance–chrominance representation before this processing takes place.

Either Huffman or arithmetic techniques can be used for entropy coding in any of the JPEG modes of operation (except in the *baseline* system, where Huffman coding is mandatory). A Huffman coder compresses a series of input symbols by assigning short code words to frequently occurring symbols and long code

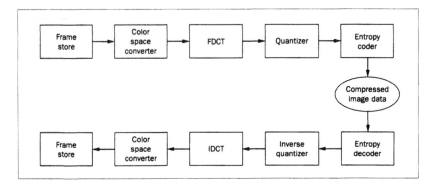

Figure 11.1   Sequential DCT codec.

words to improbable symbols [5, 6]. The output of an arithmetic coder is a single real number. After initialization to a range of 0 to 1, the probability of each input symbol is used to restrict the range of the output number further. Unlike a Huffman coder, an arithmetic coder does not require an integral number of bits to represent an input symbol. As a result, arithmetic coders are usually more efficient than Huffman coders [7, 8]. For the JPEG test images, Huffman coding (using fixed tables) resulted in compressed data requiring, on average, 13.2% more storage than arithmetic coding.

11.2.2 JPEG Operating Parameters and Definitions

A number of parameters related to the source image and the coding process may be customized to meet the user's needs. In this section, we discuss some of the important variable parameters and their allowable ranges. Also, as an aid to the algorithm descriptions in the following sections, we define some JPEG terms and present the hierarchical structure of the compressed data.

Parameters   An image to be coded using any JPEG mode may have from 1 to 65,535 lines and from 1 to 65,535 pixels per line. Each pixel may have from 1 to 255 components (only 1 to 4 components are allowed for progressive mode). The operating mode determines the allowable precision of the component. For the DCT modes, either 8 or 12 bits of precision are supported (only 8-bit precision is allowed for *baseline*). Lossless mode precision may range from 2 to 16 bits. If a DCT operating mode has been selected, the quantizer precision must be defined. For 8-bit component precision, the quantizer precision is fixed at 8 bits. Twelve-bit components require either 8- or 16-bit quantizer precision.

Data Interleaving To reduce the processing delay or buffer requirements, up to four components can be interleaved in a single scan (for progressive mode, only the dc scan may have interleaved components). A data structure called the *minimum-coded unit* (MCU) has been defined to support this interleaving. An MCU consists of one or more data units, where a data unit is a component sample for the lossless mode, and an 8×8 block of component samples for the DCT modes. If a scan contains only one component, then its MCU is equal to one data unit. For multiple component scans, the MCU for the scan consists of interleaved data units. The maximum number of data units per MCU is 10. As an interleaving example, consider an International Radio Consultative Committee (CCIR) 601 digital image in which the chrominance components are subsampled 2:1 horizontally. For a DCT coder, a CCIR-601 MCU could consist of two Y blocks, followed by a C_R block and a C_B block, where Y is the luminance of the image and C_R and C_B are proportional to the two color differences $(R - Y)$ and $(B - Y)$, respectively.

Marker Codes JPEG has defined a number of 2-byte marker codes to delineate the various sections of a compressed data stream. All marker codes begin with a byte-aligned hexadecimal "FF" byte, making it easy to scan and extract parts of the compressed data without actually decoding it. Because it is possible to create a byte-aligned hexadecimal FF byte within the entropy-coded data, the coder must detect this situation and follow the "FF" byte with a zero byte. When the decoder encounters the hexadecimal "FF00" combination, the zero byte must be removed.

Compressed-Image Data Structure At the top level of the compressed data hierarchy is the *image* (see Fig. 11.2). A nonhierarchical mode image consists of

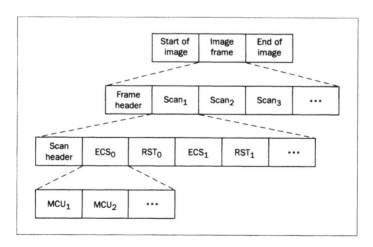

Figure 11.2 Structure of compressed-image data.

a *frame* surrounded by "start of image" (SOI) and "end of image" (EOI) marker codes. There will be multiple *frames* in a hierarchical mode image. Within a frame, a start of frame (SOF) marker identifies the coding mode to be used. The SOF marker is followed by a number of parameters [2], and then by one or more *scans*. Each scan begins with a header identifying the components to be contained within the scan, and more parameters. The scan header is followed by an entropy-coded segment (ECS). An option exists to break the ECS into chunks of MCUs called *restart intervals*, $(RST_0, RST_1,$ etc.). The restart interval structure is useful for identifying select portions of a scan and for recovery from limited corruption of the entropy-coded data. Quantization and entropy-coding tables may either be included with the compressed-image data or communicated separately.

11.2.3 Sequential DCT Mode

The sequential DCT mode offers excellent compression ratios, while maintaining image quality. A subset of the sequential DCT capabilities has been identified by JPEG for a *baseline system*. All DCT-based JPEG implementations are required to include baseline capability. This requirement should help to ensure interoperability between codecs from different vendors. Restrictions on the baseline system related to sample and quantizer precision were pointed out in the "Parameters" paragraph. One further restriction should be noted: although a full sequential DCT coder may employ either Huffman or arithmetic entropy coding, a baseline coder can use only Huffman coding. In addition, only two ac and two dc tables may be used per scan (up to four sets of tables are allowed for full sequential mode).

The following paragraphs describe the processing steps for a baseline coder. A decoder is formed by reversing the coder steps.

DCT Mode and Quantization   All JPEG DCT-based coders begin the coding process by partitioning the input image into nonoverlapping 8×8 blocks of component samples. After level-shifting the 8-bit samples so that they range from -128 to +127, the blocks are transformed to the frequency domain using the FDCT [9, 10]. The equations for the forward and inverse discrete cosine transforms are given by

$$\text{FDCT:} \quad F(u, v) = \frac{1}{4} C(u) C(v) \sum_{x=0}^{7} \sum_{y=0}^{7} f(x, y)$$
$$\cos(\frac{\pi u(2x + 1)}{16}) \cos(\frac{\pi v(2y + 1)}{16}) \qquad (11.1)$$

IDCT: $f(x, y) = \dfrac{1}{4} \displaystyle\sum_{u=0}^{7} \sum_{v=0}^{7} C(u)C(v)F(u, v)$

$$\cos(\frac{\pi u(2x + 1)}{16})\cos(\frac{\pi v(2y + 1)}{16}) \qquad (11.2)$$

where

$$C(u), C(v) = \begin{cases} \frac{1}{\sqrt{2}} \text{for } u, v = 0 \\ 1 \quad \text{otherwise.} \end{cases}$$

The DCT concentrates most of the energy of the component samples' block into a few coefficients, usually in the top left corner of the DCT block. The coefficient in the immediate top left corner is called the *dc coefficient* because it is proportional to the average intensity of the block of spatial domain samples. The ac coefficients corresponding to increasingly higher frequencies of the sample block progress away from the dc coefficient.

The next step in the process, quantization, is the key to most of the JPEG compression. A 64-element quantization matrix, where each element corresponds to a coefficient in the DCT block, is used to reduce the amplitude of the coefficients and to increase the number of zero-value coefficients. The quantization and dequantization is performed according to (11.3) and (11.4), respectively.

$$Fq(u, v) = \text{round}\left[\frac{F(u, v)}{Q(u, v)}\right], \qquad (11.3)$$

$$R(u, v) = Fq(u, v)Q(u, v). \qquad (11.4)$$

A carefully designed quantization matrix will produce high compression ratios while introducing negligible "visible" distortion [11]. Up to four quantization matrices are allowed by JPEG. The standard does not mandate quantization matrices, but includes a set that gives good results for CCIR-601 type images. Many JPEG implementations control the compression ratio (and output image quality) by using a *q-factor*, which is usually just a scale factor applied to the quantization matrices.

dc Coefficient Entropy Coding   Greater compression efficiency can be obtained if a simple predictive method is used to entropy code the dc coefficient separately from the ac coefficients. Recall that the dc coefficient corresponds to the average intensity of the component block. Adjacent blocks will probably have similar average intensities. It is, therefore, advantageous to code the *differences* between the dc coefficients of adjacent blocks rather than their values. Each differential dc value is coded using a variable-length code (VLC) and a variable-length integer (VLI). The VLC corresponds to the size, in bits, of the VLI, while the VLI gives the amplitude of the differential dc value.

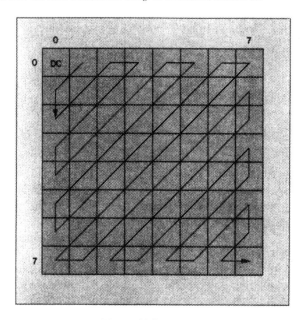

Figure 11.3   Zigzag scan.

Zigzag Scan and ac Coefficient Entropy Coding   After they have been quantized, the coefficient blocks usually contain many zero-value ac coefficients. If the coefficients are reordered, using the zigzag scan illustrated in Fig. 11.3, there will be a tendency to have long runs of zeroes. Only the nonzero ac coefficients are entropy coded. As in the dc coefficient coding, a VLC–VLI pair results from coding an ac coefficient. However, the ac VLC corresponds to two pieces of information: the number of zeroes (run) since the last nonzero coefficient, and the size of the VLI following the VLC.

11.2.4 Progressive DCT Mode

A progressive DCT mode has been defined by JPEG to satisfy the need for a fast decoded picture when a low-bandwidth medium separates a coder and decoder. By partially encoding the quantized DCT coefficients in multiple scans, the decoded image quality builds progressively from a coarse level to the quality attainable with the quantization matrices. Either spectral selection, successive approximation, or a combination of the two is used to code the quantized coefficients.

Spectral Selection   In this method, the quantized DCT coefficients of a block are first partitioned into nonoverlapping bands along the zigzag block scan. The bands are then coded in separate component scans. Before an ac coefficient band of a component may be coded, its dc coefficient must be coded. The dc coefficients from as many as four components may be interleaved in a single scan. Interleaving is not permitted for ac bands because of the introduction of an efficient means for coding contiguous blocks of zero-valued coefficients. From 1 to 32,767 blocks can be coded with a single VCL–VLI combination called an *end-of-band code*.

Successive Approximation   With this method, the precision of the coefficients is successively increased during multiple scans. Following a scan for a specified number of most significant bits of the quantized coefficients, subsequent scans increase the precision in increments of one bit until the least significant bits have been coded.

11.2.5 Lossless Mode

The lossless mode was defined for applications in which output pixels from a decoder must be identical to the input pixels to the coder. The compression ratios achievable with the lossless mode, typically around 2:1, are much smaller than those afforded by the lossy modes. This method is similar to the one used to code the dc coefficients in the DCT-based modes, but the predictor is selectable from one of seven choices, as shown in Table 11.1. Samples a, b, and c in the table correspond to neighbors of the sample x to be predicted. Figure 11.4 illustrates the prediction neighborhood. Entries 1 to 3 in Table 11.1 are used for one-dimensional predictive coding, and 4 through 7 form two-dimensional predictors. Entry 0 identifies differential coding for the hierarchical mode. As in the dc coefficient entropy coding described earlier, differences between the actual and predicted values are entropy coded.

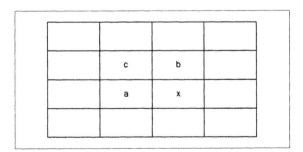

Figure 11.4   Prediction neighborhood.

Table 11.1   Lossless mode predictors.

Selection	Prediction value
0	No prediction
1	a
2	b
3	c
4	$a + b - c$
5	$a + ((b - c)/2)$
6	$b + ((a - c)/2)$
7	$(a + b)/2$

11.2.6 Hierarchical Mode

In the hierarchical mode, an image is coded as a succession of increasingly higher resolution frames. This "pyramidal" approach offers an alternative to the previously described methods for achieving progression. It also allows decoders with different resolution capabilities to use the same compressed data stream.

The first coded frame is created by reducing the resolution of the input image by a power of 2 in one or both dimensions, and then processing the lower resolution image using one of the lossy or lossless techniques of the other operating modes. Subsequent frames are formed by up-sampling the decoded image by a factor of 2 in the dimension(s) having reduced resolution, subtracting the up-sampled image from the input image at the same resolution, and coding the difference. "Missing" pixels in the unsampled image are filled in using linear (or bilinear) interpolation. This process continues until the decoded image has the same resolution as the full-resolution input image. After that, one or more full-resolution difference images may be coded. A hierarchical decoder may abort the decoding process after it has decoded a frame that provides the desired resolution.

Any coding methods described in the other three modes of operation may be used to code the hierarchical mode frames, with the following restrictions:

1. If a lossy method is chosen, all but the last frame must be coded using that method. A lossless method may be used optionally to code the last frame.
2. If a lossless method is chosen, all frames must be coded with that method.
3. The same entropy-coding technique (Huffman or arithmetic) must be used for all frames.

The hierarchical coding–decoding process is not symmetrical. Indeed, a hierarchical coder must also include the greater part of a decoder. However, a hierarchical decoder is only more complex than a nonhierarchical decoder in that it must provide a way to upsample and add. This increased complexity may be justified, given the flexibility afforded in matching the decoder to the application. This type of codec is well suited for "one-to-many" applications, as in a number of decoders (possibly having different resolution capabilities) accessing a database of images precoded by a hierarchical coder.

11.3 Videoconferencing Standards H.261

11.3.1 From the JPEG Standard to H.261 (P×64)

From an algorithmic point of view, the extension from JPEG, intraframe DCT coding, to H.261, motion-compensated DCT video coding, is a rather natural one. Historically, H.261 was developed long before JPEG. In December 1984, CCITT Study Group XV (Transmission Systems and Equipment) established a Specialists' Group on Coding for Visual Telephony. The development of this video transmission standard for low-bit-rate ISDN services has gone through several stages. At the beginning, the goal was to design a coding scheme for a transmission rate of $m \times 384$ bit/sec channels, where m was between 1 and 5. Later, $n \times 64$ kb/sec transmission rates (n from 1 to 5) were considered. However, by late 1989, the final CCITT Recommendation H.261 [12] was made for a $p \times 64$ kb/sec video codec, where p is between 1 and 30.

In fact, the H series of audiovisual teleservices is a group of standards (or recommendations) consisting of H.221, frame structure; H.230, frame synchronous control; H.242, communication between audiovisual terminals; H.320, systems and terminal equipment; and H.261, video codec. Audio codecs at several bit rates have also been specified by other CCITT Recommendations, such as G.725. In this chapter, we concentrate on the H.261 video codec specification.

Both JPEG baseline and H.261 codecs use DCT and VLC techniques. The major difference between the JPEG compression scheme and H.261 is that JPEG codes each frame individually, whereas H.261 performs interframe coding. In H.261, block-based motion compensation is performed to compute interframe differences, which are then DCT coded. Here, the picture data in the previous frame can be used to predict the image blocks in the current frame, as shown in Fig. 11.5. As a result, only differences, typically of small magnitude, between the displaced previous block and the current block have to be transmitted.

There are several interesting characteristics or design considerations in H.261.

1. It defines essentially only the *decoder*. However, the *encoder*, which is not completely and explicitly specified by the standard, is expected to be compatible with the decoder.

2. Because H.261 is designed for real-time communications, it uses only the closest previous frame as prediction to reduce the encoding delay.

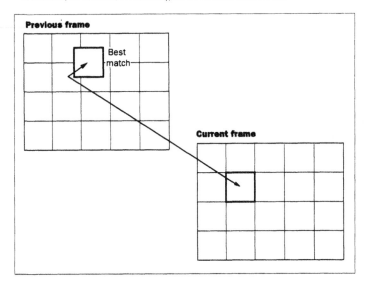

Figure 11.5   Block motion compensation.

3. It tries to balance the hardware complexities of the encoder and the decoder, since they are both necessary for a real-time videophone application. Other coding schemes, such as vector quantization (VQ), may have a rather simple decoder, but a very complex encoder.

4. H.261 is a compromise between coding performance, real-time requirement, implementation complexity, and system robustness. Motion-compensated DCT coding is a mature algorithm, and after years of study, quite general and robust in that it can handle various types of pictures.

5. The final coding structures and parameters are tuned more toward low-bit-rate applications. This choice is logical, because selection of the coding structure and coding parameters is more critical to codec performance at very low bit rates. At higher bit rates, the less-than-optimal parameter values do not affect codec performance very much.

11.3.2 Decoder Structures and Components

H.261 specifies a set of protocols that every compressed bit stream has to follow, and a set of operations that every standard compatible decoder must be able to perform. The actual hardware codec implementation and the encoder structure can vary drastically from one design to another. In a few places, user-defined bit streams may be inserted into the standard bit stream. We will first explain briefly the data structure in an H.261 bit stream and then the functional elements in an H.261 decoder.

The compressed H.261 bit stream [12] contains several layers (see Fig. 11.6). They are *picture* layer, *group of blocks (GOB)* layer, *macroblock (MB)* layer, and *block* layer. The higher layer consists of its own header and a number of instances of the lower layer data.

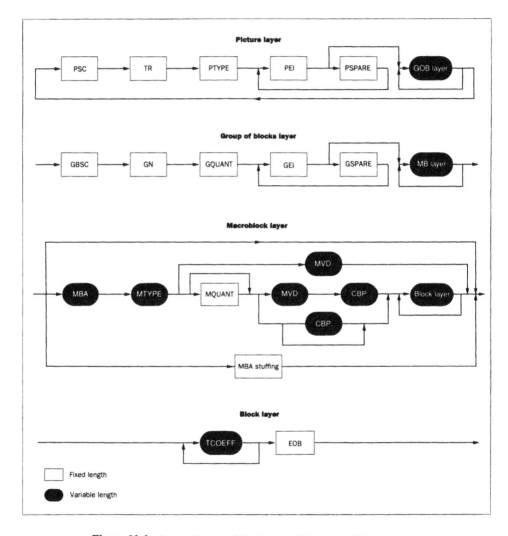

Figure 11.6   Syntax diagram of the video multiplex coder [12].

Only two picture formats—common intermediate format (CIF) and quarter-CIF (QCIF)—are allowed. CIF pictures are made of three components: luminance Y and color differences C_B and C_R, as defined in CCIR Recommendation 601. The CIF picture size for Y is 352 pixels per line by 288 lines per frame. The two-color difference signals are subsampled to 176 pixels per line and 144 lines per frame. Figure 11.7 shows the sampling pattern of Y, C_B, and C_R. The picture aspect ratio is 4(horizontal):3(vertical), and the picture rate is 29.97 noninterlaced frames per second. All standard codecs must be able to operate with QCIF; CIF is optional.

A picture frame is partitioned into 8 lines by 8 pixels image blocks. The so-called MB is made of 4 Y blocks, one C_B block, and one C_R block at the same location, as shown in Fig. 11.8a. Figure 11.8b contains 33 MBs grouped into a GOB. Therefore, one CIF frame contains 12 GOBs and one QCIF frame contains 3 GOBs, as shown in Fig. 11.8c.

In a compressed bit stream, we start with the picture layer. Its header contains:

- Picture start code (PSC), a 20-bit pattern;
- Temporal reference (TR), a 5-bit input frame number.
- Type information (PTYPE), such as CIF/QCIF selection;
- User-inserted bits.

Then, a number of GOB layer data follow.

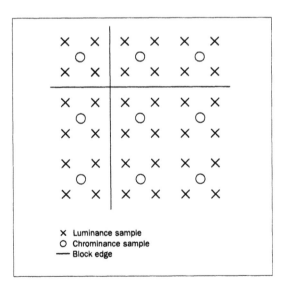

Figure 11.7   Relative positioning of luminance and chrominance samples.

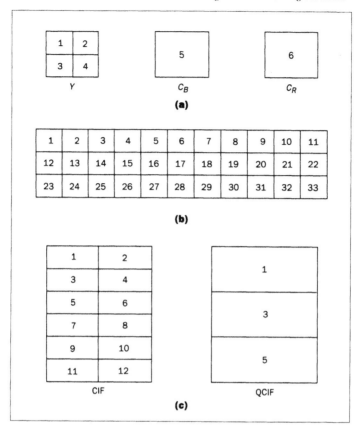

Figure 11.8   Successive arrangement of (a) blocks in a macroblock, (b) macroblocks in a GOB, and (c) GOBs in a picture.

At the GOB layer, a GOB header contains:

- Group of blocks start code (GBSC), a 16-bit pattern;
- Group number (GN), a 4-bit GOB address;
- Quantizer information (GQUANT), quantizer step size normalized to lie in the range 1 to 31;
- User-inserted bits.

Next come a number of MB layer data. An 11-bit stuffing pattern can be inserted repetitively right after a GOB header or after a transmitted macroblock.

At the MB layer, the header contains:

- Macroblock address (MBA), location relative to the previously coded MB;
- Type information (MTYPE), 10 types in total;
- Quantizer (MQUANT), normalized quantizer step size;
- Motion vector data (MVD), the differential displacement;
- Coded block pattern (CBP), the coded block location indicator.

The lowest layer is the block layer, consisting of quantized transform coefficients (TCOEFFs), followed by the end of block (EOB) symbol.

Not all header information need be present. For example, at the MB layer, if an MB is not motion compensated (as indicated by MTYPE), MVD does not exist.

Figure 11.9 is a functional diagram of a typical H.261 decoder. The received bit stream is first kept in the receiver buffer. The VLC decoder decodes the compressed bits and distributes the decoded information to the elements that need that information. The VLC tables are given in the standard.

There are essentially four types of MBs:

- Intra, original pixels are transform coded;
- Inter, the difference pixels (with zero motion vectors) are coded;
- Inter with motion compensation (MC), the displaced (nonzero motion vectors) differences are coded;
- Inter MC with filter, the displaced blocks are filtered by a predefined filter, which may help reduce visible coding artifacts at very low bit rates.

Certain MB types in this list allow the optional transmission of MQUANT and TCOEFF information. The received MTYPE information controls various switches at the decoder to produce the right combination.

A single-motion vector (horizontal and vertical displacement) is transmitted for one inter MC microblock; that is, the four Y blocks, one C_B, and one C_R block all share the same motion vector. The range of motion vectors is ±15 pixels

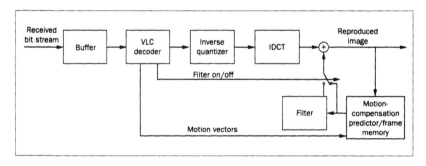

Figure 11.9   A typical H.261 decoder.

with integer values. Using both MVD and MTYPE information, the predictor can choose the right pixels for prediction.

The transform coefficients of either the original or the differential pixels are ordered according to the zigzag scanning pattern in Fig. 11.10. These transform coefficients are selected and quantized at the encoder, and then variable length coded. Just as with JPEG, successive zeros between two nonzero coefficients are counted and called a *RUN*. The magnitude of a transmitted nonzero quantized coefficient is called a *LEVEL*. The most likely occurring combinations of (RUN, LEVEL) are encoded with the standard supplied VLC tables. The other combinations are coded with a 20-bit word consisting of a 6-bit ESCAPE code, 6 bits RUN, and 8 bits LEVEL. EOB is appended to the last nonzero coefficient, indicating the end of a block.

The inverse quantizer or the reconstruction process for all the coefficients other than the intra dc is defined by the following formula: If QUANT is odd,

$$\mathrm{REC} = \mathrm{QUANT} \times (2 \times \mathrm{LEVEL} + 1), \quad \text{for LEVEL} > 0,$$
$$\mathrm{REC} = \mathrm{QUANT} \times (2 \times \mathrm{LEVEL} - 1), \quad \text{for LEVEL} < 0;$$

if QUANT is even,

$$\mathrm{REC} = \mathrm{QUANT} \times (2 \times \mathrm{LEVEL} + 1) - 1, \quad \text{for LEVEL} > 0,$$
$$\mathrm{REC} = \mathrm{QUANT} \times (2 \times \mathrm{LEVEL} - 1) + 1, \quad \text{for LEVEL} < 0,$$

where REC is the reconstructed value of a quantized coefficient. Almost all the reconstruction levels are odd numbers to reduce problems of mismatch between

Increasing cycles-per-picture width →

Increasing cycles-per-picture height ↓

1	2	6	7	15	16	28	29
3	5	8	14	17	27	30	43
4	9	13	18	26	31	42	44
10	12	19	25	32	41	45	54
11	20	24	33	40	46	53	55
21	23	34	39	47	52	56	61
22	35	38	48	51	57	60	62
36	37	49	50	58	59	63	64

Figure 11.10 Transmission order for transform coefficients.

encoders and decoders from different manufacturers. The intra-dc coefficient is uniformly quantized with a fixed step size of 8, and coded with 8 bits.

The standard requires a compatible inverse DCT (IDCT) to be close to the ideal 64-bit floating point IDCT. H.261 specifies a measuring process for checking a valid IDCT. The peak error, mean error, and mean square error between the ideal IDCT and the IDCT under test have to be less than certain small numbers given in the standard.

A few other items are required by the standard. One of them is the image-block updating rate. To prevent mismatched IDCT error and channel error propagation, every MB should be intra-coded at least once in every 132 transmitted picture frames. The contents of the transmitted bit stream must meet the requirements of a *hypothetical reference decoder* (HRD). For CIF pictures, every coded frame is limited to fewer than 256 kb; for QCIF, the limit is 64 kb. The HRD receiving buffer size is $B + 256$ kb, where $B = 4 \times R_{max}/29.97$ and R_{max} is the maximum connection (channel) rate. At every picture interval (1/29.97 sec), the HRD buffer is examined. If at least one complete coded picture is in the buffer, then the earliest picture data are removed from the buffer and decoded. The buffer occupancy, right after the preceding data have been removed, must be less than B.

11.3.3 Encoder Constraints and Options

Figure 11.11 shows a typical encoder structure. For the purpose of this discussion, the elements inside a standard compatible encoder can be classified, based on their functionalities, into two categories:

1. The *basic coding operation* units, such as motion estimator, quantizer, transform, and variable word length encoder (VLE).

2. The *coding parameter decision* units, such as the coding control in Fig. 11.11. These units select the parameter values of the basic operation units, including motion vectors, quantization step size, and frame rate.

Although H.261 does not explicitly specify a standard encoder, most basic operation elements are strongly constrained by the standard. However, other crucial elements, such as the parameter decision unit, are still open to the design engineers. We briefly outline our observations here.

The VLE implements the VLC H.261 tables. The forward DCT is not specified, but it is expected that the forward DCT inside the encoder matches the decoder IDCT, and this forward DCT should be able to match its own IDCT.

Because the inverse quantizer (IQ) is defined at the decoder, variations of the encoder quantizer are quite limited. From a theoretical viewpoint, however, it is not necessary for the decision levels of the encoder quantizer to be in the middle of two reconstruction levels. Also, encoder designers determine the criterion (a fixed or an adaptive threshold, for example) for selecting transform coefficients.

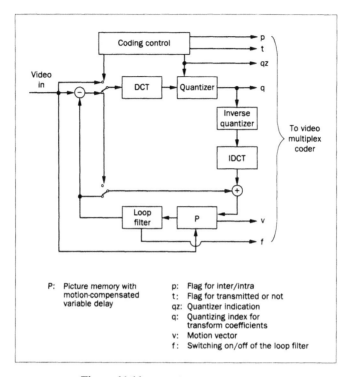

Figure 11.11   A typical H.261 encoder.

If motion compensation is selected, the motion estimator must be able to produce one motion vector for the entire MB. Block-matching motion estimation is used to produce such a motion vector; there can be several variations, such as hierarchical motion estimation [13]. Because of the HRD model required by the standard, the encoded output bits of every frame must be regulated carefully. For example, successive frames producing small numbers of bits may violate the HRD requirement.

Although individual basic coding elements may affect the overall coding performance, the most critical and global influence on the encoder performance comes from the parameter decision units. The encoder must make several decisions such as

- How many frames should be transmitted, or conversely, how many should be skipped?
- What MTYPE should each macroblock use?
- What is the proper quantization step size?

- How do we control the buffer fullness so that it does not produce long delay and does not violate the HRD requirements?

Also, it is important to keep the hardware simple for practical applications. Many issues discussed here should be investigated before a complete solution can be implemented.

11.4 Moving Picture Experts Group

The Moving Picture Experts Group (MPEG) has designed an international standard [14–17] for the compression of digital audio and video transmission. The MPEG first-phase (MPEG-1) video compression standard, aimed primarily at coding video for digital storage media, at rates of 1 to 1.5 Mb/sec, is well suited for a wide range of applications at a variety of bit rates. The standard mandates real-time decoding and supports features to facilitate interactivity with the stored bit stream. It specifies a syntax for only the bit stream and the decoding process; sufficient flexibility is allowed for encoding complexity. Encoders can be designed for optimal tradeoff of performance versus complexity, depending on the specific application.

MPEG was chartered by the ISO to standardize a coded representation of video and audio suitable for digital storage media, such as compact disc-read-only memory (CD-ROM), digital audio tape (DAT), etc. The group's goal, however, has been to develop a *generic* standard, one that can be used in other digital video applications, such as telecommunication. The MPEG standard has three parts:

- Part 1 concerns the synchronization and multiplexing of video and audio.
- Part 2 concerns video.
- Part 3 concerns audio.

An overview of the video portion of the MPEG standard follows.

11.4.1 Requirements of the Standard

Uncompressed digital video requires an extremely high transmission bandwidth. Digitized North American Television Standards Committee (NTSC) resolution video, for example, has a bit rate of approximately 100 Mb/sec. With digital video, compression is necessary to reduce the bit rate to suit most applications. The required degree of compression is achieved by exploiting the spatial and temporal redundancy present in a video signal. However, the compression process is inherently lossy, and the signal reconstructed from the compressed bit stream is not identical to the input video signal. Compression typically introduces artifacts into the decoded signal.

The primary requirement of the MPEG video standard is that it should achieve the highest possible quality of the decoded video at a given bit rate. In addition to picture quality, different applications stipulate additional requirements. For instance, multimedia applications require the ability to access, i.e., decode, any video frame in a short time. The ability to perform fast search directly on the bit stream—forward and backward—is extremely desirable if the storage medium has "seek" capabilities. Most applications require some degree resilience to bit errors. It is also useful to be able to edit compressed bit streams directly while maintaining decodability. A variety of video formats should be supported.

11.4.2 Compression Algorithm Overview

References [15] and [17] describe the basic algorithms and syntax of the MPEG standard, and reference [17] details video coding using this standard. Here, we present the background and the basic information necessary for understanding this standard.

Exploiting Spatial Redundancy   The compression approach of MPEG video uses a combination of the ISO JPEG (still image) and CCITT H.261 (videoconferencing) standards. Because video is a sequence of still images, it is possible to compress or encode a video signal using techniques similar to JPEG. Such methods of compression are called *intraframe coding techniques*, where each frame of video is individually and independently compressed or encoded. Intraframe coding exploits the spatial redundancy that exists between adjacent pixels of a frame.

As in JPEG and H.261, the MPEG video-coding algorithm employs a block-based two-dimensional DCT. A frame is first divided into 8×8 blocks of pixels, and the two-dimensional DCT is then applied independently on each block. This operation results in an 8×8 block of DCT coefficients in which most of the energy in the original (pixel) block is typically concentrated in a few low-frequency coefficients. A quantizer is applied to each DCT coefficient that sets many of them to zero. This quantization is responsible for the lossy nature of the compression algorithms in JPEG, H.261, and MPEG video. Compression is achieved by transmitting only the coefficients that survive the quantization operation and by entropy coding their locations and amplitudes.

This standard allows the quantization operation to achieve a higher level of adaptation, a key factor in achieving good picture quality. Reference [18] details the relevant details of a quantizer adaptation scheme applicable within this context.

Exploiting Temporal Redundancy   Many of the interactive requirements discussed earlier can be satisfied by intraframe coding. However, as in H.261, the quality achieved by intraframe coding alone is not sufficient for typical video signals at bit rates around 1.5 Mb/sec. Temporal redundancy results from a high degree of correlation between adjacent frames. The H.261 algorithm exploits this redundancy by computing a frame-to-frame difference signal called the *prediction error*. In computing the prediction error, the technique of motion compensation is employed to correct for motion. A block-based approach is adopted for motion compensation, where a block of pixels, called a *target block*, in the frame to be encoded is matched with a set of blocks of the same size in the previous frame, called a *reference frame*. The block in the reference frame that "best matches" the target block is used as the prediction for the latter; i.e., the prediction error is computed as the difference between the target block and the best matching block. This best matching block is associated with a motion vector that describes the displacement between it and the target block. The motion vector information is also encoded and transmitted along with the prediction error. The prediction error itself is transmitted using the DCT-based intraframe encoding technique summarized previously. In MPEG video (as in H.261), the block size for motion compensation is chosen to be 16×16, representing a reasonable tradeoff between the compression provided by motion compensation and the cost associated with transmitting the motion vectors.

Bidirectional Temporal Prediction   Bidirectional temporal prediction, also called *motion-compensated interpolation*, is a key feature of MPEG video. In bidirectional prediction, some of the video frames are encoded using two reference frames, one in the past and one in the future. A block in those frames can be predicted by another block from the past reference frame (*forward prediction*), or from the future reference frame (*backward prediction*), or by the average of two blocks, one from each reference frame (*interpolation*). In every case, the block from the reference frame is associated with a motion vector, so that two motion vectors are used with interpolation. Motion-compensated interpolation for a block in a bidirectionally predicted frame is illustrated in Fig. 11.12. Frames that are bidirectionally predicted are never themselves used as reference frames.

Bidirectional prediction provides a number of advantages. The primary one is that the compression obtained is typically higher than can be obtained from forward prediction. To obtain the same picture quality, bidirectionally predicted frames can be encoded with fewer bits than frames using only forward prediction. However, bidirectional prediction introduces extra delay in the encoding process, because frames must be encoded out of sequence. Further, it entails extra encoding complexity because block matching (the most computationally intensive encoding procedure) has to be performed twice for each target block, once with the past reference and once with the future reference.

11.4.3 Features of the Bit-Stream Syntax

The MPEG video standard specifies the *syntax* of the bit stream and also the decoder; i.e., the standard specifies how this bit stream is to be parsed and decoded

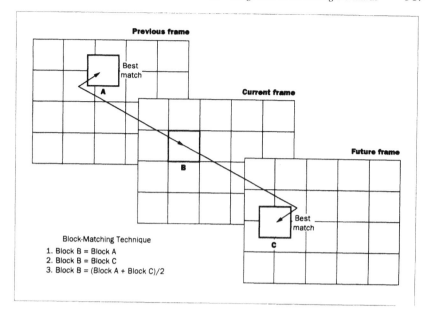

Figure 11.12 Motion-compensated interpolation.

to produce a decompressed video signal. However, a specific encoding method is not mandatory; different algorithms can be employed at the encoder so long as the resulting bit stream is consistent with the specified syntax. For example, the details of the block-matching procedure are not part of the standard. This is also true in H.261.

The bit-stream syntax should be flexible to support the variety of applications envisaged for the MPEG video standard. To this end, the overall syntax is constructed in several layers, each performing a different logical function. The outermost layer is called the *video sequence* layer, which contains basic parameters such as the size of the video frames, the frame rate, the bit rate, and certain other global parameters. A wide range of values is supported for all these parameters.

Inside the video sequence layer is the group of pictures (GOP) layer, which provides support for random access, fast search, and editing. A sequence is divided into a series of GOPs, where each GOP contains an intracoded frame (I-frame) followed by an arrangement of (forward) predictive-coded frames (P-frames) and bidirectionally predicted, interpolative-coded frames (B-frames). Figure 11.13 shows a GOP example with six frames, 1 to 6. This GOP contains I-frame 1, P-frames 4 and 6, and B-frames 2, 3, and 5. The encoding and transmission order of the frames in this GOP is shown at the bottom of Fig. 11.13. B-frames 2 and 3 are encoded after P-frame 4, using P-frame 4 and I-frame 1 as reference. We note

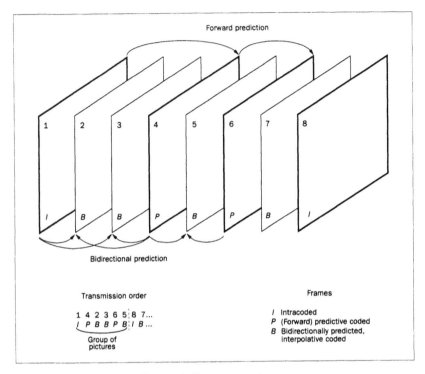

Figure 11.13 Group of pictures.

that B-frame 7 in Fig. 11.13 is part of the next GOP because it is encoded after I-frame 8. Random access and fast search are enabled by the availability of the I-frames, which can be decoded independently and serve as starting points for further decoding. The MPEG video standard allows GOPs to be of arbitrary structure and length. The GOP layer is the basic unit for editing an MPEG video bit stream.

The compressed bits produced by encoding a frame in a GOP constitute the *picture layer*. The picture layer first contains information on the type of frame that is present (I, P, or B) and the position of the frame in display order. The bits corresponding to the motion vectors and the DCT coefficients are packaged in the *slice* layer, the *macroblock* layer, and the *block* layer. Here, the block is the 8 × 8 DCT unit, the macroblock is the 16 × 16 motion-compensation unit, and the slice is a string of macroblocks of arbitrary length running from left to right and top to bottom across the frame. The slice layer is intended to be used for resynchronization during the decoding of a frame, in the event of bit errors. Prediction registers used in the differential encoding of motion vectors are reset at the start of a slice. It is again the responsibility of the encoder to choose the length of each slice.

Figure 11.14 shows an example in which slice lengths vary throughout the frame. In the macroblock layer, the motion vector bits for a macroblock are followed by the block layer, which consists of the bits for the DCT coefficients of the 8 × 8 blocks in the macroblock. Figure 11.15 shows an MPEG video encoder and decoder. The different layers in the syntax and their use are illustrated in Table 11.2.

In demonstrations of MPEG video at a bit rate of 1.2 Mb/sec, noninterlaced frames of size of 352 pixels by 240 lines at a frame rate of 29.97 per second have been used, with 2:1 color subsampling both horizontally and vertically. This resolution is roughly equivalent to one field of an interlaced NTSC frame. The quality achieved by the MPEG video encoder at this bit rate has often been compared to that of VHS. Although the MPEG video standard was originally intended for operation in the neighborhood of this bit rate, a much wider range of resolution and bit rates is supported by the syntax. The MPEG video standard thus provides

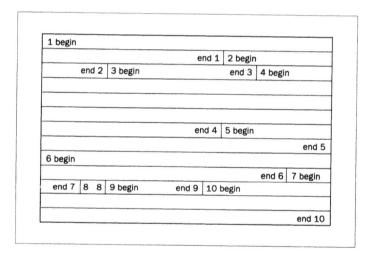

Figure 11.14   Possible arrangement of slices in a 256 × 192 picture.

Table 11.2   Layers of MPEG video bit-stream syntax.

Syntax layer	Functionality
Sequence layer	Context unit
Group of pictures layer	Random access unit: video coding
Picture layer	Primary coding unit
Slice layer	Resynchronization unit
Macroblock layer	Motion compensation unit
Block layer	DCT unit

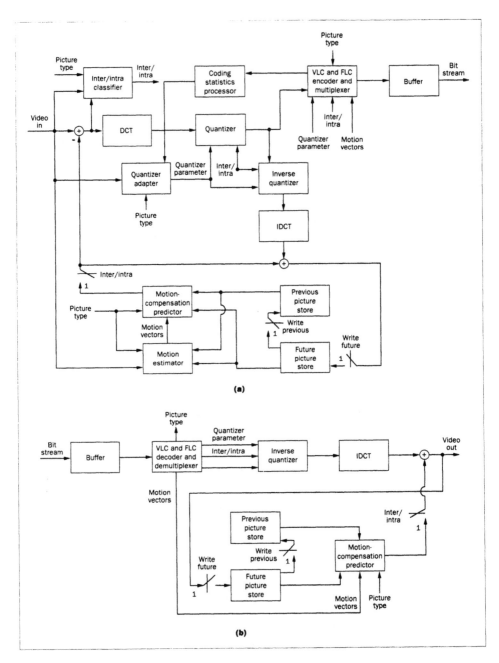

Figure 11.15 A typical (a) MPEG-1 encoder and (b) MPEG-1 decoder [17].

a generic bit-stream syntax that can be used for a variety of applications. The MPEG-Video International Standard ISO/IEC 11172-2 provides all the details of the syntax, complete with informative sections on encoder procedures that are outside the scope of the standard [14].

11.4.4 The MPEG Second-Phase Standard

Currently, the second phase of MPEG (MPEG-2) is in progress. This phase is aimed at coding the video signals created by CCIR 601; e.g., 720 pixels, 480 lines, 30 frames per sec, 2:1 interlace at bit rates of 2 Mb/sec, or higher.

The first-phase standard, MPEG-1, focused on coding of single-layer (nonscalable) video of progressive format. The MPEG-2 standard is addressing issues of improved functionality by using scalable video coding. To initiate technical work for this phase of the MPEG standard, a worldwide video coding competition was held at Kurihama, Japan, in November 1991. Nearly 30 international organizations submitted a video-coding scheme to this contest. Immediately after this competition, a collaborative phase of work began and, thus far, has resulted in a compromise scheme that retains most of the best features of the best performing schemes.

In the MPEG-2 standard, the main improvements in nonscalable coding result from emphasis on interlaced video. Various forms of frame/field motion-compensated predictions have been adapted to increase the coding efficiency. Frame/field DCT coding and quantization have also been adapted. All optimization experiments are being performed at bit rates between 4 and 9 Mb/sec.

The MPEG-2 standard is also addressing scalable video coding for a range of applications where video needs to be decoded and displayed at a variety of resolution scales. Among the noteworthy applications of interest are multipoint video conferencing, window display on workstations, video communications on asynchronous transfer mode networks, and high-definition television (HDTV) with embedded standard TV.

In scalable video coding, which can be achieved in the *spatial* or the *frequency* domain, it is assumed that, given an encoded video bit stream, decoders of various complexities can decode and display appropriate-size replicas of the original video. A scalable video encoder and the corresponding highest resolution decoder are likely to have increased complexity compared to single-layer encoder–decoder. However, this increase in complexity may be well justified in applications where increased functionality and error resilience are important.

11.5 Conclusion

Image coding standards are crucial to the robust growth of visual services in communication and computer systems. Without them, communication between terminals and systems becomes extremely inconvenient and costly. In the absence of standards, economies of scale in the manufacture of user devices, board systems, and VLSI chips may be lost.

The JPEG, P×64, and MPEG standards provide compression algorithms for all types of images that might be carried on multimedia services. With the onset of inexpensive chips, high-speed communication, and large capacity disk storage, all elements needed for rapid growth are in place.

References

[1] J. R. Allen *et al.*, "VCTV: A video-on-demand market test." *AT&T Tech. J.*, vol. 72, pp. 7–14, Jan–Feb. 1993.

[2] *Digital Compression and Coding of Continuous-Tone Still Images, Part 1: Requirements and Guidelines*, ISO/IEC DIS 10918-1, 1991.

[3] *Digital Compression and Coding of Continuous-Tone Still Images, Part 2: Compliance Testing*, ISO/IEC CD 10918-2, 1991.

[4] A. N. Netravali and B. G. Haskell, *Digital Pictures: Representation and Compression*. New York: Plenum Press, 1988.

[5] D. A. Huffman, "A method for the construction of minimum-redundancy codes." *Proc. IRE*, vol. 40, pp. 1098–1101, Sept. 1952.

[6] J. Amsterdam, "Data compression with Huffman coding." *BYTE*, vol. 11, pp. 99–108, May 1986.

[7] G. G. Langdon, Jr., "An introduction to arithmetic coding." *IBM J. Res. Develop.*, vol. 28, pp. 135–149, March 1984.

[8] I. H. Witten, R. M. Neal, and J. G. Cleary, "Arithmetic coding for data compression." *Commun. ACM*, vol. 30, pp. 520–540, June 1987.

[9] N. Ahmed, T. Natarajan, and K. R. Rao, "Discrete cosine transform." *IEEE Trans. Comput.*, vol. 23, pp. 90–93, Jan. 1974.

[10] R. J. Clarke, *Transform Coding of Images*. Orlando, FL: Academic Press, 1985.

[11] H. Lohscheller, "A subjectively adapted image communication system." *IEEE Trans. Commun.*, vol. COM-32, pp. 1316–1322, Dec. 1984.

[12] CCITT, *Recommendation H.261—Video Codec for Audiovisual Services at px64 kbit/sec*, Geneva, Aug. 1990.

[13] M. Bierling and R. Thoma, "Motion compensating field interpolation using a hierarchically structured displacement estimator." *Signal Processing*, vol. 11, pp. 387–404, Dec. 1986.

[14] *Information Technology—Coding of Moving Pictures and Associated Audio for Digital Storage Media up to about 1.5 Mbit/sec*, ISO/IEC 11172.

[15] D. J. LeGall, "MPEG: A video compression standard for multimedia applications." *Commun. ACM*, vol. 34, pp. 47–58, April 1991.

[16] R. K. Jurgen, "Digital video." *IEEE Spectrum*, vol. 29, pp. 24–30, March 1992.

[17] A. Puri, "Video coding using the MPEG-1 compression standard." In *Proc. Int. Symp.: Society for Inform. Display*, Boston, May 1992, pp. 123–126.

[18] A. Puri and R. Aravind, "Motion-compensated video coding with adaptive perceptual quantization." *IEEE Trans. Circuits Syst. Video Technol.*, vol. 1, pp. 351–361, Dec. 1991.

Chapter 12

Hybrid High-Definition Television

Y. Ninomiya
Science and Technical Research Laboratories
NHK (Japan Broadcasting Corporation)
Tokyo, Japan

Because HDTV is a new system, signal processing is mostly digital. In this sense, all HDTV systems are analog and digital hybrids. In this chapter, we will take a hybrid HDTV system to mean something narrower: an HDTV transmission system with digital signal processing but transmitted on analog lines. It is quite natural that hybrid system signals can be transmitted in digital form, but in this case the coding efficiency may be less than in other systems that are not transmitted on analog lines. Many transmission media are suitable only for analog signals. Also, in many cases we should select an analog transmission system not only to match with existing systems but also because of the existing technology and industry.

There are two hybrid HDTV transmission systems: MUSE and HD-MAC. To broadcast HDTV, these two systems are the only ones in the world already developed, and it has been decided to use only them for practical HDTV broadcasts. These two systems will be described in this chapter. HD-MAC and MUSE were developed for European and Japanese HDTV broadcasting, respectively. They belong to the same category and are technically similar, but their scanning parameters differ. Naturally, the studio systems used both in Europe and Japan meet

CCIR Recommendation 709, but scanning parameters have not been specified in the recommendation.

Digital methods are key technologies for HDTV. Throughout the signal processing, from studio to receiver, the HDTV system is supported by digital technologies. The essential point about HDTV is that its broadcast system is completely supported by digital technology. All signal processing, both in MUSE and HD-MAC, is done in digital form. In the United States, the leading candidates for ATV (advanced television) include systems using digital bit-rate reduction coding, including the DigiCipher.

The use of VLSIs is indispensable for making digital equipment reasonably cheaply and of an appropriate size. In the case of broadcast systems, it is still essential to reduce the price of receivers by mass production. The introduction of VLSIs is the only way to achieve this.

12.1 Definition and Standard for High-Definition Television

HDTV is defined as a wide screen television system in which the horizontal and vertical resolution are both almost double that of current standard television including NTSC and PAL.

As an international standard for HDTV, there is a CCIR Recommendation on HDTV studio standards, Rec. 709 [1]. Unfortunately, this is not perfect. There are no specifications for scanning parameters or frame rate. The parameters specified are summarized in Table 12.1.

Table 12.1   Specified parameters in Rec. 709 of CCIR.

Opto-electric transfer characteristics
Chromaticity coordinates
Reference white
Aspect ratio
Samples per active line
Sampling lattice
Nonlinear precorrection of primary signal
Derivation of luminance and color-difference signals
Levels of video and synchronizing signals
Synchronizing signal formats
Some digital representations

Although some digital representation formats are specified, this recommendation is fundamentally an analog standard. As this recommendation is not perfect in the practical sense, standards for establishing practical systems have been developed. Now two HDTV systems are being practically used. The main differences between these systems are the scanning parameters and frame rates. One is the 1125/60 system used in Japan and to a degree in the United States and other is the 1250/50 system used in Europe. Naturally, both studio systems meet CCIR Recommendation 709, in which the scanning parameters have not yet been specified.

12.1.1 1125/60 System

The 1125/60 system is the standard for Japanese HDTV broadcasts. The main parameters of this system are shown in Table 12.2. The standards for this system are BTA-S001 in Japan and SMPTE-240M in the United States, but the content of both are the same. A digital format of this system is also specified as BTA-S002 and SMPTE-260M. The MUSE system has been developed to broadcast the HDTV signal of this standard.

12.1.2 1250/50 System

The 1250/50 system was developed in Europe, mainly as the Eureka 95 project. At this moment we have no established or published standard for this system. A specification in an annex of a draft modified version of CCIR Recommendation 709 is the only published standard, but it is not known that this is exactly the same as the Eureka internal standard that will be used to develop HDTV hardware. The main parameters of this system are shown in Table 12.3. The HD-MAC system was developed to broadcast HDTV signals in this system.

Table 12.2   Important parameter values
of the 1125/60 system.

Aspect ratio	16:9
Interlace ratio	2:1
Frame rate	60.0
Total number of scan lines	1125
Number of active scan lines	1035
Sampling frequency	74.25 MHz

Table 12.3   Important parameter values
of the 1250/60 system.

Aspect ratio	16:9
Interlace ratio	2:1
Frame rate	50
Total number of scan lines	1250
Number of active scan lines	1152
Sampling frequency	72.0 MHz

12.1.3 Other Systems

There are some systems, proposed mainly in the United States: the 1080/2:1 (active lines) system, the 720/1:1 (active lines) are examples. Establishing standards for these systems have almost been finished in SMPTE.

12.2 Basic Construction of a Hybrid High-Definition Television System

In Fig. 12.1, the conceptual construction of a hybrid system is shown. The transmission medium is analog, but all signal processing is digital. The basic technology used for signal processing is almost the same as for a digital HDTV transmission system. Hybrid HDTV and digital HDTV use the same technology. But when we examine them in detail, there are important differences. A comparison is summarized in Table 12.4.

Considering the hardware, it is important that hybrid systems need A/D and D/A converters. But this is not quite essential, because A/D or D/A converters are also used quite frequently even in digital transmission. For example, we need them in waveform equalization systems and soft decision receiving systems.

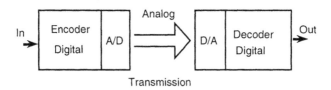

Figure 12.1   Basic construction of a hybrid system. In the hybrid system, A/D (analog to digital) and D/A (digital to analog) converters are indispensable.

Table 12.4   Differences between hybrid and digital systems.

Item	Hybrid	Digital
Coding using the characteristics of vision	M	U
Level processing	M	M
Spatio–temporal frequency domain processing	M	U
Time-domain processing	M	M
Prediction coding	X	M
Transform coding	D	U
Statistical redundancy deduction	U	M
Entropy coding	X	M
Digital transmission media	U	M
Analog transmission media	M	X

Key:
M, mainly applied; U, applicable; D, difficult to apply; X, impossible or very difficult to apply.

12.3 Types of Hybrid Systems

Although only MUSE and HD-MAC have been developed and used in practice, an overview of several categories of hybrid systems may be useful for better understanding this technology. In Table 12.5 some possible categories of hybrid HDTV are shown.

Category 1 is the most primitive system. The essence of this method is matching the filter characteristic to human visual perception and eliminating redundant frequency components. Theoretically, a simple two-dimensional filter or fixed spatio–temporal filter system is possible. But if we want to keep reasonable picture quality, a motion-adaptive spatio–temporal filter must be used.

Category 2 has the merit of higher resolution than category 1. If we can solve the problem of how to transmit sampled values on an analog line, the subsampling

Table 12.5   Categories of hybrid HDTV systems.

Category	Technology
1	Filtering
2	Subsampling
3	Level companding
4	Transform
5	Other

system is the most practical of all. Not only MUSE and HD-MAC, but also almost all hybrid HDTV transmission systems that have been experimented on or proposed are subsampling systems. Subsampling is the main technology to be described in this chapter.

There were some systems belonging to category 3 in the past. The merits of these were that they can be achieved with purely analog circuits. But efficiency was not so good. In the future we expect new coding systems in this category because digital signal processing will be more common.

Transform coding schemes belonging to category 4 have the potential of being transmitted on analog lines. But practical systems of this kind have not yet been proposed.

12.4 Multidimensional Sampling

For a simple example, we analyze the two-dimensional (2-D) sampling of a picture and its aliasing effect. In Fig. 12.2 we see a diagonal sampling pattern and the carrier.

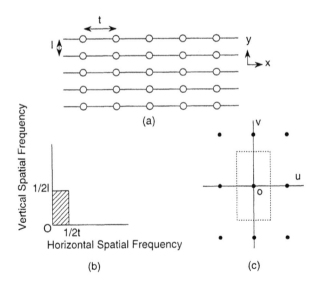

Figure 12.2 An example of a sampling pattern and carrier: (a) sampling pattern, (b) aliasing free area of the sampling pattern, (c) inverse space of the sampling pattern in the frequency domain.

The carriers can be calculated as follows. The periodic function, such as shown in Fig. 12.2a, is

$$s = f((x - t) \bmod t, (y - l) \bmod l). \tag{12.1}$$

Using a delta function for the sampling, the sampling pattern is written as

$$s = \delta((x - t) \bmod t, (y - l) \bmod l) = \delta((x - t) \bmod t) \cdot \delta((y - l) \bmod l). \tag{12.2}$$

Because the delta function is localized at the original point, (12.2) becomes

$$s = \sum_{j=-\infty}^{+\infty} \delta(x - jt) \cdot \sum_{k=-\infty}^{+\infty} \delta(y - kl). \tag{12.3}$$

The Fourier transform of this gives the inverse space. Because it is a perfect periodic function, the Fourier transform is a delta function and this gives carrier distribution:

$$F(s) = \frac{1}{t} \sum_{m=-\infty}^{+\infty} \delta(U - \frac{m}{t}) \cdot \frac{1}{l} \sum_{n=-\infty}^{+\infty} \delta(V - \frac{n}{l}). \tag{12.4}$$

This is shown in Fig. 12.2c. We can have an aliasing free area, as shown in Fig. 12.2. The aliasing free area is given as a space in which the frequency space can be covered periodically without overlap. The shape of the aliasing free area is not unique, but volumes of all aliasing free areas of a sampling pattern are the same.

The same procedure can be used to analyze aliasing free areas for more complicated sampling patterns.

12.5 Subsampling System

A good and simple example of subsampling is the interlaced scanning widely used in television systems. This is categorized as one-dimensional (1-D) and spatio–temporal subsampling using a fixed pattern.

In Fig. 12.3 the interlaced sampling pattern is shown in the real domain and the spatio–temporal frequency domain. The reason we call it 1-D is that the sampling is in one direction in space. If we also consider the temporal direction it should be called 2-D.

A television system transmits projected images of real moving objects in a 2-D plane. The signals can be treated in a 3-D space. Horizontal, vertical, and time are the three coordinate axes. The picture is sampled by field rate in the time axis. In the spatial plane, it is sampled along the vertical axis but is continuous along the horizontal axis.

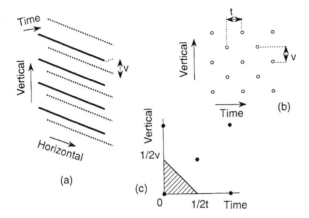

Figure 12.3   Interlaced scanning: (a) interlaced scanning in spatio–temporal domain, (b) section of the scanning in time and vertical space, (c) the carrier distribution and aliasing free area.

In the case of 2:1 interlaced system, as in Fig. 12.3, the signal is field-offset subsampled along the vertical axis. When we see a section of the vertical and time axes, it is in the so-called quincunx pattern as in Fig 12.3b.

Originally the interlaced scanning system was introduced to reduce the frame rate of television signals to take advantage of a characteristic of human visual perception. For human vision, large area flicker is much more visible than small area flicker. This means that vision takes in an integrated picture in the time domain for a small area. Thus for still pictures, the interlace system has a resolution of $1/v$, that is, twice the resolution of the line spacing in the field.

But for moving pictures it produces an aliasing component. Conventional television suffers from this aliasing effect. If we want to avoid aliasing, we must limit the input signal to aliasing free areas. Figure 12.4 shows examples of aliasing free areas, that is, areas transmissible by the quincunx subsampled system.

If we use type 1 for still pictures and type 2 for moving pictures as a motion adaptive scheme, we can avoid aliasing. As can be seen from Fig. 12.4, it is clear that the vertical resolution of a moving picture is less than that of still pictures. But human vision is less sensitive to resolution for moving pictures than for still pictures. In this sense, this system matches human vision well. Type 3 area (triangle shaped) is the most reasonable to use when we do not want to use an adaptive scheme.

Sophisticated systems such as MUSE and HD-MAC use a much more complicated subsampling system. But the basic principle is almost the same as described here. The general configuration of a subsampling system is shown in Fig. 12.5.

A prefilter limits the input signal to aliasing free areas, as in Fig. 12.4. The transmission characteristics of subsampling and resampling should meet the

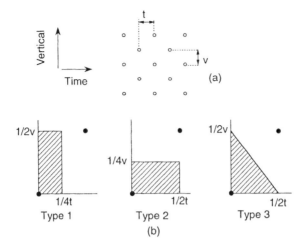

Figure 12.4 Aliasing free areas: (a) sampling pattern, (b) several aliasing free areas of the sampling pattern (black spots are carriers).

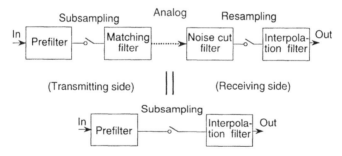

Figure 12.5 Configuration of a subsampling system. If total characteristics of the matching filter, analog line, and noise cut filter meet the distortion free condition and resampling timing is proper, the signal system is the same as shown in the bottom diagram.

distortion free conditions for sampled value transmission. We will discuss this later. If the distortion free condition is kept, the signal after resampling is the same as just after subsampling. Thus the total system is equivalent to the lower diagram in Fig. 12.5.

The prefilter and interpolation filter should have the same passband. Prefilters and interpolation filters should be used selectively when an adaptive transmissible area scheme is adopted.

12.6 Sampled Value Transmission

Sampled values can be transmitted on an analog line without distortion if the SNR of the line is good and the amplitude and phase frequency characteristics meet the distortion free condition.

Figure 12.6 shows the signal flow of a sampled value transmission. The transmission line is bandlimited. The received waveform of a single sample is smoothed, as shown in Fig. 12.6. The resampling pulse has the same period as the original sampling. If the resampling timing meets the zero cross points, there is no crosstalk between samples. The received waveform is the impulse response of the transmission line. If the frequency response of the line is $g(f)$, this is given by the Fourier transform of it as

$$F(i\omega) = \int_{-\infty}^{+\infty} g(s)\,e^{-i\omega f}\,df. \tag{12.5}$$

The constant is neglected in (12.5).

If the line is phase linear, the transmission characteristics are symmetric about the origin. The impulse response is given by the cosine term only in this case. The impulse response is

$$I(u) = \int_{-\infty}^{+\infty} g(f)\cos(uf)df. \tag{12.6}$$

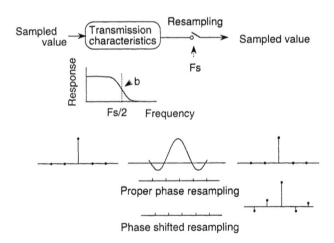

Figure 12.6 Sample value transmission. The top diagram shows basic flow of a sampled value transmission system. If the transmission characteristic is skew symmetric and resampling timing is proper, there will be no intersample contamination. If the resampling phase is shifted, the distortion will occur even if the transmission characteristic is skew symmetric.

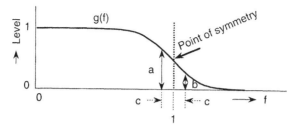

Figure 12.7 Distortion free characteristic. The axes of this diagram are normalized: 1 in the horizontal axis corresponds to the Nyquist frequency of the sampling; the response should be 0.5 at this point. And the characteristic should be skew symmetric around this point.

We resample on the points of $u = n\pi$ ($n \in Z$). If $I(u) = 0$ for all n except $n = 0$, the transmission is distortion free.

If the characteristic is skew symmetric as shown in Fig. 12.7, the integral is zero. This is a sufficient condition but not a necessary one, but it is the most natural.

12.7 Adaptive Subsampling

Almost all practical systems of relatively high performance are adaptive multidimensional subsampling systems. The sampling pattern or transmissible area is adaptively switched depending on the motion or spatial characteristic of the picture.

There are several types of switching methods. These are summarized in Table 12.6.

Type 1 is the most stable, but its coding efficiency is relatively low. In this system, we can adaptively switch pixel by pixel. Picture degradation caused by mismatch between switching in the encoder and decoder of this type of system is not severe. Adaptive switching in the encoder and decoder can be independent. So there is no need for a switching control signal.

Table 12.6 Types of adaptive subsampling methods.

Type	Sample density	Sub-sampling pattern	Transmissible area
1	Fixed	Fixed	Adaptive
2	Fixed	Adaptive	Adaptive
3	Adaptive	Adaptive	Adaptive

Type 2 is rather more efficient, but it is difficult to switch pixel by pixel. A switching mismatch between the encoder and decoder produces a severe degradation of the picture. The switching decision must be met exactly. So we must send a control signal from the encoder to the decoder. This means that switching must be blockwise, because the information rate for sending pixelwise control signals is too great. Blockwise control is too coarse and may produce block distortion.

Type 3 is most efficient in theory, because its statistical characteristics can be used effectively. But if a switching mismatch occurs, the degradation of the picture is almost catastrophic. Therefore, control must be blockwise.

Several proposals for this type of coding scheme have been published. The TAT system is an example, and hardware experiments for conventional television have been done [2]. However, there are no practical systems as yet.

12.8 System Requirements

We must consider system requirements when we design a transmission system. Almost all hybrid HDTV systems have been developed for HDTV broadcasts. Like other TV systems, HDTV is bound by basic technical requirements, the most basic and important of which are shown in Table 12.7. The HDTV broadcasting systems like HD-MAC and MUSE have been developed from these. To meet these requirements, both systems have been designed as hybrid HDTV systems. Although these requirements are of a general nature, we shall discuss them briefly.

12.8.1 Nationwide Reception

Even where it is not stipulated by law, using a system that can be received in only a limited area is not desirable as a basic means of broadcasting. It is also important

Table 12.7   Important requirements for HDTV broadcasting.

1. Reception capability throughout the country
2. Availability of low-cost receivers
3. A certain quality under inferior conditions
4. The best picture quality
5. Good match with other media
6. Reasonable compatibility with existing systems
7. Relatively simple signal format
8. Easy to monitor
9. Good match with current industrial technology

from the economic viewpoint. We need as many subscribers as possible because the production cost of HDTV programs is enormous.

12.8.2 Receiver Price

The concept here is the same as earlier. The need to buy an expensive receiver would seriously hamper the popularization of a system. The receivers, therefore, must be as cheap as possible.

12.8.3 Reception under Inferior Conditions

This is always important in broadcast systems. In FM transmission, for example, a picture of a certain quality should be receivable even with a low C/N (carrier to noise ratio), below the threshold. Unlike fixed communications, broadcasts are received by an unspecified number of receivers, which in general have no technical support. Consequently, in terms of hardware design or the operation of the hardware, it is not possible to receive transmissions under ideal conditions at any given time.

12.8.4 Good Picture Quality

Naturally, good picture quality is desirable, but the important thing is overall picture quality. In the case of broadcasts, we are expected to be able to transmit a wide range of pictures. This is completely different from conference TV systems. Good picture quality must in general be maintained for signals of widely varying quality. It should be noted that picture quality is often determined by the worst case. The most important factor affecting picture quality is the effects of transmission noise and bandwidth.

12.8.5 Matching with Other Media

The spread of VCRs into homes is a remarkable phenomenon, and a similar situation is expected for HDTV. The "package" should be taken into account when considering a complete HDTV broadcast system. The same may be said for CATV systems.

12.8.6 Compatibility with Existing Systems

This is rather complex. Philosophies on compatibility are different for HD-MAC and MUSE. Details of the compatibility problem will be described in the next section, but clearly HDTV broadcasting must have some degree of compatibility with conventional television.

12.8.7 Simple Signal Format

The development of a broadcast system is not undertaken by a single research laboratory; rather, it is done by cooperation and competition among numerous institutions and manufacturers. A too complex system would, therefore, result in a generally less efficient development, as it would tend to block mutual understanding and hamper cooperation.

The cost of a receiver for a simple system would also, naturally, be lower. Moreover, experience has shown that a simple, flexible form allows more room for improvement at a later stage.

12.8.8 Monitoring

Monitoring is an important requirement for a broadcast system. Monitoring is easier when the signal format is closer to the original signal (this means the content of the transmitted signal).

12.8.9 Match for the Industry

As broadcasting is a very large and open system, it is indispensable to get the support of industry. This is clear if we consider the price of receivers. Nobody can sell reasonably cheap receivers without a technical industrial base.

12.9 Compatibility

The range of alternatives in the compatibility problem is very wide. They range from the situation in which HDTV broadcasts can be received by conventional, that is, PAL, SECAM, or NTSC, receivers to one in which they cannot be received without a receiver designed exclusively for the purpose. In Table 12.8, several grades of compatibility are summarized. The former alternative corresponds to Grade 1 or 2, and the latter corresponds to Grade 6 or 7.

Grade 1 and 2 are very difficult because in this case, the system requirements would spoil the picture quality of HDTV. This can be understood by the fact that fully compatible systems with NTSC were rejected in discussions of ATV in the United States. However, from the business viewpoint, it is quite dangerous to adopt a system that matches existing systems poorly. It is desirable to use common hardware with conventional systems as far as possible.

The HD-MAC system belongs to Grade 4, and the MUSE belongs to Grade 5. Most digital television systems will be of Grade 7. The greatest difference between design philosophies for HD-MAC and MUSE is in compatibility. They are

Table 12.8   Grades of compatibility.

Grade	Compatibility
1	Fully compatible with an existing system without any problem in quality and without any modification of receivers.
2	Compatible with an existing system with tolerable problems of quality and without any modification of receivers.
3	Compatible with a new system that has the same scanning system as an existing system without any problems in quality.
4	Compatible with a new system that has the same scanning system as an existing system with tolerable problems of quality.
5	The signal can be received by an existing system with simple additional equipment, such as a converter.
6	The signal can be received by an existing system with major modifications.
7	A new receiver is required in practice.

Table 12.9   Compatibility design of HD-MAC and MUSE.

	HD-MAC	MUSE
To which?	MAC family	NTSC
Grade	4 (direct)	5 (simple converter)
To terrestrial	No	Same as preceding
Assumed BS system	MAC family	NTSC

summarized in Table 12.9. These are derived from differences in the broadcasting satellite (BS) of conventional television. The Japanese BS system uses an NTSC signal, but it was decided that the European BS system should use the MAC family including D2-MAC [3]. Theoretically, the MAC signal matches satellite FM modulation better because it is a component type and has no subcarrier. But it needs a special receiving decoder because it is not directly compatible with conventional terrestrial signals, including PAL and SECAM. On the other hand, when NTSC, PAL, or SECAM is used for satellite broadcasting, the video signal can be decoded by conventional receivers for terrestrial reception.

Table 12.10   Technical commonality level.

Level	Technology need
1	Needs no new category of technology
2	Needs no new category of technology except in the encoder and decoder
3	Needs new category of technology

12.10 Commonality with the Existing System

Compatibility is a matter of commonality of signal format between a new system and an existing system. Commonality of technology in both systems should also be considered.

In this case, different from compatibility, we are thinking about HDTV reception. The question is: To what extent can we use technology supporting existing systems for the new system? Levels of commonality are summarized in Table 12.10.

Naturally technical innovation is important for the progress of society. But, on the other hand, state-of-the-art technology is essential for the easy introduction of a new system.

12.11 Concepts of MUSE and HD-MAC

Before describing the two systems in detail, technical points and the design concept of MUSE [4] and HD-MAC [5] are summarized. Because both systems were developed for HDTV broadcasts and should meet the requirements described in Section 12.8, they have many common features. The fundamental purpose of both systems is the satellite broadcasting of HDTV. FM modulation, bandwidth compression by subsampling, and sampled value transmission are important common technical points. The major factors of these systems are summarized in Table 12.11.

Both of these systems were developed for the practical broadcasting of HDTV, and easy introduction has been considered. The technical commonality of the systems is Level 2. Encoder and decoder technology was new, but other components of reception facility, including antenna and tuner, were developed with almost the same technology as conventional broadcast satellite systems.

The scanning parameters do not coincide, as described already, but the number of lines and frames does not affect the technical principles of the system if compatibility is not taken into consideration. In this sense, a 1250/50 version of

Table 12.11   Major factors of MUSE and HD-MAC.

	MUSE	HD-MAC
Media	BS	BS
Transmission	Analog sampled value	Analog sampled value
Bandwidth	8.1 MHz	10.125 MHz
Modulation	FM	FM
RF bandwidth	27 MHz	27 MHz
Subsampling	Field/frame	Field/frame
Scanning format	1125/60 2:1	1250/50 2:1
Audio multiplex	Digital, baseband	Digital, baseband
Control signal	Digital	Digital DATV system
Motion compensation	Fieldwise	Blockwise
Sampling pattern	Fixed	Switched, blockwise
Adaptive control	Pixelwise (Type 1)	Blockwise (Type 2)
Chrominance	Line sequential	Line sequential
Compatibility	Grade 5	Grade 4
Commonality	Level 2	Level 2

MUSE or a 1125/60 version of HD-MAC are possible. But this does not make sense because HD-MAC was developed to keep direct compatibility with MAC family, of which the scanning is 1250/50.

The big difference is that MUSE uses a fixed sampling pattern but HD-MAC uses two sampling patterns chosen according to the grade of motion. This relates to the blockwise control of HD-MAC.

12.12 MUSE System

The MUSE system was developed for HDTV broadcasting via satellite [4] and has been adopted as the Japanese HDTV broadcasting system for broadcast satellites. It is very flexible, and many experiments have shown that MUSE can also be used for AM, including VSB systems, CATV, and package media like VCR or videodisc.

MUSE is being used in daily HDTV broadcasts, which have already been going on for more than two years on an experimental basis and which expanded into nearly full scale test broadcasts late in 1992. HDTV receivers including an LSI MUSE decoder [6] are already being sold in Japan.

12.12.1 Technical Basis of MUSE

The most important technical bases developed for MUSE are as follows:

1. Bandwidth compression using multiple subsampling and motion compensation.
2. Analog sampled value transmission including an automatic equalization system.
3. Synchronization that provides an accurate resampling phase. This uses a positive synchronization signal, which eliminates a synchronization pulse loss of 3 dB.
4. Efficient nonlinear emphasis of gain 9.5 dB.
5. Quasi-constant luminance principle system applied to signal processing and transmission. In this way, crosstalk between the chrominance (C) and the luminance (Y) signal can be considerably reduced, and the SNR for highly color-saturated pictures improved. This leads to a lower necessary received C/N.
6. A baseband multiplex system for audio and independent digital data. The signal format is the same as for the video signal.

12.12.2 Principle of the MUSE System

The MUSE system can be transmitted through an analog channel with the technical bases just listed. The fundamental basis is dot-interlaced sampling. In the MUSE system, the picture is processed according to its motion. For moving parts of the picture, a line-offset subsampling system is used, and for stationary parts a frame- and field-offset subsampling system. The resultant sampling patterns are the same. MUSE is of type 1 in Table 12.7. This is important, because motion detection, needed to switch between moving-area processing and stationary-area processing, must be done independently in the encoder and decoder. If the sampling patterns were different, the degradation of picture quality would be severe when motion detection was not the same in the encoder and decoder. When the sampling patterns are the same, the evil effects of a mismatch in motion detection are not so great.

The sampling patterns for a stationary part of the Y signal are

- Original sampling; 48.6 MHz orthogonal;
- First subsampling; 24.3 MHz field offset; and
- Second subsampling; 16.2 MHz frame offset and line offset.
- For moving parts, only the second subsampling is used; The sampling system is shown in Fig. 12.8.

The transmissible areas in two-dimensional spatial frequency space are shown in Fig. 12.9. To match with the horizontal blanking period of a studio HDTV signal, a slight time compression of 12:11 is inserted in the input of the MUSE

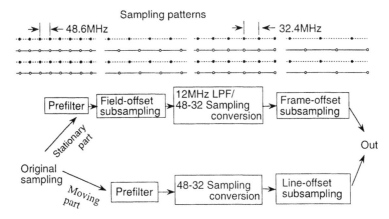

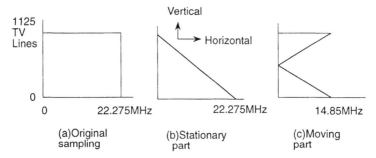

Figure 12.8   Subsampling system of MUSE. The top is a sampling system for the stationary part. In the case of a moving part, the second from the left is not need.

Figure 12.9   Transmissible areas of MUSE subsampling. Prefilters before the subsampling should have pass areas corresponding to these transmissible areas.

encoder. The limit of the horizontal passband for a moving area is 14.85 and not 16.2 for this reason. Prefilters used in real hardware are not the same as in Fig. 12.9. The horizontal values should be multiplied by 12/11. That is, 24.3 MHz and 16.2 MHz should be used instead of 22.275 MHz and 14.85 MHz, respectively.

As in Fig. 12.9, the resolution of the picture is less than for the original sampling, both for moving parts and stationary parts. But we should consider that, in the original sampling, the diagonal resolution is $\sqrt{2}$ times of that in the horizontal and vertical directions. Even if the human visual system were isotropic, this high diagonal resolution would be useless, and it is known that the modulation transfer function (MTF) of human vision is less in the diagonal direction. We may conclude that the MUSE system is a good match to the human visual system.

For moving parts, the resolution is less because temporal interpolation cannot be used. But again we know that the resolution of human vision is much less for moving objects. Thus motion blur is not objectionable.

In the case of a camera pan or tilt, blurring and a dirty-window effect may be noticeable because the eye can track moving objects and because the viewer knows that the real object is not moving. Motion compensation is introduced to reduce the blurring and the dirty-window effect. We can avoid blurring produced by spatial interpolation by using motion-compensated temporal interpolation, which can be achieved by shifting the read-out addresses in the field memory and frame memory used for interpolation. We can avoid a sticking dirty-window effect by the same method.

MUSE, therefore, is a sampled value transmission system. It is important not only to have good amplitude and group delay response, as in other waveform transmission systems, but also to have accurate resample timing.

It is well-known that if the resampling phase is accurate, distortion free transmission can be achieved if the transmission characteristic has a -6dB roll-off at half the sampling frequency and a phase linear response, as described before. An automatic waveform equalization system has been developed to keep the distortion free condition.

12.12.3 Encoding and Decoding

Figure 12.10 is a block diagram of an encoder, and Fig. 12.11, that of a decoder. The construction of MUSE encoders and decoders is not unique: there can be many designs. These diagrams are examples, and have been somewhat simplified to aid understanding.

At the input of the encoder, the RGB HDTV signals are put into linear form by an antigamma process. By using a matrix, the signals are converted into Y and two chrominance components (R-Y, B-Y). The video signal is first time compressed by 11/12, so after compression, the original sampling frequency is 48.6 MHz. The chrominance signals and luminance signal are time-domain multiplexed by the TCI (time compressed integration) encoder.

The signal is processed in two ways: for the stationary parts of a picture and for the moving parts. For the moving parts, the processing is rather simple. A diamond prefilter is used for line offset subsampling. The shape of the filter is shown in Fig. 12.9c. In this case, sampling frequency conversion is needed because the original sampling frequency is 48.6 MHz.

For stationary parts, the processing filter is of a diamond type, of which the first quadrant is shown in Fig. 12.9b, and is followed by field-offset subsampling. The signal is field-offset subsampled at 24.3 MHz. The higher frequencies are eliminated by a 12 MHz low-pass filter, and the sampling frequency is converted into 32.4 MHz.

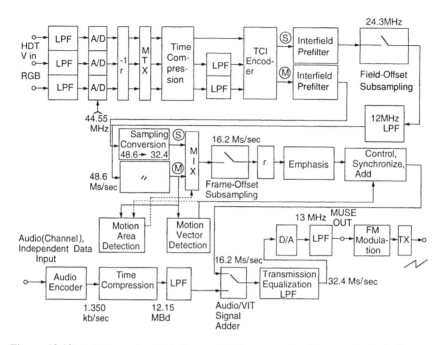

Figure 12.10 MUSE encoder block diagram. This is an example of the encoder block diagram. There would be many ways to construct an encoder.

These two signals, that is, the signals for the moving parts and the stationary parts, are combined pixelwise according to the degree of motion. The signal is subsampled again at 16.2 MHz, the sampling frequency of the MUSE signal.

This MUSE signal is a linear signal. To reduce the visibility of noise in dark areas, the signal is processed nonlinearly, and this is called *transmission gamma*. After preemphasis, this signal, the control signal, the synchronization signal, and the digital audio signal are all multiplexed into a single format.

The audio signal is multiplexed in the vertical blanking period in ternary 12.15 MBaud format. The transmission capacity is 1350 kb/sec, which means that four channels of high-quality audio and about 100 kb/sec of independent digital data can be carried.

In the decoder, these processes are inverted. A block diagram of an example decoder is given in Fig. 12.11. The demodulated MUSE signal is fed to the A/D converter. The resampling phase of the A/D converter is important, as described earlier. After demultiplexing the audio, synchronization, and control signals, the video signal is interpolated.

The interpolation process has two branches, as in the encoder. One is for moving parts and the other for stationary parts. For moving parts, the signal is interpolated

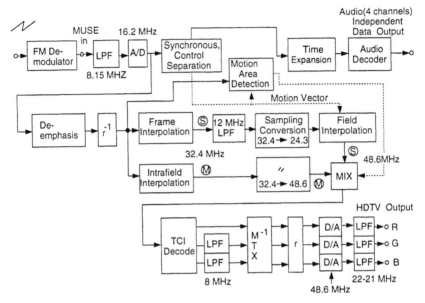

Figure 12.11 MUSE decoder block diagram. This is an example of the decoder block diagram. There would be many ways to construct a decoder.

by the intrafield interpolator (spatial interpolation). For stationary parts, the signal is first interpolated by the frame interpolator, which gives the signal a sampling rate of 32.4 MHz. After elimination by a 12 MHz low-pass filter, the sampling frequency is converted into 24.3 MHz to carry out field interpolation. Motion compensation is applied to both frame and field interpolation [7]. The motion vector is detected in the encoder and transmitted as part of the control data.

These two branches of the signal are combined pixelwise according to the degree of motion. The signal is then fed to the TCI decoder, from which the Y and two components of the C signals are extracted. The inverse matrix converts the signals into RGB format, and at the last stage, they are corrected for display gamma.

The transmission signal format of the MUSE is shown in Fig. 12.12.

12.13 HD-MAC System

A European research project, the Eureka 95 project, whose objective is to bring HDTV to the home of the consumer starting in 1992, has developed the HD-MAC system. An essential demand for European HDTV broadcasting is compatibility with the MAC–packet family, which is the agreed-on European television standard for direct broadcast by satellite. That is, both a high-definition receiver on the

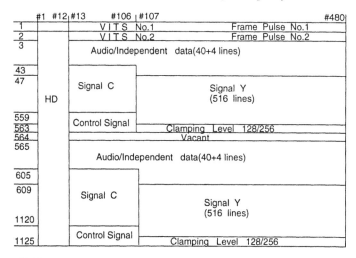

Figure 12.12   Transmission signal of MUSE. All signals are multiplexed in time domain. The sampling frequency and phase are uniform throughout the total signal.

higher line standard and a normal definition receiver on a 625-line standard must be able to receive HD-MAC broadcasting.

A 1250-line standard, which is optimally related to the 625-line MAC emission standard, has been adopted. The HDTV (1250 lines, 50 field/sec, 2:1 interlaced) signal is coded into a video signal and an additional digital control signal, which is transmitted in the vertical blanking interval. After bandwidth reduction, the luminance and chrominance signals are multiplexed according to the MAC–packet family standard. This allows normal MAC receivers to display an HD-MAC signal as a MAC-like signal. The displayed quality depends on the signal format and the construction of the MAC receiver. Some artifacts can be observed in HD-MAC reception by current D2-MAC receivers. The grade of impairment has been examined by the EBU (European Broadcasting Union). Unfortunately, the results have not been published at the time of writing.

HD-MAC is a type 2 system in Table 12.7. The digital control signal is specially designed for HD-MAC receivers and conveys all the information required to reconstruct the HDTV signal. This is called the *DATV signal*. The DATV allows the HD-MAC receiver to act as a complete slave of the encoder, so that reliable motion adaptive video processing can be done in the decoder. Motion detection can be done in the encoder better than in the decoder. The coincidence of motion adaptive control in the encoder and decoder is perfect if the DATV transmission has no error. The maximum available bit rate for the DATV channel of an HD-MAC signal is about 1 Mb/sec. This means that motion adaptive control must be blockwise, not pixelwise.

In the Eureka 95 project, many candidates for the HD-MAC system were studied over two years. This is why there seem to be several versions of HD-MAC.

The sound and data signals of HD-MAC are the same as the standard specification for the MAC–packet family.

12.13.1 Technical Basis of HD-MAC

As described previously, the basic technology of HD-MAC has many things in common with MUSE. Naturally, the practical construction of the HD-MAC system differs from MUSE, even if it uses field- and frame-offset subsampling, like MUSE. But one point is unique. The system was designed for compatibility with the MAC family. To have a common signal format for both 1250- and 625-line systems, two consecutive intrafield lines of the 1250-line system are transmitted as one of the 625-line system, as shown in Fig. 12.13 [3].

This technique is called *line shuffling*. With this method, HD-MAC can carry a 1250-line signal on a 625 scanning format. Figure 12.13 is an explanatory figure for the 80 msec mode of sampling to be described later. The left-hand side of the picture shows the original HDTV scanning pattern. In Fig. 12.13, two successive fields are shown. The solid lines are scanning lines of the current field, and the broken lines are of another field. As shown in Fig. 12.13, same sampled signals are moved to another line. We transmit signals originating in two lines on a single line of the transmission format. This is the MAC transmission format.

In an HD-MAC receiver, the samples are separated and moved into the original positions. In the case of MAC reception, band limitation by the MAC receiver acts as a vertical low-pass filter.

Unlike MUSE, HD-MAC uses blockwise motion adaptive control. This has three modes with different sampling methods. The motion compensation system

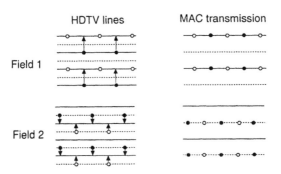

Figure 12.13 Line shuffling technique of HD-MAC. By this method HD-MAC keeps compatibility with the MAC reception.

of HD-MAC is also blockwise. Control signals are assigned to each block and transmitted through the DATV channel.

12.13.2 Principle of the HD-MAC System

Bandwidth reduction in HD-MAC systems is done by spatio–temporal sub-Nyquist sampling, adaptive to the picture content. Three different picture-refresh frequencies provided for three different velocity ranges.

In the stationary mode (velocity range; 0–0.5 sample/frame), the refresh interval is 80 msec. In the slowly moving mode (0.5–12 samples/frame), the refresh interval is 40 msec. In the moving mode (more than 12 samples/frame), the refresh interval is 20 msec. These three modes correspond to the three subsampling patterns, as shown in Fig. 12.14. The numbers correspond to the field numbers, and the positions of the number show the sampling points. The sampling patterns are in the 1250-line system, and the original sampling frequency is 54 MHz. The numbered points are sampled points of the numbered fields.

The transmissible areas of the sampling modes in the two-dimensional frequency domain are shown in Fig. 12.15. In HDTV scanning, the selection of the mode is done blockwise in blocks of 16 lines and 16 samples. For the 40 msec mode, motion compensation is used. Motion compensation is also blockwise. The motion vector and mode-switching information are transmitted through the DATV channel, which is a baseband digital multiplexed channel in the vertical interval of the video signal.

Line shuffling, described previously, is used only in the luminance 80 and 20 msec modes of the luminance signal.

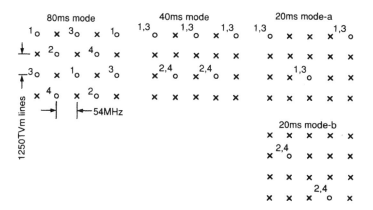

Figure 12.14 Sampling system of HD-MAC. There are four sampling patterns. In the 20 msec mode, two patterns will be used depending on the phase of sampling and motion vector.

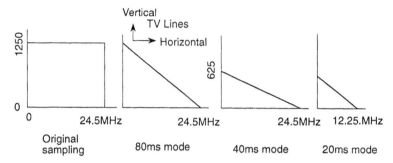

Figure 12.15   Transmissible areas of HD-MAC. Prefilters before the subsampling should have pass areas corresponding to these transmissible areas.

12.13.3 Encoding and Decoding

Figure 12.16 is a block diagram of an encoder, and Fig. 12.17 is one of a decoder. In Figs. 12.16 and 12.17, only the luminance parts are shown, but the encoding and decoding block diagrams for chrominance signals are not much different. Of course, many important constituent elements, including input and output matrices, audio signal processors, and the DATV encoder and decoder are not included in Figs. 12.16 and 12.17, but these are not essential technically.

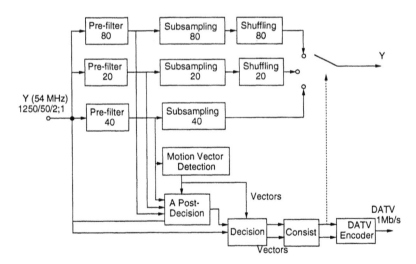

Figure 12.16   Block diagram of HD-MAC luminance encoder. This diagram shows only the luminance part. The DATV signal is combined with the video signal as shown in Fig. 12.18.

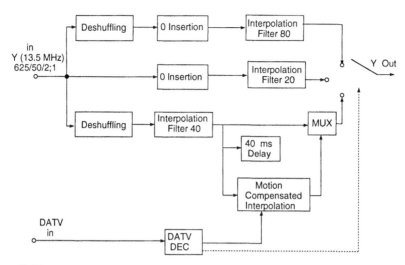

Figure 12.17   Block diagram of HD-MAC luminance decoder. This diagram shows only the luminance part.

In the encoder, the video signal is fed to three branches; that is, the 80 msec mode, 40 msec mode, and 20 msec mode. In each branch, the signal is processed by a prefilter corresponding to the transmissible areas shown in Fig. 12.15. The output of the 40 msec prefilter is used for motion vector detection for motion compensation in the decoder. Which branch should be used is decided using the motion vector, which shows the range of motion.

In each branch, the prefiltered signal is subsampled and in the 80 and 20 msec modes the signals are shuffled. The output signal to be transmitted is chosen from these three outputs according to the decision for selecting the mode blockwise. The result of this decision is sent though the DATV channel, and the video signal, audio signal, and DATV signal are combined into a single MAC signal for emission.

In the decoder, these processes are inverted. After the separation of the luminance, chrominance, audio, and DATV signals, the luminance signal is fed to three branches again. In the 80 msec and 20 msec branches, the signal is deshuffled. After that, the signals are interpolated according to the transmissible areas of the modes. In Fig. 12.17, the interpolation processes are shown as divided into zero insertion and low-pass filtering. For the 40 msec mode, motion compensation is done using the motion vectors sent through the DATV channel. The output of the decoder is chosen from the outputs of the three branches according to the control signals sent through the DATV channel blockwise.

The transmission signal format of the HD-MAC is almost the same as the D2-MAC signal, as shown in Fig. 12.18.

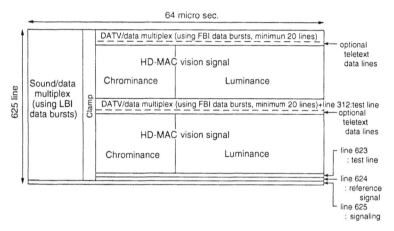

Figure 12.18   HD-MAC signal format. All signals are multiplexed in the time domain. This format is almost the same as the D2-MAC signal except for the DATV signal.

References

[1] CCIR Rec. 709, "Basic parameter values for the HDTV standard for the studio and for international program exchange," 1990.

[2] M. Tanimoto *et al.*, "TAT bandwidth compression system for high-definition television." *J. ITE Japan*, vol. 39, p. 934, 1985.

[3] P. Bernard, "HD-MAC coding scheme and compatibility." In *Proc. Third HDTV Workshop, HDTV Broadcast.*, 1989, 4-2.

[4] Y. Ninomiya *et al.*, "An HDTV broadcasting system utilizing a bandwidth compression technique—MUSE." *IEEE Trans. Broadcast.*, vol. 33, p. 130, 1987.

[5] F. W. P. Vreeswijk and M. R. Haghiri, "HD-MAC coding for MAC compatible broadcasting of HDTV signals." In *Proc. Third HDTV Workshop, HDTV Broadcast.*, 1989, 1-1.

[6] M. Abe *et al.*, "LSI family for MUSE decoder." *IEEE Trans. Consumer Electron.*, vol. 36, pp. 684, 1990.

[7] Y. Ninomiya and Y. Ohtuska, "A motion compensated interframe coding scheme for television pictures." *IEEE Trans. Commun.*, vol. COM-30, p. 20, 1982.

Chapter 13

Video Communications Technologies I: Narrowband Transmissions

L. F. Chang
Bellcore
Red Bank, New Jersey

and

T. R. Hsing
Bellcore
Morristown, New Jersey

13.1 Introduction

Communications through visual means dominate most of our daily lives. Statistically speaking, about 70% of the communication in human beings activity mechanism belongs to visual communication category. Coupling the rapid progress of signal processing and VLSI technology with many world-wide standards such as H.261 and MPEGs, visual communications have become more feasible than ever, both technically and economically. At present, there is an increasing demand for visual communications through relatively narrowband channels. For example, one current goal that poses a challenge to the visual communication community is to produce various video codecs, ranging from a few kb/sec to a few Mb/sec that will allow transmission of dynamic visual information over narrowband networks such as ordinary telephone lines and mobile communications channels.

Just a few years after inventing the telephone, Alexander Graham Bell patented twisted pair wiring (1881). Since then, the worldwide subscriber loop plant has grown to become one of the world's most valuable technological assets. High-performance transmission over subscriber loops has historically required considerable engineering to accommodate the wide range of loop makeups encountered in the distribution plant. Recently, self-adaptive filtering techniques have been applied to transmission over subscriber loops. This technology, which had previously been too complex and expensive for widespread use, can now be integrated into single VLSI devices that make transmission over the loop plant much more robust and cost effective. As we move forward into the age of optical fiber, we should view the copper plant as an asset to aid evolution, rather than as a liability preventing progress. Continuing advances in digital signal processing, parallel processing, and VLSI technology will enable higher bit rates to be achieved. The embedded copper will continue to make up a significant portion of the local loop plant well into the next century, even under the most aggressive fiber deployment scenarios. We believe that the BCC's ownership of the embedded plant can have positive, synergistic consequences relating to the deployment of fiber-in-the-loop (FITL) systems.

The other side of narrowband transmission technologies, namely, wireless transmission, has experienced rapid growth in the past few years. Wireless radio communications provide freedom from a tether, from the need for endpoint wiring. This attribute has offered a new means of communication for people on the move or away from wired communication terminals. Especially, the following factors have led to the feasibility of future personal communications systems (PCSs): (1) evolution in various technologies such as availability of portable data terminals, (2) increased market demand for cellular telephones and cordless phones, and (3) numerous regulatory activities such as the FCC's allocation of PCS spectrum between 1.85 and 2.2 GHz for new telecommunications technologies and applications in the United States and WARC'92's (World Administrative Radio Conference) decision to designate the spectrum around 2 GHz for future public land mobile telecommunications services (FPLMTSs).

It is envisaged that a future PCS featuring low-power handheld multimedia terminals will provide voice and moderate rate data and may further provide video communications services. Currently, per-channel bit rates for various wireless access technologies range from 8 kb/sec to 32 kb/sec. The 32 kb/sec technologies are expected to have compatible voice quality to wireline services. In the United States, several bodies consisting of manufacturers and service providers have started the standard process for low-bit-rate voice and moderate data rate applications for PCS. Internationally, PCS-related activities are continuing in ITU-T and ITU-R. In addition, ITU-T and MPEG are looking into standards for very low-bit-rate visual telephony. MPEG4, a standard aimed at video applications for radio channels with bit rates around 10 kb/sec has been initiated [1]. All these activities indicate the wide range of interest and commitment for future PCS services.

In this chapter, we will review the current research and development activities on the transport media such as wireline loop transmission and wireless radio transmission. Compared with a fiber-based network, it provides a narrowband communication that can cover certain spectrum visual communications services ranging from a few kb/sec to a few Mb/sec. The first part is devoted to the wireline loop transmission technologies, while the second part will focus on wireless.

13.2 Wireline Loop Transmission

13.2.1 The Local Loop Environment and Its Characteristics

The local exchange plant is a resource that began growing over a century ago and has provided basic telephone service ever since. It now represents one of the major components of capital investment in the public network. It is reported, under the FCC-established Uniform System of Accounts, as comprising over 25% of the regional companies' assets, or over $50 billion.

Initially, cables were made up of individual copper conductors wrapped with paper for insulation. The insulated conductors are twisted together into pairs and pairs are grouped into binder groups of typically 25–50 pairs. Binder groups are combined into cables, ranging in sizes from 100 pairs to 2500 pairs. These cables typically emanate from a central office (CO) in underground conduits or ducts. As the cables branch out to businesses and residences they become smaller and aerial suspension becomes more common. Direct burial, without supporting ducts, is also common, particularly in suburban and rural areas. Over the past several decades, various mechanisms including polyethylene insulated conductor (PIC), pressurization, or use of a nonconductive jelly to fill the gaps between copper pairs have been developed to increase the efficiency or reduce the cost of cable maintenance.

New technology has increased the potential digital information-carrying capacity of wire pairs to the point where certain leading-edge broadband services can be provided to customers well in advance of direct fiber access [2]. Recent research indicates that it will be feasible to transmit 1.5 Mb/sec over the majority of existing copper loops for distances up to 18 kilofeet [3]. With recent progress in digital video compression, for instance, such data rates should be adequate for several important classes of video services. We can also envisage other services that can profitably use such data rates. The ability to rapidly deploy new broadband services to sparsely located customers using the existing wires should have a number of positive effects on FITL deployment [4]: affordable broadband services in the near term will build and increase the demand for full-capacity fiber; the local exchange carriers (LECs) will, more immediately, be viewed as viable suppliers of broadband; the costs of FITL will become more manageable since broad-based service offerings can be made without the commitment to "fiberize" all of the serving area; and the appearance of a large volume of new digital traffic in the loop will accelerate the need for fiber.

As shown in Fig. 13.1, the loop plant widely used today for voice (and data modem) access consists of individual copper wire pair connections from a CO (or remote electronics site) to customer locations. Of the over 130 million pairs in service today, about 28% are equipped with loading inductors and are not directly usable as new digital subscriber lines (DSLs). The majority of unloaded loop plant is less than 18 kilofeet in length, so this is a good working limit for ubiquitous residential service. Loops that are in carrier serving areas include loops up to 9 kilofeet if they are 26 gauge, 12 kilofeet if they are 24 gauge, and in between if they are of mixed gauge. This is important because performance improves significantly if the application of new DSL technology can be limited to these shorter lengths. A few typical loop makeups of the plant are shown in Fig. 13.2. The path leading to the customer is typically not just a simple pair of wires. Circuits are frequently made up of various gauges of wire spliced together. Gauge changes result in impedance discontinuities that tend to distort and reflect high-speed digital pulses. Additionally, bridged taps, which are open-circuited branches placed along a main line for provisioning flexibility, also echo signals. The result is not only that received signal pulses must be amplified but also that significant reshaping must be done to reverse the damage done by transmission. The exact degree of reshaping is unique to every wire pair.

In the 1970s, efforts began to improve the bandwidth of the loop plant. One important aspect of these efforts was to introduce the carrier serving area, or CSA, administration guidelines. These guidelines are aimed at shortening the distribution of loop lengths, both eliminating the longer loops and the use of load

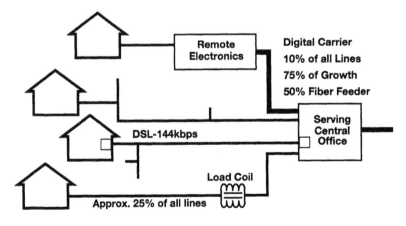

Figure 13.1   Existing loop plant.

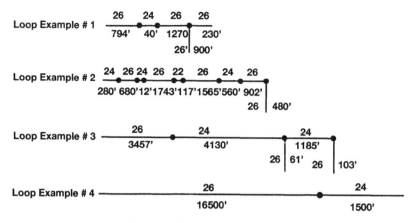

Figure 13.2 Typical loop makeups.

coils to position the loop plant to provide new digital services at speeds in the region of 64 kb/sec. The guidelines also restrict the working length of loops with bridged taps and limit bridged taps and excessive mixing of gauges. Administration of the loop plant using the CSA guidelines began in the early 1980s, particularly with increasing deployment of digital loop carrier (DLC) systems. DLC fueled a trend to place electronics in the distribution network. These electronics are located between the local office and the customer, requiring some type of environmental enclosure such as a hut, a vault, or a cabinet. Among the driving forces behind DLC deployment are the desire to use the plant more efficiently and to shorten the lengths of voice frequency copper loops, making the loop plant more amenable to new digital services. In addition, recent DLC systems use fiber in the feeder network, those facilities from the local office to the remote electronics site.

Although the operating telephone companies are aggressively moving toward fiber deployment in the distribution network, complete penetration may take several decades, particularly to smaller businesses and residences. Thus, there is motivation to use the embedded base of copper facilities effectively during the transition from copper to fiber.

13.2.2 Loop Impairments

Figure 13.3 shows the frequency responses of the sample loops of Fig. 13.2. As can be seen, these responses drop off rapidly at higher frequencies. For voice service, simple equalizers can be used to flatten the channel. If we wish to transmit data, we also need to be concerned with the phase response. A copper loop provides a dispersive communications channel; the phase response is not strictly linear with frequency. Due to the bandwidth-limited nature and the nonlinear characteristics

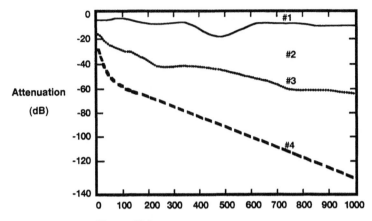

Figure 13.3   Loop frequency responses.

of the phase response, the received pulse at the receiver will be spread out. In a pulse train, the smeared out pulses will overlap with each other, a phenomenon referred to as *intersymbol interference* (ISI). The pulses are also considerably attenuated. The receiver equalizer and timing recovery circuits are faced with the task of compensating for these degradations.

In the access network, there is strong economic motivation to minimize the number of pairs needed to provide service. This is why two-wire transmission has been used for voice service, where transmission occurs simultaneously in both directions over the same copper loop. This is accomplished with a hybrid balance network, which separates the two directions of transmission using a four-port balanced bridge. If the impedances are balanced, a two-wire to four-wire conversion is effected. The same principle is used in data transmission, where pulse trains are simultaneously transmitted in both directions over a single loop in a full duplex fashion. One of the impairments resulting from this technique is residual echo, in which some of the transmitting energy leaks into the receiving path since the hybrid network cannot be perfectly balanced to the line in practice.

Aside from echo and ISI, additional impairments will affect the transmission performance. They are impulse noise, crosstalk, and radio frequency interference (RFI). Impulse noise is created primarily by several mechanisms in which transients are induced in a loop by nearby sources such as dial pulses, switching equipment in the office, and current surges in heavy equipment such as elevators. As the name implies, impulse noise tends to be sporadic with short pulses of energy, on the order of microseconds.

Crosstalk generally comes from the capacitive coupling between different pairs of transmission lines. Within a binder group of a cable there are usually several active pairs. Since these pairs are in close physical proximity over long distances, coupling takes place and the pairs crosstalk into each other. There are two main types of crosstalk to consider. Far-end crosstalk (FEXT) will cause less problems since it encounters large coupling loss and line attenuation; Near-end crosstalk (NEXT) can often cause a large problem.

Telecommunication equipment, by nature of its application in the telecommunication network, may be exposed to one or more sources of electromagnetic energy. In high-bit-rate transmission over copper lines, impairment due to electromagnetic interference needs to be considered. Signals from high-bit-rate services on unshielded drop wires and in-house wires may fall within the frequency bands assigned to aeronautical safety communication channel, maritime service or radio navigation such as Loran, and commercial AM broadcast. One concern is RFI to these licensed services due to electromagnetic emissions from any high-bit-rate transmission system. The other concern is that this kind of system must have the immunity to conducted RFI that is present at the drop wire and in-house wires.

13.2.3 Existing Systems

Historically, high-speed transmission over wire pairs has required considerable special engineering. T1 carrier systems that provide 1.544 Mb/sec digital transmission rates, which have been used in the interoffice plant for many years, require detailed design for proper operation. Carefully separated pairs are used for each direction of transmission. Repeaters are needed at approximately 6000 ft. intervals. Although such systems are often used to provide DS1 (1.544 Mb/sec) service to business customers, they are really not an appropriate vehicle for broadly deployed residential services.

Although it is hoped that FITL installation will bring widespread broadband applications to residential customers toward the end of this decade, there is an urgent need to explore new technologies for immediate applications to meet the requirements of those customers who have wideband service needs prior to such time as the copper is replaced by fiber. While an evolution of copper plant to fiber plant seems inevitable, and fiber is the media of the next century, any increasing demand for wider bandwidth services through a loop plant will need to be satisfied by increasing the efficiency of the existing ubiquitous copper loop plant. Even as fiber systems are deployed, copper wire will still be used, in many cases, for the last portion of the connection to the terminal.

13.2.4 ISDN Basic Access

Due to recent advances in electronics and digital signal processing, it has become possible to transmit a higher bandwidth from voice communication through high-speed data communications over longer distance in the loop plant. The ISDN basic access system [5] was the first embodiment of the digital subscriber line.

This system uses adaptive digital signal processing for both pulse restoration and cancellation of transmit/receive echoes. It provides 144 kb/sec (2B + D) transport in both directions over up to 18 kilofeet of the existing unloaded wire plant. The appearance of the first commercial chips in 1990 proved that sophisticated, automatic line conditioning is feasible in a compact, affordable realization. The planned deployment of ISDN was a major motivation for having the video compression techniques standardized at the transmission rate of $P \times 64$ kb/sec ($P = 1$ to 30), using the ITU-T "Recommendation H.261 : Video Codec for Audiovisual Service at px64 kb/s" [6]. For residential applications, a visual telephony service can be offered today over ISDN-based local access loops by using H.261 standards.

13.2.5 High-Bit-Rate Digital Subscriber Line

With the advance in high-speed digital signal processing and VLSI technology, we can now use high-bit-rate digital subscriber line (HDSL) technology [7] to transmit high-speed digital data through the existing copper loop plant. HDSL is a scheme that uses two pairs to provide two-way 1.544 Mb/sec transport (DS1) to cover the full CSA range. To reduce the impairments caused by NEXT and cable losses, a "full duplex" architecture was adopted. Two pairs of copper wires (as in the case of digital loop carrier (T1) lines) are needed in this architecture. The use of echo cancellation allows each pair to carry 784 kb/sec with full duplex transmission. This system was originally envisaged as a rapidly deployable vehicle for business DS1 services. With the large demand for 1.5 Mb/sec in the commercial arena, a system that provides the equivalent capability as the T1 technology, but uses existing wire pairs with no special engineering is expected to fill an important market need. Since T1 repeaters are avoided, cost savings will result. The customer interface is as with conventional T1.

One of the prime objectives of HDSL technology is to provide a transparent replacement for a T1 line, and several telephone companies are now conducting field trials. It is understood that HDSL will be a transition technology supporting the availability of ubiquitous DS1 service while fiber penetration accelerates.

13.2.6 Asymmetrical Digital Subscriber Line

Several potential services are asymmetric in terms of the bandwidth required in each direction of transmission. These include the delivery of video services at the DS1 rate, many multimedia applications, and high-speed access to database services and educational networks. Because of these and similar applications, the asymmetric digital subscriber line (ADSL) technology was proposed [8] to provide repeaterless unidirectional 1.544 Mb/sec transport capability from network to subscriber, along with a standard bidirectional POTS or ISDN basic rate access channel or both. A low-rate digital control channel will also be available in the reverse direction over the same single, nonloaded twisted copper pair. One feasible technique for the ADSL configuration that we have investigated uses a quadrature amplitude modulation (QAM) passband transmission scheme [9] to transmit information at approximately 1.6 Mb/sec, forming a high-rate channel from the network to the customer. In this configuration, the transmission bands from approximately 50–500 kHz are being considered for the high-rate channel. A low-rate return channel from the customer to the network to provide signaling and control information is expected to occupy a portion of the frequency band that is below the high-rate channel. This architecture results in three separate signal components appearing simultaneously on one wire pair: the POTS operating at 0–4 kHz baseband, the low-speed control channel operating above the voice band, and the high-speed signal operating above that. Several modulation schemes and bandwidth allocations are being studied. The tradeoffs are complex, involving issues of the severe loop loss at the extremes of the nonloaded plant, spectral compatibility [10] with other services in the plant, impulse noise at the residence, coupling signaling transients into the digital receivers, and RFI [11].

Compared with HDSL, ADSL offers several advantages in a residential setting. It can operate over one loop pair whereas HDSL will require two loop pairs. In addition, ADSL structures can potentially cover the entire nonloaded loop plant whereas HDSL is restricted to the CSA range. This increase in capability comes about because self-NEXT, the limiting impairment in bidirectional systems such as the HDSL, does not exist for ADSL. All of the ADSL high-speed transmitters are at the network ends and the receivers are at the customer ends of the loops. Instead of being NEXT-limited, the ADSL architecture will be limited primarily by signal loss, crosstalk from other services in the binder group, RFI, and impulse noise. The ADSL would be ideally suited for residential applications requiring an asymmetric data flow. In the residential environment, the HDSL characteristics could be potentially restrictive, because many residential customers are beyond 12 kilofeet, and the use of multiple pairs per living unit would exhaust facilities in some areas. On the other hand, the present DS1 (and HDSL) service environment is made up almost exclusively of business customers that are expected to be within 12 kilofeet of a remote terminal.

13.2.7 ADSL Transceiver Structure and Potential Impairments

Generally, there are three approaches to implement an ADSL transceiver: carrier-less AM/PM (CAP) [12], discrete multitone technique (DMT) [13], and QAM [8]. Theoretically speaking, those three techniques should have a similar performance if all of them have the same signal processing functions implemented. For example, a block diagram of a candidate QAM-based ADSL transceiver for simulation purposes [8] is shown in Fig. 13.4. A minimum mean-square-error (mmse) decision feedback equalizer (DFE), comprised of a fractionally-spaced feed forward filter (FFF) and a feedback filter (FBF), are included in the receiver. Optimal tap weight coefficients for the DFE are obtained by minimizing the mse between the signal at the input to the detector and the detected DFE output symbol.

Some of the potential ADSL impairments include cable and receiver thermal noise, FEXT or NEXT from other services on the cable, RFI, and impulse noise. The receiver noise is generated primarily by amplifiers in the front-end of the receiver. The cable and receiver noise is usually modeled as zero-mean, additive white Gaussian noise (AWGN), and is referred to as the *background noise* (BGN). To date, we have been using a BGN power spectral density (PSD) of -140 dBm/Hz, as a safe engineering implementation value.

Impulse noise is expected to be one of the dominant impairments for an uncoded ADSL system. The impact of NEXT will depend on the nature of the digital signals in the other cable pairs in the same binder group. It has been shown [9] that if basic access DSLs, HDSLs, or T1 lines are in the same binder group as ADSL, then NEXT could potentially become a very significant impairment. The NEXT crosstalk model has a magnitude squared transfer function that can be expressed in the following manner:

$$|H_{NEXT}(f)|^2 = K_{NEXT}(f/f_0)^{1.5}, \tag{13.1}$$

where $f_0 = 20$ kHz, $K_{NEXT} = 10^{[-(76-6\log(N))/10]}$, and N is the number of crosstalkers. The coupling path for the generation of FEXT can be expressed as

$$|H_{FEXT}(l, f)|^2 = |H_{dis}(l, f)|^2 K_{FEXT} l_{coup} f^2, \tag{13.2}$$

where $|H_{dis}(l, f)|^2$ is the disturbing loop transfer function, l_{coup} is the loop length in feet, f is the operating frequency in Hz, and the coupling constant K_{FEXT} has a typical value of 8×10^{-20} for the 1% equal-level 49-disturber crosstalk. Studies

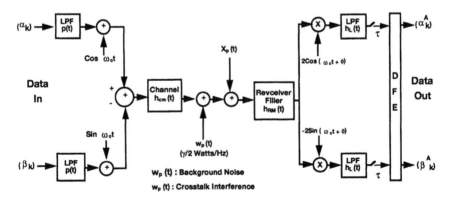

Figure 13.4   A functional block diagram of a passband ADSL QAM transceiver.

have shown that self-FEXT and BGN are not expected to be limiting impairments for 1.5 Mb/sec ADSL.

In the simulation model, it is assumed that the impairment sources are additive on a power basis and statistically independent. They are also modeled as stationary Gaussian stochastic processes. As such, their respective correlation matrices can be derived and used to obtain optimal tap coefficients for the DFE. With a judicious choice of the ADSL operating frequency band, RFI is not expected to be a limiting impairment for a QAM-based ADSL system in general [10].

13.2.8 Next Generation ADSL

Currently about 75 million lines in the U. S. nonloaded loop plant are potential sites for ADSL-based video services. Using the MPEG-I decoder, only one channel can be supported using current ADSL technology. From the video compression point of view, we expect that full NTSC video quality will be achievable at about 3–4 Mb/sec. To provide a full range of video services, offering multiple channels and better picture quality, obviously more bandwidth is required. The needed bandwidth will eventually be provided by fiber, but the current penetration rate of fiber in the feeder plant (primarily in conjunction with the deployment of DLC in the residential environment) is still not large enough to immediately provide ubiquitous service. This gives us the motivation to explore the technical feasibility of a next generation ADSL, providing a digital transport capability that is significantly higher than 1.5 Mb/sec over shorter ranges.

Based on an uncoded QAM passband signaling scheme with perfect timing and carrier recovery, our preliminary simulation results [14] indicate that bit rates of 3–4 Mb/sec may be possible over CSAs (see Fig. 13.4). Compared with the ADSL technology at 1.544 Mb/sec, there will be additional difficulties such as increased signal loss and crosstalk at higher frequencies of operation, spectral compatibility with yet another system operating in the same cable, impulse noise, and RFI to or from AM broadcasting. Research efforts for this higher rate architecture are just beginning now.

If this higher transport capability were provided to next generation ADSL system, it would allow a simultaneous viewing of multiple channels with MPEG-I within a home [15] or at least one channel of NTSC-quality video with the coming MPEG-II standard. The standards effort is now underway, examining compression algorithms for digital video at nominal bit rates of higher than 3 Mb/sec. A working draft of a standard is expected to be completed by the end of 1993.

13.2.9 Discussion

Currently several technologies (including ordinary POTS, voice band modem, T1, and ISDN basic access) are available for providing access using the embedded copper loop plant. Advances in VLSI technology have made ubiquitous per-line DSL technology economically feasible. Continuing advances in both VLSI implementation and signal processing algorithms are now making it possible to provide the HDSL for a repeaterless T1 capability within CSAs, ADSL for a 1.5 Mb/sec transport capability over the nonloaded copper loop plant, and the next generation ADSL for a much higher transmission rate (> 1.544 Mb/sec) over shorter loop ranges. Although more research is needed to better understand the loop and customer cable and plant properties at these higher rates, clearly there is the potential for copper-based public network transport capabilities at rates above DS1. Coupled with the advances in video compression techniques and the recent standards activities in ITU-T, these bit rates will enable LECs to provide video communication and services using highly compressed digital video, while allowing the LECs to use their existing embedded loop plant to enter the video market and establish a base of customers. Demand for much higher quality video and multichannel services will grow. That will then accelerate penetration of fiber in the feeder network first, and eventually fiber in the distribution network supporting fiber-to-the-curb (FTTC) systems.

In the public network, there is a natural synergy between the penetration of fiber into the distribution plant and reuse of the remaining copper at higher rates using DSL technologies. These higher rate technologies are referred to as *very high-bit-rate digital subscriber lines* (VHDSL). Penetration of electronics into the distribution plant began with DLC systems in the 1980s, increasingly fed by fiber in the feeder network. The next stage of network evolution will be driven by FTTC architecture, which, as the name implies, will bring remote electronics even closer to the customer. Initially these systems will provide conventional metallic service interfaces. The service area will be considerably smaller than a CSA. This means that VHDSL technologies operating over distances of 1000 feet can be considered, at bit rates in the range of 10 Mb/sec.

As we have seen, the use of unshielded twisted pairs leads to crosstalk between pairs in the same binder group. Impulse noise also couples into and between pairs. When installing a new transmission capability, network engineers must be concerned with the noise environment that the new transmission system will have to withstand and also the noise levels that the new system will couple into other existing services in the same binder group. To simplify provisioning and engineering, it is highly desirable that all systems in the public network be spectrally compatible, that is, they are neither excessively impaired by crosstalk or impulse noise and do not excessively crosstalk into other systems.

The introduction of new transmission capabilities must be carefully plan-ned in terms of the transmission spectra and the power levels of the systems. Studies indicate that the ISDN basic access DSL, the dual duplex HDSL, and the ADSL can coexist with each other and with digital data service and T1 carrier.

At the higher rates, spectrum compatibility becomes more difficult because of increased coupling and potential radiation to and from systems outside of the cable on which they are implemented. Now, in addition to the relatively short range interference among pairs within a binder group, radiation will travel farther and must conform to radiation limitations set by FCC rules. In addition, at rates of 10 Mb/sec and above over unshielded pairs, commercial aviation, and maritime broadcasting may actually couple into the pairs, creating yet another source of noise. However, these spectrum management issues can be successfully controlled with proper system design by using robust digital filters.

The copper distribution plant can be used to support the evolution of the public network toward a fiber-based broadband network. The use of DSL technology will facilitate this evolution, allowing copper facilities to be reused at even higher bit rates over shorter distances.

ISDN basic rate access is now emerging to provide ubiquitous 144 kb/sec digital capability over the nonloaded loop plant. HDSL will provide repeaterless T1 capability within CSAs. It is expected that the ADSL transport technology concept will emerge to deliver one-way transmission of 1.544 Mb/sec or more from the network to the residence to deliver compressed video and asymmetric data. As FTTC and other broadband architectures are deployed, copper pairs can provide even higher bit-rate access capabilities over shorter distances.

Advances in VLSI technology made ubiquitous per-line DSL technology economically feasible in the 1980s. Continuing VLSI advances will allow DSL technology to prove in at higher bit rates. These advances will allow the telephone companies to take advantage of their embedded base of copper facilities to complement the deployment of fiber.

13.3 Wireless Radio Transmission

13.3.1 System Overview

Wireless radio communication provides freedom from a tether, from the need for endpoint wiring. This attribute has offered a new means of communication for people on the move or away from wired communication terminals. The potential to use this for visual communication, in addition to speech, must be investigated.

Figure 13.5 illustrates one suggested overview system architecture of a future PCS [16]. A multimedia terminal accesses the existing wired network via a radio link. A radio port (RP) serves as an entry/exit to the network. The radio port control unit (RPCU) [17] provides management and control functions between the radio port and the network. The access manager (AM) supports several RPCUs, by querying remote databases for visiting users, assisting in network setup and

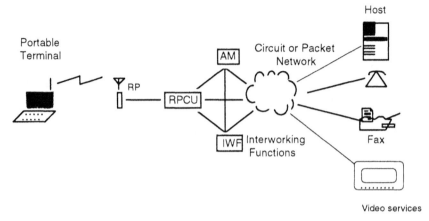

Figure 13.5   Architecture for future PCS system.

delivery, and coordinating link transfer between RPCUs. The interworking functions (IWF) [18–20] module is an important element in the system architecture to support wireless data/video communications. For various data/video applications, it changes signals from the wireless system to a form suitable for the connecting wireline network. For example, a service interworking over wireline voiceband modems could rely on a modem pool in the IWF. These modems may be shared by many RPCUs. Signals from RP to RPCU may be carried over T1, HDSL, or DSL by multiplexing the signals from several RPs. In this architecture, the RP acts as a radio modem and the RPCU performs higher layer signaling and control functions.

In another architecture, the RPCU functions are placed with those of the RP to form an intelligent base station (IBS). Otherwise, the systems are similar.

For each RP or IBS, a segment of radio frequency spectrum is assigned to serve a certain coverage area. The size of the coverage area depends on whether service is provided indoors or outdoors, in an urban or rural area. Since the radio spectrum resource is limited, the same segment of the radio frequency is reused by other RPs as long as they are separated from one another by sufficient distances so that the signal to cochannel interference ratio (S/I) is acceptable to the system. Thus, for spectrum efficiency, the total allocated radio spectrum is divided into separate frequency segments, which are assigned to various sets of RPs. This is the concept of frequency reuse [21]. Frequency reuse has been employed not only in cellular telephony and future PCS services but also in entertainment broadcasting services. Figure 13.6 illustrates the concept of frequency reuse on a perfectly regular grid of RPs.

One major task in employing a frequency reuse concept is to perform frequency planning for an entire working system so that the cochannel interference is minimized. Two approaches can be used to realize frequency planning. One is based

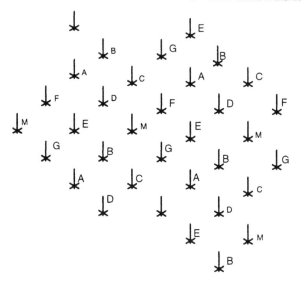

Figure 13.6   An arrangement of Radio Ports. The letter indicates its corresponding carrier frequency.

on centralized computations of frequencies that require location-specific propagation models and high computing power. The other is a self-organizing frequency assignment [22] scheme that calls for each RP to measure interference from other RPs and autonomously select the frequency with the lowest interference. One can also adopt a scheme called dynamic channel assignment (DCA) [23] that dynamically attempts to minimize mutual interference among all the active calls. In this approach, the frequency reuse pattern varies from call to call and it completely removes the requirement for a stable and structured frequency reuse pattern in a service area; however, synchronization among all the RPs is required.

The remainder of this chapter will focus on the description of the characteristics of the radio channel, discussions of the multiple access techniques, and some major issues involved in providing video communication for the future personal communications system.

13.3.2 Radio Channel Characteristics

The radio channel impairments are hard to control and unpredictable. In general, signals transmitted over the radio channel are subject to two types of signal variations; namely, small- and large-scale signal variations [24, 25]. The small-scale signal variations are due to multipath signal propagation. Between each transmitter/receiver pair, many scattered or reflected signals exist due to signal reflections from walls, ceilings, or other objects, resulting in multipath signal propagation. The signals in each path may have different time delay and encounter different path attenuation. Thus, the multipath propagation medium introduces time delay spread and amplitude fluctuation on the received signal. Time delay spread degrades system performance by causing intersymbol interference (ISI). ISI leads to an irreducible bit error rate for typical modulation schemes (e.g., BPSK, QPSK, QAM) [26, 27]. The impairment caused by ISI is usually called *frequency-selective fading*. The instantaneous signal amplitude fluctuations are due to the additions of the multipath signals. In the typical case, the statistics of the received signal amplitude fit a Rayleigh distribution [6, 28], and the phase of the received signal is uniformly distributed between 0 and 2π. Errors occur in bursts when the received signal envelope fades below some noise related threshold, and the length of the erroneous bursts depends on the speed of the time variations of the channel. For the vehicular radio channel, the time variation is relatively faster than the portable radio channel. That is, the signal amplitude may vary from symbol to symbol for the former and remains constant for several symbol durations for the latter.

The effect of multipath propagation varies from environment to environment and can be characterized by a power delay profile $p(t)$. Some measured power delay profiles in mobile [29, 30] and portable [31, 32] radio environment have been documented by Cox and Leek and Devasirvatham. Figure 13.7 shows an average power delay profile obtained from measurements made in New York City.

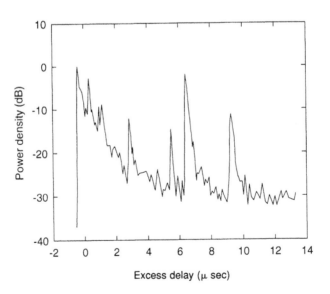

Figure 13.7 Average power received as a function of time delay.

Figure 13.8 shows a typical fading envelope for a small area. One important parameter in the frequency-selective fading channel is the width of the power delay profile, normally called the *root mean-square* (rms) delay spread τ, defined as the square root of the second central moment. The rms delay spread of a radio channel limits the maximum transmission rate of a digital system. Typically, the rms delay spread ranges from a few tenths of a microsecond to several microseconds from portable to mobile propagation environments. In the mobile propagation environments where the rms delay spread significantly degrades the system performance, techniques such as channel equalization [33], i.e., picking up a "main signal" and canceling the echoes; or a Rake receiver [34], which takes advantage of all the multipath, may be employed to combat the effect of multipath propagation. In the portable propagation environment (short transmission range, in-building, and low-height antennas), the multipath delay spread is generally a few tenths of a microsecond and the system can support reasonable transmission rates (a few hundreds kb/sec to Mb/sec) without equalization.

In addition to the effect of the multipath signal propagation on the instantaneous signal power, the local mean signal (average of the instantaneous signal over a distance of several wavelengths) power experiences other type of signal variations, the large-scale signal variations. Factors affecting the large-scale signal variations include path attenuation due to the distance between the transmitter and the receiver, terrain obstructions, and antenna height. The impact of these factors on the receiving signal can not be mathematically analyzed and are resolved mostly by measurement or experimental data. Measurements done in the 800–900 MHz band in the large buildings and residential areas have indicated that (1) the

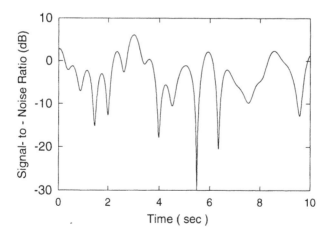

Figure 13.8 A typical Rayleigh fading envelope, at 1.8 GHz for a speed of 1 mile per hour and an average receiver input signal-to-noise ratio of 0 dB.

path attenuation of the received signal level is approximately proportional to d^{-4}, where d is the distance between the base station and the portable; (2) the local mean signal is a log-normal distributed random variable with a mean that varies with distance. This type of signal variation excluding the effect of path attenuation is called *shadow fading*.

Diversity techniques are believed to be the most efficient way to mitigate radio channel impairments. The concept of diversity is to form two or more noncorrelated signals and then obtain a selection or a combination of the "best" received signal that is robust to interference, multipath delay spread, and shadow fading. There are two types of diversity; namely, macroscopic and microscopic. *Macroscopic diversity* is implemented by means of automatic link transfer, to mitigate large-scale signal variations or shadowing. *Microscopic diversity* is usually implemented at the radio link level to mitigate small-scale multipath signal variations.

Antenna, frequency, and time diversity are used to provide noncorrelated signals. Interleaving with error-correcting codes can be viewed as a type of time diversity. The effectiveness of time diversity is determined by whether the overall interleaving span can completely decorrelate the received signal. For PCSs, since the portable set is either stationary or moving slowly, the overall interleaving delay required to randomize the slowly varying signal may be excessive. For some PCS applications such as voice communication, this excessive delay (which can be as much as a second) is not tolerable, thus time diversity alone is inefficient. Antenna diversity, using physically separated antennas, antennas with different polarization, different angles, or different patterns, is currently used in some existing cellular systems and is proposed for the future PCS system. Antenna diversity is widely used and attractive because it requires no radio spectrum and thus is spectral efficient. Frequency diversity is accomplished by sending the signal at different frequencies. Frequency hopping and direct sequence spread spectrum modulation techniques are examples. A system using a spread spectrum multiple access technique benefits from the effect of frequency diversity. The feasibility of implementing various diversity techniques also relies on the multiple access approaches adopted in the PCS system. The next section discusses various approaches for wireless multiple access and duplexing methods.

13.3.3 Duplexing and Multiple Access Techniques

The duplexing method is crucial to the overall system design in two-way (full duplex) transmission systems. Systems using time division duplexing (TDD) multiplex the uplink (portables to RP) and the downlink (RP to portables) transmissions together in the same radio frequency channel; i.e., the uplink and downlink transmissions are operated in the same carrier frequency in a ping-pong fashion. TDD requires only one frequency band, however, the transmission rate in each direction is only half of the radio channel rate. The uplink performance of a TDD system with asynchronous ports transmissions is very sensitive to the height and power differences between the portable and the radio port. The uplink interferences are not only from the cochannel portables but also from cochannel port transmissions. Therefore, if the power difference between the portables and the ports is large or if the antenna heights are not the same, the uplink performance is dominated by the strong port-to-port interference and can be degraded by up to 15 dB or more [35]. In this case, time synchronization among ports is required to maintain good frequency reuse efficiency.

Frequency division duplexing (FDD) uses a pair of frequencies for uplink and downlink transmission separately. Therefore, it requires two distinct frequency bands. However, it allows complete freedom for the transmitter and the receiver to operate independently. FDD avoids the need for port synchronization since ports receive and transmit at different frequencies; one port's transmission will not interfere with another port's reception.

It appears obvious that one of the criteria for the selection of duplexing methods depends on the power difference of the portables and the ports. Because of the allocations available in Europe, CT-2 and DECT (digital European cordless telephone) systems adopted TDD. This is acceptable because these systems are targeted inside buildings and homes, ports and portables are about the same height, resulting in the uplink experiencing similar interference levels from cochannel ports and portables. Thus, time synchronization is not required to minimize port-to-port interference. On the other hand, for mobile radio systems, such as the European global system for mobile (GSM) for pan-European vehicular digital cellular mobile radio, and IS-54 (the North American digital cellular radio system), FDD is employed. The reason is the strong port-to-port interference due to the large power differences between ports and portables. Therefore, FDD is adopted to avoid time synchronization among ports and to double the equipment utilization.

In addition to the duplexing method, the multiple access scheme is also a major issue in engineering the system architecture. Much of the debate in the industry regarding the appropriate technology for the next generation cellular or future PCS system has been on the multiple access techniques. Currently, three multiple access technologies have been discussed. They are frequency division multiple access (FDMA), time division multiple access (TDMA), and Spread-Spectrum-Multiple-Access (SSMA).

In an FDMA communication system, radio spectrum is divided into nonoverlapping frequency segments; each user is assigned an exclusive frequency segment for communicating with the RP. To implement FDMA requires bandpass filters at the output of the transmitter and input of the front-end receiver to ensure the orthogonality of the carrier frequencies. An FDMA system is a one-channel-per-carrier system, multiple radio receivers are required for implementing antenna diversity. Apart from the added complexity for achieving antenna diversity, the bandwidth of the allocated frequency is fixed, resulting in a fixed user transmission rate. Therefore, to accommodate various transmission rates in an FDMA communications system, simultaneous use of multiple frequency segments or a flexible bandwidth assignment scheme has to be implemented, yielding higher complexity in the link design. The capacity of an FDMA system is the total number of frequency segments within the allocated band. Notice that, in a stable and structured frequency-reuse system, FDMA is adopted for frequency assignment for different RPs. FDMA is also used in radio and TV broadcasting, in the current U.S. analog cellular system, and in the Northern Telecom CT-2 Plus system.

In a TDMA system, users communicating with the same RP transmit data on the same carrier frequency but at different times. The transmission timing is segmented into frames, each frame is further divided into time slots. A user with a transmission rate equivalent to the time slot duration (i.e., basic transmission rate) uses that time slot in every frame to communicate with the RP. For subrate transmission such as a half or a quarter of the basic rate, a time slot is shared by two (half rate) or four (quarter rate) users, and each individual user would communicate with the RP using the sharing time slot in every other frame (half rate) or every fourth frame (quarter rate), respectively. For a multiple basic rate transmission, multiple time slots per frame are assigned to carry the user information. Thus, a TDMA radio communications system can flexibly adapt to various transmission rates without much complexity.

In addition to providing a flexible transmission rate, a TDMA radio system also has several other advantages; for example, (1) time during a TDMA frame when a portable is not required to transmit or receive can be used to assess the quality of other frequency channels and to measurements for microscopic diversity selection; (2) fewer radio transceivers are required for a given number of user circuits at RP; (3) power control and port synchronization are not required (but, adding the power control and port synchronization to the TDMA system reduces cochannel interference and as a result improves spectrum efficiency of the system). However, a TDMA system does require frame–time slot synchronization and guard times shared among various users.

Forms of TDMA/FDD have been used in the GSM system and the IS-54 North America digital cellular standard. In the future low-power PCS, DECT, Japanese Handy-Phone, and CT-3/DCT-900 will all use TDMA/TDD because of local allocation rules; WACS (wireless access communications system proposed by Bellcore) will use TDMA/FDD as the access/duplexing technique. The WACS is a low delay, low overhead system with 2.5-msec frame duration. In this system, digital signal processing techniques implemented in the receiver permit robust symbol timing and carrier recovery without bit overhead and without performance degradation [36]. Furthermore, a combined time slot synchronization and error detection scheme is implemented in the WACS to perform time slot synchronization with only 2 or 3 bits overhead [37].

Code division multiple access (CDMA) and frequency hopped multiple access (FHMA) are based on spread spectrum technology. In a CDMA communication system, a unique binary spreading pseudo random code (PN code) is assigned for each call to each user, and all active users share the same frequency spectrum at the same time. The signal of each user is separated from the others at the receiver by using a correlator keyed with the associated selected binary spreading code. Since all users share the same frequency spectrum, all other users' signals contribute to the interference level in the system. Power control is essential in a CDMA system where the near-far problem arises (transmitters near a receiver generate overwhelming interference relative to those far from the receiver). If power control can be performed *perfectly*, the overall interference for the weakest users can be minimized. As a result, CDMA system capacity in terms of the number of simultaneous users that can be handled in a given system bandwidth can be maximized. With proper design, CDMA is more robust to the multipath delay spread environment than the other access schemes because of the use of the multiple PN correlators at the receiver to provide path diversity (if PN correlators can capture all the signal power from each individual delay path). It also does not need frequency coordination among all the RPs since all RPs share the same frequency band.

A CDMA proposal is being processed for a U.S. digital cellular standard [38], and a modification of this proposal is also targeted as the future PCS services. Although CDMA has the aforementioned advantages, the system [38] requires systemwide time synchronization among all RPs, coordination of precise power control among RPs from multiple service providers, and higher implementation complexity than FDMA or TDMA. The maximum user bit rate for the aforementioned proposed system [38] is 9.6 kb/sec for 1.25 MHz spreading bandwidth. Increasing the bit rate to more than 64 kb/sec per user to allow low-bit-rate video service can be done either by using more than a 10 MHz spreading bandwidth (which implies significant complexity to be added to the overall radio link design and would require large contiguous bandwidth spectrum allocations to be available) or by reducing the spreading gain, which implies significant capacity reduction.

In an FHMA communication system, the available channel bandwidth is divided into a large number of contiguous frequency slots. The instantaneous carrier frequency for a user is pseudo-randomly hopped to various frequency slots at different time intervals according to the frequency hopping pattern assigned to the user. In an FHMA communications system, frequency diversity is provided inherently except for a low rms delay spread propagation environment. Thus, power control also needs to be implemented to account for a deep fade in the low rms delay spread portable propagation environments. Similar to the CDMA system, frequency planning may not be required because all users hop over the same bandwidth, but system-wide synchronization is required to reduce overall interference. In the GSM system, slow frequency hopping over the TDMA frame is implemented to enhance the radio link performance.

The choice of various wireless access and duplexing technologies for a wireless communications system is determined by the service environments and its applications. For various wireless communications environments, some technology compromises among issues such as complexity, quality, coverage area, capacity, and spectrum efficiency are needed [39].

13.3.4 Video Communications over a Digital Radio Channel

It is envisaged that a multimedia portable may access to a digital video database containing entertainment programs, education material, or other isochronous data. However, video data have to be highly compressed to get by with limited storage space and limited radio transmission bandwidth. Therefore, the portable has to support video data decompression. The portable is a low-power unit and is expected to sustain a long battery recharge time, hence, it is crucial to design the compression techniques such that the decompression will take minimum number of operations with minimum power consumption.

Currently, using the best video compression/decompression scheme, it is estimated that 3–4 Mb/sec data transmission rate is required to support full motion video with entertainment quality. This implies that, for example, with a QPSK (quadrature phase-shift keying) modulation scheme, at least 1.5 MHz per user is needed. For the emerging PCS frequency band, this much bandwidth per user may not be readily available. However, in the future if a separate radio spectrum is allocated for high-bit-rate data service usage or a low-bit-rate (less than 256 kb/sec) video codec is available for full motion video, then full motion video entertainment programs will be well supported by the PCS system. Therefore, in the initial deployment of the PCS system providing multimedia communication services, the most probable image or video applications will be in low-bit-rate services such as inquiry information, educational programming, videophone, or two-way videoconferencing. The ITU-T H.261 standard is for video transmission of videophone and videoconferencing signals at transmission bit rate of P × 64 kb/sec, where P is an integer ranging from 1 to 30. A preliminary study for sending video over a two-way point-to-point radio link using H.261 and DECT has been presented [40]; however, issues such as video quality and delay problems due to error control scheme were not addressed.

The H.261 standard is designed for wireline transmission where the channel characteristics are much more benign than those of the wireless transmission. In ISO, standard processes for MPEG4 have started. The MPEG4 standard is aimed at applications for very low-bit-rate video with bit rates around 10 kb/sec.

It is clear that video compression techniques are crucial in making wireless video transmission feasible. The "best" compression algorithm for wireless video transmission will likely be quite different from the one chosen for the present compression standards. Compression algorithms for wireless video communications need to consider the following issues that pertain to the PCS system: delay, portability of the terminal and robustness to the radio channel impairments. The conventional approach for designing a digital communications system is to optimize the source coder and channel coder separately. The design of channel coding for digital broadcasting channel to portable or mobile receivers has been investigated [41, 42]; however, an optimum video codec for wireless communications remains to be demonstrated.

If the video coder is to be optimized separately from the channel coder, video delay will be the primary concern since the visual quality depends on the end-to-end delay to a great extent. The delay estimation of the present chosen standard includes mainly the video signal processing delay of the compression technique. For wireless transmission, extra delay is likely to occur due to error recovery techniques (e.g., error detection plus an automatic repeat request protocol), and privacy–authentication algorithms employed in the system. (Authentication is needed to ensure that service is not obtained fraudulently; the privacy of the communication is provided to avoid eavesdropping since the radio channel is more susceptible to eavesdroppers than the wireline channel.) Therefore, for a given visual quality criterion, the allowable delay budgeted for the compression/decompression technique designed for wireless application has to be less than that of the wireline transmission. In addition to this delay, during automatic link transfer (ALT), a capability of the PCS system to reroute the call to different RPs when the radio link quality deteriorates due either to the movement of the user or the changes in the radio environment, extra degradation to the visual quality may be presented since during this period the radio link may be momentarily interrupted. Therefore, the video codec should be designed to handle missing frames intelligently, so that appearance of jerky motion can be avoided during the ALT process (typically ranging from 20 msec to several seconds).

An alternative approach is to adopt combined source coding and channel coding [43, 44] in the system. The basic concept is that different features of the digital signal to be transmitted may have different error protection needs due to the characteristics of the source coding algorithm. For example, the coding process may result in certain bits having more "significance" than others; in that case, the more significant bits should be more heavily coded. In addition, error-correcting coding can be further made to adapt to the channel error conditions by using side information containing the channel state. This feature is of particular interest to the multipath radio transmission channel since the channel characteristics vary considerably in time. A rate-compatible punctured convolutional code is one of the proposed channel coding schemes to achieve this purpose [45]. However, the added complexities of the channel encoder and decoder require further evaluation.

13.3.5 Discussion

The design of a high-quality, low-delay, low-complexity video code for personal communications is a challenging task. Several factors influence the design, such as the multiple access scheme or the duplexing method to be employed in the system. Of course, one critical point in the design is to be robust to the radio channel errors. In this chapter, we did not intend to solve the problem; instead, we considered two different approaches. For either approach, one should always bear in mind that asymmetric compression/decompression schemes that require minimum operations and circuitry at the portable are highly desirable. This results in minimum power consumption by the portable unit and leads to a successful personal communications terminal for multimedia services.

13.4 Conclusion

In this chapter, we have reviewed developments in the wireline loop and wireless radio transmissions technologies and discussed related issues of video communications over the two transmission media. Technology evolution plays an important role in the feasibility of supporting video communications over both wireline loop and radio channels. It is likely that continued advances in VLSI technologies will lead to the success of future multimedia communication services either through wireline local loop or wireless radio communications systems.

References

[1] ISO-IEC/JTC1/SC29/WGII MPEG ad hoc group, "Project description for very low bit-rate A/V coding." Draft, Nov. 1992.

[2] D. L. Waring, J. W. Lechleider, and T. R. Hsing, "Digital subscriber line technology facilitates a graceful transition from copper to fiber." *IEEE Commun. Mag.*, vol. 29, pp. 96–104, March 1991.

[3] T. R. Hsing, C.-T. Chen, and J. A. Bellisio, "Video communications and services in the copper loop." *IEEE Commun. Mag.*, vol. 31, pp. 62–68, Jan. 1993.

[4] D. L. Waring, D. S. Wilson, and T. R. Hsing, "Fiber upgrade strategies using high-bit-rate copper technologies for video delivery." *IEEE J. Lightwave Technol.*, vol. 10, pp. 1743–1750, Nov. 1992.

[5] "ISDN basic access digital subscriber line." *Bellcore Technical Reference TR-TSY-000393*, no. 1, May 1988.

[6] M. L. Liou, "Overview of the Px64 kbit/s video coding standard." *Commun. ACM*, vol. 34, pp. 59–63, April 1991.

[7] Special issue on high-speed digital subscriber lines, *IEEE J. Selected Areas Commun.*, vol. 9, Aug. 1991.

[8] D. W. Waring, "The asymmetrical digital subscriber line (ADSL): A new transport technology for delivering wideband capabilities to the residence." In *IEEE Conf. Proc. Globecom'91*, Phoenix, Dec. 1991, pp. 1979–1986.

[9] M. Barton, "On the performance of an asymmetrical digital subscriber lines QAM transceiver." In *IEEE Conf. Proc. Globecom'91*, Phoenix, Dec. 1991, pp. 2002–2006.

[10] K. Sistanizadeh, "Spectral compatibility of asymmetrical digital subscriber lines (ADSL) with basic rate DSLs, HDSLs, and T1 lines." In *IEEE Conf. Proc. Globecom'91*, Phoenix, Dec. 1991, pp. 1969–1971.

[11] M. Barton, "Electromagnetic interference performance of QAM ADSL." *ANSI T1E1.4 Contribution*, T1E1.4/92, May 18, 1992.

[12] M. Sorbara, J. J. Werner, and N. A. Zervos, "Carrierless AM/PM." *AT&T Contribution*, T1E1.4/90–154, Sept. 24, 1990.

[13] J. S. Chow, J. C. Tu, and J. M. Cioffi, "A discrete multitone transceiver system for HDSL applications." *IEEE J. Selected Areas Commun.*, vol. 9, pp. 895–908, Aug. 1991.

[14] M. Barton, L. Chang, and T. R. Hsing, "Very high-speed digital transport capability of a passband asymmetric digital subscriber line." In *IEEE Conf. Proc. Globecom'92*, Florida, Dec. 1992.

[15] D. J. LeGall, "MPEG: A video compression standard for multimedia applications." *Commun. ACM*, vol. 34, pp. 46–57, April 1991.

[16] D. C. Cox, "Universal digital portable radio communications." *Proc. IEEE*, vol. 75, pp. 436–477, April 1987.

[17] TA-NWT-001313, "Generic criteria for version 0.1 wireless access communications systems (WACS)," Issue 1 (Bellcore), July 1992, and Supplement 1, Nov. 1992.

[18] F. Hillebrand, "Implementation of data and telematic services in GSM PLMNs." In *Conf. Proc. Third Nordic Seminar on Digital Land Mobile Radio Commun.*, Copenhagen, Sept. 1988.

[19] D. Weissman, A. H. Levesque, and R. A. Dean, "Interoperable wireless data." *IEEE Commun. Mag.*, pp. 68–77, Feb. 1993.

[20] L. F. Chang, D. J. Harasty, and A. R. Noerpel, "Circuit based data transport for a wireless access communications system." In *IEEE Conf. Proc. ICUPC'93*, Ottawa, Canada, Oct. 1993.

[21] V. H. MacDonald, "The cellular concept." *Bell Syst. Tech. J.*, vol. 58, pp. 15–41, Jan. 1979.

[22] J. Chuang, "Autonomous adaptive frequency assignment for TDMA portable radio systems." *IEEE Trans. Vehicle Technol.*, pp. 627–635, Aug. 1991.

[23] D. Cox and D. Reudink, "A comparison of some channel assignment strategies in large scale mobile communications systems." *IEEE Trans. Commun.*, vol. COM-20, pp. 190–195, April 1972.

[24] W. C. Jakes, *Microwave Mobile Communications*. New York: J. Wiley and Sons, 1974.

[25] W. C. Y. Lee, *Mobile Communications Engineering*. New York: McGraw-Hill, 1982.

[26] J. B. Anderson, S. L. Lauritzen, and C. Thommesen, "Statistics of phase derivatives in mobile communications." In *IEEE VTC'86 Proc.*, May 1986.

[27] J. C.-I. Chuang, "The effects of time delay spread on portable radio communications channels with digital modulation." *IEEE J. Selected Areas Commun.*, vol. 5, pp. 879–889, June 1987.

[28] P. A. Bello, "Characterization of randomly time-variant linear channels." *IEEE Trans. Commun. Syst.*, vol. COMM-11, pp. 360–393, Dec. 1963.

[29] D. C. Cox and R. P. Leck, "Distributions of multipath delay spread and average excess delay for 910-MHz urban mobile radio paths." *IEEE Trans. Antennas Propagat.*, vol. AP-23, pp. 206–213, 1975.

[30] D. C. Cox and R. P. Leck, "Correlation bandwidth and dealy spread multipath propagation statistic for 910 MHz urban mobile radio channels." *IEEE Trans. Commun.*, vol. COM-23, pp. 1271–1280, 1975.

[31] D. M. J. Devasirvatham, "Time delay spread measurements of wide-band radio signals within a building." *Electron. Lett.*, vol. 20, pp. 950–951, Nov. 1984.

[32] D. M. J. Devasirvatham, "Time delay spread and signal level measurements of 850 MHz radio waves in building environments." *IEEE Trans. Antennas Propagat.*, vol. AP-34, pp. 1300–1305, Nov. 1986.

[33] S. U. H. Qureshi, "Adaptive equalization." *Proc. IEEE*, vol. 73, pp. 1349–1387, Sept. 1985.

[34] G. L. Turin, "Introduction of spread-spectrum antimultipath techniques and their application to urban digital radio." *Proc. IEEE*, vol. 68, pp. 328–353, March 1980.

[35] J. C.-I. Chuang, "Performance limitations of TDD wireless personal communications with asynchronous radio ports." *Electron. Lett.*, vol. 28, pp. 532–533, March 1992.

[36] N. R. Sollenberger *et al.*, "Architecture and implementation of an efficient and robust TDMA frame structure for digital portable communications." *IEEE Trans. Vehicle Technol.*, pp. 250–260, Feb. 1991.

[37] L. F. Chang and N. R. Sollenberger, "Performance of a TDMA portable radio system using a block code for burst synchronization and error detection." *IEEE Trans. Commun.*, Jan. 1993.

[38] "Wideband spread spectrum digital cellular system dual-mode mobile station-base station compatibility standard," *Proposed EIA/TIA Interim Standard*, proposed by Qualcomm, April 1992.

[39] D. C. Cox, "Wireless network access for personal communications." *IEEE Commun. Mag.*, pp. 96–115, Dec. 1992.

[40] A. Heron and N. MacDonald, "Video transmission over a radio link using H.261 and DECT." in *Fourth Int. Conf. on Image Processing and its Applications*, Mastricht, the Netherlands, April 1992.

[41] M. Alard and R. Lassalle, "Principles of modulation and channel coding for digital broadcasting for mobile receivers." *EBU (European Broadcasting Union) Technical Review*, No. 224, Aug. 1987.

[42] J. F. Helard and B. Le Floch, "Trellis coded orthogonal frequency division multiplexing for digital video transmission." In *IEEE Conf. Proc. Globecom'91*, Dec. 1991.

[43] D. J. Goodman and C.-E. Sundberg, "Combined source and channel coding for matching the speech transmission rate to the quality of the channel." *Bell Syst. Tech. J.*, vol. 62, pp. 2017–2036, Sept, 1983.

[44] R. V. Cox, J. Hagenauer, N. Seshadri, and C.-E. Sundberg, "A sub-band coder designed for combined source and channel coding." In *IEEE Conf. Proc. ICASSP'88*, New York, April 1988, pp. 235–238.

[45] J. Hagenauer, "Rate-compatible punctured convolutional codes (RPCU codes) and their applications." *IEEE Trans. Commun.*, vol. COM-36, pp. 389–400, April 1988.

Chapter 14

Video Communications Technologies II: Broadband Cable Television Transmissions

W. I. Way
Department of Telecommunication Engineering
National Chiao-Tung University
Hsinchu, Taiwan, R.O.C.

14.1 Introduction

All network providers wish their networks can eventually provide complete broadband services; i.e., services that include voice, data, image, and video. Public switching networks carriers, due to their slow growth in plain-old-telephone-service (POTS)-only revenue, wish to incorporate data and video services in their networks. Cable TV (CATV) operators, with their cable penetration having already reached a saturation level of 65%, have a strong intention to move into POTS and data business to improve their current financial situation. Local-area-network (LAN) operators, knowing their plain data-based revenue cannot last long, are eager to incorporate real-time traffic such as video and POTS, or the so-called multimedia information. It is clear that any company that can provide broadband services to its customers at low cost and at the right time will financially benefit significantly, and this has to do with what transmission media can be economically used. It is clear that coaxial cables, which were initially used to transport multichannel video signals, are inherently broadband for an appreciable distance. This fact, together with the significant improvement in the capacity of many video distribution systems that are composed of optical fiber trunks and coaxial cable branches, suggests a near-term feasible solution from both the economic and technology points of view. The technologies involved in such systems will be the focus of this chapter.

Another point that will be discussed is the evolutionary strategy of broadband systems. If we take a look at those successfully deployed broadband systems, ironically, most of them are not the so-called broadband integrated-service digital network (BISDN) that has been advocated by many telecommunication leaders in the past 10 to 20 years. There is a tremendous gap between what those fancy networks intend to provide and what customers really need today. Early field trials on sending plain old telephone service via optical fibers without knowing how to evolve to broadband economically is one example, the slow acceptance of narrow-band ISDN due to its low bits/sec per dollar is another example. Therefore, only those that design their networks with careful considerations on evolutionary strategies can succeed in deploying broadband systems realistically. This will be the second part of our discussion.

14.2 Coaxial Cable Distribution of Video Signals

A coaxial cable distribution system is a *broadband* system with a *limited distance*. For example, a 1000-feet, 0.5"-aluminum cable can have a total bandwidth of 500 MHz with an attenuation loss of about 14 dB. Furthermore, the cable attenuation loss is proportional to the square root of signal frequency. What this means is that, for wider bandwidth transmission, more amplifiers will have to be used for a given transmission distance. As more amplifiers are used, thermal noise and nonlinear distortion accumulations become a problem. This problem, in combination with other problems such as electromagnetic interference, signal reflections due to impedance mismatch of components, and poor amplifier reliability, makes optical fiber trunk distribution a much better alternative, as will be discussed in Sections 14.3 and 14.4. Despite their limitations, coaxial cables have reached 65% of U.S. subscribers to deliver mainly broadcast CATV signals.

14.2.1 Basic Tree-and-Branch Coaxial Cable System

Figure 14.1 shows a basic tree-and-branch coaxial cable video distribution system. The high-quality video signals for broadcasting are assembled at the headend. The video sources can be from satellites, microwaves, off-air VHF/UHF, or from locally originated programming. The distribution system can be generally separated into

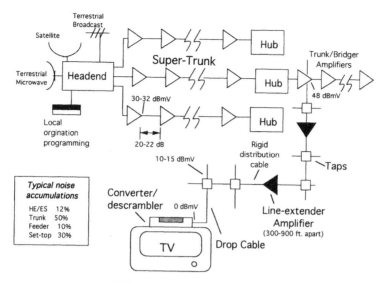

Figure 14.1   Coaxial cable tree-and-branch systems.

four parts; i.e., super-trunk, feeder, subscriber drop, in-home wiring, and customer premises equipment (CPE). In the super-trunk portion, a maximum of 20–30 amplifiers can be cascaded with 20–22 dB budget allocated between amplifiers. Therefore, a typical spacing between trunk amplifiers is about 2000 ft, although an actual spacing depends on the number of video channels to be distributed. Output power of each trunk amplifier is typically around 30–32 dBmV. In the feeder portion, bridger amplifiers are used not only to forward deliver signals, but also to split the signals to a number of neighborhood distribution cables (see Fig. 14.1). Therefore, the output power of a bridger amplifier is typically as high as 48 dBmV. Due to this high output power, nonlinear distortions in the bridger amplifiers and in the distribution line-extender amplifiers are higher than those in the trunk amplifiers. To avoid nonlinear distortion accumulation, only one or two line-extenders are used (typically spaced between 300 and 900 ft) in the distribution portion. Taps in Fig. 14.1 are used to split the signals to drop cables, with a tapping power level of about 10–15 dBmV. About 0 dBmV signal level is expected to be received at the set-top converter box.

Major noise contributions to the entire coaxial cable system are from the trunk and set-top box [1]. However, most nonlinear distortions are contributed by bridger amplifiers and line extenders [2]. Therefore, by replacing the coaxial trunk cables with optical fibers, the noise accumulation problem may be overcome, but the nonlinear distortions due to bridger amplifiers and line extenders may still exist.

14.2.2 Video Modulation Formats for Cable Transmission

Currently, there are two major modulation formats for coaxial CATV distribution; i.e., amplitude-modulated–vestigial-sideband (AM-VSB), whose spectrum is shown in Fig. 14.2, and frequency-modulated (FM) video signals. Each AM-VSB video signal requires only 6 MHz of bandwidth for transmission, but also requires a high signal-to-noise ratio (SNR) which is typically more than 50–55 dB for trunk applications and more than 40–45 dB to the customer site. In a multichannel AM-VSB system, the nonlinear distortion requirements are also very stringent due to the susceptibility of AM-VSB video signals to peaky composite second-orders (CSOs) and composite triple beats (CTBs) that are generated by the multiple video carriers. Typical numbers for trunk systems are that both CSO and CTB should be lower than the video carrier by 65 dB. Note that the aural carrier in an AM-VSB channel is usually lower than the video carrier by about 17 dB to reduce their nonlinear distortion effects. U.S. spectrum allocations for 77 channels of AM-VSB signals (up to 550 MHz) are shown in Fig. 14.3. FM video signals are used mostly in super-trunk systems, which are typically around 20 miles [1]. FM video signals require more bandwidth for transmission but are much less susceptible to noise and nonlinear distortions than AM-VSB signals. Depending on the range of FM deviation, an FM video signal typically requires

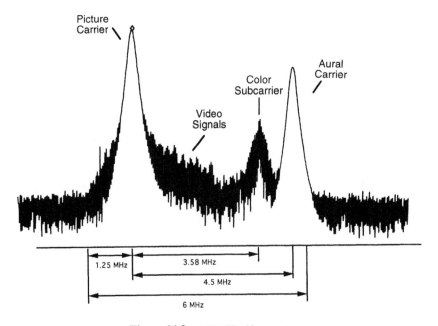

Figure 14.2   AM-VSB video spectrum.

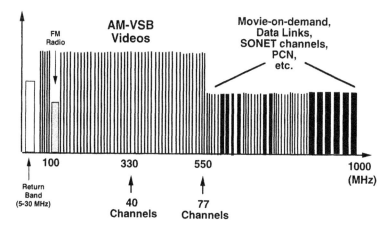

Figure 14.3 A 1 GHz spectrum planning of CATV networks. The current broadcast signals can be increased to 77 channels (up to 550 MHz). The remaining spectrum can be used for various communication purposes by using modulation techniques such as *m*-ary QAM.

only about 17–24 dB of carrier-to-noise ratio (CNR) in order to achieve an SNR more than 56 dB (end-to-end broadcast quality [3]).

Presently, systems or equipment used for analog video signals such as AM-VSB or FM are much less expensive than those used for digital video signals. However, digital video signals are bound to be the future modulation formats because of its versatility for applications such as video on demand or video storage and search. The current trend is toward using digital video signals in CATV systems to transport compressed digital video signals such as MPEG-1 or MPEG-2 [2], and the compressed video signals will be up-converted to frequency range of 550–1000 MHz, as shown in Fig. 14.3, by using multilevel quadrature-amplitude-modulation (QAM) technique [4, 5]. Many companies are currently working toward manufacturing CATV set-top boxes that are based on QAM and MPEG. Multilevel *m*-ary QAM modulation technique is one of the so-called spectral-efficient modulation techniques, since the spectral efficiency can be increased, theoretically, from the conventional 1 bit/sec/Hz to $\log_2 m$ bits/sec/Hz. For example, a 16-QAM modulation can enable a typical 6 MHz AM-VSB video bandwidth accommodate a total of $6 \times \log_2 16 = 24$ Mb/sec digital video signals. If the MPEG-2 video signal bit rate is 6 Mb/sec, then a total of $24/6 = 4$ digital video signals can now be accommodated in one conventional analog channel, and therefore the channel capacity is increased by four times. The theoretical CNR requirement for a bit error rate (BER) of 10^{-9} in the case of 16-QAM is about 23 dB and about 28 dB in the case of 64-QAM. Therefore, as the number of QAM levels increases, the spectral efficiency increases at the expense of higher signal susceptibility to random amplitude noise and hence a higher CNR is required. By compromising between system capacity and cost/ease of transmission, one may find that 16- or 32-QAM signals are the currently most popular format.

14.3 From Coaxial to Optical Fiber Cable Television: Current Status

Starting in 1988–1989, the CATV industry in United States began to deploy optical fibers in their trunk systems. Before that, only a few hundred miles of optical fiber cables had been installed in U.S. cable systems [6]. The slow acceptance of fiber optics resulted from mainly two factors: (1) the cable industry had already wired most nonrural areas with coaxial cable systems, as evidenced by the 65% penetration rate, and (2) early fiber optic technology was unsuitable for transmitting multichannel video signals due to its high cost. The technological advancement in making highly linear and low-noise lasers in the 1989 time frame suddenly boosted the optical fiber deployment in the trunk systems. The laser diodes with superior linearity and low-noise performance enable the cost-effective transmission of up to about 40 channels of AM-VSB analog video signals at that time frame. Even though the cost of a single laser diode may be relatively high, its cost is shared by 10,000 to 20,000 subscribers, who use the same trunk system. In addition, the analog NTSC video format is compatible with the 2 million TV sets that already exist, and no additional cost needs to be spent on digital decoders. Other advantages of using optical fiber trunk systems are as follows:

1. Improved signal-to-noise ratios (SNRs) and linearity performance due to the bypass of multiple cascaded electrical amplifiers in a conventional coaxial cable system. The problems of noise and nonlinear distortion accumulations in a multiple-amplifier coaxial trunk system can be avoided.

2. Easier to implement two-way communications due to the fact that (1) an overlaid optical fiber link can go from the headend to the subscriber site and the original coaxial cable system can be reversed for upstream communications, (2) a typical optical cable accommodates multiple optical fibers, and (3) the wavelength-division-multiplexing (WDM) technique, which transmits multiple wavelengths in a single fiber can be employed.

3. Less susceptible to signal ingress and egress.

4. Lower maintenance cost because most of the optical fiber trunk system is passive, and the only active components that have to be placed in the fields are those optical receivers and a few stages of electrical amplifiers in the hub or subhub locations (see Fig. 14.1). Therefore, there are much fewer active components to experience extreme environmental variations.

To have a smooth transition in replacing coaxial cables by optical fiber cables from an economic point of view, there are several evolutionary stages as will be described.

14.3.1 The Fiber Backbone

This is a system [6] essentially overlaying existing coaxial cable trunk lines with optical fiber cables. Thus, a direct optical fiber path is established from the headend to a remote hub (see Fig. 14.1) that consists of O/E conversion and power supply equipment. From that point on, the existing coaxial cables are still used, with some amplifiers being reversed in direction to distribute video signals. Some spans of trunk cables between hubs can be abandoned or used as a standby, or the trunk amplifiers can now be reversed for transporting upstream signaling for pay-per-view or video-on-demand applications.

14.3.2 Fiber-to-the-Feeder Systems

As a step further from the backbone system, the feeder portion of coaxial cables is replaced by optical fibers, as shown in Fig. 14.4. Unlike the backbone system, which is a star, tree, and branch system, the fiber to the feeder is a double-star, tree-and-branch system. As optical systems penetrate deeper in the subscriber loop, a better signal quality can be achieved at the expense of higher cost of installation. Several technological options are also shown in Fig. 14.4, where we see that, in the headend, one can use either multiple lasers or a power optical amplifier, and in the hub, one can use either another set of laser diodes or an in-line optical amplifier. Basically, economics determines the optimum choice.

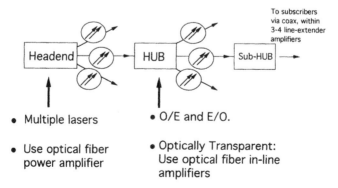

Figure 14.4   A fiber-to-the-feeder system and its associated technology options.

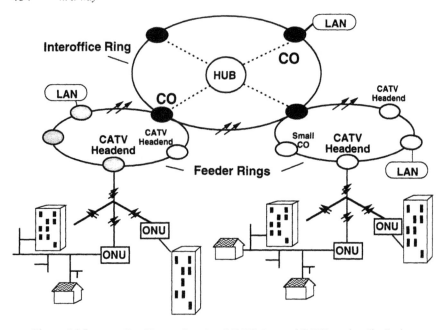

Figure 14.5   Overall architecture based on SONET rings and CATV star-bus distribution.

14.3.3 Ring-Configured Networks

In the third evolutionary stage, CATV's multiple system operators (MSOs) are planning to link multiple headends via a synchronous optical network (SONET) ring architecture as shown in Fig. 14.5. The SONET rings are self-healing if fiber breaks or one of the headends burns out. Network survivability is important as the capacity of optical-fiber based CATV networks increases and as critical services (no disruptions allowed) such as POTS and data are added onto the network.

14.4 Optical Fiber Distribution of Video Signals

Generally speaking, video signal transmission/distribution via optical fiber systems can be classified into four categories: (1) single analog video channel, (2) single (compressed or uncompressed) digital video channel, (3) multiple analog video channels, and (4) multiple baseband time-division-multiplexed (TDM) video channels. Category (1) is the least expensive system, which is suitable for short-distance applications such as remote teaching. Low-cost O/E components such as LED transmitters and connectorized photo-detectors can be used in this kind of applications. Many commercial products are available and not many technical problems remain; therefore, we will not discuss further regarding this part. As for category (2), the most common form is via DS3 (45 Mb/sec) channel through public network. Lower bit rates than T1 (1.544 Mb/sec) for applications such as video conferencing may not require the use of optical fibers. Our discussions will focus mainly on applications for (3) and (4) which require more rigorous technical considerations on system design. The main optical components to be discussed include the use of semiconductor laser diodes and external modulators.

14.4.1 Overall System Considerations

The overall system considerations are concentrated on subcarrier-multiplexed (SCM) lightwave systems that can transport both analog and digital video signals and baseband TDM digital systems, respectively.

Subcarrier-Multiplexed Lightwave Systems   A basic configuration of an SCM system [7–9] is shown in Fig. 14.6. A number of baseband analog or digital signals are first frequency division multiplexed (FDM) using local oscillators (LOs) of different radio frequencies. The up-converted signals are then combined to drive a high-speed light source, which can be either a laser diode or an external modulator, as shown in Fig. 14.7. The LO frequencies are the so-called subcarriers in contrast to the optical carrier frequencies. At the receiver site, a user can receive any one of the FDM channels by tuning a local oscillator and down-convert the RF or microwave signals to baseband or IF frequencies, similar to the way we tune in radio or TV channels. SCM systems have an advantage over TDM baseband digital systems in that services carried by different subcarriers are independent

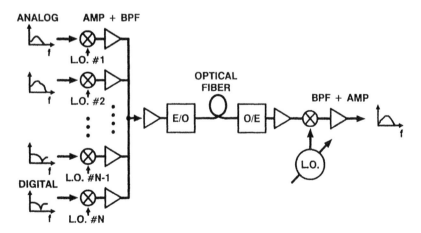

Figure 14.6   A basic subcarrier-multiplexed (SCM) lightwave system configuration.

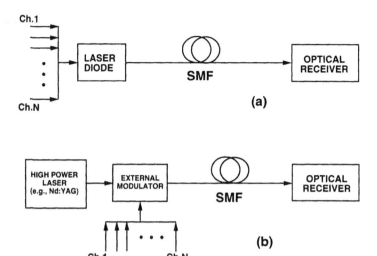

Figure 14.7   Two basic methods of transporting multichannel CATV signals.

of each other and require no synchronization. In addition, SCM systems are presently more cost effective than high-capacity TDM lightwave systems, which are attractive features for near-term deployment in broadband subscriber loop systems.

In Fig. 14.7b, we see that the use of an external modulator, which is usually a LiNbO$_3$-based modulator [10], requires the presence of a high power solid-state laser such as a Nd:YAG laser. A high-power (> 100 mW into single-mode fiber) Nd:YAG laser can overcome the insertion loss associated with the external modulator, which is typically 5 dB or more, and also exhibit an appropriate energy transition after a 800 nm laser diode-array pump into a Nd:YAG rod, which corresponds to a wavelength of 1319 nm [10].

The two most important design considerations in an SCM system are the control of intensity noise and nonlinear distortions. We will review the essence of these two aspects in the next two paragraphs.

Intensity Noise and Degradation Factors   Intensity noise embedded in a received signal directly degrades the signal quality. The basic unit used for intensity noise is relative intensity noise (RIN), which is defined as

$$\text{RIN} = < i^2 > / I_p^2 \quad \text{(dB/Hz)}, \tag{14.1}$$

where $< i^2 >$ is the noise current spectral density, and I_p is the detected dc photocurrent. The noise current spectral density is normalized by I_p^2 so that it is independent of the received optical power or photocurrent. A number of various sources contribute to the RIN of the entire system, and they are listed as follows:

1. *Intrinsic intensity noise.* Intrinsic intensity noise of a laser diode arises as a result of amplification of the quantum fluctuations in the electron and photon population caused by stimulated and spontaneous emission processes and by absorption and radiation losses [11, 12]. The static RIN decreases as the bias current increases according to $(I_b/I_{th} - 1)^{-3}$ [11, 13], where I_b is the bias current, and I_{th} is the threshold current. The dynamic RIN displays a resonance peak at the relaxation oscillation (RO) frequency, and therefore for a fixed RF modulation amplitude, CNR is the lowest near the RO frequency. Since the RO frequency increases as the bias current increases according to $(I_b/I_{th} - 1)^{1/2}$, one should carefully adjust the bias to push the RO frequency away from the frequency range of interest.

The RIN noise of a Nd:YAG laser that is used in an external modulation system is extremely low and was measured to be in the range of -165 to -170 dB/Hz [14], which is at least 10 dB better than a low-noise DFB laser. The peak of the relaxation oscillation typically occurs around 200 KHz, but can be reduced by using a feedback control technique [10].

2. *Direct-optical-reflection induced intensity noise.* Optical light reflected back into the active region of a laser diode that is caused by fiber discontinuities (e.g., fiber connectors or splices) within the coherence length of a laser diode can severely degrade the intensity noise performance. Here the coherence length of a laser diode is inversely proportional to the linewidth of a laser diode [8]. Two measured results on quantifying RIN degradation due to different levels of direct optical reflections are shown in Fig. 14.8 for both Fabry–Perot lasers [13] and DFB lasers [8]. The horizontal axis in each diagram, feedback power ratio (FPR), is essentially the product of the reflectivity of fiber discontinuity and the square of the coupling efficiency from the laser into the SM fiber (in and out). The results show that, in general, the FPR should be lower than -70 dB to achieve a RIN < -150 dB/Hz, which is required for multichannel AM-VSB CATV systems. It is for this reason that an optical isolator (with >30–40 dB isolation) is required at the transmitter end (either within or outside the transmitter package).

3. *Multiple-reflection induced intensity noise.* In spite of an extremely low-noise laser with an optical isolator, a high-RIN level near the low-frequency region may still be observed. One of the main factors is the so-called interferometric noise due to multiple reflections along the system. The interferometric noise is caused by the interference of two light beams arriving at the photodetector at different instants of time (with a time difference of τ), and is a result of converting laser phase noise to extra intensity noise [15–18]. In practical lightwave systems, interferometric noise is due mainly to multiple reflections from fiber connectors. It has been shown that multiple-reflection induced intensity noise is insignificant

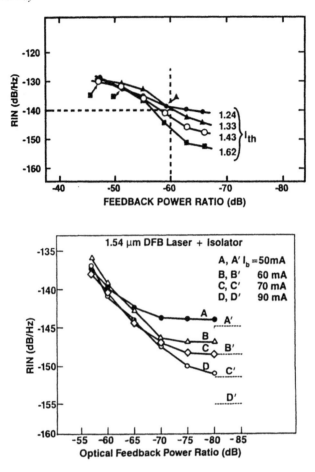

Figure 14.8 RIN versus optical feedback: showing that a laser diode's intensity noise is highly sensitive to optical reflections.

when the laser linewidth is either very narrow or very broad, and the worst case occurs at $\Delta\nu \cdot \tau \cong 0.2$, where $\Delta\nu$ is the laser linewidth. For a typical DFB laser with linewidth 10 MHz $< \Delta\nu <$ 100 MHz, and a path-cord/jumper cable with a length of about 1–10 m, we have $0.05 < \Delta\nu \cdot \tau < 5$. Therefore, multiple-reflection induced RIN for a typical DFB laser linewidth and a typical jumper cable length is close to the worst case condition. On the other hand, an external modulator system that employs a Nd:YAG laser is essentially immune to the multiple-reflection induced intensity noise. This is because the linewidth of a Nd:YAG laser is below tens of kilohertz [19].

Multiple-reflection induced intensity noise can also be induced by "distributed" reflectivities such as Rayleigh backscattering [20, 21], which is caused by small inhomogeneities in the local refractive index of the fiber. A RIN noise caused by this phenomenon was observed to be as high as about -145 dB/Hz for the frequency range close to TV Ch. 2, when the fiber length was 20 km.

Both of these phenomena will be enhanced when the optical amplifiers that provide bidirectional gain for the multiple reflections are used [18]. Unless all discrete fiber discontinuities can be completely eliminated, and when the fiber length is short enough to avoid Rayleigh backscattering-induced intensity noise, optical isolators before and after the optical amplifiers are critical to the system performance.

4. *Mode-partition noise and fiber-dispersion induced intensity noise.* In an SCM system when a multilongitudinal-mode (MLM) laser is used, the phenomenon of mode-partition noise [22] in combination with fiber dispersion will rapidly decrease the available channel CNR [23]. The degradation of CNR due to this effect is proportional to both fiber dispersion and modulating frequency.

5. *Modal noise.* Modal noise is generally observed in multimode fiber systems, caused by interference between light rays traveling different paths in a multimode fiber [24]. This is most commonly observed when a coherent light goes through a multimode fiber discontinuity and experiences spatial filtering of those coherent fiber modes.

6. *Stimulated Brillouin-scattering induced intensity noise.* In an externally modulated system, substantial excess noise can be observed due to Brillouin scattering for injected single-frequency laser power beyond a few to several milliwatts [25]. This essentially sets an upper limit on the maximum power of a Nd:YAG laser that can be launched to a long fiber (e.g., 10–20 km).

Nonlinear Distortions   As in conventional RF and microwave systems, nonlinear distortions include harmonics, intermodulation, and cross-modula-tions. An SCM lightwave system that carries a number of hybrid up-converted analog and digital channels must be carefully designed so that these nonlinear distortions can be minimized. In a multiple-CATV-channel system, the dominant second-order nonlinear distortions are caused by the sum of type $A - B$ or $A + B$, where A and B stand for two arbitrary RF signals; this is the so-called composite-second-order (CSO) term. The dominant third-order distortions are due to the summation of beats due to three distinct spectral components, or the so-called composite-triple-beat (CTB) term. In the following, we will separately discuss the nonlinear distortions due to static nonlinearity (for laser diodes and external modulators) and dynamic laser nonlinearity:

1. *Static laser nonlinearity.* Static nonlinearity can be observed from the light versus bias $(L - I)$ curve, or its slope dL/dI. The optical power ratio of a

second-order nonlinear distortion term to the fundamental carrier is proportional to d^2L/dI^2; the optical power ratio of a third-order nonlinear distortion and a fundamental carrier is proportional to d^3L/dI^3 [7]. Due to the significant progress in improving the laser linearity performance in the last couple years, a number of papers studied the fundamental laser linearity performance according to an idealized $L - I$ curve and the so-called clipping phenomenon [26–28]. A system experiment that closely matches the fundamental clipping limit was an 80-channel AM-VSB, 20-km system with CNR = 51 dB, CSO = -60 dBc, and a received optical power of 3 mW [29].

The nonlinearity of a Mach–Zehnder modulator is essentially frequency independent, and the received photocurrent versus driving voltage can be modeled by a cosine function [30]. The modulator is usually biased at the inflection point of this cosine function so that the second-order distortions vanish [10, 31]. This bias point must be stabilized to maintain the minimum second-order distortions. The third-order nonlinearity due to the cosine function characteristics can be corrected by using a predistortion circuit before driving the modulator [10] or by using a modulator with dual parallel Mach–Zehnder interferometers [32]. Both methods have been successfully applied to several commercial products. A quasi-feedforward technique [10, 33] can also be used, but it may be more costly to implement.

2. *Dynamic laser nonlinearity.* In addition to static nonlinearity, a laser diode also exhibits a dynamic nonlinearity that is intrinsic to the nonlinear photon–electron interaction mechanism. Using a small-signal analysis of the laser rate equations, Lau and Yariv [34] first showed that the calculated two-tone third-order intermodulation distortions from a GaAs laser match well with experimental data. Their results were later extended to InGaAsP lasers [35]. Although the close-form analysis is restricted to two closely spaced microwave carriers, the result shows clearly the frequency-dependent characteristics. The analysis shows that the nonlinear distortions become larger when the modulating frequencies are closer to the resonance frequency of the laser. Because of this frequency-dependent characteristics of a laser diode, the resonance frequency is usually higher than 7 or 8 GHz for a laser diode used for transporting multichannel AM-VSB signals (with modulating frequencies lower than 550 MHz currently). The results in [35] also showed that the second-order nonlinear distortions are proportional to the square of the optical modulation index (OMI), and the third-order nonlinear distortions are proportional to the fourth power of OMI.

3. *Fiber-dispersion induced nonlinearity.* A large fiber dispersion can occur when a 1.3 μm conventional optical fiber is used together with a 1.55 μm source. This large fiber dispersion, when combined with significant frequency chirping of a laser diode, will produce unacceptable second-order nonlinear distortions [36, 37]. It has been shown that, due to the frequency chirp of a typical 1.55 μm DFB laser diode and about 3 to 4 km of conventional single-mode fiber, the CSO can exceed

the limit of -60 dBc. This sets a severe limitation on the transmission distance of such a system. Electrical or optical dispersion compensation techniques [38] can be used to improve this limitation. It should be noted this problem does not exist in an external modulation system, in which case a solid-state or semiconductor laser is running as a continuous wave.

4. *Erbium-doped fiber amplifier-gain-slope induced nonlinearity.* Erbium-doped fiber amplifiers (EDFAs) have been proven to be very useful in extending the transmission distance or the total power budget of a CATV transmission system [39]. For in-line or pre-EDFAs, negligible nonlinear distortions are generated from the EDFAs. However, significant second-order nonlinear distortions can be generated when a gain-saturated power EDFA is used to transmit multichannel AM-VSB signals [39]. This is mainly because there is spectral-gain slope in a saturated amplifier, and when the unavoidable frequency chirp of a laser diode passes through this spectral gain slope, an undesirable FM-to-AM conversion occurs. The undesirable converted AM is the source of the high CSO observed in such systems. Again, optical gain equalization techniques can be used to compensate for these distortions. Note that external modulation systems do not have this kind of problem.

Baseband TDM Digital Lightwave Systems   Owing to economics considerations, multichannel TDM digital video signal distribution at baseband most likely will be used in the public networks that are responsible for transporting multichannel video signals from one location to another. If we take a look of the current development and deployment status of the synchronous optical network, it is obvious that in the near future SONET will be the main vehicle for transporting these TDM video signals. The high speed (622 Mb/sec or 2.488 Gb/sec) and the self-healing characteristics of a SONET network make it a promising network for all video signals that have to be moved around a metropolitan area network (MAN) or a wide area network (WAN); e.g., for the exchange of video programs, or for specialized advertisements.

14.5 Conclusion

Cross-ownership actions between telephone companies and CATV companies have been taking place intensively in United States recently. This can be viewed as due to the inherent broadband feature of CATV networks and the associated potential revenue attract the giant telephone companies. Enabling broadband or multimedia communications via CATV networks will become a major factor to increase the operations revenue. The broadcast entertainment programs of CATV systems will not be the only revenue source. In this chapter, however, we have reviewed only the optical fiber and coaxial cable transmission technologies for broadcasting CATV systems; i.e., one-way point to multipoint broadcast systems. Although the technologies that we reviewed here can also be applied to two-way communications in CATV networks, it is important to know that the hardware technologies are just our stepping stone to a complete two-way communication systems. In the future, we should use the transmission technologies and architectures covered in this chapter as a base and pay more attention to the network aspects such as network communication protocols, video-on-demand storage and access techniques, network management software, etc. Only when the network software issues are resolved can we have a complete broadband CATV-based network and provide complete broadband services.

References

[1] L. Thompson, R. Pidgeon, and F. Little, *IEEE LCS 1*, pp. 26, 1990.
[2] ISO-IEC JTC1/SC29/WG11 CD11172-2.
[3] Electronic Industries Association (EIA) standard RS-250-B, 1976.
[4] K. Feher, *Digital Communications: Microwave Applications*. Englewood Cliffs, NJ: Prentice-Hall, 1981, Chapter 6.
[5] J. G. Proakis, *Digital Communications*. New York: McGraw-Hill, 1983, Chapter 4.
[6] J. A. Chiddix and D. M. Pangrac, *NCTA 37th Annual Convention and Exposition*, 1988, p. 73.
[7] R. Olshansky, V. A. Lanzisera, and P. M. Hill, *IEEE J. Lightwave Technol.*, vol. 7, p. 1329, 1989.
[8] W. I. Way, *IEEE J. Lightwave Technol.*, vol. 7, p. 1806, 1989.
[9] T. E. Darcie, *IEEE J. Selected Areas Commun.*, vol. 8, p. 1240, 1990.
[10] M. Nazarathy *et al.*, *IEEE J. Lightwave Technol.*, vol. QE-11, p. 82, 1993.
[11] Y. Yamamoto, *IEEE J. Quantum Electron.*, vol. QE-20, p. 34, 1984.
[12] Y. Yamamoto, S. S. Saiko, and Y. Mukai, *IEEE J. Quantum Electron.*, vol. QE-20, p. 47, 1984.
[13] K. Sato, *IEEE J. Quantum Electron.*, vol. QE-19, p. 1380, 1983.
[14] R. B. Childs and V. A. O'Byrne, *IEEE J. Selected Areas Commun.*, vol. 8, p. 1369, 1990.
[15] K. Petermann and E. Weidel, *IEEE J. Quantum Electron.*, vol. QE-17, p. 1251, 1982.
[16] R. W. Tkach and A. R. Chraplyvy, *J. Lightwave Technol.*, vol. 4, p. 1711, 1986.
[17] J. Gimlett and N. K. Cheung, *J. Lightwave Technol.*, vol. 7, p. 888, 1989.
[18] W. I. Way *et al.*, *IEEE Photon. Technol. Lett.*, vol. 2, p. 360, 1990.
[19] L. G. Katzovsky and D. A. Atlas, *IEEE J. Lightwave Technol.*, vol. 8, p. 294, 1990.
[20] A. F. Judy, In *ECOC'89 Technical Digest*, 1989, p. 486.
[21] S. Wu and A. Yariv, *Appl. Phys. Lett.*, vol. 59, p. 1156, 1991.
[22] K. Ogawa, *IEEE J. Quantum Electron.*, vol. QE-17, p. 849, 1982.
[23] G. Meslener, In *Optical Fiber Commun. Conf. Technical Digest* 1992, p. 22.
[24] R. E. Epworth, *Laser Focus*, p. 109, 1981.
[25] X. P. Mao *et al.*, *IEEE Photon. Technol. Lett.*, vol. 4, p. 287, 1992.
[26] A. A. M. Saleh, *Electron. Lett.*, vol. 25, p. 776, 1989.
[27] N. J. Frigo and G. E. Bodeep, *IEEE Photon. Technol. Lett.*, vol. 4, p. 781, 1992.
[28] C. J. Chung and I. Jacobs, *IEEE Photon. Technol. Lett.*, vol. 4, p. 289, 1992.
[29] M. Tanabe *et al.*, Technical Digest, *LEOS Summer Topical Meeting on Broadband Analog Optoelectronics—Devices and Systems*, 1990, p. 18.
[30] B. H. Kolner and D. W. Dolfi, *Appl. Optics*, vol. 26, p. 3676, 1987.

[31] G. E. Bodeep and T. E. Darcie, *IEEE Photon. Technol. Lett.* vol. 1, p. 401, 1989.

[32] J. L. Brooks *et al.*, *IEEE J. Lightwave Technol.*, vol. 11, p. 34, 1993.

[33] L. S. Fock and R. S. Tucker, *Electron. Lett.*, vol. 27, p. 1298, 1991.

[34] K. Y. Lau and A. Yariv, *Appl. Phys. Lett.*, vol. 45, p. 1034, 1984.

[35] T. E. Darcie, R. S. Tucker, and G. J. Sullivan, *Electron. Lett.* vol. 21, p. 665, 1985.

[36] E. E. Bergmann, C. Y. Kuo, and S. Y. Huang, *IEEE Photon. Technol. Lett.*, vol. 3, p. 59, 1991.

[37] M. R. Phillips *et al.*, *IEEE Photon. Technol. Lett.*, vol. 3, p. 481, 1991.

[38] C. Y. Kuo, In *Optical Fiber Commun. Conf. Postdealine Paper Digest*, 1992, p. 340.

[39] K. Kikushima *et al.*, *Topical Meeting on Optical Amplifiers and their Applications*, 1990, paper WB1.

Chapter 15

VLSI for Video Coding

P. Pirsch
Institut für Theoretische Nachrichtentechnik und Informationsverarbeitung
Universität Hannover
Hannover, Germany

Architectures for VLSI implementation of video coding algorithms have been studied. These algorithms exhibit very high signal processing demands that call for architectural structures with large number of concurrent operations and data access. Optimization of architectures has to consider the achievable throughput rate and the overall silicon area. VLSI implementations according to function as well as the software-oriented approach will be discussed. As examples of the function-oriented approach, the architectures of dedicated circuits for filtering, discrete cosine transform, and block matching will be evaluated. Software-oriented implementations by multiprocessor systems will be also presented. Heterogeneous multiprocessor systems as a combination of dedicated modules and programmable modules offer compact implementation with restricted flexibility to modifications of algorithms by software changes.

15.1 Introduction

Due to the increasing availability of digital transmission channels and digital recording equipment, video communication services with new features are under discussion or will be introduced in the near future. Essential for the introduction of new services is low cost and equipment that can be easily handled and is small in size. To reduce the transmission cost, source coding methods are applied. Examples of source coding schemes are predictive, transform and interpolative coding [1, 2]. By application of simple source coding schemes, only a small bit-rate reduction can be achieved. Higher reduction factors require sophisticated source coding schemes, such as a hybrid coder.

Video system definition has to consider the specification of transmission standards. The CCITT has standardized a hierarchy of digital channels. Starting from the basis channel of ISDN (64 kbit/sec), also several of these channels are planned for bearer services (H0 = 384 kbit/sec, H1 = 1563 or 1920 kbit/sec). Higher levels in the hierarchy of bearer services with bit rates of approximately 32 Mbit/sec (H3) and 140 Mbit/sec (H4) await final definition. In addition to a hierarchy of synchronous transmission channels with fixed rate, an asynchronous transfer mode (ATM) with flexible bit rates is envisaged for the future.

The bit rates according to the CCITT hierarchy are planned for video transmission. Because of high picture quality requirements, TV transmission between studios is under discussion with 32 Mbit/sec and HDTV with 140 Mbit/sec [3]. To have worldwide transmission for videophone and videoconference services, bit rates in the range of 64 kbit/sec to 1920 kbit/sec are envisaged [4]. Recently new activities are directed to store motion video on compact disc (CD). The motion picture expert group (MPEG) of ISO proposes a coding scheme with about 1.2 Mbit/sec for video [5].

In addition to storage on CD, the proposed MPEG standard should be also used for interactive video services and processing on new high-performance workstations. Because of the expected progress for future high-density recording media the MPEG is now working on new standards with about 9 Mbit/sec for broadcast video and about 45 Mbit/sec for HDTV. The video coding systems just discussed in most cases employ hybrid coding schemes for bit-rate reduction. Hybrid coding is a combination of motion-compensated prediction and transform coding of the prediction error. A block diagram of such a hybrid coder is shown in Fig. 15.1.

The evolution of the standard TV results in systems with higher spatial resolution (HDTV). Subband coding is frequently proposed for bit-rate reduction of HDTV [6, 7]. By 2-D analysis filter banks the original signal is split into several bands of smaller bandwidth. The low-frequency band could be coded by a hybrid coding scheme according to Fig. 15.1, where the high-frequency bands are adaptively quantized and coded by variable length codes. After decoding at the receiver reconstruction of the HDTV signal is performed by 2-D synthesis filter banks.

The envisaged mass application of the discussed video services calls for terminal equipment of low manufacturing cost and small size. For this reason a video codec should be implemented with the least number of chips. Therefore, high complexity in terms of logic functions per chip is requested that will result in large

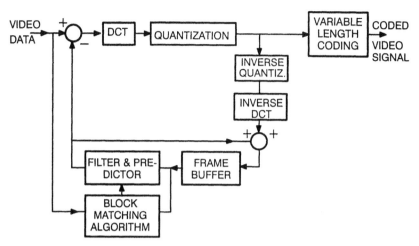

Figure 15.1 Hybrid coding scheme (transmitter side).

silicon area. Manufacturing these large area chips has several limiting factors. A few factors follow:

- Yield (chip size),
- Number of transistors (design geometries),
- Power dissipation (package),
- Number of pins (package),
- Design tools (data volume),
- Testability.

Large area chips have large complexity in terms of basic logic functions, which will increase the time for design, verification, and test. New CAD tools give the designer the possibility to cope with high complexity VLSI components [8, 9]. These CAD tools support all different levels of system description.

The complexity of the circuit description is decreased by the application of hierarchical design methodologies. The design efforts and design errors will be reduced by regular and modular architectures. Design of memory parts and special function blocks is becoming inexpensive due to their high regularity and modularity. The building block concept simplifies the VLSI design by use of high-level synthesis tools. As the number of blocks per chip increases, the number of interconnections between the blocks also increases. The silicon area for interconnection can be restricted by architectures with local connections between the blocks.

The silicon area can be minimized and the performance optimized by a full custom design style. A major drawback of the hand-honed chips is the time for specification, design, and fabrication turnaround. Improvement of these points can be achieved by a semicustom design style [10]. Standard cell design uses predesigned blocks of common height but variable width for basic logic functions. Standard cells can be combined with parameterized full custom macros of high regularity as RAMs, ROMs, and PLAs.

Gate array design is based on premanufactured chips containing an array of transistors. Customization is accomplished by laying down the interconnect paths between the transistors for creating the needed logic functions. Semicustom design is supported by a completely specified database and cell library provided by the vendor and a set of CAD tools.

The alternatives in design styles offer differences between silicon utilization, design efforts, and fabrication turnaround time. For large production volume the design efforts are less important. Hence, full custom design is frequently used. For moderate production volume and system prototypes semicustom is preferred. Due to extreme performance requirements, full custom design is sometimes also used for moderate production volume.

Despite differences in specification and design of the described design styles, the efficiency of a particular implementation can be defined by the quotient between performance and cost [11]. Performance can be measured in signal processing capability per time unit. For a fixed algorithm the throughput rate of data specifies performance. Considering that the manufacturing cost of integrated circuits is mainly proportional to the silicon area, a simplified measure for efficiency becomes

$$\eta = \frac{1}{AT} \tag{15.1}$$

with T as time period specifying throughput rate. With the efficiency according to (15.1) alternative implementations can be compared on the logical level as well as the architectural level. Here the silicon area and time period could be exchanged reciprocally. By taking just the silicon area as a cost factor, the impact of silicon technologies, device testing, chip housing, chip mounting, power dissipation, etc. are not considered. Modification of the efficiency according to (15.1) is reported by considering the impact of yield [12] and power dissipation [13]. The drawback of the improved efficiency measure is that comparison of architectures are more complex. For this reason comparison based just on a cost factor of silicon area is still frequently applied. There is an essential influence of the algorithm onto the architecture. For this reason the derivation of best suited architectures for given video coding algorithms will be further treated.

Characteristics of algorithms related to the required hardware resources and the needed concurrency in operations and parallel access are presented in Section 15.2. Then the two major implementation possibilities will be exemplified. Direct implementation by circuits dedicated to specific functions will be explained for subband coding, DCT, and block matching in Section 15.3. Implementations by programmable multiprocessor systems are shown in Section 15.4. Heterogeneous systems as a combination of dedicated function oriented modules and programmable processors are included.

15.2 Required Parallelism of Video Coding Algorithms

Of primary concern are the hardware resources required for implementation of the signal processing tasks. Hardware resources are the number of logic gates, the memory capacity, and the communication bandwidth for I/O as well as between the modules. A key element in programmable processors is the arithmetic logic unit (ALU). Such an ALU can perform almost all operations. The execution of many operations requires just one clock cycle. Examples are addition, subtraction, compare, negate, and logical XOR.

ALU operations such as multiplication, division, and general shift require several cycles. To achieve higher efficiency, frequently dedicated hardware for implementation of specific operations is applied. Even if hardware implementations of operations result in large variations in number of logic gates and processing time, a simplified processor model will be used to determine a figure of merit for hardware expense. For this simplified processor model each operation is assumed to require the same amount of logic and processing time. In principle an average of all needed operations weighted by the frequency of occurrence has to be considered. For this simple processor model the computational rate would be a figure of merit for the required hardware resources of the operational part. The computational rate specifies the needed number of operations per time unit.

The computational rate is proportional to the source rate which is the product of image size and frame rate:

$$R_C = R_S \cdot n_{\text{OP}} \tag{15.2}$$

with n_{OP} as mean number of operations per sample, R_S as source rate in samples per second, and R_C as computational rate in operations per second. It should be noted that difference in the number of bits per operand is not considered in (15.2).

The relation (15.2) is obvious for low-level tasks that are performed in the same manner for each sample or a group of samples. The number of operations of high-level tasks is data dependent. In that case the computational rate can be determined as only an average of typical image material. The number of operations per sample for three important low-level tasks is listed in Table 15.1. This table shows that there are several alternative algorithms for implementations of the same tasks with different computational rates due to the different number of operations

Table 15.1   Mean number of operations per sample. $N \times N$
window size; p, maximum displacement.

Task/algorithm	Operations per sample
2-D DCT	
Dot-product with basis images	$2N^2$
Matrix–vector multiplication	$4N$
Fast DCT (Lee)	$4 \log N - 2$
Block Matching	
Full search	$3(2p + 1)^2$
Conjugate direction search	$3(2p + 3)$
Modified 2-D log search	$3(1 + 8\lceil \log p \rceil)$
2-D FIR Filter	
2-D convolution	$2N^2$
Separable filter	$4N$
Separable symmetrical filter	$3N$

per sample. The wording is here such that task is the general description of the processing scheme whereas algorithm is used for the accurate specification with sequence of operations etc.

In addition to the operative part, the hardware expense for implementation consists of the memory and the interconnection between all modules. The memory requirements are influenced by the multiple access to original image data and intermediate results. The interconnect bandwidth depends on the frequency of communication between memory modules and the operative part. Table 15.2 shows the average number of accesses per sample n_{ACC} and the memory capacity, under the assumption that the operative part contains just one register. Similar to (15.2) the access rate R_{ACC} becomes

$$R_{ACC} = R_S \cdot n_{ACC}. \tag{15.3}$$

Table 15.2 shows also large variations for the alternatives of the computation schemes. Because computational rate, access rate, and memory capacity are rough indicators for the required hardware expense, preference for implementation is for that scheme with the smallest measures. For more accurate comparison of the hardware expense, detailed investigations of architectures are required.

Sophisticated processing schemes as the hybrid coding scheme shown in Fig. 15.1 cannot be described with a few relations. High-complexity processing schemes have to be defined in a hierarchical manner. On the top, a processing scheme can be specified by a block diagram as given in Fig. 15.1. By piecewise refinement, further details of data transfers and computations are derived.

Table 15.2 Number of access per sample and memory capacity. $N \times N$ window
size; p, maximum displacement; K, number of samples per line.

Task / Algorithm	Number of access per sample	Memory capacity
2-D DCT		
Dot-product with basis images	$2N^2 + 1$	$N^4 + N^2$
Matrix–vector multiplication	$4N + 2$	$2N^2 + N$
Fast DCT (Lee)	$10 \log N - 6$	$N^2 + 2N$
Block Matching		
Full search	$4(2p + 1)^2$	$(N + 2p)^2 + N^2$
Conjugate direction search	$4(2p + 3)$	$(N + 2p)^2 + N^2$
Modified 2-D log search	$4(1 + 8\lceil \log p \rceil)$	$(N + 2p)^2 + N^2$
2-D FIR Filter		
2-D convolution	$2N^2 + 1$	$K(N - 1) + N^2$
Separable filter	$4N + 2$	$K(N - 1) + 2N$
Separable symmetrical filter	$5N + 2$	$K(N - 1) + 2N$

Depending on the level of hierarchy, algorithms could be defined on groups of data such as arrays or vectors, single data (word level), or even on the bit level. The computations and data dependencies of algorithms can be described either by recurrence equations, program notations, or dependence graphs.

It is essential for processing of continuous data that the algorithms are periodically defined over a basic interval. In case of a hybrid coding scheme this interval is a macroblock. Almost all tasks of this coding scheme are defined on a macroblock of 16×16 luminance pels and $2 \times 8 \times 8$ chrominance pels. The arrangement of tasks in a specific processing sequence is forming a functional space as depicted in Fig. 15.2 [14]. The knowledge of the required parallelism is essential for VLSI implementation with real-time capability. Let a processing element PE be a hardware unit that offers processing of one operation in a time interval T_{OP}. A figure of merit for the needed number of parallel operating PEs (n_{PE}) can be determined by

$$n_{\mathrm{PE}} = R_C \cdot T_{\mathrm{OP}} \tag{15.4}$$

with T_{OP} as average time for one operation. With at present available technologies T_{OP} is possible on the order of 20 ns. From this follows that n_{PE} is in the range from 2 to 443 depending on the source rate (Table 15.3). Because of (15.2) n_{PE} is proportional to the source rate. In Table 15.3 the source rate of several image formats is listed. For a complete hybrid coding scheme n_{OP} is on the order of 160 for the encoder and 40 for the decoder.

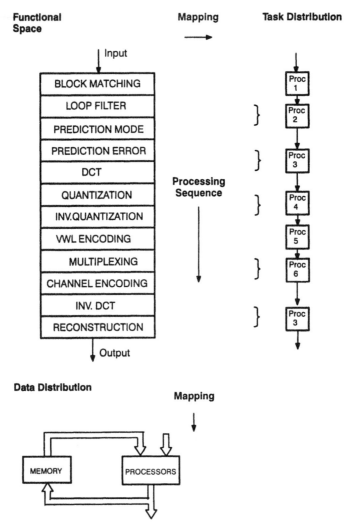

Figure 15.2   Functional space of the hybrid coder and mapping to multiprocessor systems exploiting data and task distribution. Reprinted from [14] with permission of Kluwer Academic Publishers.

What follows is that one processor is not sufficient to provide the required computational rate. In particular for systems with high source rate, extensive parallelism is needed by use of multiple processors. There are basically two multiprocessor arrangements. By projection of the functional space in the direction of the processing sequence (see Fig. 15.2), each processor has to perform all defined

Table 15.3   Source rate of several image formats and number of PEs for hybrid
coding. Net rate without blanking intervals; Y, luminance; C, chrominance.

Name	Image size (active area)	Frame rate in Hz	Source rate in Msamples/sec	Number of PEs
QCIF	Y: 176×144 C: 88×72	10	0.4	2
CIF	Y: 352×288 C: 176×144	30	4.6	19
CCIR 601	Y: 720×576 C: 360×576	25	20.7	83
HDTV	Y: 1920×1152 C: 960×1152	25	110.6	443

operations in the specified sequence. This mapping results in a processor with
time-dependent processing according to the sequence of functions. The feedback
memory is needed for the storage of intermediate results. Parallel processing is
possible by assigning to each processor a subsection of an image. Here the fact
is exploited that image segments can be almost independently processed. The
smallest segment for independent processing in case of the hybrid coding scheme
is the macro block. A block diagram of a multiprocessor system exploiting this
data distribution is shown in Fig. 15.3.

Mapping of functional blocks orthogonal to the processing sequence (see
Fig. 15.2) results in processors dedicated to one or a group of dedicated functions. The tasks of a coding scheme are distributed over several processors. This

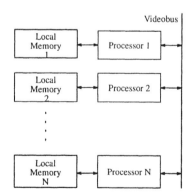

Figure 15.3   Multiprocessor system exploiting data distribution.

approach can be interpreted as pipelining on the macroblock level. A block diagram of a multiprocessor system exploiting task distribution is shown in Fig. 15.4.

In addition to the operative part, limitations of performance can be caused by the data access. Figure 15.5 shows a simple processor model consisting of an operative part with n_{PE} parallel processors, a memory part, and a bus for connection between both. The needed number of bus lines is dependent on the access rate. The number of accesses of the operative part is on the same order as the number of operations. Considering one large external memory for storage of video data and intermediate results, the number of parallel access lines becomes very large. Because the memory access time T_{ACC} is limited, the required buswidth becomes

$$n_{BUS} = R_{ACC} \cdot T_{ACC}. \qquad (15.5)$$

The access time of static memory is at present on the order of 30 ns. With (15.3) and n_{ACC} of about 200 the buswidth becomes 27 bytes for the CIF format and 124 bytes for the CCIR 601 format. Taking into consideration that the bus access rate is influenced mainly by multiple access of image source data and intermediate results,

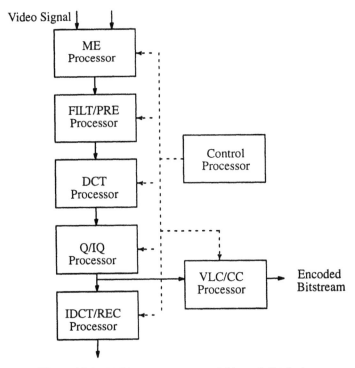

Figure 15.4 Multiprocessor system exploiting task distribution.

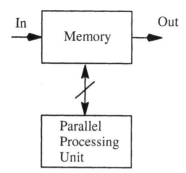

Figure 15.5   Simple general processor model with memory separated from the processor unit.

the bus access rate can be essentially reduced by assigning to each processor a local memory. By storing in the local memory all source data and intermediate results for processing of one macro block, the video bus access rate can be restricted to the order of the source rate. This would require storing the search area for block matching in the local memory. The preceding discussion indicates that a shared memory concept is not advisable for video coding applications whereas a distributed memory concept is appropriate because of local access conditions.

From the preceding results it follows that hardware structures with a concurrency in operations according to n_{PE} of (15.4) are needed. This concurrency can be achieved by parallel processing and pipelining. There are several alternative architectures to achieve a specified concurrency. Because pipelining structures store intermediate results in registers, this effects also the access and memory requirements.

To achieve high silicon efficiency, optimization of architectures according to (15.1) is requested. A high throughput rate for the smallest expense of silicon area is desired for the envisaged throughput rate. A realization according to the task distribution can be optimized by adapting each processor architecture to the specific algorithm. In most cases the algorithms of the subtasks are very regular. The algorithms can be expanded to multidimensional dependence graphs with nodes describing the operations and edges between the nodes describing the data dependencies. For such regular algorithms as filtering, linear transform, and block matching almost all nodes are identical and based on a small number of different operations. The data dependencies can be restricted to spatially neighboring nodes (localization). Mapping regular algorithms onto processor arrays is known from the literature [15]. Implementation of the image coding tasks by devices dedicated to the specific algorithms will be discussed in the following. Independent optimization of each processor could result in devices that section require, at input or output, internal memories for intermodule communication of macroblock size.

This is caused by the macroblock pipelining. By matching the data access of neighboring processors, the internal memories can be essentially reduced. Restriction up to one register is possible. Therefore, a global view of the complete hardware arrangement is advisable for optimization.

15.3 Key Components for Function-Oriented Implementations

A function-oriented realization can be very attractive because it has the potential to achieve the smallest size possible by being tailored to the requirements of the application. When mapping the operational parts of a video processing scheme to the blocks of a function-oriented realization, the data transfer between blocks has to be taken into account. Considerable hardware expenses for formatting data sequences might occur if the output from any one block is not matched to the input sequence of the successive blocks. The specific features of the algorithm influence the architecture, and this will have an impact on the overall hardware complexity. As discussed in the previous section measures such as computational rate, access rate, and memory capacity will give the system designer indications which algorithms are to be preferred from the hardware point of view.

An appropriate optimization criterion is the minimization of the total required silicon area under consideration of I/O constraints. Here the tradeoff between areas devoted to computation, storage, and interconnection has to be taken into account. In general, there are several alternative architectures which contribute differently to the three mentioned parts. It should be recognized that there is a strong interaction between the algorithms and the most appropriate architecture. In most cases specialized architectures considering all a priori known features of algorithms are the best solution. There are a few exceptions, where more general architectures are better because of high regularity, which offers cell connections by abutment and reduced effort for specification and design. Cell connection by abutment reduces considerably the silicon area for connection.

In this section special realizations of three widely used algorithms from the video coding area, namely, subband filtering, DCT, and block matching, will be discussed. These algorithms have been selected because of the high requirements they establish. Distinct hardware structures are examined that achieve the high performance requirements in a small silicon area.

15.3.1 Filter Banks for Subband Coding

In particular for bit-rate reduction of HDTV signals, subband coding is proposed. Key components for implementation of subband coding schemes are bandsplitting filter banks at the transmitter side and band-synthesis filter banks at the receiver side. At the transmitter side the video signal is separated into several bands by filtering. Sampling in each band is decimated according to the bandwidth. Then each band is coded according to the individual properties, and the code words are transmitted to the receiver. After decoding at the receiver the video signal is synthesized by band-interpolation filters.

Due to the transition between pass- and stopbands, subsampling at the transmitter side will cause aliasing effects. The alias cancellation necessary for good picture quality can be achieved by appropriate choice of the parameters of the coder and decoder filter banks [16, 17]. Filter characteristics that incorporate aliasing error cancellation are quadrature mirror filters (QMFs) and conjugate quadrature filters (CQFs) [18]. Both are finite impulse response (FIR) filters with a nonrecursive structure. QMFs offer a linear phase behavior. A disadvantage of QMFs is the small ripples in their overall frequency transfer function. In contrast, CQFs allow a perfect reconstruction of the original signal, if undistorted transmission of the subband signals is provided.

Computational Part   Because of the high source rate, processing of HDTV signals results in high computational rates that can be enabled only by intensive pipelining and parallel processing. Therefore, FIR filters are preferred because of their simplicity for pipelining-intensive realizations. Images are two-dimensional from nature. For this reason 2-D filtering is applied. For simplicity in the following, 2-D filters with $N \times N$ coefficients are considered. The convolution of the image data $x(\cdot)$ with an impulse response $h(\cdot)$ is given by

$$y(i, j) = \sum_{m=0}^{N-1} \sum_{n=0}^{N-1} h(m, n)x(i - m, j - n). \tag{15.6}$$

Using separable filter characteristics the number of multiplications of a 2-D filter with N^2 coefficients can be reduced from N^2 to $2N$. The filtering process is then

$$\begin{aligned} \tilde{y}(i, j) &= \sum_{m=0}^{N-1} h(m)x(i - m, j), \\ y(i, j) &= \sum_{n=0}^{N-1} h(n)\tilde{y}(i, j - n). \end{aligned} \tag{15.7}$$

A disadvantage of filters along the horizontal and vertical dimensions is that only orthogonal sampling patterns and not quincunx sampling patterns are supported.

Considering that filtering according to (15.7) requires two operations (MUL, ADD) per filter tap, the number of operations per sample of a separable subband filter bank with K bands results in

$$n_{op} = K \cdot 2 \cdot 2N. \tag{15.8}$$

By utilization of the fact that filters are linear and can be factorized into a product of filter sections a cascade implementation is possible. Figure 15.6 shows a cascade implementation for the case of $K = 4$. This treelike filter structure essentially reduces the number of operations down to

$$n_{op} = \log K \cdot 2 \cdot 2N. \tag{15.9}$$

The decimation process at the bandsplitting filters suppresses samples that have been determined by the expensive filter process. This is not very efficient. By implementing the filters in polyphase structures, only partial results that are needed for the transmitted samples are determined. For polyphase filter structures, the filter is split into several parallel filter parts. The impulse response of the filter parts is determined by a subsampled impulse response of the complete filter. If the number of phases corresponds to the decimation factor, each part of the filter processes only a certain phase component of the subsampled signal. By transition of the decimation process from filter output to input, the filter phases can operate with reduced clock frequency without affecting the function. This is depicted for a 2:1 decimation filter in Fig. 15.7. In this specific case the filter phases are given by filter parts with odd and even indices, respectively. The polyphase structure operates with the output sample rate, which reduces the computational rate by a factor of 2.

Filter banks of separable filters can be implemented by low-pass and high-pass filters. For QMFs the high-pass and low-pass filters have fixed relations on their coefficients [16]. The transfer function H_1 of the high-pass filter is a mirror image of the low-pass filter H_0:

$$H_1(z) = H_0(-z). \tag{15.10}$$

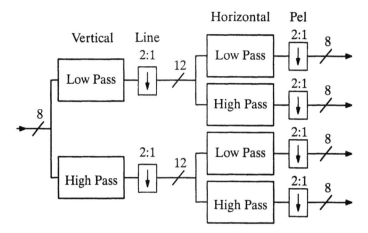

Figure 15.6   Cascade implementation of an analysis subband filter bank.

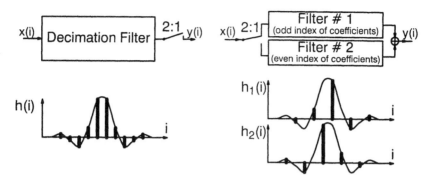

Figure 15.7 Principle of polyphase filter structure for decimation.

With (15.10) follows for FIR filters

$$H_0(z) = a_0 + a_1 z^{-1} + a_2 z^{-2} + a_3 z^{-3} + \ldots$$
$$H_1(z) = a_0 - a_1 z^{-1} + a_2 z^{-2} - a_3 z^{-3} + \ldots \qquad (15.11)$$

The filter characteristics G_0, G_1 for synthesis at the receiver can be directly derived from the filter H_0, H_1 at the transmitter. Considering the conditions for aliasing compensation [16, 18],

$$G_0(z) = H_1(-z) \quad = H_0(z),$$
$$G_1(z) = -H_0(-z) = -H_1(z). \qquad (15.12)$$

The filter coefficients for high-pass and low-pass filters of QMF filter banks differ only in the sign of every second coefficient. This can be utilized for implementation. The filter functions of the high-pass and low-pass filters can be split into two partitions. The regular change of the sign in (15.11) results in a structure (Fig. 15.8) similar to polyphase filters. According to the sign change, filter taps with even and odd indices are in separate partitions of the filter. The output values of the low-pass filter are the sum of both filter partitions. Due to the QMF characteristics, only one additional subtracter is necessary for realization of the high-pass filter within a filter bank, since the difference of both partitions gives the output values for the corresponding signal. This reduces the hardware expense by nearly 50%. A similar structure exists for QMF synthesis filter banks.

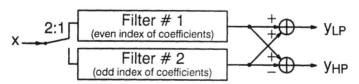

Figure 15.8 Structure of a QMF decimation filter (analysis side).

The results above can be applied to 2-D filters. Figure 15.9 shows the derived 2-D filter structure of the analysis filter for the case $K = 4$. The incoming data are first distributed into K parallel operating filter paths. Each filter path operates at a clock rate reduced by $1/K$ compared to the incoming sample rate. The overall number of operations for QMF subband filters results in

$$n_{op} = \log K \cdot N. \tag{15.13}$$

This result indicates that the specific knowledge of algorithms can be utilized for architectures with minimum number of operations per sample, which will also result in hardware savings for the filter arithmetic.

The main operations for filtering are multiplication and addition. The analysis preceding counts both as one operation even if a multiplication has a much larger hardware expense than an addition. As a rough estimate, one multiplier has the expense of $m - 1$ adders, if m is the number of multiplicator bits. For video applications fixed point representation of the coefficients with 8 and 9 bits are required to fulfill the needed stopband attenuation. Depending on the sign bit, 9 or 10 bit multipliers are needed. For a given filter characteristic the hardware expense for realization can be essentially reduced by an implementation with fixed coefficients instead of programmable filter coefficients. Fixed coefficients can be implemented by hardwired shifts and adders where the number of adders is specified by the number of nonzero binary digits. Figure 15.10 shows a typical impulse response of an FIR low-pass filter. Just by considering a specific amplitude characteristic it is obvious that the coefficients apart from the center have a decay in magnitude that will result in several leading zeros in binary representation. The savings in number of adders by the decay of the coefficients magnitude are on the order of 50%. Integers can be represented by binary numbers with digits 0 and 1. As an extension, signed digit representation is possible with digits -1, 0, and 1. Recoding of binary numbers to canonical signed digits (CSDs) provides a representation

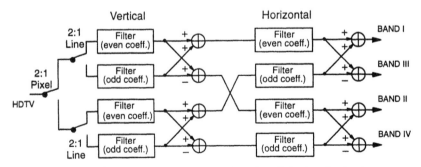

Figure 15.9   A 2-D QMF analysis filter bank for splitting into four bands.

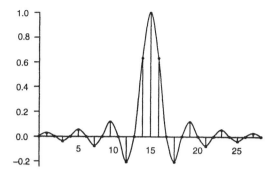

Figure 15.10 Typical impulse response of an FIR low-pass filter.

with at least 50% zeros [19]. From this it follows that CSD coding of coefficients for typical low-pass characteristics will result in at most 25% nonzero digits. The filter characteristics reported in [7] confirm this fact. For implementation of the filter arithmetic, addition, subtraction, and hardwired shifts of several samples are needed. Carry–save adder trees are very appropriate for this [20]. The sequence of additions can be performed according to either the sequence of coefficients or the bit planes of all coefficients. By a bit plane grouping (least bit plane first) of the filter coefficients it is possible to have a nearly constant word length for all adders. Also early rounding and truncation of word length is supported.

I/0 Scheme and Internal Memories The operational part of two-dimensional $N \times M$ filters has to be fed in parallel with $N + M$ samples. To lower the I/O bandwidth to a filter circuit, internal on-chip memories are used. For filters along scanning direction (horizontal), a register chain according to the number of taps can provide all needed samples in parallel. For vertical filters a sequence of several line delays performs the same function. A disadvantage of this approach is the large amount of on-chip memory. But the good news is that the I/O bandwidth is restricted to the video source rate. There are alternative architectures with an exchange between I/O bandwidth and internal memory. The amount of internal memory can be lowered at the expense of I/O bandwidth. The total delay needed is specified by the interface between video input and the filter operational part. A reduction of the internal memory will increase the capacity for external memory. Complete line delays on-chip are frequently preferred because of simple control and direct feeding with the video sample rate.

The 2-D filters are implemented as a sequence of two 1-D filters. To avoid visible artifacts, the results coming out of the first filter cannot be truncated to 8 bits. At least a 12 bit representation for intermediate filter output is needed. Resulting from the large line length of 1920 active pels for HDTV signals, line delays form a major part of subband filter banks. For this reason the vertical filter with the line delays should be placed in front of the horizontal filter. In this case the data stored in the line delays are 8 bit words instead of 12 bit.

The line delays can be implemented by shift register arrays and RAM circuits. The advantage of shift register arrays is the simple control, but the disadvantage is the high power dissipation. By parallel implementation of shift registers and the use of a multiphase clocking scheme the power dissipation and silicon area can be considerable reduced [21, 22]. The input multiplexers of the analysis filters (Fig. 15.9) result in a reduction of clock frequency in the filter sections. Horizontal down-sampling by a factor of 2 requires the expansion of every second sample. In hardware realization, this can be easily achieved by registers with a clock enable or by different clocking of registers. Vertical down-sampling by a factor of two requires the expansion of the data of one line over the period of two lines. This kind of data formatting can be offered by first-in–first-out buffers (FIFOs). The easiest implementation of FIFOs is possible with double-sized memories and a read/write in a ping-pong mode. The memory capacity in the FIFOs can be lowered to the minimal amount by implementation as a rate conversion structure based on a combination of multiplexers and synchronous clocked shift registers of different length [22].

Prototype VLSI Chip for Subband Filter Banks   Prototype chips for a 10 × 14 subband filter bank have been designed for a 1.2 μm two metal layer CMOS process [22]. Because of the high transistor complexity of the filter bank, an implementation with two identical chips has been considered. The application of the polyphase structure allows an easy partitioning of the filter bank into two identical chips. The HDTV signal is multiplexed twice at the input of the subband filter. This corresponds to the parallel polyphase structure of the architecture that is used to divide a filter bank into two parts.

Pel-by-pel multiplexed chrominance signals result in the same clocking rate and line length as for the luminance signal. Inserting in the horizontal filter part switchable delay parts one chip can perform operations for both luminance and chrominance. Two chips have been designed: one as analysis filter, the other as synthesis filter. Four identical chips form a complete filter bank for luminance and chrominance.

Figure 15.11 shows the floor plan of both chip types. The line delays and FIFOs for vertical filtering and subsampling occupy the largest part of the chip area. Due to the dedicated filter arithmetic with multioperand adders, the arithmetic units are relatively small in terms of chip area. The block horizontal filter includes memory and arithmetic part. Both chip types have about 450,000 transistors on 90mm^2. They have been manufactured and tested. Future CMOS technologies with smaller geometry will allow one chip realization of one filter bank.

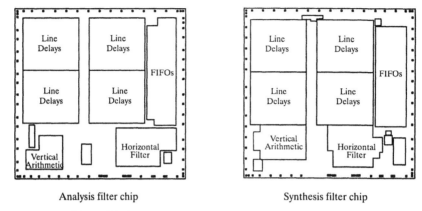

Analysis filter chip   Synthesis filter chip

Figure 15.11   Floor plans of analysis and synthesis filter bank ICs.

15.3.2 Discrete Cosine Transform

The purpose of transform coding is to convert a sequence of statistically dependent samples into an array of relatively independent and energy compacted coefficients. Because of high energy compaction, the discrete cosine transform (DCT) is the most frequently applied transform coding scheme [23]. A DCT circuit has to perform continuously the transformation of blocks $\mathbf{X}$ of $N \times N$ image samples to $N \times N$ transform coefficients $y(u, v)$. In analogy to a Fourier series expansion of periodic functions, the block $\mathbf{X}$ can be represented as a weighted sum of basis functions. In the Fourier series the basis functions are $\cos(\cdot)$ and $\sin(\cdot)$ functions and the weights are determined by scalar products of the original function with the basis functions. For 2-D transformations the basis functions are described as basis images. Each block $\mathbf{X}$ can be represented as a linear combination of a set of basis images weighted by the transform coefficients $y(u, v)$:

$$\mathbf{X} = \sum_{u=0}^{N-1} \sum_{v=0}^{N-1} y(u, v) \cdot \Phi_{uv}. \tag{15.14}$$

The transform coefficients $y(u, v)$ are determined by the dot product between $\mathbf{X}$ and Φ_{uv}. The dot product is a 2-D scalar product.

$$\begin{aligned} y(u, v) &= \mathbf{X} \odot \Phi_{uv} \\ &= \sum_{i=0}^{N-1} \sum_{j=0}^{N-1} x(i, j) \cdot \phi_{uv}(i, j). \end{aligned} \tag{15.15}$$

The DCT implementation according to (15.15) requires N^2 dot products between the block $\mathbf{X}$ and basis images. This counts to N^4 multiplications and additions. Also storage of N^2 basis images of size N^2 is needed. This shows that DCT implementation according to (15.15) is not very efficient. Considering that, the basis images ϕ_{uv} can be determined as a product of two basis vectors:

$$\Phi_{uv} = \Phi_u \cdot \Phi_v^T. \tag{15.16}$$

N basis vectors Φ_u form an $N \times N$ matrix $\mathbf{C}$. Therefore, the DCT can be alternatively rewritten by the following matrix product:

$$\mathbf{Y} = \mathbf{C} \cdot \mathbf{X} \cdot \mathbf{C}^T. \tag{15.17}$$

This can be seen as a sequence of 1-D transform, transposition of N^2 intermediate results and a second 1-D transform:

$$\mathbf{Y} = \mathbf{C} \cdot (\mathbf{C} \cdot \mathbf{X}^T)^T. \tag{15.18}$$

The matrix multiplication according to (15.18) requires $2N^3$ multiplications and additions. Thus the computational effort of (15.18) is smaller by a factor $N/2$ when compared with (15.15). Also the storage requirements are reduced. A memory for N^2 elements of $\mathbf{C}$ and a transposition memory of size N^2 is needed.

A VLSI realization according to (15.18) has been proposed by Totzek, Matthiesen, and Noll [24]. A basic processing element in this structure has to contain a multiplier followed by an adder for accumulation. Employing the idea of a linear systolic array for implementation of the matrix–vector multiplication a high-speed circuit with high regularity is possible. The multiplications and accumulations can be realized by carry–save adder trees. Intensive pipelining allows high clock rates. The devised chip incorporates mainly two linear arrays with 8 PEs each, a ROM for the coefficients of the matrix $\mathbf{C}$, and a RAM block for the transposition of an 8×8 block (Fig. 15.12). The computational part of this chip consists of 16 multipliers and accumulators. In a 1.5 micron CMOS technology the IC requires

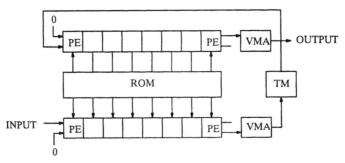

Figure 15.12   DCT circuit based on matrix–vector multiplication (VMA = vector merging adder, TM = transposition memory).

a silicon area of about 92 mm^2 and contains 284,000 transistors. The achievable maximal sample rate is at least 45 MHz.

As an alternative to multiplier based realizations, a distributed arithmetic can be used, where partial precalculated results are stored in ROMs. Let the transformation of a vector **x** containing N values into a vector **y** given by

$$\mathbf{y} = \mathbf{C} \cdot \mathbf{x}. \tag{15.19}$$

Considering a twos complement code with m bits each vector **x** can be described by a sum of bit plane vectors $\mathbf{x}_r$:

$$\mathbf{x} = -\mathbf{x}_{m-1} \, 2^{m-1} + \sum_{r=0}^{m-2} \mathbf{x}_r \, 2^r. \tag{15.20}$$

Splitting the matrix **C** into N row vectors $\mathbf{c}(n)$ each value $y(n)$ has to be determined by

$$\begin{aligned} y(n) &= \mathbf{c}(n) \, \mathbf{x} \\ &= -\mathbf{c}(n) \, \mathbf{x}_{m-1} \, 2^{m-1} + \sum_{r=0}^{m-2} \mathbf{c}(n) \, \mathbf{x}_r \, 2^r. \end{aligned} \tag{15.21}$$

Let F_n be the scalar product between the row vector $\mathbf{c}(n)$ and a column vector $\mathbf{x}_r$. The precalculated results of F_n can be stored in a ROM that will be addressed by the bit patterns of bit plane vectors $\mathbf{x}_r$. This allows the design of a PE as depicted in Fig. 15.13, which generates the values $y(n)$ based on table look-up by a ROM and bit serial processing. A 2-D transform requires $2N$ ROMs and a transposition RAM. For $N = 8$ the number of transistors for a ROM implementing the function F_n is on the order of that of a multiplier and accumulator. For this reason the chip size for $N = 8$ is for the distributed approach on the same order as for the matrix–vector implementation based on multipliers. The throughput rate for the distributed approach is smaller because the number of cycles depends on the word width m and not on N. The number of transistors for the ROM is growing exponentially by 2^N and not linearly with N. For this reason special additional measures to reduce the number of transistors for table look-up are reported [25]. The idea of the distributed arithmetic has been implemented in a chip performing 16×16 DCT for video signals with up to 15 MHz [25].

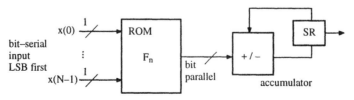

Figure 15.13   PE for DCT based on distributed arithmetic.

By consideration of the special characteristic of the coefficients of the matrix C, architectural structures can be derived with reduced computational and storage requirements. The basis vectors of the 1-D DCT are given by

$$\phi_u(i) = b(i)\cos[(\pi/N)(i + 1/2)u], \quad i = 0, 1, \ldots, N - 1$$
$$u = 0, 1, \ldots, N - 1$$
$$b(i) = \begin{cases} \sqrt{1/N} & i = 0 \\ \sqrt{2/N} & i = 1, 2, \ldots, N - 1. \end{cases} \tag{15.22}$$

By taking advantage of the periodicity in the basis vectors, more efficient algorithms with smaller number of multiplications can be derived. These structures are referred to as fast cosine transforms (FCTs). Lee [26] has proposed an algorithm that requires $N/2 \log N$ multiplications and $3N/2 \log N - N + 1$ additions to perform a 1-D DCT. A signal flow graph (SFG) of the FCT according to Lee is shown in Fig. 15.14. It contains an alternating sequence of data permutation stages and arithmetic stages. In an arithmetic stage the basic operation includes an addition and a subtraction followed by a multiplication.

A direct realization of the SFG would need an excessive amount of silicon area even if fabricated in an advanced technology. However, realization examples can be derived by using projections within the graph leading to sufficiently small implementations. Arterie *et al.* [27] have presented a realization that uses a projection in the direction of word length to achieve a smaller silicon area. The realized chip consists of mainly a transposition memory, a parallel to serial and a serial to parallel converter, and an operative part. The operative part is a direct mapping of the SFG of the FCT into silicon and employs digit serial techniques. Thus every addition, subtraction, and multiplication is assigned to a physical operator that receives its operands as a series of 2-bit groups. A chip has been developed that supports 2-D DCT and IDCT with $N = 4$, $N = 8$, and $N = 16$ for sample rates up to 13.5 MHz [27]. It has been fabricated in a 1.25 micron CMOS technology

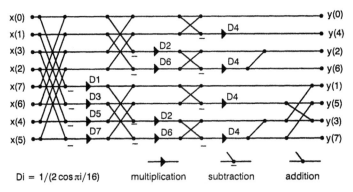

$Di = 1/(2\cos\pi i/16)$ multiplication   subtraction   addition

Figure 15.14   SFG of the FCT based on Lee [26].

and contains about 114,000 transistors in a die area of 40 mm². The advantage of a hardware structure according to the SFG of Fig. 15.14 is the reduced expense for the computational part and internal memory. A disadvantage is the complex data access that results in large amount of silicon area for interconnect.

In addition to the reduction of the hardware expense by a transfer from bit parallel techniques in the direction of bit serial techniques, spatial projection techniques also can be applied. A vertical projection in the SFG of Fig. 15.14 can also be employed. Through vertical projection, operations are mapped to PEs, called *butterfly PEs*, that generate two result values from two input values. One result is the sum of the two input values while the other result is their difference multiplied by a coefficient. Between the PEs a delay and commutator network is needed for shuffling the data sequences. After the vertical projection, a subsequent horizontal projection can be envisioned to generate a realization example with one PE (Fig. 15.15). A multiport RAM is needed for the data access and storage in the predefined sequence. As well as butterfly PEs rotator PEs can be used as a basis element for DCT implementations. The rotator PE is more complex than the butterfly PE because four multipliers and two adders are needed. But the number of rotators is smaller. Ligtenberg and O'Neill [28] have reported on a DCT chip based on one rotator.

The projection techniques allow adaption of the number and type of PEs to the computational requirements by offering the right amount of concurrent processing. A DCT implementation with one PE is appropriate for low source rates as the CIF format. Broadcast video rates need at least four concurrent operating PEs. Combinations of the discussed schemes have been also implemented. The first stage of the fast algorithm (additions and subtractions) have been combined with proceeding parallel inner product stages of multipliers and accumulators [29]. Others combine the first stage of the fast algorithms with the distributed arithmetic [25, 30]. Combinations of the DCT with subtracters for prediction error determination and adders for sample reconstruction as needed in prediction loops have been also reported [29, 30].

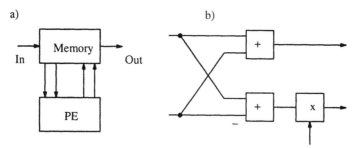

Figure 15.15   (a) DCT realization based on one PE; (b) butterfly PE.

15.3.3 Block Matching Algorithm

Motion estimation is required to improve prediction of moving objects. Block matching is a simple scheme where motion is determined for small rectangular objects [31]. Here the actual frame is divided uniformly into reference blocks of size $N \times N$ pels. Every reference block is compared with candidate blocks from a search area in the previous frame. The offset between the best matching candidate block and its reference block specifies the displacement vector $\mathbf{v} = (v_i, v_j)$. In general the mean of the absolute differences is used as a matching criterion. The search can be limited to a maximum displacement p in both directions if the maximum motion of objects is assumed to be limited (Fig. 15.16). The block matching algorithm is then given by

$$s(n, m) = \sum_{i=1}^{N} \sum_{j=1}^{N} |x(i, j) - y(i + m, j + n)|, \quad -p \le m, n \le p, \quad (15.23)$$

$$u = \min_{m,n} \{s(m, n)\}, \tag{15.24}$$

$$\mathbf{v} = (m, n)|_u. \tag{15.25}$$

The block matching algorithm requires the calculations of $(2p + 1)^2$ sums according to (15.23) and a subsequent detection of the minimum sum to estimate the displacement vector from a full search of all candidate blocks. The computational rate is directly proportional to the number of candidate blocks. A full search considers $(2p + 1)^2$ candidate blocks. Search strategies offer an essential reduction of the computational rate by reducing the number of investigated candidate blocks. The disadvantage of the search strategies is the enlarged control overhead and a

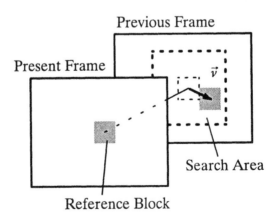

Figure 15.16   Motion estimation based on block matching.

much less regular data flow. For this reason, until now realized block matching chips have been based on full search techniques.

Realization of the block matching algorithm requires two types of basic PEs. Both are depicted in Fig. 15.17. One basic PE generates the absolute difference between two pels x and y from reference and candidate blocks, respectively, which is then accumulated for all N^2 pels within a block. Fig. 15.17a exemplifies an efficient implementation of the magnitude operation by using the MSB of the difference value to trigger bitwise inversion in the XOR operator as well as increment in the subsequent adder. In a second basic PE, the minimum sum is searched among the accumulated sums $s(m, n)$ and the corresponding displacement vector is detected. The first type of PE will be denoted as AD (absolute value of differences and accumulation) and the second as M (determination of minimum). The required processing power for real-time implementation can be achieved by a certain number of PEs of type AD and of type M. The numbers of PEs type M can in principle be smaller than the number of PEs type AD. Based on the computational rate the ratio could be $1/N^2$.

According to (15.4) the total number of PEs should be on the order of

$$n_{PE} = (2p + 1)^2 \cdot R_S \cdot T_{PE} \qquad (15.26)$$

with T_{PE} as processing time of one PE. Because motion estimation is determined by the luminance component alone, the source rates given in Table 15.3 have to be reduced accordingly. The evaluation of (15.26) indicates that an extensive concurrency by pipelining and parallel processing is needed. Systolic architectures are very appropriate for this. Because of the associativity of additions and minimum search, there are several alternative arrangements of the PEs to perform the calculations. A systematic design has to follow a methodology similar to those of Kung [15]. The dependence graph of the algorithm has to be specified. By

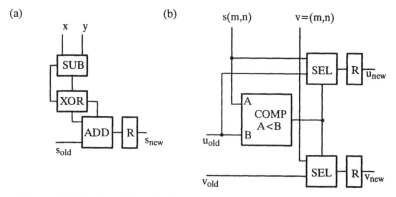

Figure 15.17   Basic PEs for the block matching algorithm: (a) PE type AD; (b) PE type M.

assignment of a schedule and multiple projections, several alternative SFG can be derived [32]. Figure 15.18 shows a 1-D systolic array that offers the computational power for the CIF image format with a reduced frame rate. The CCIR image format will require 2-D arrays with N^2 or even $N \times (2p + 1)$ PEs of type AD [32]. The systolic arrays need adequate data transport to the boundaries of the array. A total of $2 \times N^2 \times (2p + 1)^2$ pels from reference block and search area have to be transported into the array during calculation of one displacement vector. The corresponding data rate cannot be transferred across the IC boundaries due to a limitation of pin count. To reduce the I/O rate at the IC boundaries, local memories have to be considered. Having stored the reference block and the search area in two local memories operating as double-sized buffers in a ping-pong mode, the I/O rate can be restricted to $3 \times R_s$.

In the literature several designs for block matching chips are reported [29, 30]. Frequently 1-D processor arrangements with local memories based on register arrays according to the proposal of Yang, Sun and Wu [33] have been applied.

15.4 Programmable Multiprocessor Systems

A software-oriented implementation is very attractive because it incorporates flexibility to accommodate a wide variety of application schemes and it allows modifications of algorithms by software changes. To improve the overall silicon efficiency according to (15.1), special multiprocessor architectures have been developed that incorporate the data path and data access adapted to the required class of algorithms. With respect to data and control flow, architectures of multiprocessors are generally classified as single instruction, multiple data stream (SIMD) or multiple instruction, multiple data stream (MIMD). Block diagrams of generic SIMD and MIMD architectures are shown in Fig. 15.19. An interconnection network is provided for communication between the processing units. To restrict the I/O bandwidth for operand transport, memories local to the processing units are

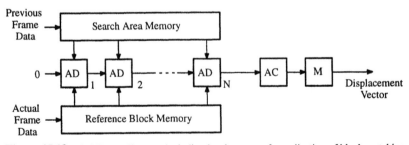

Figure 15.18 A 1-D systolic array including local memory for realization of block matching.

SIMD

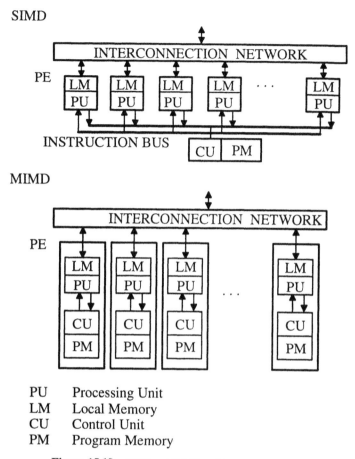

PU   Processing Unit
LM   Local Memory
CU   Control Unit
PM   Program Memory

Figure 15.19   SIMD and MIMD multiprocessor systems.

needed for video applications. The appropriate size of the local memories depends on the kind of algorithms and is driven by multiple access to image data of local image segments. Because most image processing algorithms allow independent processing of image segments, the PEs will be distributed over the image space (see Section 15.2). In this case, image data have to be distributed to the PEs according to a specification of image segments. The need for communication between the PEs can be avoided by matching the local memory size to the maximum requirements of the set of algorithms to be considered.

SIMD multiprocessors are very efficient for applications where the employed algorithms allow identical operations for many parallel data streams. Most low-level algorithms with a fixed time-dependent processing sequence are of this kind.

The efficiency of SIMD decreases for algorithms with result-dependent alternative parts. During operation of alternatives, parts of the PEs have to be disabled and idle, while the remaining PEs perform the operations. This loss of efficiency for data-dependent processing can be reduced for MIMD architectures because independent processing of all PEs is possible. The efficiency of SIMD can be also improved by extension to autonomous SIMD (ASIMD). The controlling of ASIMD offers restricted local decisions for the execution of instructions and data access. ASIMD provides for a large class of algorithms with irregular control flow good efficiency.

Implementation of multiprocessors with standard RISC or standard signal processors will result in very large hardware arrangements. Smaller hardware is possible by adaptation to the specific signal processing algorithms and restriction of flexibility by software changes to some extend.

The literature reported on several multiprocessor systems specifically designed for video signal processing [34–40]. The structures, basic elements, and technologies are different. For this reason the overall performance and the needed silicon area using advanced technologies are very difficult to compare. A first attempt to achieve a unified performance measure for multiprocessor architectures independent of technology constraints has been investigated in [41].

15.4.1 Improvement of Multiprocessor Performance

Increase of silicon efficiency of multiprocessors can be achieved by processor structures matched to the algorithms to be processed. Algorithms can be specified in different hierarchical levels. Adaptation of hardware to the algorithms can be performed on different levels accordingly. On the lowest level the data path of RISC-type signal processors should be extented. An analysis of the algorithms indicates that arithmetic operations as

$$\begin{aligned} & \sum |a - b| \\ & \sum (a - b)^2 \\ & \sum (a - b)c \end{aligned} \tag{15.27}$$

frequently occur and should be supported. A pipeline of ALU, array multiplier, accumulator, and shifter is appropriate and offers several operations per clock cycle. Also result control by shift, round, and clipping should be implemented by additional hardware. The efficiency of an instruction pipeline can be increased by considering special measures for data prefetching and instruction prefetching to avoid pipeline hazards. Hardware implemented loop counter and special instructions like

```
if   a > b
        goto        LABEL
else
```

or

```
if  a /= b
          c = a + b
else
          c = a - b
```

are further increasing the efficiency of data dependent operations.

Homogeneous multiprocessor systems with identical processing elements (PE) have the advantage of modularity and regularity, which is favorable under consideration of scalability, design efforts, and redundancy implementation. The appropriate parallelization strategy for this kind of multiprocessor system is data distribution (see Fig. 15.3).

Compact implementation with some restriction to programmability can be achieved by combining the programmable part with function-oriented modules for specific tasks. This will be a parallelization by a combination of task and data distribution. Candidates for dedicated implementation are tasks with high performance requirements such as block matching, DCT, and IDCT. Also data-dependent tasks such as quantization and variable length coding and decoding are favorable for dedicated implementation. The dedicated modules could be assigned to each PE by offering scalability and data distribution on the highest hardware level [42]. The alternative would be to assign to the complete multiprocessor system hardware arrangements of dedicated modules [39, 40]. In this case task distribution is given on the highest hardware level.

In addition to modules dedicated to just one task, more flexible modules are possible that allow implementation of a class of algorithms. A few control lines cause the change of operation. A possible example would be a coprocessor for convolution like kernel operations as specified in (15.27) [36]. The combination of programmable sections with dedicated modules results in more complex systems that offer higher silicon efficiency as long as the increase in throughput rate is higher than the increase in silicon area.

15.4.2 Design Examples

Two design examples of multiprocessor systems for video applications will be discussed. Both systems have identical PEs on the highest level. The general programmable section within each PE is extended by function-oriented modules. Independent macroblock processing of the PEs is achieved by local memory of sufficient size and an appropriate control strategy. All PEs are connected via busses.

In the first design example, each PE consists of a RISC-type processor and a low-level coprocessor (Fig. 15.20). The RISC-type processor performs medium-level

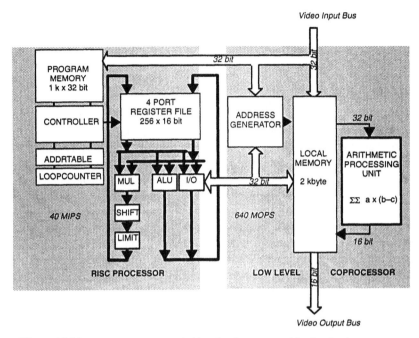

Figure 15.20   Processing element of video signal processor with a low-level coprocessor.

tasks and controlling. The instruction set of the RISC processor is extended by additional hardware. Fast multiplication is offered by an array multiplier.

The low-level coprocessor consists of an arithmetic processing unit (APU), local memory (LM), and address generation unit (AGU). The APU is adapted to window-based low-level algorithms and can operate in parallel 4 data streams (Fig. 15.21). It performs by pipelining a sequence of ADD/SUB, MUL, accumulation over a window, and shift/limit for normalization of sums. For the access to two operands, a register file of 128×8 bits is included. The APU based on data paths with 8 bit representation. Nevertheless, the mode shifters between multiplier and accumulator enable the APU to perform 16 bit operation in two consecutive clock cycles. The high flexibility micro-controlled AGU is generating address sequences for the LM and performs static and dynamic control of the APU processing.

Considering a clock rate of 40 MHz a peak performance of 640 MOPS is achieved in the 8 bit mode. The computational rate is half for the 16 bit mode. Simulations of the architecture have shown that 6 PEs are sufficient for a complete video codec according to H.261 video telephone standard. Three PEs are required for the MPEG-1 decoder. A 0.8 micron CMOS technology allows implementation of a complete PE on 1 chip.

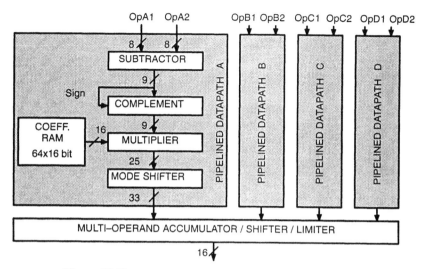

Figure 15.21   Arithmetic processing unit of a low-level coprocessor.

In the second design example a programmable unit is accompanied by two application-specific modules. Taking the computational rate as a criterion a block matching module and a DCT/IDCT module are to be used. Figure 15.22a shows a sequential schedule for such an arrangement. The sequential schedule considers the data dependencies between the modules. Having the start of each macroblock as a synchronization barrier, a large amount of dummy cycles reduces efficiency of this structure. The amount of dummy cycles can be essentially reduced by interleaved processing of several macroblocks. Because of the complex data dependencies, about four macroblocks have to be processed concurrently and the size of the local memory has to be increased accordingly. Higher efficiency can be achieved by extending the DCT module to a block-level coprocessor. This module processes in addition to DCT/IDCT, variable thresholding quantization, inverse quantization, and loop filtering. As indicated by the schedule in Fig. 15.22b, the data dependencies between the modules are improved, which results in smaller local memories and a much simpler interleaved processing. A reduction of idle times can be achieved by parallel processing of different macroblocks based on an interleaved schedule as depicted in Fig. 15.22c. The resulting processor architecture is depicted in Fig. 15.23.

The block matching based on full search provides the best picture quality. Due to the regularity and predefined data access of full search, an implementation based on a systolic array architecture is recommended (see Section 15.3.3). The task prediction and reconstruction require access to the frame memory. Since these

a) BM, DCT, PU (sequential schedule)

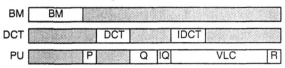

b) BM, BLC, PU (sequential schedule)

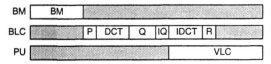

c) BM, BLC, PU (interleaved schedule)

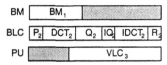

Figure 15.22   Schedule of a heterogeneous processor with three modules (BM, block matching module; DCT, DCT, module; PU, programmable processing unit; BLC, block level coprocessor).

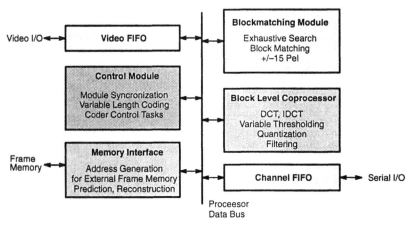

Figure 15.23   Processor architecture overview.

tasks require low computational power, they can be also implemented within the frame memory interface instead of the block level coprocessor (BLC) module.

An efficient implementation of the DCT on the BLC can be achieved by employment of fast algorithms (see Section 15.3.2). The data path of the BLC has to contain a butterfly structure in regular and transpose forms. A shuffle network based on delay commutators is needed for appropriate data access. The numerical behavior of the fast DCT requires an increase of the internal word width to 18 bits. To provide the additional task quantization, inverse quantization, and filtering, the multiplier has to be extended to a multiply–accumulate unit, and a dedicated unit for variable thresholding has to be included. The BLC architecture is depicted in Fig. 15.24.

The remaining programmable module is a RISC-type of controller, consisting of program controller, interrupt handler, data path, and local data memory. By provision for a dedicated unit for variable length coding and decoding the overall efficiency can be improved.

Assuming a 0.6 micron CMOS process, the required silicon area for one processor according to Fig. 15.23 is estimated to 75 mm^2. One processor is sufficient to perform source coding and decoding for video telephone signals as specified by H.261. A video codec as defined by MPEG-1 requiring three processors.

15.5 Conclusion

The envisaged low-cost implementation of video codecs requires specification and design of application-specific VLSI circuits. The total hardware complexity of these circuits is influenced by the algorithms, the applied architectures, and the circuit techniques. Characteristics of algorithms related to the hardware expense are computational rate, access rate, and memory capacity. From the computational rate, a figure of merit of the required parallelism can be derived. There are several alternatives to map algorithms onto multiprocessor structures. For hybrid coding schemes, task distribution as well as data distribution is possible. The efficiency

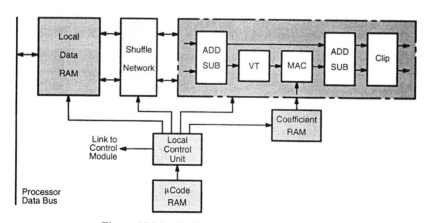

Figure 15.24   Block-level coprocessor architecture.

of multiprocessor systems can be improved by adapting the data path of the processing elements to the type of operation and their frequency of occurrence. For high efficiency, each processing element will contain a set of function specific modules matched to specific subtasks. The appropriate scheduling of the function-specific modules results in a task distribution within each processing element that operates on subsets of data according to data distribution.

The normal design flow is from algorithms to architectures. The impact of algorithms on hardware complexity is determined by the best suited architecture, which is derived by selection from several alternative architectures. Providing the algorithm designer with characteristics of appropriate architectures, algorithms can be modified accordingly. Therefore, high efficiency given by high signal processing performance and low hardware expense requires an interactive process between algorithm and architecture design.

References

[1] N. S. Jayant and P. Noll, *Digital Coding of Waveforms.* Englewood Cliffs, NJ: Prentice Hall, 1984.

[2] A. N. Netravali and B. G. Haskell, *Digital Pictures: Representation and Compression.* New York: Plenum Press, 1988.

[3] CMTT, Draft Revision of Recommendation 723, "Transmission of Component-Coded Digital Television Signals for Contribution-Quality Applications at the Third Hierarchical Level of CCITT Recommendation G 702," Document CMTT/BL/5-E, May 1992.

[4] CCITT Study Group XV, Recommendation H.261, "Video codec for audiovisual services at px64 kbit/sec," Report R 37, Geneva, July 1990.

[5] ISO-IEC JTC1/SC2/WG11, MPEG/Document 176 Rev. 2, 1990 and Document 0159, 1992.

[6] T.-C. Chen, P. E. Fleischer, and S.-M. Lei, "A subband scheme for advanced TV coding in BISDN applications." *Third Int. Workshop on HDTV,* Turin, Italy, 1989.

[7] U. Pestel and K. Grüger, "Design of HDTV subband filterbanks considering VLSI implementation aspects." *IEEE Trans. Circuits Syst. Video Technol.,* vol. 1, pp. 14–21, March 1991.

[8] Synopsys, Inc., Reference Manual, Version 2.2a, Technical Report, 1991.

[9] F. Catthoor *et al.,* "Architectural synthesis for medium and high throughput signal processing with the new CATHEDRAL environment." In *High-Level Synthesis,* ed. R. Camposano and W. Wolf. Hingham, MA: Kluwer Academic Publishers, 1991, pp. 27–54.

[10] C. K. Erdelyi *et al.,* "Custom and semi-custom design." In *Design Methodologies,* ed. S. Goto. Amsterdam: Elsevier Science Publishers, 1986, pp. 3–41.

[11] R. Jain, A. C. Parker, and N. Park, "Predicting system-level area and delay for pipelining and nonpipelining designs." *IEEE Trans. Computer-Aided Design,* vol. CAD-11, pp. 955–965, Aug. 1992.

[12] P. Pirsch, W. Gehrke, and R. Hoffer, "A hierarchical multiprocessor architecture for video coding applications." In *Proc. ISCAS'93,* Chicago, May 1993, pp. 1750–1753.

[13] T. G. Noll and E. de Man, "Pushing the performance limits due to power dissipation of future ULSI chips." In *Proc. ISCAS'92,* May 1992, pp. 1652–1655.

[14] P. Pirsch, "VLSI architectures for digital video signal processing." In *Computer Systems and Software Engineering*, ed. P. Dewilde and J. Vandewalle. Hingham, MA: Kluwer Academic Publishers, 1992, pp. 65–99.

[15] S. Y. Kung, *VLSI Array Processor*. Engleswood Cliffs, NJ: Prentice-Hall, 1988.

[16] P. P. Vaidyanathan, "Quadrature mirror filter banks, m-band extensions and perfect-reconstruction techniques." *IEEE ASSP Mag.*, pp. 4–20, July 1987.

[17] R. E. Crochiere and L. R. Rabiner, *Multirate digital signal processing*. Englewood Cliffs, NJ: Prentice-Hall, 1983.

[18] J. J. T. Smith and T. P. Barnwell, "Exact reconstruction techniques for tree-structured subband coders." *IEEE Trans. Acoust., Speech, Signal Processing*, vol. ASSP-34, pp. 434–441, June 1986.

[19] K. Hwang, *Computer Arithmetic, Principles, Architectures, and Design*. New York: J. Wiley and Sons, 1979.

[20] T. G. Noll, "Carry-save architectures for high-speed digital signal processing." *J. VLSI Signal Processing*, vol. 3, pp. 121–140, 1991.

[21] P. Pirsch, K. Grüger, and M. Winzker, "VLSI architectures of two-dimensional filters for HDTV coding." In *Proc. ISCAS '92*, San Diego, May 1992, pp. 1648–1651.

[22] K. Grüger, M. Winzker, W. Gehrke, and P. Pirsch, "VLSI realization of 2-D HDTV subband filterbanks with on-chip memories and FIFOs." In *ESSCIR'92*, Sept. 1992, pp. 319–322.

[23] W. K. Pratt, *Digital Image Processing*. New York: J. Wiley and Sons, 1978.

[24] U. Totzek, F. Matthiesen, and T. Noll, "DCT-bausteine für die codierung von HDTV-signalen." *Mikroelektronik*, vol. 5, pp. 124–127, 1991.

[25] M.-T. Sun, T.-C. Chen, and A. M. Gottlieb, "VLSI implementation of a 16x16 discrete cosine transform." *IEEE Trans. Circuits Syst.*, vol. CAS-36, pp. 610–617, 1989.

[26] B. G. Lee, "A new algorithm to compute the discrete cosine transform," *IEEE Trans. Acoust., Speech, Signal Processing*, vol. ASSP-32, pp. 1243-1245, 1984.

[27] A. Artieri *et al.*, "A one chip VLSI for real time two-dimensional discrete cosine transform." In *Proc. IEEE Int. Symp. Circuits Syst.*, Helsinki, 1988, pp. 701–704.

[28] A. Ligtenberg and J. H. O'Neill, "A single chip solution for an 8 by 8 two dimensional DCT." In *Proc. IEEE Int. Symp. Circuits Syst.*, Philadelphia, 1987, pp. 1128–1131.

[29] P. A. Ruetz, P. Tong, D. Bailey, D. Luthi, and P. Ang, "A high-performance full-motion video compression chip set." *IEEE Trans. Circuits Syst. Video Technol.*, vol. 2, pp. 111–122, June 1992.

[30] H. Fujiwara, M. L. Liou, M. T. Sun, K. M. Yang, K. Maruyama, K. Shomura, and K. Oyama, "An all-ASIC implementation of a low bit-rate video codec." *IEEE Trans. Circuits Syst. Video Technol.*, vol. 2, pp. 123–134, June 1992.

[31] H. G. Musmann, P. Pirsch, and H.-J. Grallert, "Advances in picture coding." *Proc. IEEE*, vol. 73, pp. 523–548, 1985.

[32] T. Komarek and P. Pirsch, "Array architectures for block matching algorithms." *IEEE Trans. Circuits Syst.*, vol. CAS-36, pp. 1301–1308, Oct. 1989.

[33] K.-M. Yang, M. T. Sun, and L. Wu, "A family of VLSI designs for the motion compensation block matching algorithm." *IEEE Trans. Circuits Syst.*, vol. CAS-36, pp. 1317–1325, Oct. 1989.

[34] Th. Micke, D. Müller, and R. Heiss, "ISDN-bildtelefon-codec auf der Grundlage eines Array-Prozessor-IC." *Mikroelektronik*, vol. 5, pp. 116–119, 1991.

[35] P. Weis, "Video-parallelprozessor zur Bild-codierung und -verarbeitung auf einem Chip." *Mikroelektronik*, vol. 5, pp. 112–115, 1991.

[36] K. Gaedke, H. Jeschke, and P. Pirsch, "A VLSI based MIMD architecture of a multiprocessor system for real-time video processing applications," *J. VLSI Signal Processing*, vol. 5, pp. 159–169, 1993.

[37] Y. Suzuki *et al.*, "Single board video codec using VLSIs for 64/128 kbit/s CIF video." In *Proc. Int. Workshop on 64 kb/s Coding of Moving Images,* paper 3–1, Rotterdam, Sept. 1990.

[38] T. Nishitani, "Parallel video signal processor configuration based on overlap-save technique and its LSI processor element: VISP." *J. VLSI Signal Processing*, vol. 1, pp. 25–34, 1989.

[39] S. K. Rao *et al.*, "A real-time Px64/MPEG video encoder chip." In *ISSCC'93*, Feb. 1993, pp. 32–33.

[40] D. Brinthaupt *et al.*, "A video decoder for H.261 video teleconferencing and MPEG stored interactive video applications." In *ISSCC'93*, Feb. 1993, pp. 34–35.

[41] H. Jeschke, K. Gaedke, and P. Pirsch, "Multiprocessor performance for real-time processing of video coding applications." *IEEE Trans. Circuits Syst. Video Technol.*, vol. 2, pp. 221–230, June 1992.

[42] P. Pirsch, W. Gehrke, and R. Hoffer, "A hierarchical multiprocessor architecture for video coding applications." In *Proc. ISCAS'93*, Chicago, May 1993, pp. 1750–1753.

Index